Methods in Molecular Biology

Series Editor
John M. Walker
School of Life and Medical Sciences
University of Hertfordshire
Hatfield, Hertfordshire, AL10 9AB, UK

For further volumes:
http://www.springer.com/series/7651

Peptide Self-Assembly

Methods and Protocols

Edited by

Bradley L. Nilsson

Department of Chemistry, University of Rochester, Rochester, NY, USA

Todd M. Doran

Department of Medicinal Chemistry, Institute for Translational Neuroscience, University of Minnesota, Minneapolis, MN, USA

Humana Press

Editors
Bradley L. Nilsson
Department of Chemistry
University of Rochester
Rochester, NY, USA

Todd M. Doran
Department of Medicinal Chemistry
Institute for Translational Neuroscience
University of Minnesota
Minneapolis, MN, USA

ISSN 1064-3745 ISSN 1940-6029 (electronic)
Methods in Molecular Biology
ISBN 978-1-4939-7809-0 ISBN 978-1-4939-7811-3 (eBook)
https://doi.org/10.1007/978-1-4939-7811-3

Library of Congress Control Number: 2018940502

Printed on acid-free paper

This Humana Press imprint is published by the registered company Springer Science+Business Media, LLC part of Springer Nature.
The registered company address is: 233 Spring Street, New York, NY 10013, U.S.A.

Preface

Peptide and protein self-assembly phenomena are endemic in biological systems. The regulated self-assembly and disassembly of cytoskeletal and extracellular matrix proteins form the bedrock of cellular structure and function. The dysfunctional assembly of proteins and peptides underlies a growing number of amyloid pathologies that include Alzheimer's disease, Parkinson's disease, and bovine spongiform encephalopathy. These disorders are marked by the aberrant self-assembly of peptides into one-dimensional cross-β amyloid fibrils. It has also been recognized that amyloid aggregates are not only related to disease states but have also evolved to be functional biomaterials. This recognition has inspired extensive efforts to develop novel biomaterials derived from natural and designed self-assembling peptides. This volume attempts to capture modern methods that span the breadth of the exciting and expanding field of peptide self-assembly. The contributions to this volume are divided into three categories: (1) Methods for the interrogation of the structure and function of self-assembled peptides, (2) Functional peptide materials with an emphasis on β-sheet peptide materials, and (3) Methods for the investigation of pathological amyloid.

Correlation of the emergent properties of self-assembled peptides with the structure of the self-assembled material and the mechanism of self-assembly has been exceptionally challenging. The first section of this volume is focused on methods that have been developed to enable studies that establish these correlations and thus deepen our understanding of self-assembled peptides in the fields of biomedicine and biomaterials. The contributions in this first section include discursive reviews that provide a survey of spectroscopic methods that have been adapted for the study of self-assembled peptides and a more focused discussion of the expanding use of solid-state NMR as a powerful method to enhance structural understanding of self-assembled peptide materials. Methods for the application of theoretical computational approaches to the understanding of peptide self-assembly are presented as are novel spectroscopic methods that improve functional analysis of peptide self-assembly phenomena.

The design and development of functional self-assembled biomaterials is a rapidly burgeoning field of study. The second section of this volume focuses on methods for the design, synthesis, and application of self-assembled peptide materials. This volume focuses primarily on β-sheet assemblies, although collagen materials are also represented. The applications of these materials are diverse and include hydrogels for tissue engineering, fibril assemblies in vaccine development, self-assembled peptides with novel photophysical properties, and cell-penetrating materials for delivery of therapeutic agents. Collectively, these methods focus on both the synthetic and operational aspects of the novel materials.

The understanding of amyloid self-assembly has also been a significant research challenge that has received intense attention. Specifically, it has been difficult to elucidate how the self-assembly of dysfunctional amyloid is related to the observed functional cytotoxicity of the assemblies due to their complicated self-assembly pathways, which include metastable oligomeric intermediates as well as mature cross-β fibrils. It is clear that the properties of the folding intermediates and the ultimate fibril assemblies are distinctive. In the final section of this volume are presented methods focused on structural and mechanistic analyses of pathological amyloid systems that provide novel ways to assess function of the various possible aggregates as well as to determine how the structure of these materials correlates to function/dysfunction in the biological context.

Rochester, NY, USA *Bradley L. Nilsson*
Minneapolis, MN, USA *Todd M. Doran*

Contents

Contributors

ROJELIO ELIAS ALVARADO • *Department of Pharmacology & Toxicology, University of Texas Medical Branch, Galveston, TX, USA*

AIDA ATTAR • *Department of Neurology, David Geffen School of Medicine, University of California at Los Angeles, Los Angeles, CA, USA; Brain Research Institute, University of California at Los Angeles, Los Angeles, CA, USA*

ANGELO BELLA • *National Physical Laboratory, Middlesex, UK*

GAL BITAN • *Department of Neurology, David Geffen School of Medicine, University of California at Los Angeles, Los Angeles, CA, USA; Brain Research Institute, University of California at Los Angeles, Los Angeles, CA, USA; Molecular Biology Institute, University of California at Los Angeles, Los Angeles, CA, USA*

C. BRENT CHESSON • *Department of Pharmacology & Toxicology, University of Texas Medical Branch, Galveston, TX, USA; Sealy Center for Vaccine Development, University of Texas Medical Branch, Galveston, TX, USA*

V. CASTELLETTO • *Department of Chemistry, University of Reading, Reading, UK*

SAKETH CHEMURU • *Department of Structural Biology and Pittsburgh Institute for Neurodegenerative Diseases, University of Pittsburgh School of Medicine, Pittsburgh, PA, USA; Department of Physiology and Biophysics, Case Western Reserve University, Cleveland, OH, USA*

HYEWON CHUNG • *Department of Ophthalmology, Konkuk University Medical Center, Konkuk University School of Medicine, Seoul, South Korea*

JOEL H. COLLIER • *Department of Surgery, University of Chicago, Chicago, IL, USA; Department of Biomedical Engineering, Duke University, Durham, NC, USA*

MICHAEL L. CONNOLLY • *Biogelx Ltd, Biocity Scotland, Lanarkshire, Scotland, UK*

VINCENT P. CONTICELLO • *Department of Chemistry, Emory University, Atlanta, GA, USA*

ELLIOT J. CROOKS • *Department of Biochemistry and Cell Biology, Stony Brook University, Stony Brook, NY, USA*

HONGGANG CUI • *Department of Chemical and Biomolecular Engineering and Institute for NanoBiotechnology (INBT), Johns Hopkins University, Baltimore, MD, USA; Department of Oncology and Sidney Kimmel Comprehensive Cancer Center, Johns Hopkins University School of Medicine, Baltimore, MD, USA*

MATTHEW J. DALBY • *Centre for Cell Engineering, Institute of Molecular, Cell & Systems Biology, College of Medical, Veterinary & Life Sciences, Joseph Black Building, University of Glasgow, Glasgow, UK*

LOUIS DERIDDER • *Department of Chemistry and Biomolecular Science, Potsdam, NY, USA*

JING DI • *Department of Neurology, David Geffen School of Medicine, University of California at Los Angeles, Los Angeles, CA, USA*

NINA DINJASKI • *Department of Biomedical Engineering, Tufts University, Medford, MA, USA*

KATELYNNE E. DOHERTY • *Department of Chemistry, State University of New York at Binghamton, Binghamton, NY, USA*

HE DONG • *Department of Chemistry and Biomolecular Science, Potsdam, NY, USA*

TODD M. DORAN • *Department of Medicinal Chemistry, Institute for Translational Neuroscience, University of Minnesota, Minneapolis, MN, USA*

BRENDAN ELMORE • *Department of Chemistry and Biomolecular Science, Potsdam, NY, USA*
YUAN GAO • *School of Chemical and Biomolecular Engineering, Georgia Institute of Technology, Atlanta, GA, USA*
LAURA GOLDIE • *Biogelx Ltd, Biocity Scotland, Lanarkshire, Scotland, UK*
I. W. HAMLEY • *Department of Chemistry, University of Reading, Reading, UK*
HUIFANG HAN • *Department of Surgery, University of Chicago, Chicago, IL, USA*
MHAIRI M. HARPER • *Biogelx Ltd, Biocity Scotland, Lanarkshire, Scotland, UK*
MINAKO HOSHI • *Institute of Biomedical Research and Innovation, FBRI, Kobe, Japan; Department of Anatomy and Developmental Biology, Graduate School of Medicine, Kyoto University, Kyoto, Japan*
RUNDONG HU • *Department of Chemical & Biomolecular Engineering, The University of Akron, Akron, OH, USA*
DANTING HUANG • *School of Chemical and Biomolecular Engineering, Georgia Institute of Technology, Atlanta, GA, USA*
WENWEN HUANG • *Department of Biomedical Engineering, Tufts University, Medford, MA, USA*
BENJAMIN C. HUDSON • *School of Chemical and Biomolecular Engineering, Georgia Institute of Technology, Atlanta, GA, USA*
ELEANORE J. IRVINE • *Biogelx Ltd, Biocity Scotland, Lanarkshire, Scotland, UK*
YOSHITAKA ISHII • *Department of Chemistry, University of Illinois at Chicago, Chicago, IL, USA; School of Life Science and Technology, Tokyo Institute of Technology, Yokohama, Japan*
VINEETHA JAYAWARNA • *Biogelx Ltd, Biocity Scotland, Lanarkshire, Scotland, UK*
BINBO JIANG • *Department of Chemical & Biomolecular Engineering, The University of Akron, Akron, OH, USA; College of Chemical and Biological Engineering, Zhejiang University, Zhejiang, Hangzhou, China*
TAO JIANG • *Department of Chemistry, Emory University, Atlanta, GA, USA*
TEJASWINI S. KALE • *Department of Chemistry, Johns Hopkins University, Baltimore, MD, USA*
DAVID L. KAPLAN • *Department of Biomedical Engineering, Tufts University, Medford, MA, USA*
KARUNAKAR KAR • *Department of Structural Biology and Pittsburgh Institute for Neurodegenerative Diseases, University of Pittsburgh School of Medicine, Pittsburgh, PA, USA; School of Life Sciences, Jawaharlal Nehru University, New Delhi, India*
FRANK-GERRIT KLÄRNER • *Faculty of Chemistry, University of Duisburg-Essen, Essen, Germany*
RAVI KODALI • *Department of Structural Biology and Pittsburgh Institute for Neurodegenerative Diseases, University of Pittsburgh School of Medicine, Pittsburgh, PA, USA; Department of Chemistry and Biochemistry, Duquesne University, Pittsburgh, PA, USA*
JIN MO KOO • *Department of Chemical and Biomolecular Engineering and Institute for NanoBiotechnology (INBT), Johns Hopkins University, Baltimore, MD, USA*
IVAN V. KORENDOVYCH • *Department of Chemistry, Syracuse University, Syracuse, NY, USA*
SAMUEL A. KOTLER • *Biophysics Program, University of Michigan, Ann Arbor, MI, USA*
ANGUS LAU • *Tanz Centre for Research in Neurodegenerative Diseases, Department of Biochemistry, University of Toronto, Toronto, ON, Canada*
HEATHER H. C. LAU • *Tanz Centre for Research in Neurodegenerative Diseases, Department of Biochemistry, University of Toronto, Toronto, ON, Canada*
ZSOFIA LENGYEL • *Department of Chemistry, Syracuse University, Syracuse, NY, USA*

David Lightbody • *Biogelx Ltd, Biocity Scotland, Lanarkshire, Scotland, UK*
Sara Linse • *Department of Biochemistry and Structural Biology, Lund University, Lund, Sweden*
Yi-An Lin • *Department of Chemical and Biomolecular Engineering and Institute for NanoBiotechnology (INBT), Johns Hopkins University, Baltimore, MD, USA*
Yonglan Liu • *Department of Chemical & Biomolecular Engineering, The University of Akron, Akron, OH, USA*
Xiaomeng Lu • *Biophysics Program and Department of Chemical and Biological Engineering, University of Wisconsin-Madison, Madison, WI, USA*
Buyong Ma • *Basic Science Program, Leidos Biomedical Research, Inc., Cancer and Inflammation Program, National Cancer Institute, Frederick, MD, USA*
Jie Ma • *Department of Chemical & Biomolecular Engineering, The University of Akron, Akron, OH, USA; State Key Laboratory of Pollution Control and Resource Reuse, School of Environmental Science and Engineering, Tongji University, Shanghai, China*
Ravinder Malik • *Department of Neurology, David Geffen School of Medicine, University of California at Los Angeles, Los Angeles, CA, USA*
Dan McElheny • *Department of Chemistry, University of Illinois at Chicago, Chicago, IL, USA*
Jennifer J. McManus • *Department of Chemistry, Maynooth University, Maynooth, Co. Kildare, Ireland*
Pinaki Misra • *Department of Structural Biology and Pittsburgh Institute for Neurodegenerative Diseases, University of Pittsburgh School of Medicine, Pittsburgh, PA, USA; Department of Biochemistry and Molecular Biology, Mayo Clinic College of Medicine, Rochester, MN, USA*
Judith J. Mittag • *Fakultät für Physik and CeNS, Ludwig-Maximilians-Universität, Munich, Germany*
Smita Mukherjee • *Department of Structural Biology and Pittsburgh Institute for Neurodegenerative Diseases, University of Pittsburgh School of Medicine, Pittsburgh, PA, USA; The Wharton School, University of Pennsylvania, Philadelphia, PA, USA*
Regina M. Murphy • *Biophysics Program and Department of Chemical and Biological Engineering, University of Wisconsin-Madison, Madison, WI, USA*
Gayatri Nair • *Department of Neurology, David Geffen School of Medicine, University of California at Los Angeles, Los Angeles, CA, USA*
Bradley L. Nilsson • *Department of Chemistry, University of Rochester, Rochester, NY, USA*
Ruth Nussinov • *Basic Science Program, Leidos Biomedical Research, Inc., Cancer and Inflammation Program, National Cancer Institute, Frederick, MD, USA; Sackler Institute of Molecular Medicine, Department of Human Genetics and Molecular Medicine Sackler School of Medicine, Tel Aviv University, Tel Aviv, Israel*
Tiernan T. O'Malley • *Laboratory for Neurodegenerative Research, Center for Neurologic Diseases, Brigham and Women's Hospital and Harvard Medical School, Boston, MA, USA*
Anant K. Paravastu • *School of Chemical and Biomolecular Engineering, Georgia Institute of Technology, Atlanta, GA, USA*
Ruxi Qi • *Department of Physics, State Key Laboratory of Surface Physics, Key Laboratory for Computational Physical Sciences (MOE), Fudan University, Shanghai, P. R. China*
Wei Qiang • *Department of Chemistry, State University of New York at Binghamton, Binghamton, NY, USA*

Joachim O. Rädler • *Fakultät für Physik and CeNS, Ludwig-Maximilians-Universität, Munich, Germany*
Ayyalusamy Ramamoorthy • *Biophysics Program, University of Michigan, Ann Arbor, MI, USA; Department of Chemistry, University of Michigan, Ann Arbor, MI, USA*
Vincent Raussens • *Faculty of Science, Laboratory of Structure and Function of Biological Membranes, Center for Structural Biology and Bioinformatics, Université Libre de Bruxelles, Brussels, Belgium*
Baiping Ren • *Department of Chemical & Biomolecular Engineering, The University of Akron, Akron, OH, USA*
Stephen M. Richardson • *Faculty of Biology, Medicine and Health, Division of Cell Matrix Biology and Regenerative Medicine, School of Biological Sciences, University of Manchester, Manchester, UK*
Evan K. Roberts • *School of Chemical and Biomolecular Engineering, Georgia Institute of Technology, Atlanta, GA, USA*
Jai S. Rudra • *Department of Pharmacology & Toxicology, University of Texas Medical Branch, Galveston, TX, USA; Sealy Center for Vaccine Development, University of Texas Medical Branch, Galveston, TX, USA*
Caroline M. Rufo • *Department of Chemistry, Syracuse University, Syracuse, NY, USA*
Jean-Marie Ruysschaert • *Faculty of Science, Laboratory of Structure and Function of Biological Membranes, Center for Structural Biology and Bioinformatics, Université Libre de Bruxelles, Brussels, Belgium*
Maxim G. Ryadnov • *National Physical Laboratory, Middlesex, UK*
Emiliana De Santis • *National Physical Laboratory, Middlesex, UK*
Thomas Schrader • *Faculty of Chemistry, University of Duisburg-Essen, Essen, Germany*
Joshua E. Shaw • *Faculty of Biology, Medicine and Health, Division of Cell Matrix Biology and Regenerative Medicine, School of Biological Sciences, University of Manchester, Manchester, UK*
Michael Shaw • *National Physical Laboratory, Middlesex, UK*
Junfeng Shi • *Brandeis University, Waltham, MA, USA*
Steven O. Smith • *Department of Biochemistry and Cell Biology, Stony Brook University, Stony Brook, NY, USA*
Carolina Mora Solano • *Department of Surgery, University of Chicago, Chicago, IL, USA*
Hao Su • *Department of Chemical and Biomolecular Engineering and Institute for NanoBiotechnology (INBT), Johns Hopkins University, Baltimore, MD, USA*
Karen Taylor • *Department of Neurology, David Geffen School of Medicine, University of California at Los Angeles, Los Angeles, CA, USA*
Edmond Teng • *Department of Neurology, David Geffen School of Medicine, University of California at Los Angeles, Los Angeles, CA, USA; Greater Los Angeles Healthcare System, Veterans Hospital, West Los Angeles, CA, USA*
John D. Tovar • *Department of Chemistry, Johns Hopkins University, Baltimore, MD, USA; Department of Materials Science and Engineering, Johns Hopkins University, Baltimore, MD, USA; Institute for NanoBioTechnology, Johns Hopkins University, Baltimore, MD, USA*
Rein V. Ulijn • *CUNY Advanced Science Research Center (ASRC), City University of New York, New York, NY, USA*
Dominic M. Walsh • *Laboratory for Neurodegenerative Research, Center for Neurologic Diseases, Brigham and Women's Hospital and Harvard Medical School, Boston, MA, USA*

JOEL C. WATTS • *Tanz Centre for Research in Neurodegenerative Diseases, Department of Biochemistry, University of Toronto, Toronto, ON, Canada*

GUANGHONG WEI • *Department of Physics, State Key Laboratory of Surface Physics, Key Laboratory for Computational Physical Sciences (MOE), Fudan University, Shanghai, P. R. China*

YI WEN • *Department of Surgery, University of Chicago, Chicago, IL, USA; Department of Biomedical Engineering, Duke University, Durham, NC, USA*

RONALD WETZEL • *Department of Structural Biology and Pittsburgh Institute for Neurodegenerative Diseases, University of Pittsburgh School of Medicine, Pittsburgh, PA, USA*

YILING XIAO • *Department of Chemistry, University of Illinois at Chicago, Chicago, IL, USA*

BING XU • *Brandeis University, Waltham, MA, USA*

DAWEI XU • *Department of Chemistry and Biomolecular Science, Potsdam, NY, USA*

LIJIAN XU • *Department of Chemical & Biomolecular Engineering, The University of Akron, Akron, OH, USA; College of Life Sciences and Chemistry, Hunan University of Technology, Zhuzhou, China*

DAN YUAN • *Brandeis University, Waltham, MA, USA*

MINGZHEN ZHANG • *Department of Chemical & Biomolecular Engineering, The University of Akron, Akron, OH, USA*

JIE ZHENG • *Department of Chemical & Biomolecular Engineering, The University of Akron, Akron, OH, USA*

NING ZHOU • *Brandeis University, Waltham, MA, USA*

MARTINE ZILIOX • *Department of Biochemistry and Cell Biology, Stony Brook University, Stony Brook, NY, USA*

Part I

Fundamental Methods for Characterizing Peptide Self-Assembly

Chapter 1

Methods to Characterize the Nanostructure and Molecular Organization of Amphiphilic Peptide Assemblies

V. Castelletto and I. W. Hamley

Abstract

Methods to characterize the nanostructure and molecular organization of aggregates of peptides such as amyloid or amphiphilic peptide assemblies are reviewed. We discuss techniques to characterize conformation and secondary structure including circular and linear dichroism and FTIR and Raman spectroscopies, as well as fluorescence methods to detect aggregation. NMR spectroscopy methods, especially solid-state NMR measurements to probe beta-sheet packing motifs, are also briefly outlined. Also discussed are scattering methods including X-ray diffraction and small-angle scattering techniques including SAXS (small-angle X-ray scattering) and SANS (small-angle neutron scattering) and dynamic light scattering. Imaging methods are direct methods to uncover features of peptide nanostructures, and we provide a summary of electron microscopy and atomic force microscopy techniques. Selected examples are highlighted showing data obtained using these techniques, which provide a powerful suite of methods to probe ordering from the molecular scale to the aggregate superstructure.

Key words Peptides, Conformation, Secondary structure, Characterization methods, X-ray diffraction, Fluorescence spectroscopy, Circular dichroism, Linear dichroism, FTIR spectroscopy, NMR, Light scattering, Small-angle X-ray scattering (SAXS), Small-angle neutron scattering (SANS), Atomic force microscopy (AFM), Electron microscopy

1 Introduction

Peptides are one of the key classes of biomolecules with many roles in nature. Peptides are short chains of amino acids linked via amide bonds. Amino acids are the natural building blocks which make up both proteins and peptides. Amino acids consist of an amine, carboxylic acid, as well as distinctive side chains. There arc approximately 20 different amino acids, with the same number of different side chains, which give rise to their unique characteristics and variability. Differences in the amino acid side chains allow them to be grouped into different categories according to their chemical composition such as hydrophobic/hydrophilic/aliphatic/aromatic/ or neutral/positively/negatively charged [1]. The numerous interactions between amino acids allow for complex self-

Bradley L. Nilsson and Todd M. Doran (eds.), *Peptide Self-Assembly: Methods and Protocols*, Methods in Molecular Biology, vol. 1777, https://doi.org/10.1007/978-1-4939-7811-3_1,

assembled nanostructures to develop. These types of interactions include hydrogen bonds, electrostatic interactions, hydrophobic interactions, aromatic interactions (π-π stacking), and van der Waals forces. The varying properties of amino acids can therefore be manipulated in the design of peptides to control self-assembly to produce novel functional biomaterials.

There are two main classes of amphiphilic peptides; those are [1] designed purely peptidic systems with amphiphilic properties arising from sequences of hydrophobic and hydrophilic (charged) residues, and [2] peptides modified by attachment of hydrophobic lipid chains, these are termed peptide amphiphiles (PAs) or lipopeptides.

This chapter is focussed on methods to characterize the molecular conformation and self-assembly (leading to defined nanostructures) of amphiphilic peptides, with a particular focus on amyloid and sequenced peptides and lipopeptides. Other classes of peptide such as coiled coils may have amphiphilic properties (helix facial amphiphilicity contributes to coiled-coil formation), but this is discussed elsewhere [2–4]. Research on the self-assembly of amphiphilic peptides has been the subject of several reviews [5–13].

This chapter is organized as follows. The first sections deal with methods to probe the structure of amphiphilic peptide assemblies through X-ray diffraction (XRD, Subheading 1) and the determination of the critical aggregation concentration through fluorescence probe methods (Subheading 2). Then spectroscopic methods to characterize peptide secondary structure are considered including circular and linear dichroism (CD, LD; Subheading 3), Fourier transform infrared spectroscopy (FTIR), Raman spectroscopy (Subheading 4), and NMR spectroscopy (Subheading 5). We then turn to methods to investigate fibrillization and the morphology of amphiphilic peptide self-assemblies. Subheading 6 is concerned with light scattering, Subheading 7 with small-angle scattering methods, and Subheading 8 with microscopic imaging techniques. The chapter concludes with some brief summarizing remarks.

2 X-ray Diffraction

With a very few exceptions, amyloid peptides and lipopeptides do not crystallize, prohibiting single crystal X-ray diffraction. This is because fibrillar assemblies are by their nature noncrystalline, one- or two-dimensional arrays with a low degree of molecular ordering. A very few studies have succeeded in obtaining single crystal XRD patterns from small amyloid peptide fragments [14–16].

Peptide fibrils typically contain bundles of β-sheets with backbones orthogonal to the fiber axis, in the so-called "cross-β" structure [17–21]. Figure 1 shows a representative "cross-β"

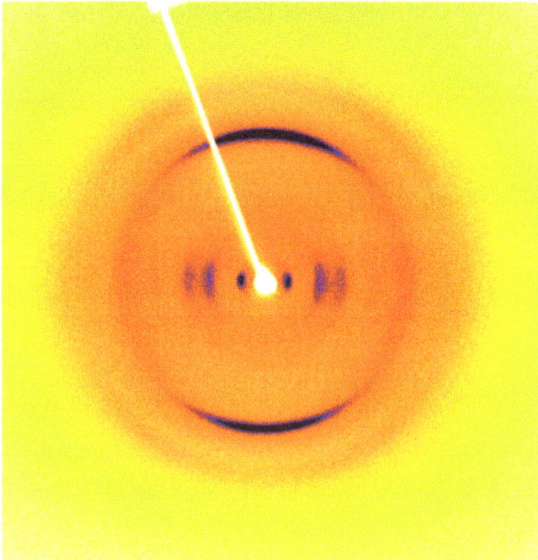

Fig. 1 Fiber XRD pattern obtained from a dried stalk of amyloid peptide YYKLVFFC [22]

XRD pattern from a peptide YYKLVFFC [22] containing a key aggregation sequence (KLVFF) from the amyloid β peptide, which itself is the subject of a detailed review [23]. This is obtained from a stalk dried from a concentrated solution at the end of a wax-coated capillary [24]. The signature in an X-ray scattering pattern of the cross-β structure is a 4.7 Å meridional reflection corresponding to the spacing between peptide backbones and at least one equatorial reflection at 8–12 Å which is broader and which corresponds to the stacking periodicity of β sheets (the range of values reflects different side chain dimensions). In the case of the pattern in Fig. 1 there are additional equatorial reflections. In general, multiple such reflections come from molecular and β-sheet periodicities [20]. It is possible to estimate unit cell parameters by indexing cross-β fiber XRD patterns and this can be done using automated software such as CLEARER [25]. Most X-ray diffraction work that showed cross-β structure has been performed on dried films. Since hydration can affect the structure of β-sheet containing fibrils, Squires and coworkers have examined whether the cross-β structure is retained in flow-aligned solution [26]. The wide angle X-ray scattering (WAXS) results for a synthetic peptide (fragment of transthyretin) and hen egg white lysozyme indeed showed the same pattern as for a dried sample, indicating that the cross-β structure is present in solution and is not an artifact caused by dehydration. Fibrils may be formed by other classes of peptides such as those adopting coiled-coil conformations but these have distinct XRD patterns [27, 28].

As mentioned above, fiber XRD is performed on dried samples, in the form of films or "stalks," the latter being dried threads of solution. Other methods of alignment include use of stretch frames or cryo-loops, the latter producing a dried flat film or "mat" [29]. Stalks may also be dried in the presence of a magnetic field to improve alignment [30, 31].

3 Fluorescence Methods Including Dye Staining and Critical Aggregation Concentration Determination

Congo red staining provides a method that, by definition, identifies amyloid [32]. Under polarized light, amyloid samples exhibit green birefringence when stained with Congo red. Thioflavin T is a fluorescent dye widely used to study amyloid formation. Excitation at 450 nm produces fluorescence at 482 nm [33, 34].

Peptide amphiphiles behave in some respects like conventional amphiphilic molecules (surfactants, lipids) [35] in that they are observed to undergo aggregation (often into fibrils) above a critical aggregation concentration (*cac*). The critical aggregation concentration may be detected via fluorescence methods. Pyrene is a fluorescent probe molecule which is used to locate the *cac* for conventional amphiphiles, as its fluorescence is sensitive to the local hydrophobic environment [36, 37]. It has also successfully been used to determine the *cac* for several PA systems [38–46]. Pyrene fluorescence measurements of *cac* usually involve determination of the ratio of third and first vibronic band intensities, I_{373}/I_{383} or I_1/I_3, although sometimes I_1 itself shows discontinuities at the same concentration [45, 47]. For conventional amphiphiles, I_1/I_3 decreases at the critical micelle concentration, from a typical value around 1.7–1.8 to a lower value 1.0–1.2 [see, for example [48, 49]], although this is not always observed for lipopeptides, in particular if they contain an aromatic residue which contributes to UV absorbance. Figure 2 shows a representative *cac* curve obtained from concentration-dependent pyrene fluorescent measurements for a lipopeptide studied by our group [41]. Pyrene derivatives such as 1-pyrene carboxylic acid offer lower toxicity than pyrene itself and have been successfully used in amyloid peptide cac determination [50].

In contrast to the use of pyrene, the fluorescence of Thioflavin T (ThT) is dependent on the formation of amyloid-like structures (β-sheet fibrils) [33, 51] and has been used for amyloid fibril-forming peptides. We have recently shown that ThT can be used to determine the *cac* of lipopeptide PAs, the value obtained being similar to that obtained from pyrene fluorescence techniques [40, 44, 45]. Other fluorescent dyes such as Nile red have been used to determine the *cac* of amyloid-forming peptides [52, 53] but have not yet been widely employed in studies of PA systems. The fluorescent probe DPH [1,6-diphenyl-1,3,5-hexatriene] has also been used to locate the cac of lipopeptides [54].

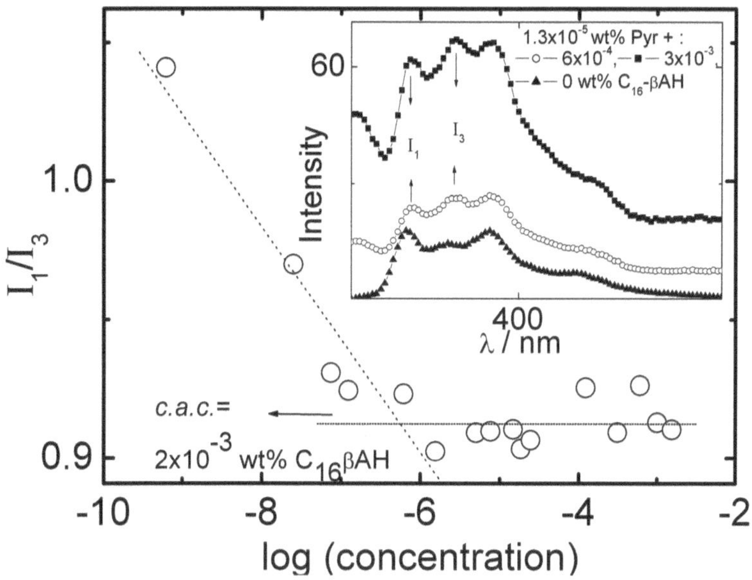

Fig. 2 Determination of critical aggregation concentration (*cac*) for a lipopeptide C_{16}-βAH [carnosine dipeptide headgroup, βA: beta-alanine] from pyrene fluorescence ratio I_1/I_3 (representative fluorescence emission spectra shown in inset) [41]

In principle, other physicochemical methods sensitive to the state of molecular aggregation, such as light scattering [55, 56] and proton NMR solubility measurements [57, 58], may also be used to determine the *cac*. The *cac* will, as for conventional surfactants, depend on the hydrophile-lipophile balance (HLB) although in addition for PAs it will be influenced by hydrogen bonding as well as electrostatic interactions. A detailed model for this is still lacking and this presents an interesting future challenge for researchers.

The kinetics of fibril formation by amyloid peptides and PAs may be also followed using techniques well established for amyloid peptides such as ThT fluorescence [23, 32].

Förster resonance energy transfer (FRET) has been used to investigate the binding of dye to amyloid fibrils [59].

4 Circular and Linear Dichroism

Circular dichroism (CD) refers to the differential absorption of right- and left- circularly polarized light. Electronic CD in the UV region of the spectrum is sensitive to the presence of chiral groups in organic molecules and is a primary technique to characterize protein secondary structure. The usual method is based on "fingerprinting" of features in the 190–250 nm far UV region [60]. Figure 3 shows schematic spectra in this regime for some common secondary structures. Data in the near UV region can provide

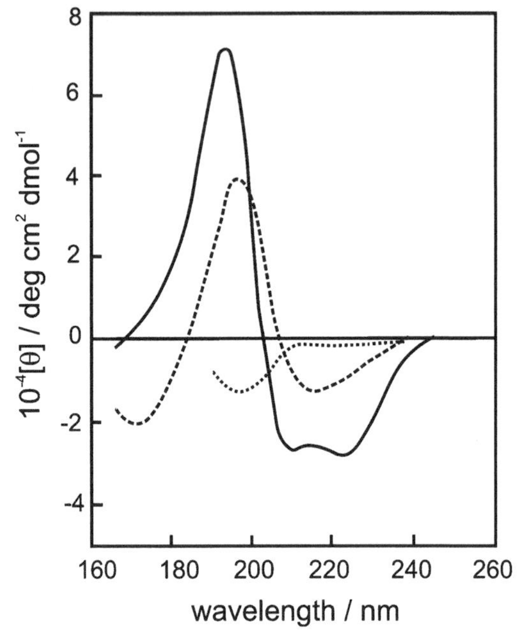

Fig. 3 Schematic CD spectra for peptides with common secondary structures. Solid line: α-helix, dashed line: β-sheet, dotted line: disordered conformation. Adapted from [60]

information on the conformation of peptides containing aromatic residues with absorption features in the 250–310 nm range. In the far UV region, characteristic minima are observed in the absorption spectra at (approximately) 208 and 222 nm (α-helix) or 216–220 nm (β-sheet) or a broad weak minimum in the range 195–200 nm (disordered, sometimes known as random coil conformation). The polyproline II (PPII) conformation is characterized by a deep minimum in a CD spectrum at around 190–205 nm, along with a broad positive maximum at around 210–220 nm [61–64]. Examples showing the comparison between CD spectra for PPII and disordered conformation can be found in the literature [63, 65]. The CD spectra for these structures are characterized by typical values of the molar (or alternatively mean residue) ellipticity [60, 66], as well as the position of the maxima/minima. In fact, CD spectra should be normalized in this way (Eq. 1) for this reason, and also to facilitate comparison between samples at different concentration or measurements in different path length cells. The molar ellipticity is given by

$$[\theta] = \frac{\theta}{10cl} \tag{1}$$

Here $[\theta]$ is the molar ellipticity (in units deg cm^2 dmol^{-1}), θ is the measured CD signal amplitude in millidegrees (mdeg), c is the molar concentration, and l is the path-length of the cell in cm.

CD spectra for proteins are usually analyzed using algorithms based on databases compiled for proteins for which the X-ray crystal structure is known [60]. This permits an accurate determination of secondary structure content which can be used to "calibrate" CD spectra. Several algorithms are available based on various databases. Most consider only larger proteins although there are limited reference data sets (and curve fitting programs) for shorter peptides [67]. This type of analysis is of little use for amphiphilic peptides such as amyloid peptides and lipopeptides for which individual residues (especially aromatic residues) and specific conformations (e.g., turns) can dominate the CD spectrum.

Linear dichroism refers to the differential absorbance of (UV) light parallel and perpendicular to a reference axis. It can be used to provide information on the alignment of extended objects resulting from peptide self-assembly. Since amyloid fibrils and nanotapes are highly anisotropic, they can orient under flow or other fields and this can be probed using LD which in particular provides information on the orientation of the peptide backbone and of chromophores such as aromatic residues [22, 66, 68, 69].

5 FTIR and Raman Spectroscopy

Since a full quantum chemical analysis of the vibrational modes of peptides or proteins is not usually possible, analysis of FTIR data relies on "fingerprinting" or peak fitting methods [70, 71]. These are prone to uncertainty associated with the overlap of features in the spectra [70–72] although in our experience clear features of secondary structures such as β-sheets can be resolved by performing measurements in transmission mode on a modern FTIR instrument, with sufficiently concentrated samples in D_2O in narrow path-length cells.

The amide I region of the FTIR spectrum is the primary one used to identify peptide secondary structures as it is associated with amide group stretching modes, primarily C=O stretch but with a contribution from CN and NH deformations. A band typically in the 1620–1640 cm^{-1} range is associated with β-sheets [70–74]. A band in the typical range 1648–1657 cm^{-1} is associated with α-helix structure while disordered peptides give a peak in the typical range 1642–1650 cm^{-1} [71, 72, 75]. The narrow intense band observed for some peptides at 1675–1695 cm^{-1} is usually ascribed to antiparallel β-sheet structure, although there is still controversy around this assignment [75], for example, ab initio calculations [76] suggest that it may be due to vibrational coupling of residues. Tables of specific side-group FTIR bands are available [75, 77]. Figure 4 shows representative FTIR spectra for two amphiphilic alanine-rich peptides studied by our group. A peak near 1705 cm^{-1} may be assigned to carbonyl stretch, e.g., in acidic

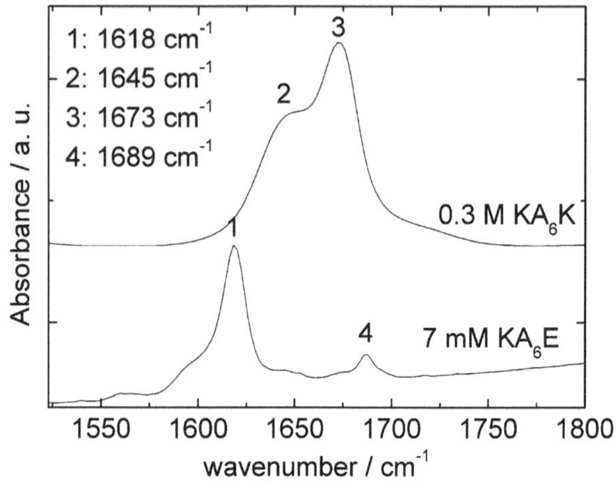

Fig. 4 FTIR spectra measured for a disordered peptide (0.3 M KA_6K) and a peptide under conditions where β-sheet formation is observed (7 mM KA_6E) [128]

side residues or from the C terminus [43, 75, 78, 79]. All band positions mentioned above are slightly downshifted if measurements are performed in D_2O [75], this then being referred to as an amide I′ spectrum. In fact, measurements in D_2O are beneficial since water absorption features around 1650 cm^{-1} can be avoided. Our group always perform measurements in D_2O solutions. In our experience, peptide concentration must be 0.5 wt% or more typically to obtain a reliable signal. It is also quite common to perform FTIR on dried samples in the form of films for transmission mode measurements or via ATR (attenuated total reflectance) measurements.

A peak in the amide I region of an FTIR spectrum at 1673 cm^{-1} is due to trifluoroacetic acid (TFA, a common component of peptide cleavage media and/or HPLC eluents) bound to peptide cations [70, 74, 80–82]. Similar band positions are observed in peptide Raman spectra, although this technique is much more rarely employed.

The amide II band around 1550 cm^{-1} mainly results from the N–H bending vibrations which are responsive to deuteration (the deuterium from D_2O exchanges positions with hydrogen from the N–H bond). As a consequence the amide II' band is shifted by approximately 100 cm^{-1} to 1450 cm^{-1} in D_2O.

Isotope labeling of peptides enables specific features in the FTIR spectrum to be resolved as bands associated with deformation modes of labeled residues will be shifted [83–86]. Typically, isotopic substitution using ^{13}C-labeled peptide splits amide I bands for β-sheets into higher and lower frequency peaks. The magnitude of the wavelength shift depends on the extent of perturbation of the ^{12}C carbonyl coupling induced by ^{13}C substitution and strongly coupled β-sheets are generally sensitive to such isotopic substitution [84]. This can be used to infer information on the registry of β-strands [83, 84].

FTIR can be extended to study linear dichroism (polarized FTIR) on aligned samples with isotope labeling [85, 87] and can be adapted to study vibrational circular dichroism (VCD) [73]. VCD is analogous to (UV) electronic CD, but evidently extended to the IR mode, sensitive to bond vibrations [88, 89]. The analogous technique using Raman scattering is Raman Optical Activity (ROA) [88]. Polarized Raman spectroscopy can provide information on the orientation of specific residues such as aromatic residues [22].

6 NMR Spectroscopy

Solid-state NMR (ssNMR) has provided much detail on amyloid fibril structure. Various high resolution experiments employing magic angle spinning may be performed using ^{13}C or ^{15}N labeled peptides [90–93]. For example, the Aβ(16–22)] peptide KLVFFAE has been found to adopt an antiparallel β-sheet arrangement through solid-state NMR [94], and isotope-edited FTIR on labeled peptides (combined with ssNMR) [83, 84] and computer simulation [95]. In another example, solid-state NMR on aligned films of labeled variants of the surfactant-like peptide A_6K including measurement of ^{2}H and ^{15}N spectra for peptides labeled with these isotopes at specific alanine residues provided information on peptide β-strand alignment within nanotubes (based on line-shape analysis) [86].

Homo-nuclear and hetero-nuclear 2D and 3D NMR spectroscopy may be performed. These techniques enable information on interatomic distances and torsion angles to be obtained for isotopically labeled peptides in the dried state [96, 97].

Hydrogen/deuterium exchange techniques have also been employed to probe structural changes during amyloid formation [96]. Exchange rates provide information on structural rearrangements of sub-segments of a protein or peptide during folding, unfolding, or fibrillization. This method has also been used to provide a 3D structure for the Amyloid β peptide Aβ42 [98].

Solution ^{1}H NMR has been used to probe the aggregation state of many peptides. For example, in collaboration with the Miravet and Escuder groups, we examined the degree of ionization of the C terminus of a model amyloid peptide as a function of concentration [65]. In collaboration with this group, ^{1}H NMR solubility measurements have been used to obtain solubility phase diagrams and related diffusion measurements have enabled the critical micelle concentration of a lipopeptide to be obtained [58]. Specific chemical information, for information on metal ion binding to peptides, can also be obtained from solution ^{1}H NMR [99].

7 Dynamic and Static Light Scattering

In dynamic light scattering (DLS, sometimes known as photon correlation spectroscopy), analysis of the intensity autocorrelation function of scattered light from a particle undergoing translational or rotational diffusion enables determination of a diffusion coefficient, and hence via the Stokes-Einstein equation, the effective hydrodynamic radius. There are several approximate methods to relate this quantity to the fibril length and diameter [100, 101].

Static light scattering (SLS) can be used to obtain the molecular weight of peptide aggregates and also to provide an indication on particle shape, via measurements of the angular dependence of the scattered intensity. The angular dependence of the scattered light from amyloid fibrils in an SLS measurement can provide information on the conformation of amyloid fibrils, for example, whether stiff, flexible, or branched. Murphy and coworkers have applied linear and branched wormlike chain models to describe static light scattering data from Aβ alone and also from mixtures of Aβ with Aβ fragments [including Aβ(16–20)]] [102]. They also reported a branched wormlike chain form factor [102, 103]. We have used this method to fit SLS data from Aβ fragment peptide KLVFF [104].

Dynamic light scattering has been used to monitor the formation of amyloid protofilaments and fibrils [105–108] in the sense that it can distinguish between monomers, oligomers, and larger structures. However, since DLS provides information based on the assumption of spherical particles (in the case of the usual Stokes-Einstein) equation, it cannot reliably be used to obtain the dimensions of highly anisotropic particles such as amyloid fibrils and is therefore not discussed in further detail in this chapter.

8 SAXS and SANS

Small-angle X-ray scattering (SAXS) and small-angle neutron scattering (SANS) are powerful complementary techniques to probe the nanostructure of self-assembled objects. SAXS can be performed in the laboratory or at a central synchrotron facility, whereas SANS is performed using either a spallation or nuclear reactor as the source of neutrons. In either technique, the dependence of the scattered intensity at low angle or low wavenumber $q = 4\pi \sin\theta/\lambda$ (2θ is the scattering angle and λ is the wavelength) provides information on the interaction between self-assembled objects via the structure factor, and also on the shape and scattering density profile

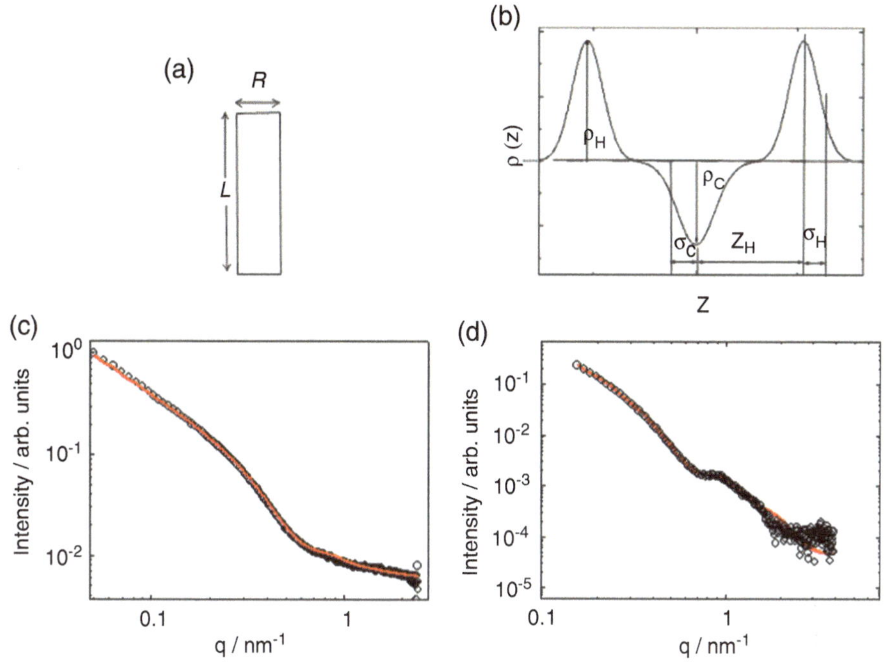

Fig. 5 Showing parameters for SAXS/SANS form factor fitting, and representative fits. (**a**) Parameters for a uniform cylinder model, (**b**) Gaussian bilayer electron density profile with parameters indicated, (**c**) representative fit to a uniform cylinder form factor, SAXS data for amyloid peptide KLVFF, [104] (**d**) representative fit to a Gaussian bilayer form factor, SAXS data for a C_{16} lipopeptide with a lumican-derived peptide headgroup [129]

(from the form factor). Since X-rays are scattered by electrons, a SAXS form factor depends on the electron density profile, whereas SANS provides information on the scattering length density profile (this depends on the nuclear scattering factors). In dilute solution, only form factor scattering is obtained and this is used to obtain the dimensions of peptide fibrils, lipopeptide tapes, etc.

For amyloid fibrils, a uniform [109] or core-shell [110] cylinder form factor can be used to fit SAXS or SANS data. Figure 5 shows the parameters for model uniform cylinder and (lipo)peptide bilayer form factor fits, along with representative fits to data obtained with such models. A core-shell model is required when there is significant contrast difference between the core of the peptide fibril and the exterior, for instance, in the case of PEG-peptides with a PEG corona [110, 111]. For nanotapes formed in particular (but not exclusively) by lipopeptides we have found that a form factor developed for lipid bilayers can successfully be employed. This so-called Gaussian bilayer form factor comprises three Gaussian functions, one of which represents the electron density (in the case of SAXS) of the lipid-chain rich core (negative amplitude), the other two being (positive amplitude) Gaussians

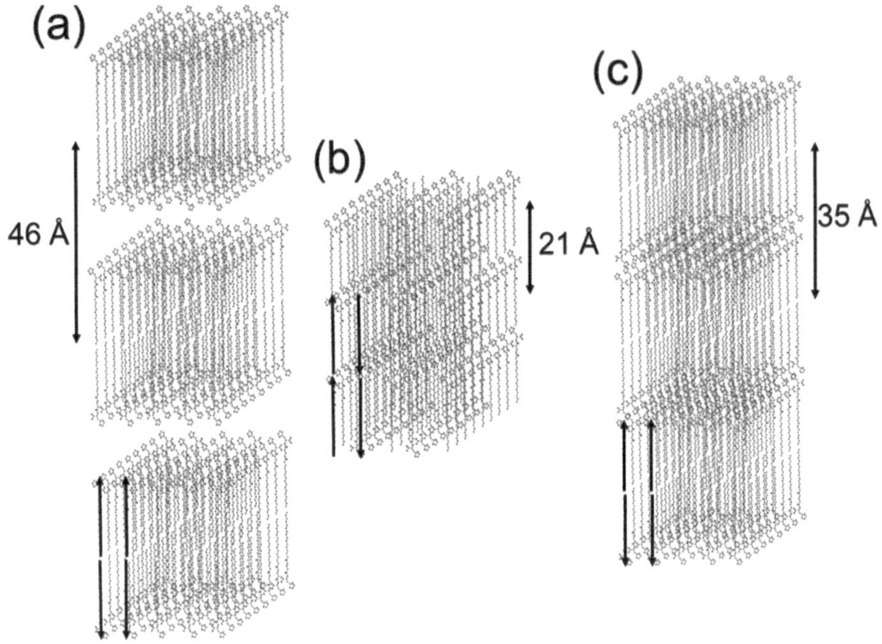

Fig. 6 Different layer structures observed for lipopeptide C_{16}-βAH [41]. (**a**) Hydrated bilayer, (**b**) monolayer, (**c**) dehydrated bilayer

representing the electron density of the charged head-groups (Fig. 5b). The corresponding parameters are shown in Fig. 5b and the form factor equations and definitions can be found in the original paper where this form factor was reported [112], or our papers utilizing this form factor [41, 43, 113]. We have found that the software SASfit [114] is very useful in fitting form factor data from amphiphilic peptide assemblies, as it incorporates many different form factors including those described above and many others, and it also has a powerful least-squares fitting algorithm. It also incorporates several model structure factors, and in some cases for lipopeptide bilayer systems we have used the Caillé structure factor [115] to represent multi-bilayer ordering [41]. In several cases, we have observed multiple polymorphisms of lipopeptide layer structures such as those shown in Fig. 6 [41].

The remarkable assembly of amphiphilic peptides into nanotubes has also been observed, and data have been fitted with corresponding form factors [113, 116]. If the nanotube radius $R \gg t$, where t is the tube wall thickness (e.g., of a bilayer), then the Gaussian bilayer form factor used above can be used to fit the high q portion of the small-angle scattering data [117].

Lipopeptides and amphiphilic peptides can also self-assemble into spherical micelles, SAXS data from which has been fitted using a core-shell (two electron density level) sphere form factor [47, 58, 118–120].

9 Atomic Force Microscopy and Electron Microscopy

These methods provide direct imaging of peptide fibrils. Atomic force microscopy (AFM) which is a scanning probe microscopy provides images of the surface. To date the method has mainly been used to image fibrils from solutions dried onto planar solid substrates such as mica. A nice recent review on the study of amyloid fibrils by AFM is available [121]. Recent developments include sophisticated analysis software which can be used to quantify the properties of amyloid fibrils including conformation, stiffness, orientation, etc. [122–124]. Measurement of fiber mechanical properties using peak force microscopy with an AFM instrument has now yielded values of the elastic modulus for many amyloid fibril systems [121, 125].

Electron microscopy includes transmission electron microscopy (TEM) of dried films on carbon grids (these are usually stained with heavy-metal-containing compounds to enhance contrast) or scanning electron microscopy (SEM) of surfaces. Cryo-techniques are often preferred as a method to avoid artifacts caused by slow drying. Indeed, we have found cryo-TEM to be the best imaging technique for amphiphilic peptide nanostructures. A review of the topic is available [126]. In our experience, SEM will provide an image of a "fibrillar network" morphology for many peptide structures which does not resolve the individual nanostructure elements on the 1–10 nm lengthscale, which is possible using cryo-TEM. Figure 7 shows a representative cryo-TEM image of nanotapes formed by self-assembly of a β-sheet-forming peptide.

Fig. 7 Cryo-TEM image obtained from a 2 wt% solution of peptide GNNDESNISFKEK [130]

In some cases, amphiphilic peptide assemblies such as fibrils or nanotapes are large enough to be imaged by optical microscopy—see for example [127].

10 Summary

In summary, a variety of methods are available to investigate the self-assembly of amphiphilic peptides. Often it is first of interest to determine the critical aggregation concentration, if present, from concentration-dependent fluorescent probe intensity measurements. Spectroscopic methods can be used to probe molecular conformation (NMR), chiral ordering (CD), and hydrogen bonding (FTIR). Molecular packing can be studied using fiber XRD which is appropriate to the typical extended fibrillar and tape structures formed by β-sheet amphiphilic peptide assemblies. Small-angle scattering is an in situ technique to determine the morphology and dimensions and scattering density profile of self-assembled structures. We generally complement SAXS or SANS data with cryo-TEM images or AFM images to provide a comprehensive characterization of nanostructure.

Acknowledgements

This work was supported by EPSRC Platform Grant reference EPSRC EP/L020599/1. We thank our collaborators for their vital contributions to several aspects of our ongoing research, especially in the fields of AFM imaging (Prof Raffaele Mezzenga and Dr. Jozef Adamcik, ETH Zürich) and cryo-TEM imaging (Prof Janne Ruokolainen and his team at Aalto University, Finland).

References

1. Creighton TE (1993) Proteins. Structures and molecular properties. W.H. Freeman, New York
2. Woolfson DN (2005) The design of coiled-coil structures and assemblies. Adv Protein Chem 70:79–112
3. Woolfson DN, Bartlett GJ, Bruning M, Thomson AR (2012) New currency for old rope: from coiled-coil assemblies to α-helical barrels. Curr Opin Struct Biol 11:432–441
4. Apostolovic B, Danial M, Klok HA (2010) Coiled coils: attractive protein folding motifs for the fabrication of self-assembled, responsive and bioactive materials. Chem Soc Rev 39:3541–3575
5. Santoso SS, Vauthey S, Zhang S (2002) Structure, function and applications of amphiphilic peptides. Curr Opin Colloid Interface Sci 7:262–266
6. Zhang SG (2003) Fabrication of novel biomaterials through molecular self-assembly. Nat Biotechnol 21:1171–1178
7. Zhao X, Zhang S (2004) Fabrication of molecular materials using peptide construction motifs. Trends Biotechnol 22:470–476
8. Löwik DWPM, van Hest JCM (2004) Peptide based amphiphiles. Chem Soc Rev 33:234–245
9. Cavalli S, Kros A (2008) Scope and applications of amphiphilic alkyl- and lipopeptides. Adv Mater 20:627–631
10. Versluis F, Marsden HR, Kros A (2010) Power struggles in peptide-amphiphile nanostructures. Chem Soc Rev 39:3434–3444

11. Cui HG, Webber MJ, Stupp SI (2010) Self-assembly of peptide amphiphiles: from molecules to nanostructures to biomaterials. Pept Sci 94:1–18
12. Hamley IW (2011) Self-assembly of amphiphilic peptides. Soft Matter 7:4122–4138
13. Dehsorkhi A, Castelletto V, Hamley IW (2014) Self-assembling amphiphilic peptides. J Pept Sci 20:453–467
14. Nelson R, Eisenberg D (2006) Recent atomic models of amyloid fibril structure. Curr Opin Struct Biol 16:260–265
15. Nelson R, Eisenberg D (2006) Structural models of amyloid-like fibrils. Adv Protein Chem 73:235–282
16. Nelson R, Sawaya MR, Balbirnie M, Madsen AO, Riekel C, Grothe R, Eisenberg D (2005) Structure of the cross-β spine of amyloid-like fibrils. Nature (London) 435:773–778
17. Eanes ED, Glenner GG (1968) X-ray diffraction studies on amyloid filaments. J Histochem Cytochem 16:673–677
18. Kirschner DA, Abraham C, Selkoe DJ (1986) X-ray diffraction from intraneuronal paired helical filaments and extraneuronal amyloid fibers in Alzheimer disease indicates cross-β conformation. Proc Natl Acad Sci U S A 83:503–507
19. Sunde M, Blake CCF (1997) The structure of amyloid fibrils by electron microscopy and X-ray diffraction. Adv Protein Chem 50:123–159
20. Serpell LC (2000) Alzheimer's amyloid fibrils: structure and assembly. Biochim Biophys Acta 1502:16–30
21. Makin OS, Serpell LC (2005) Structures for amyloid fibrils. FEBS J 272:5950–5961
22. Hamley IW, Castelletto V, Moulton CM, Rodriguez-Perez J, Squires AM, Eralp T, Held G, Hicks M, Rodger A (2010) Alignment of a model amyloid peptide fragment in bulk and at a solid surface. J Phys Chem B 114:8244–8254
23. Hamley IW (2012) The amyloid beta peptide: a chemist's perspective. Role in Alzheimer's and fibrillization. Chem Rev 112:5147–5192
24. Morris K, Serpell L (2010) From natural to designer self-assembling biopolymers, the structural characterization of fibrous proteins and peptides using fibre diffraction. Chem Soc Rev 39:3445–3453
25. Makin OS, Sikorski P, Serpell LC (2007) CLEARER: a new tool for the analysis of X-ray fibre diffraction patterns and diffraction simulation from atomic structural models. J Appl Crystallogr 40:966–972
26. Squires AM, Devlin GL, Gras SL, Tickler AK, MacPhee CE, Dobson CM (2006) X-ray scattering study of the effect of hydration on the cross-β structure of amyloid fibrils. J Am Chem Soc 128:11738–11739
27. Papapostolou D, Smith AM, Atkins EDT, Oliver SJ, Ryadnov MG, Serpell LC, Woolfson DN (2007) Engineering nanoscale order into a designed protein fiber. Proc Natl Acad Sci U S A 104:10853–10858
28. Banwell EF, Abelardo ES, Adams DJ, Birchall MA, Corrigan A, Donald AM, Kirkland M, Serpell LC, Butler MF, Woolfson DN (2009) Rational design and application of responsive alpha-helical peptide hydrogels. Nat Mater 8:596–600
29. Makin OS, Serpell LC (2004) The structure of amyloid. Fibre Diffraction Review 12:29–35
30. Sunde M, Serpell LC, Bartlam M, Fraser PE, Pepys MB, Blake CCF (1997) Common core structure of amyloid fibrils by synchrotron X-ray diffraction. J Mol Biol 273:729–739
31. Hill RJA, Sedman VL, Allen S, Williams PM, Paoli M, Adler-Abramovich L, Eaves L, Tendler SJB (2007) Alignment of aromatic peptide tubes in strong magnetic fields. Adv Mater 19:4474–4479
32. Hamley IW (2007) Peptide fibrillisation. Angew Chem 46:8128–8147
33. Levine H (1993) Thioflavine T interaction with synthetic Alzheimer's disease β-amyloid peptides: detection of amyloid aggregation in solution. Protein Sci 2:404–410
34. Wood SJ, Maleef B, Hart T, Wetzel R (1996) Physical, morphological and functional differences between pH 5.8 and 7.4 aggregates of the Alzheimer's amyloid peptide Aβ. J Mol Biol 256:870–877
35. Hamley IW (2007) Introduction to soft matter, Revised edn. Wiley, Chichester
36. Kalyanasundaram K, Thomas JK (1977) Environmental effects on vibronic band intensities in pyrene monomer fluorescence and their application in studies of micellar systems. J Am Chem Soc 99:2039–2044
37. Winnik FM (1993) Photophysics of preassociated pyrenes in aqueous polymer-solutions and in other organized media. Chem Rev 93:587–614
38. Guler MO, Claussen RC, Stupp SI (2005) Encapsulation of pyrene within self-assembled peptide amphiphile nanofibers. J Mater Chem 15:4507–4512
39. Sabate R, Estelrich J (2005) Evidence of the existence of micelles in the fibrillogenesis of

beta-amyloid peptide. J Phys Chem B 109:11027–11032

40. Castelletto V, Cheng G, Greenland BW, Hamley IW (2011) Tuning the self-assembly of the bioactive dipeptide L-carnosine by incorporation of a bulky aromatic substituent. Langmuir 27:2980–2988
41. Castelletto V, Cheng G, Stain C, Connon CJ, Hamley IW (2012) Self-assembly of a peptide amphiphile containing L-carnosine and its mixtures with a multilamellar vesicle forming lipid. Langmuir 28:11599–11608
42. Jones RR, Castelletto V, Connon CJ, Hamley IW (2013) Collagen stimulating effect of peptide amphiphile C_{16}-KTTKS on human fibroblasts. Mol Pharm 10:1063–1069
43. Castelletto V, Gouveia RJ, Connon CJ, Hamley IW (2013) New RGD-peptide amphiphile mixtures containing a negatively charged diluent. Faraday Discuss 166:381–397
44. Hamley IW, Dehsorkhi A, Castelletto V (2013) Coassembly in binary mixtures of peptide amphiphiles containing oppositely charged residues. Langmuir 29:5050–5059
45. Castelletto V, Gouveia RJ, Connon CJ, Hamley IW, Seitsonen J, Nykänen A, Ruokolainen J (2014) Alanine-rich amphiphilic peptide containing the RGD cell adhesion motif: a coating material for human fibroblast attachment and culture. Biomater Sci 2:362–369
46. Fowler M, Siddique B, Duhamel J (2013) Effect of sequence on the ionization behavior of a series of amphiphilic polypeptides. Langmuir 29:4451–4459
47. Hamley IW, Kirkham S, Dehsorkhi A, Castelletto V, Reza M, Ruokolainen J (2014) Toll-like receptor agonist lipopeptides self-assemble into distinct nanostructures. Chem Commun 50:15948–15951
48. Wilhelm M, Zhao C-L, Wang Y, Xu R, Winnik MA, Mura J-L, Riess G, Croucher MD (1991) Poly(styrene-ethylene oxide) block copolymer micelle formation in water: a fluorescence probe study. Macromolecules 24:1033–1040
49. Johnsson M, Hansson P, Edwards K (2001) Spherical micelles and other self-assembled structures in dilute aqeuous mixtures of poly (ethylene glycol) lipids. J Phys Chem B 105:8420–8430
50. Decandio CC, Silva ER, Hamley IW, Castelletto V, Liberato MS, Oliveira VX, Oliveira CLP, Alves WA (2015) Self-assembly of a designed alternating arginine/phenylalanine oligopeptide. Langmuir 31:4513–4523
51. LeVine H (1999) Quantification of β-sheet amyloid fibril structures with thioflavin T. In: Wetzel R (ed) Methods in enzymology, vol 309. Academic, San Diego, pp 274–284
52. Santra MK, Banerjee A, Krishnakumar SS, Rahaman O, Panda D (2004) Multiple-probe analysis of folding and unfolding pathways of human serum albumin - evidence for a framework mechanism of folding. Eur J Biochem 271:1789–1797
53. Jha S, Snell JM, Sheftic SR, Patil SM, Daniels SB, Kolling FW, Alexandrescu AT (2014) pH dependence of amylin fibrillization. Biochemistry 53:300–310
54. van den Heuvel M, Baptist H, Venema P, van der Linden E, Löwik D, van Hest JCM (2011) Mechanical and thermal stabilities of peptide amphiphile fibres. Soft Matter 7:9737–9743
55. Nagai A, Nagai Y, Qu HJ, Zhang SG (2007) Dynamic behaviors of lipid-like self-assembling peptide A_6D and A_6K nanotubes. J Nanosci Nanotechnol 7:2246–2252
56. Khoe U, Yang YL, Zhang SG (2008) Synergistic effect and hierarchical nanostructure formation in mixing two designer lipid-like peptide surfactants ac-A_6D-OH and ac-A_6K-NH_2. Macromol Biosci 8:1060–1067
57. Hamley IW, Nutt DR, Brown GD, Miravet JF, Escuder B, Rodríguez-Llansola F (2010) Influence of the solvent on the self-assembly of a modified amyloid beta peptide fragment. II. NMR and computer simulation investigation. J Phys Chem B 114:940–951
58. Miravet JF, Escuder B, Segarra-Maset MD, Tena-Solsona M, Hamley IW, Dehsorkhi A, Castelletto V (2013) Self-assembly of a peptide amphiphile: transition from nanotape fibrils to micelles. Soft Matter 9:3558–3564
59. Girych M, Gorbenko G, Trusova V, Adachi E, Mizuguchi C, Nagao K, Kawashima H, Akaji K, Lund-Katz S, Phillips MC, Saito H (2014) Interaction of thioflavin T with amyloid fibrils of apolipoprotein A-I N-terminal fragment: resonance energy transfer study. J Struct Biol 185:116–124
60. Kelly SM, Jess TJ, Price NC (2005) How to study proteins by circular dichroism. Biochim Biophys Acta 1751:119–139
61. Woody RW (1995) Circular dichroism. Methods Enzymol 246:34–71
62. Shi ZS, Woody RW, Kallenbach NR (2002) Is polyproline II a major backbone conformation in unfolded proteins? In: Rose GD (ed) Unfolded proteins, Advances in protein chemistry, vol 62. Academic, San Diego, pp 163–240
63. Paramonov SE, Jun H-W, Hartgerink JD (2006) Self-assembly of peptide-amphiphile

nanofibers: the roles of hydrogen bonding and amphiphilic packing. J Am Chem Soc 128:7291–7298

64. Woody RW (2009) Circular dichroism spectrum of peptides in the poly(pro)II conformation. J Am Chem Soc 131:8234–8245
65. Castelletto V, Hamley IW, Cenker C, Olsson U, Adamcik J, Mezzenga R, Miravet JF, Escuder B, Rodriguez-Llansola F (2011) Influence of end-capping on the self-assembly of model amyloid peptide fragments. J Phys Chem B 115:2107–2116
66. Bulheller BM, Rodger A, Hirst JD (2007) Circular and linear dichroism of proteins. Phys Chem Chem Phys 9:2020–2035
67. Reed J, Reed A (1997) A set of constructed type spectra for the practical estimation of peptide secondary structure from circular dichroism. Anal Biochem 254:36–40
68. Rodger A, Nordén B (1997) Circular dichroism and linear dichroism. Oxford University Press, Oxford
69. Nordén B, Rodger A, Dafforn TR (2010) Linear dichroism and circular dichroism: a textbook on polarized-light spectroscopy. Royal Society of Chemistry, Cambridge
70. Surewicz WK, Mantsch HH, Chapman D (1993) Determination of protein secondary structure by Fourier transform infrared spectroscopy: a critical assessment. Biochemistry 32:389–394
71. Stuart B (1997) Biological applications of infrared spectroscopy. Wiley, Chichester
72. Jackson M, Mantsch HH (1995) The use and misuse of FTIR spectroscopy in the determination of protein structure. Crit Rev Biochem Mol Biol 30:95–120
73. Hiramatsu H, Kitagawa T (2005) FT-IR approaches on amyloid fibril structure. Biochim Biophys Acta 1753:100–107
74. Haris P, Chapman D (1995) The conformational analysis of peptide using Fourier transform IR spectroscopy. Biopolymers 37:251–263
75. Barth A, Zscherp C (2002) What vibrations tell us about proteins. Q Rev Biophys 35:369–430
76. Kubelka J, Keiderling TA (2001) The anomalous infrared amide I intensity distribution in ^{13}C isotopically labeled peptide β-sheets comes from extended, multiple-stranded structures. An ab initio study. J Am Chem Soc 123:6142–6150
77. Krimm S, Bandekar J (1986) Vibrational spectroscopy and conformation of peptides, polypeptides and proteins. Adv Protein Chem 38:181–364
78. Bellamy LJ (1975) The infra-red spectra of complex molecules. Chapman and Hall, London
79. Castelletto V, Moulton CM, Cheng G, Hamley IW, Hicks MR, Rodger A, López-Pérez DE, Revilla-López G, Alemán C (2011) Self-assembly of fmoc-tetrapeptides based on the RGDS cell adhesion motif. Soft Matter 7:11405–11415
80. Pelton JT, McLean LR (2000) Spectroscopic methods for analysis of protein secondary structure. Anal Biochem 277:167–176
81. Gaussier H, Morency H, Lavoie MC, Subirade M (2002) Replacement of trifluoroacetic acid with HCl in the hydrophobic purification steps of pediocin PA-1: a structural effect. Appl Environ Microbiol 68:4803–4808
82. Eker F, Griebenow K, Schweitzer-Stenner R (2004) $A\beta_{1\text{-}28}$ fragment of the amyloid peptide predominantly adopts a polypropyline II conformation in acidic solution. Biochemistry 43:6893–6898
83. Liang Y, Pingali SV, Jogalekar AS, Snyder JP, Thiyagarajan P, Lynn DG (2008) Cross-strand pairing and amyloid assembly. Biochemistry 47:10018–10026
84. Mehta AK, Lu K, Childers WS, Liang S, Dong J, Snyder JP, Pingali SV, Thiyagarajan P, Lynn DG (2008) Facial symmetry in protein self-assembly. J Am Chem Soc 130:9829–9835
85. Rodríguez-Pérez J, Hamley IW, Gras SL, Squires AM (2012) Local orientational disorder in peptide fibrils probed by a combination of residue-specific ^{13}C-^{18}O labelling, polarised infrared spectroscopy and molecular combing. Chem Commun 48:11835–11837
86. Middleton DA, Madine J, Castelletto V, Hamley IW (2013) New insights into the molecular architecture of a peptide nanotube using FTIR and solid-state NMR spectroscopy combined with sample alignment. Angew Chem Int Ed Engl 52:10537–10540
87. Rodriguez-Perez J, Hamley IW, Squires AM (2011) Infrared linear dichroism spectroscopy on amyloid fibrils aligned by molecular combing. Biomacromolecules 12:1810–1821
88. Nafie LA (1997) Infrared and raman vibrational optical activity: theoretical and experimental aspects. Annu Rev Phys Chem 48:357–386
89. Keiderling TA (2002) Protein and peptide secondary structure and conformational determination with vibrational circular dichroism. Curr Opin Chem Biol 6:682–688
90. Costa PR, Kocisko DA, Sun BQ, Lansbury PT, Griffin RG (1997) Determination of

peptide amide configuration in a model amyloid fibril by solid-state NMR. J Am Chem Soc 119:10487–10493

91. Benzinger TLS, Gregory DM, Burkoth TS, Miller-Auer H, Lynn DG, Botto RE, Meredith SC (2000) Two-dimensional structure of β-amyloid(10-35) fibrils. Biochemistry 39:3491–3499
92. Siemer AB, Ritter C, Ernst M, Riek R, Meier BH (2005) High-resolution solid-state NMR spectroscopy of the prion protein HET-s in its amyloid conformation. Angew Chem Int Ed Engl 44:2441–2444
93. Heise H, Hoyer W, Becker S, Andronesi OC, Riedel D, Baldus M (2005) Molecular-level secondary structure, polymorphism, and dynamics of full-length alpha-synuclein fibrils studied by solid-state NMR. Proc Natl Acad Sci U S A 102:15871–15876
94. Balbach JJ, Ishii Y, Antzutkin ON, Leapman RD, Rizzo NW, Dyda F, Reed J, Tycko R (2000) Amyloid fibril formation by Aβ16-22, a seven-residue fragment of the Alzheimer's β-amyloid peptide, and structural characterization by solid state NMR. Biochemistry 39:13748–13759
95. Ma BY, Nussinov R (2002) Stabilities and conformations of Alzheimer's beta-amyloid peptide oligomers ($A\beta_{16\text{-}22}$, $A\beta_{16\text{-}35}$, and $A\beta_{10\text{-}35}$): sequence effects. Proc Natl Acad Sci U S A 99:14126–14131
96. Tycko R (2004) Progress towards a molecular-level structural understanding of amyloid fibrils. Curr Opin Struct Biol 14:96–103
97. Tycko R (2006) Molecular structure of amyloid fibrils: insights from solid-state NMR. Q Rev Biophys 39:1–55
98. Lührs T, Ritter C, Adrian M, Riek-Loher D, Bohrmann B, Döbeli H, Schubert D, Riek R (2005) 3D structure of Alzheimer's amyloid-β(1-42) fibrils. Proc Natl Acad Sci U S A 102:17342–17347
99. Castelletto V, Hamley IW, Segarra-Maset MD, Gumbau CB, Miravet JF, Escuder B, Seitsonen J, Ruokolainen J (2014) Tuning chelation by the surfactant-like peptide A_6H using predetermined pH values. Biomacromolecules 15:591–598
100. Lomakin A, Chung DS, Benedek GB, Kirschner DA, Teplow DB (1996) On the nucleation and growth of amyloid β-protein fibrils: detection of nuclei and quantitation of rate constants. Proc Natl Acad Sci U S A 93:1125–1129
101. Riseman J, Kirkwood JG (1950) J Chem Phys 18:512–516
102. Pallitto MM, Ghanta J, Heinzelman P, Kiessling LL, Murphy RM (1999) Recognition sequence design for peptidyl modulators of β-amyloid aggregation and toxicity. Biochemistry 38:3570–3578
103. Shen C-L, Fitzgerald MC, Murphy RM (1994) Effect of acid predissolution on fibril size and fibril flexibility of synthetic β-amyloid peptide. Biophys J 67:1238–1246
104. Krysmann MJ, Castelletto V, Kelarakis A, Hamley IW, Hule RA, Pochan DJ (2008) Self-assembly and hydrogelation of an amyloid peptide fragment. Biochemistry 47:4597–4605
105. Walsh DM, Lomakin A, Benedek GB, Condron MM, Teplow DB (1997) Amyloid β-protein fibrillogenesis. Detection of a protofibrillar intermediate. J Biol Chem 272:22364–22372
106. Kusumoto Y, Lomakin A, Teplow DB, Benedek GB (1998) Temperature dependence of amyloid β-protein fibrillization. Proc Natl Acad Sci U S A 95:12277–12282
107. Sokolowski F, Modler AJ, Masuch R, Zirwer D, Baier M, Lutsch G, Moss DA, Gast K, Naumann D (2003) Formation of critical oligomers is a key event during conformational transition of recombinant Syrian hamster prion protein. J Biol Chem 278:40481–40492
108. Modler AJ, Gast K, Lutsch G, Damaschun G (2003) Assembly of amyloid protofibrils via critical oligomers—a novel pathway of amyloid formation. J Mol Biol 325:135–148
109. Pedersen JS (1997) Analysis of small-angle scattering data from colloids and polymer solutions: modeling and least-squares fitting. Adv Colloid Interf Sci 70:171–210
110. Hamley IW, Krysmann MJ, Castelletto V, Kelarakis A, Noirez L, Hule RA, Pochan D (2008) Nematic and columnar ordering of a PEG-peptide conjugate in aqueous solution. Chem Eur J 14:11369–11374
111. Hamley IW, Krysmann MJ, Castelletto V, Noirez L (2008) Multiple lyotropic polymorphism of a PEG-peptide diblock copolymer in aqueous solution. Adv Mater 20:4394–4397
112. Pabst G, Rappolt M, Amenitsch H, Laggner P (2000) Structural information from multilamellar liposomes at full hydration: full q-range fitting with high quality X-ray data. Phys Rev E 62:4000–4009
113. Hamley IW, Dehsorkhi A, Castelletto V (2013) Self-assembled arginine-coated peptide nanosheets in water. Chem Commun 49:1850–1852
114. http://kur.web.psi.ch/sans1/SANSSoft/sasfit.html (2015)

115. Caillé A (1972) X-ray scattering by smectic-a crystals. CR Hebd Seances Acad Sci (Paris) B 274:891–893
116. Lu K, Jacob J, Thiyagarajan P, Conticello VP, Lynn DG (2003) Exploiting amyloid fibril lamination for nanotube self assembly. J Am Chem Soc 125:6391–6393
117. Hamley IW, Dehsorkhi A, Castelletto V, Furzeland S, Atkins D, Seitsonen J, Ruokolainen J (2013) Reversible helical ribbon unwinding transition of a self-assembling peptide amphiphile. Soft Matter 9:9290–9293
118. Dehsorkhi A, Hamley IW, Seitsonen J, Ruokolainen J (2013) Tuning self-assembled nanostructures through enzymatic degradation of a peptide amphiphile. Langmuir 29:6665–6672
119. Hamley IW, Dehsorkhi A, Jauregi P, Seitsonen J, Ruokolainen J, Coutte F, Chataigné G, Jacques P (2013) Self-assembly of three bacterially-derived bioactive lipopeptides. Soft Matter 9:9572–9578
120. Dehsorkhi A, Gouveia RJ, Smith AM, Hamley IW, Castelletto V, Connon CJ, Reza M, Ruokolainen J (2015) Self-assembly of a dual functional bioactive peptide amphiphile incorporating both matrix metalloprotease substrate and cell adhesion motifs. Soft Matter 11:3115–3124
121. Adamcik J, Mezzenga R (2012) Study of amyloid fibrils via atomic force microscopy. Curr Opin Colloid Interface Sci 17:369–376
122. Usov I, Adamcik J, Mezzenga R (2013) Polymorphism in bovine serum albumin fibrils: morphology and statistical analysis. Faraday Discuss 166:151–162
123. Usov I, Mezzenga R (2014) Correlation between nanomechanics and polymorphic conformations in amyloid fibrils. ACS Nano 8:11035–11041
124. Usov I, Mezzenga R (2015) FiberApp: an open-source software for tracking and analyzing polymers, filaments, biomacromolecules, and fibrous objects. Macromolecules 48:1269–1280
125. Adamcik J, Lara C, Usov I, Jeong JS, Ruggeri FS, Dietler G, Lashuel HA, Hamley IW, Mezzenga R (2012) Measurement of intrinsic properties of amyloid fibrils by the peak force QNM method. Nanoscale 4:4426–4429
126. Cui H, Hodgdon TK, Kaler EW, Abezgaous L, Danino D, Lubovsky M, Talmon Y, Pochan DJ (2007) Elucidating the assembled structure of amphiphiles in solution via cryogenic-transmission electron microscopy. Soft Matter 3:945–955
127. Castelletto V, Hamley IW, Perez J, Abezgauz L, Danino D (2010) Fibrillar superstructure from extended nanotapes formed by a collagen-stimulating peptide. Chem Commun 46:9185–9187
128. Castelletto V, Gouveia RJ, Connon CJ, Hamley IW, Seitsonen J, Ruokolainen J, Longo E, Siligardi G (2014) Influence of elastase on alanine-rich peptide hydrogels. Biomater Sci 2:867–874
129. Hamley IW, Dehsorkhi A, Castelletto V, Walter MNM, Connon CJ, Reza M, Ruokolainen J (2015) Self-assembly and collagen stimulating activity of a peptide amphiphile incorporating a peptide sequence from lumican. Langmuir 31:4490–4495
130. Castelletto V, Kirkham S, Hamley IW, Kowalczyk R, Rabe M, Reza M, Ruokolainen J (2016) Self-assembly of the toll-like receptor agonist macrophage-activating lipopeptide MALP-2 and of its constituent peptide. Biomacromolecules 17:631–640

Chapter 2

Solid-State NMR Structural Characterization of Self-Assembled Peptides with Selective ^{13}C and ^{15}N Isotopic Labels

Danting Huang, Benjamin C. Hudson, Yuan Gao, Evan K. Roberts, and Anant K. Paravastu

Abstract

For the structural characterization methods discussed here, information on molecular conformation and intermolecular organization within nanostructured peptide assemblies is discerned through analysis of solid-state NMR spectral features. This chapter reviews general NMR methodologies, requirements for sample preparation, and specific descriptions of key experiments. An attempt is made to explain choices of solid-state NMR experiments and interpretation of results in a way that is approachable to a nonspecialist. Measurements are designed to determine precise NMR peak positions and line widths, which are correlated with secondary structures, and probe nuclear spin–spin interactions that report on three-dimensional organization of atoms. The formulation of molecular structural models requires rationalization of data sets obtained from multiple NMR experiments on samples with carefully chosen ^{13}C and ^{15}N isotopic labels. The information content of solid-state NMR data has been illustrated mostly through the use of simulated data sets and references to recent structural work on amyloid fibril-forming peptides and designer self-assembling peptides.

Key words Solid-state NMR, Self-assembling peptide, Selective isotopic labeling, Dipolar redoubling, Two-dimensional NMR

1 Introduction

This chapter aims to be a practical guide to solid-state nuclear magnetic resonance (NMR)-based structural measurements of self-assembled peptide systems. We focus on well-established experiments with detection of ^{13}C or ^{15}N NMR signals, performed on isotopically labeled peptide samples produced by chemical synthesis (Fig. 1). These nuclei have low natural abundances (1% for ^{13}C and 0.4% for ^{15}N), such that isotopic labeling at specific sites makes it possible to obtain localized structural information. This information is extracted from observed effects of structure on NMR spectra and nuclear spin dynamics. Our descriptions of

Bradley L. Nilsson and Todd M. Doran (eds.), *Peptide Self-Assembly: Methods and Protocols*, Methods in Molecular Biology, vol. 1777, https://doi.org/10.1007/978-1-4939-7811-3_2, © Springer Science+Business Media, LLC, part of Springer Nature 2018

solid-state NMR methods do not assume in-depth understanding of NMR theory. However, we do not attempt to explain basic operation of an NMR spectrometer. NMR spectroscopy involves: (1) placement of samples into large static magnetic fields (~10 Tesla); (2) excitation of nuclear spins (nuclei with magnetic moments) by resonant radio frequency magnetic fields delivered via induction by a coil around the sample; and (3) detection of current induced in the coil by time-dependent nuclear magnetization. We will describe a toolkit consisting of a series of NMR experiments, each with radio frequency excitations designed to obtain specific structural information from samples with deliberate patterns of isotopic labeling. We will also mention important complementary methods, including sample preparation, computational molecular modeling, and nuclear spin simulations.

Solid-state NMR is often the method of choice for structural biology of self-assembling peptides because such samples usually lack the solubility and crystalline order necessary for solution-state NMR spectroscopy and X-ray crystallography, respectively. Solid-state NMR refers to NMR techniques that are specially designed for analysis of samples with highly constrained molecular motions, i.e., with correlation times for molecular orientation that are long on the measurement timescale (milliseconds per scan). The noncrystalline nature of self-assembled peptides further restricts experiments to those designed for random distributions of molecular orientation ("powder" samples). Solid-state NMR experiments involve hardware and methods designed to compensate for orientation-dependent effects and selectively manipulate local nuclear magnetic interactions. Observables provide information on backbone torsion angles (see Fig. 1a), proximity maps for

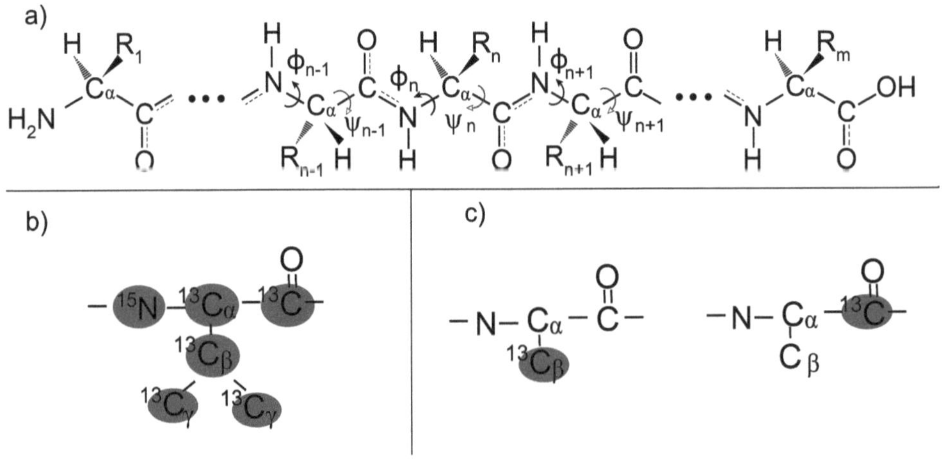

Fig. 1 (**a**) Schematic representation of a polypeptide chain. The ϕ and ψ torsion angles specify secondary structure and molecular fold. (**b**) Uniformly $^{13}C/^{15}N$-labeled valine. (**c**) Selectively ^{13}C-labeled alanine, labeled at C_β or CO. Sites labeled with ^{13}C are highlighted in gray ovals

different amino acids (residues), and three-dimensional (3D) arrangements of isotopically labeled sites. The direct observables are peak positions, peak line widths, and decays in intensity under the influence of nuclear spin–spin dipolar couplings.

In practice, data obtained by other techniques are used to complement information which is uniquely attainable by solid-state NMR. While the implementation of these techniques is beyond the scope of this chapter, understanding them can be essential to the rationalization of NMR data. Molecular dynamics computer simulations can be instrumental in understanding NMR-derived structural constraints by identifying reasonable molecular structures [1–3]. These simulations can be performed with artificial force fields inspired by NMR results [4–8]. Electron microscopy and atomic force microscopy [4, 5, 7, 9–11] can quantify structural features at larger length scales (nanometers) than are typically accessible by NMR. Quantitative measurements of fibril mass have been reported using scanning transmission electron microscopy [4, 11–14] and tilted-beam transmission electron microscopy [15]. Secondary structure (β-strands and α-helices) may be probed by Fourier transform infrared spectroscopy [16–19] and circular dichroism [20–22]. Fluorescent dyes, i.e., Thioflavin-T [23–26] and Congo Red [27–29], are effective indicators for monitoring β-sheet formation. Distinction of quaternary structure by cryogenic electron microscopy has been reported in the characterization of peptide nanofibers [13, 30–32]. Molecular folding and intermolecular packing (e.g., β-sheet stacking) can be measured using many diffraction-based techniques [33–37], H/D exchange [38, 39], electron paramagnetic resonance (EPR) [40–44], and proteolysis [45, 46].

Important background and insights are revealed by previous scientific contributions based on solid-state NMR of self-assembled peptides. For each experiment presented here, we reference precedent applications. A majority of these techniques were originally developed to study naturally occurring peptide self-assembly, which is most often associated with pathological amyloid deposition [47–52]. The study of designer peptides presents a promising new area of application for solid-state NMR [5, 7, 53]. For non-natural self-assembling peptides, solid-state NMR plays the indispensable role of "closing the design loop" by providing a means of comparing hypothesized structures with those that are actually produced. The following sections review general solid-state NMR methodologies, requirements for sample preparation, and specific descriptions of key experiments. The NMR data presented in this chapter are illustrated using simulated data sets based on previously published results.

2 Overview of Solid-State NMR Structural Investigations

2.1 Unique Characteristics of Solid-State NMR Equipment

The experiments discussed here require an NMR system (magnet, spectrometer, and probe) that is equipped for solid-state NMR measurements. As illustrated in Fig. 2, the NMR magnet establishes a uniform static magnetic field ($\vec{B}_0$) throughout the volume of the sample, while the NMR probe produces a time-dependent radio frequency magnetic field ($\vec{B}_1$). The NMR probe contains a resonant radio frequency circuit to induce $\vec{B}_1$ by means of a current delivered through a coil surrounding the sample; the coil is the only part of the probe that is drawn in Fig. 2. The experiments described here require magic-angle spinning (MAS) of samples and double-resonance (simultaneous excitation of ^{1}H and ^{13}C) or triple-resonance (simultaneous excitation of ^{1}H, ^{13}C, and ^{15}N) NMR probes. In addition, solid-state NMR experiments tend to require higher powers for radio frequency excitation compared to most solution-state NMR experiments.

MAS refers to sample rotation about an angle $\theta = \cos^{-1}\frac{1}{\sqrt{3}} \approx 54.7^{\circ}$ relative to $\vec{B}_0$ (Fig. 2) [54]. Typical MAS rotation frequencies range from 5 to 30 kHz. MAS improves spectral resolution (narrows NMR peaks) by eliminating molecular orientation-dependent effects on spin dynamics. NMR probes designed for MAS require that samples be placed inside special sample holders, called rotors, which are designed to withstand the forces involved in sample spinning. Specific probes are designed to work with specific rotors, with typical rotor diameters ranging between 2.5 and 4 mm. Probes designed for larger MAS rotor sizes make it possible to perform experiments on larger sample

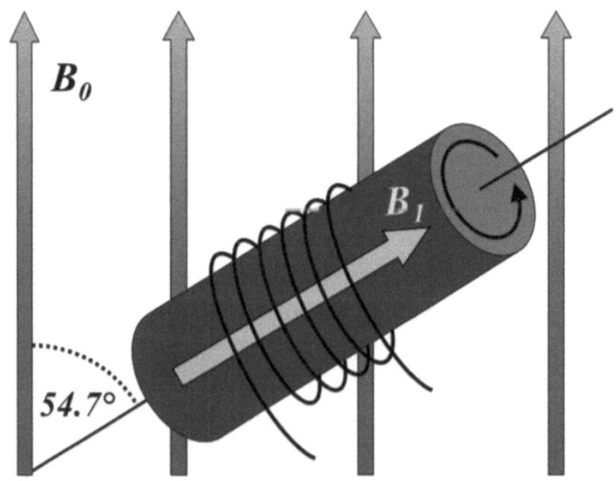

Fig. 2 Depiction of the sample (cylinder), coil, and magnetic fields involved in the NMR experiment. The sample is orientated at the magic angle relative to the static magnetic field $\vec{B}_0$. A tuned radio frequency probe circuit, to which the coil is connected, produces a time-dependent $\vec{B}_1$ field and detects the NMR signal

volumes, producing stronger NMR signals. However, larger rotors correspond to lower maximum MAS rotation rates, which in turn limit the capacity to attenuate line-broadening mechanisms.

The high radio frequency power requirements for solid-state NMR are necessitated by two special experimental features: cross polarization and decoupling. Cross polarization is a signal enhancement method in which the polarization of ^{1}H is transferred to ^{13}C or ^{15}N [55]. Decoupling is the application of high-power radio frequency excitation of ^{1}H in order to eliminate ^{1}H-^{13}C or ^{1}H-^{15}N dipolar couplings that are not fully removed by MAS [56]. Decoupling is often necessary when applying pulses to excite ^{13}C or ^{15}N spin states and when detecting ^{13}C or ^{15}N NMR signals. Probes with larger coils (to accommodate larger rotors) have lower limits on radio frequency pulse powers.

2.2 Requirements for Samples

Preparation of peptides with selective ^{13}C and ^{15}N isotopic labels is accomplished by chemical peptide synthesis using isotopically labeled amino acids that are available commercially [57]. Individual amino acids could be uniformly labeled with ^{13}C and ^{15}N (Fig. 1b), or selectively labeled (Fig. 1c). The choice of isotopic label depends on the planned NMR experiments. A widely used toolkit for solid-phase peptide synthesis is based on fluorenyl-9-methoxycarbonyl (FMOC) chemistry [58]. Peptide synthesis produces practical yields for most peptides with up to ~50 amino acids [59]. Required steps following peptide synthesis include chemical cleavage from the solid-support resin, precipitation of product from the cleavage solvent (e.g., cleavage by trifluoroacetic acid followed by precipitation in diethyl ether), and purification. Purification is usually carried out using high performance liquid chromatography. Product quality can be tested using matrix-assisted laser desorption or electrospray ionization mass spectrometry [60–62].

Preparation of samples with well-defined molecular structures is critical to the success of a solid-state NMR structural investigation. The phenomenon of peptide self-assembly is complex, often promoting structural inhomogeneity in samples. Naturally occurring amyloidogenic peptides are known to assemble along multiple energetically similar pathways to distinct amyloid fibril structures [4, 63]. The formation of multiple fibril structures by self-assembly of a single peptide is often called polymorphism, because fibrils with different molecular structures exhibit distinct morphologies (nanofiber width, width modulation, and protofilament substructure) when observed by electron and atomic force microscopy [9, 41]. Amyloidogenic peptides are known to also form non-fibrillar species that have been described as oligomers and protofibrils [64–66]. There have been fewer studies of assembly pathways for designer peptides, but it is likely that they are similarly capable of self-assembly along multiple pathways. Therefore, researchers must take special care to control molecular structure during sample preparations.

Experimental protocols designed to prepare structurally homogeneous self-assembled peptide samples rely on a combination of controlled self-assembly conditions and seeded assembly. Fibril structures in samples are known to depend on solution conditions during self-assembly, such as pH, ionic strength, access to surfaces, and solution agitation [67–70]. Fibrils also exhibit seeded propagation, such that dissolved peptide molecules can be induced to self-assemble by addition of fibril fragment solutions aliquoted from sonicated fibril hydrogels [71–73]. Furthermore, seeding results in propagation of molecular structure, such that fibrils produced by seeding have structures that match those of the fibril fragments used as seeds. When seeded self-assembly is applied repeatedly to polymorphic fibril samples, fibril polymorphs that propagate most rapidly can be purified [4, 11, 74]. In contrast to fibril studies, structural analysis of non-fibrillar aggregates of amyloid-forming peptides (oligomers and protofibrils) have not relied on seeded assembly. Instead, achievement of structural homogeneity has relied on careful production of monomeric peptide and control of subsequent assembly conditions. Since non-fibrillar aggregates are often distinguished based on their sizes, size exclusion chromatography has been used to separate specific aggregates from monomers and fibrils [64, 75]. - Environment-dependent self-assembly has been reported for the RADA16-I designer peptide [76], and evidence of polymorphism has been observed in solid-state NMR of MAX8 designer peptide nanofibers [53]. It is possible that the relative simplicity of designer peptides facilitates establishment of thermodynamic equilibrium structures rather than kinetically trapped states [77, 78].

From the spectroscopic point of view, structural stability and uniformity of the sample affect NMR sensitivity, resolution, and data interpretation. Stability is important because NMR measurements often require days to weeks of repeated scanning. Unstable samples may be especially affected by radio frequency fields that elevate sample temperature [79]. Decoupling fields tend to be the largest contributor to sample heating because of their relatively high power and long duration. Structural uniformity is also important because samples containing multiple molecular structures would exhibit multiple NMR peaks for each labeled site. Such heterogeneity would result in more NMR peaks overall, promoting reduced spectral resolution due to spectral overlap. Furthermore, since peak intensity is approximately proportional to the number of atoms contributing to each NMR signal, increased structural homogeneity promotes stronger signals. Finally, the presence of multiple distinct molecular structures in a sample complicates interpretation of experiments based on measurement of spin dynamics [53, 57].

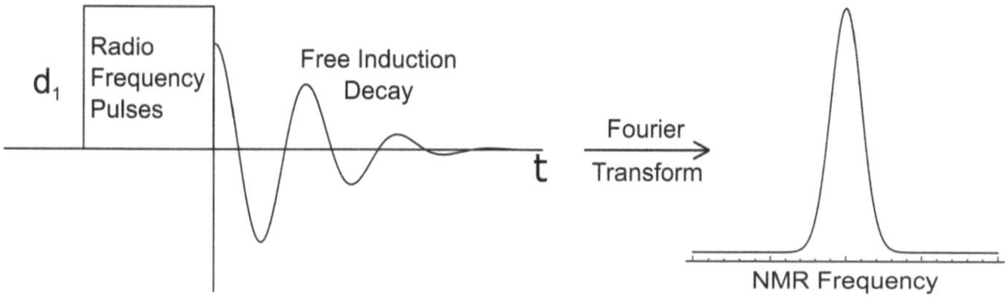

Fig. 3 Diagram of a single scan in an NMR experiment. Time-varying heights of radio frequency pulses or FID waveforms correspond to time-dependent voltage or current in the coil (not drawn to scale)

2.3 General Description of Solid-State NMR Measurements

NMR is a spectroscopic technique used to probe nuclei with magnetic moments (nuclear spins) in a magnetic field $\left(\vec{B_0}\right)$. The magnetic moment is a fundamental property of each NMR-active nucleus, such that all nuclei of a specific isotope possess the same magnetic moment. The magnetic moment for each nuclear isotope has been measured precisely and is proportional to the gyromagnetic ratio $\left(\frac{\mathrm{MHz}}{\mathrm{T}}\right)$: $\gamma_0^1\mathrm{H}^0 = 42.576$, $\gamma{}_0^{13}\mathrm{C}^0 = 10.705$, and $\gamma{}_0^{15}\mathrm{N}^0 = -4.316$ [80]. Populations of nuclear spins are perturbed by radio frequency excitation, and subsequent spin dynamics are then measured by magnetic induction (see Fig. 3). Measured current in the coil as a function of time is a decaying waveform referred to as the free induction decay (FID). Observed signals depend on local magnetic interactions and thus report on structure. Local interactions include the chemical shift and nuclear magnetic dipolar couplings. Chemical shift is a change in the effective strength of $\vec{B_0}$ experienced by each NMR-active nucleus, due primarily to magnetic field-induced distortion of orbitals of paired electrons local to each nucleus. For example, NMR signals from ^{13}C atoms at carbonyl sites are distinguishable (different frequencies) from NMR signals from ^{13}C atoms at methyl sites because of the significant differences in the local electronic orbital structures. Nuclear magnetic dipolar couplings are analogous to interactions between classical magnetic dipoles, which depend on relative position and the magnetic moments involved. If nuclei are separated by ~1 nm or less, their spin dynamics will be measurably affected by dipolar couplings.

NMR experiments involve the following general steps shown in Fig. 3:

1. An initial delay, d_1, allows nuclear spin populations in the sample to reach thermal equilibrium population distributions in the $\vec{B_0}$ field. At equilibrium, nuclear magnetization is collinear with $\vec{B_0}$ and (by convention) the *z*-axis.[1] The initial delay d_1 must be sufficiently long to achieve thermal equilibrium and is

[1] Nuclear magnetization is induced by population differences in nuclear spin states.

set based on the relaxation time T_1 of the peptide sample ($d_1 \cong 5 \times T_1$).[2]

2. Radio frequency pulses of current are applied to the coil, producing a time-varying magnetic field $\vec{B_1}(t)$ inside the sample volume (see Fig. 2). The transverse component of $\vec{B_1}(t)$ (on the *xy*-plane) induces changes in quantum mechanical nuclear spin states, resulting in detectable transverse nuclear magnetization [81–83]. A resonance condition occurs such that $\vec{B_1}(t)$ induces a change in nuclear spin state if it oscillates close to the resonant or Larmor frequency of the target nucleus.[3,4] NMR signals also depend on the pulse duration, power, and phase for $\vec{B_1}(t)$.[5] Radio frequency pulses are often described by the angle of deflection (tip angle) of nuclear magnetization they induce. For example, applying an "on-resonance" $\frac{\pi}{2}$ pulse to a sample initially at thermal equilibrium would deflect magnetization from the *z*-axis onto the *xy*-plane.[6] We express pulse powers in units of frequency, such that pulse power corresponds to $\frac{1}{\tau_{2\pi}}$, where $\tau_{2\pi}$ is the duration of a 2π pulse.[7]

3. The measured FID is a time-dependent current induced in the coil. Magnetic induction occurs because the nuclear magnetization component on the *xy*-plane will precess about the *z*-axis at the Larmor frequency (due to quantum interference effects). NMR signals decay because stochastic atomic-scale fluctuations in nuclear spin interactions destroy bulk-averaged phase coherence in NMR signal throughout the sample [81, 82]. Fourier transform of the FID produces the NMR spectrum.

4. Since NMR experiments tend to produce weak signals relative to typical noise levels in detection circuitry [85], a single application of **steps 1–3** is often insufficient to obtain desirable

---

[2] Shorter values for d_1 are sometimes used for faster scanning and better signal-to-noise, at the expense of some distortion in relative peak intensities.

[3] The energy splitting between nuclear spin energy levels is $\Delta E = h\gamma_n \cdot \left|\vec{B_0}\right|$, where h is Planck's constant and γ_n is the gyromagnetic ratio. Local nuclear magnetic interactions have very small (but measurable) effects on these energy splittings.

[4] The Larmor frequency $\nu_L = \frac{\Delta E}{h} = \gamma_n \cdot \left|\vec{B_0}\right|$. It is common to refer to a magnet in terms of its ^{1}H Larmor frequency rather than its magnetic field strength.

[5] Pulse, duration, power, and phase are independent parameters that define a radio frequency pulse. Pulse duration is the time over which the pulse is applied. Pulse power (energy input to the probe per unit time) is proportional to the second power of the voltage (or current) input to the probe. Pulse phase is the phase of the radio frequency waveform (sinusoid) for the pulse. Unless otherwise specified, the frequency of the pulse waveform is set to the spectrometer carrier frequency. In some experiments, pulse frequencies, phases, and powers can vary during the pulse.

[6] This description does not capture the implications of "off-resonance" pulses that are not perfectly at the Larmor frequency. The Bloch equations capture the effects of off-resonant pulses on nuclear spin magnetization in the absence of local spin interactions [81, 84].

[7] Pulse power in frequency units is proportional to $\left|\vec{B_1}\right|$ and the current in the coil during the pulse. Pulse power in Watts would be proportional to $\left|\vec{B_1}\right|^2$.

signal-to-noise ratios. Therefore, experiments typically require repeated scanning: **steps 1–3** are repeated and measured signals from successive scans are added together. If n repeated scans are applied for an experiment, the signal-to-noise ratio would be proportional to $\sqrt{n}$ [83].

5. NMR experiments employ phase cycling [86], since NMR is a phase-sensitive spectroscopic technique. Phase cycling is the systematic modulation of pulse and signal detection (receiver reference, ϕ_{Receiver}) phases between scans so that co-addition of data from successive scans minimizes effects of experimental imperfections (e.g., spatial inhomogeneity of $|\vec{B}_0|$) or selects for specific signal components (e.g., ^{13}C signal exclusively corresponding to transferred magnetization from ^{1}H).

The following is a list of special experimental features for solid-state NMR experiments:

- MAS is employed to eliminate most effects of chemical shift anisotropy and magnetic dipolar couplings that do not involve ^{1}H [87].
- High-power ^{1}H decoupling can be used to eliminate the effects of dipolar couplings between ^{13}C or ^{15}N and ^{1}H atoms. The effects of couplings involving ^{1}H are not usually eliminated by MAS alone at typical rotation frequencies [88–91].
- Some radio frequency excitations are designed to take advantage of the non-random nature of molecular reorientation under MAS. In contrast to liquid state molecules, which undergo uncorrelated rotations, molecules in solid samples reorient in synchrony as the sample rotates. As a result, NMR signals from ^{13}C or ^{15}N can be enhanced through transfer of polarization from the ^{1}H spin system (cross polarization) [92, 93]. In addition, techniques involving application of pulses timed in synchrony with sample rotation can temporarily reintroduce orientation-dependent spin interactions such that structural information can be extracted.

The solid-state NMR methods to probe self-assembled peptide molecular structure depend primarily on the following characteristics of nuclear spins:

- Each ^{1}H, ^{13}C, or ^{15}N nucleus is a spin-$\frac{1}{2}$ particle (total angular momentum quantum number $I_n = \frac{1}{2}$). Its behavior is described by two energy levels (quantum mechanical eigenstates) [81, 82, 94].
- In the presence of a static magnetic field $\vec{B}_0$, energy differences between spin eigenstates will scale in proportion to the strength of the magnetic field. At a specific magnitude of $\vec{B}_0$, the Larmor frequency of each type of nucleus can be predicted accurately.

For example at $\left|\vec{B_0}\right| = 11.75$ Tesla, the Larmor frequencies of ^1H, ^{13}C, and ^{15}N are 500 MHz, 126 MHz, and 51 MHz, respectively.[8]

- Radio frequency excitation near the Larmor frequency is required to induce nuclear spin transitions. Consequently, a single radio frequency pulse can only directly affect the spins of a single nuclear isotope (e.g., ^{13}C). However, heteronuclear (e.g., ^{13}C-^{15}N) spin–spin couplings could lead to effects on other spin systems.
- Long lifetimes (related to durations of FIDs) of NMR signals (ms) allow for accurate measurements of NMR frequencies. NMR signals from nuclear spins in chemically inequivalent sites are usually distinguishable by chemical shift.
- Molecular conformation and intermolecular packing affect the chemical shift. Thus, precise NMR peak positions, especially for atoms near peptide backbones, can report on secondary structure. NMR line width is also an indicator of structural order within a sample.
- Dipolar couplings between adjacent (proximities within ~1 nm) nuclei with magnetic moments perturb nuclear spin energy levels and affect spin dynamics. Dipolar couplings can therefore report on 3D arrangements of NMR-detectable atoms.
- Dipolar couplings depend strongly on distances between nuclei with magnetic moments: these couplings scale with r^{-3}, where r is the distance between a pair of nuclear spins. When more than two nuclear spins are separated by multiple interatomic distances less than ~1 nm, spin dynamics tend to be dominated by the smallest interatomic distance [5, 57].

2.4 Formats for Data Sets

Figures 4, 5, and 6 illustrate three different formats of data sets that could be obtained from the NMR experiments described in this chapter. These figures are intended to provide a basis for understanding what types of information can be extracted from NMR experiments. The radio frequency pulse sequences used to obtain these data sets will be discussed in more detail subsequently.

The most basic NMR data set is a spectrum that relates intensity to frequency, such as that shown in Fig. 4. This is a simulated ^{13}C NMR spectrum from the nanofibers of an isotopically labeled RADA16-I peptide sample [5]. For Fig. 4, all ^{13}C-labeled sites

[8] It is common to refer to static magnetic field strength in terms of the ^1H NMR Larmor frequency. For example, an 11.75 Tesla magnet would often be called a 500 MHz magnet.

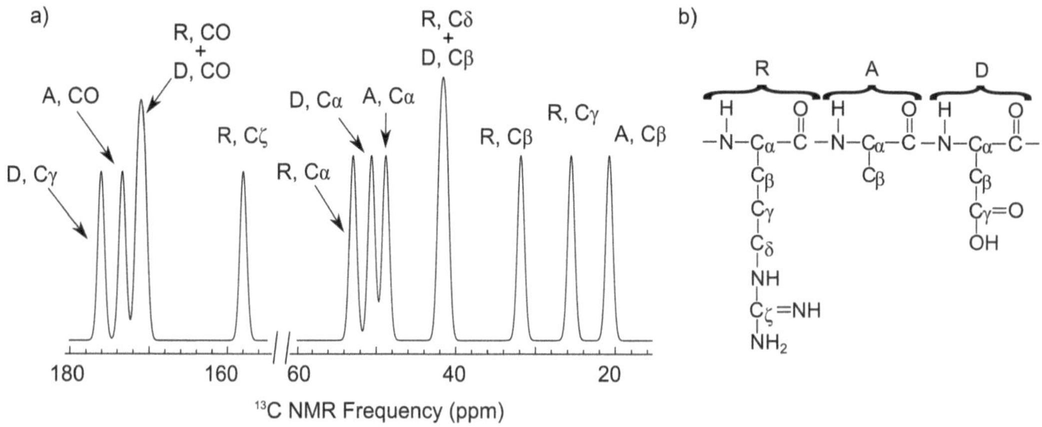

Fig. 4 (**a**) Simulated ^{13}C NMR spectrum of RADA16-I nanofiber sample labeled at R9, A10, and D11. (**b**) All-atom depiction of R, A, and D residues connected by peptide bonds

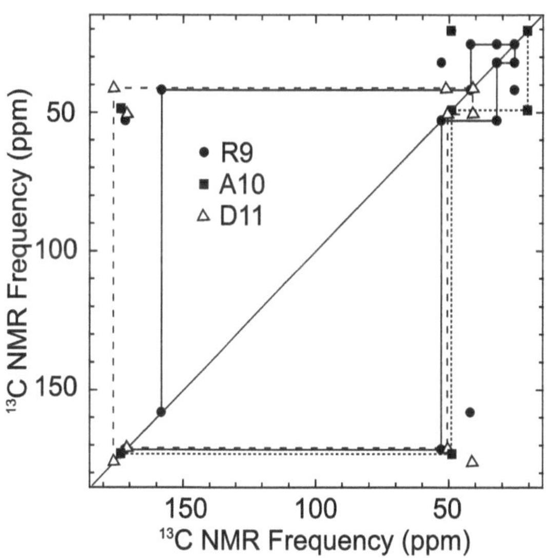

Chemical Shifts (ppm)

	CO	C_α	C_β	C_γ	C_δ	C_ζ
R9	171.4	53	32.2	25.4	42.0	157.7
A10	173.0	49.2	20.8			
D11	170.8	50.7	41.3	175.5		

Fig. 5 A simulated 2D-fpRFDR spectrum for RADA16-I nanofiber sample labeled at R9, A10, and D11. The chemical shift values shown in the table on the right are adopted from [5]. Single-bond correlations are marked by dashed lines. Solid circles, solid squares, and empty triangles indicate positions of crosspeaks within the D11, A10, and R9 residues, respectively. Since crosspeaks are expected only for directly bonded ^{13}C atoms, crosspeak patterns (highlighted with different dashed lines) are unique to each amino acid. A solid line highlights the diagonal

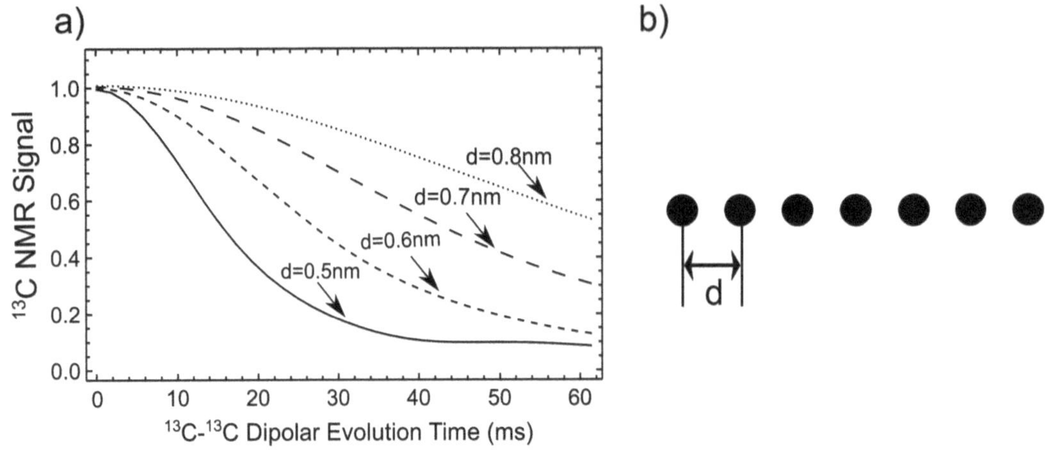

Fig. 6 (**a**) Simulated dephasing curves for PITHIRDS-CT experiments. The vertical axis corresponds to integrated intensity of an NMR peak corresponding to a selectively ^{13}C-labeled sample (e.g., with ^{13}C only at a single carbonyl or methyl site). The solid and dashed lines represent predicted decays in peak intensity as a function of time under the influence of a pattern of radio frequency pulses designed to reintroduce ^{13}C-^{13}C dipolar couplings that are normally averaged away by MAS. The dephasing curves are the results of numerical simulations of spin dynamics for 7 ^{13}C atoms arranged in a straight line separated by the constant interatomic distance, *d*. The arrangement of atoms is illustrated in (**b**)

are assumed to produce an NMR peak with equal intensity and a 1 part per million (ppm)[9] line width (full width at half maximum). The peak positions were taken from experimental data [5].[10] In this chapter, ^{13}C NMR frequencies are reported relative to the peak position for ^{13}C within tetramethylsilane (TMS) [95]. Another commonly used ^{13}C NMR standard is 4,4-dimethyl-4-silapentane-1-sulfonic acid (DSS), which has higher water solubility than TMS; to adjust ^{13}C NMR frequencies for referencing to the strongest ^{13}C NMR peak of DSS rather than that of TMS, add 1.7 ppm [96]. NMR signals from aliphatic ^{13}C nuclei appear in the 10–70 ppm region, aromatic ^{13}C signals appear in the 100–150 ppm region, and ^{13}C signals from atoms with double bonds to electronegative atoms (O or N) appear in the 160–180 ppm region. The 65–150 ppm region of Fig. 4 has been omitted because there are no aromatic groups in the RADA16-I peptide. Although each of the peaks in Fig. 4 is labeled with the corresponding ^{13}C nucleus, this spectrum alone is insufficient to determine correspondence between NMR

[9] Frequency in ppm is defined as the frequency of an NMR signal in Hz divided by a reference frequency (near the spectrometer carrier) in MHz. In ppm frequency units, many features of NMR spectra become independent of $\left|\vec{B}_0\right|$.

[10] The NMR spectra of RADA16-I nanofibers exhibited three sets of NMR peaks for each ^{13}C atom labeled within alanine residues. These spectral features indicate the coexistence of multiple distinct molecular structures. We have simplified the schematic representations of NMR data in Figs. 4 and 5 to indicate only one set of alanine ^{13}C NMR peaks.

peaks and ^{13}C-labeled sites (spectral assignments). Assignment of NMR spectra usually requires multidimensional NMR spectroscopy.

Multidimensional NMR spectroscopy is employed to probe exchange of nuclear magnetization between atoms that correspond to different NMR frequencies. Such an exchange would require magnetic interactions between atoms, and therefore spatial proximity. In multidimensional NMR spectroscopy, the number of dimensions refers to the number of frequency axes in a spectrum. Each frequency axis corresponds to the Fourier transform of NMR intensity as a function of a corresponding time domain in the experiment. Every NMR spectrum includes a direct dimension, which is the frequency determined by Fourier transform for a detected FID. Multidimensional NMR spectra are characterized by additional indirect dimensions, which correspond to Fourier transforms of NMR intensities as a function of variable time delays within radio frequency excitations. The scope of this chapter does not include NMR experiments with more than two dimensions. Figure 5 shows a sample two-dimensional (2D) exchange ^{13}C-^{13}C NMR spectrum, in which symbols emulate contours that would define intensity relative to two axes, both corresponding to ^{13}C NMR frequency. By convention, the horizontal frequency axis corresponds to the direct dimension. Off-diagonal peaks (crosspeaks) that correspond to distinct frequencies on the horizontal and vertical axes are generated through magnetization exchange between distinct ^{13}C atoms. In this particular spectrum, crosspeaks are observed between ^{13}C atoms that are directly bonded to one another, allowing determination of spectral assignments through comparison of crosspeak patterns to configurations of atoms within each labeled amino acid. Different 2D NMR techniques produce different patterns of crosspeaks, including "homonuclear" correlation experiments for which both frequency axes correspond to a single type of nucleus and "heteronuclear" correlation experiments that probe interactions between different NMR-detectable nuclei (e.g., ^{15}N and ^{13}C).

An additional scheme for solid-state NMR experiments is the quantitative measurement of decays in NMR signal intensity under the influence of dipolar recoupling pulses. By applying dipolar recoupling techniques for variable durations, one can measure a decay in NMR intensity that depends on the 3D arrangements of NMR-active atoms in the sample (^{13}C or ^{15}N). Decays can be simulated by calculating nuclear spin dynamics based on postulated atomic positions [97, 98]. As an example, Fig. 6 shows a set of simulated decay curves for normalized NMR signal intensity (peak integral) for a sample labeled with ^{13}C at a single site. The simulations employed consider seven ^{13}C atoms arranged in a straight line and separated by a constant distance [99]. For these experiments, samples are selectively ^{13}C- or ^{15}N-labeled so that measureable dipolar couplings provide specific information about molecular structures.

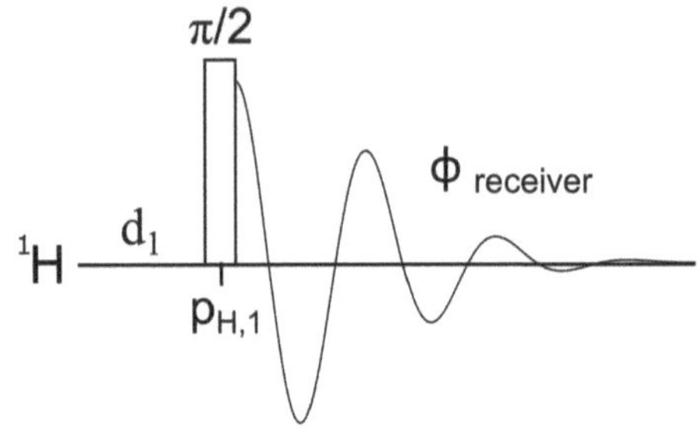

Fig. 7 Pulse sequence diagram for single pulse NMR with ^{1}H signal detection

2.5 Framework for Radio Frequency Pulse Sequences

NMR experiments are described in terms of pulse sequence diagrams such as that shown in Fig. 7. This "single pulse" experiment is the simplest pulsed NMR experiment. Symbols representing different parts of the experiment are arranged on a horizontal time line, which is labeled to specify a nucleus (^{1}H). Radio frequency pulses (rectangles) and signal detection periods (decaying waveform symbol) are performed at a carrier frequency that is set equal to the Larmor frequency of the nucleus specified to the left of the horizontal line. Although we normally employ the single pulse experiment to detect ^{1}H signal for our samples, the experiment can be used on any NMR-active nucleus. The rectangle marked $p_{\mathrm{H},\,1}$ represents a radio frequency pulse. The power, phase, and duration of the pulse are specified in Table 1. The pulse is marked with "$\pi/2$" because optimal signal is obtained when the pulse corresponds to a $\pi/2$ tip angle. The delay d_1 is necessary for the establishment of thermal equilibrium magnetization when multiple scans are performed. When multiple phases are listed in correspondence with a pulse, the list specifies how the pulse phase should be changed with each application (between scans or when a pulse appears more than once in a pulse sequence). When multiple receiver phases are listed, the list corresponds to receiver phase to be applied for successive scans. The phase lists are cyclic: phases should return to those specified at the beginning of a list after the list is completed.

An example of a double-resonance NMR experiment is shown in Fig. 8. We refer to this experiment as a ^{1}H-^{13}C cross polarization magic-angle spinning (CPMAS) experiment. The "CP" or cross-polarization pulses are aligned vertically because they require simultaneous excitation at ^{1}H and ^{13}C Larmor frequencies. These pulses induce transfer of ^{1}H magnetization to the ^{13}C spin system. The wide rectangle marked with "dec" indicates ^{1}H decoupling during acquisition of ^{13}C NMR signal (continuous wave radio frequency input to ^{1}H or a pulse sequence, depending on the decoupling method), to eliminate broadening of ^{13}C NMR peaks by ^{1}H-^{13}C dipolar couplings. The dotted "flip-back" pulse is only used for calibration of ^{13}C pulse powers; signal intensity is maximized in the absence of this pulse. Detailed parameters are specified in Table 2.

Table 1
Typical parameters for single pulse NMR experiment on ^{1}H

MAS rate: 25 kHz (rotor period, $\tau_r = 40$ μs)			
Pulse	**Power (kHz)**	**Duration (μs)**	**Phase**
$p_{H,1}(\pi/2)$	110	2.27	$x, y, -x, -y$
$\phi_{Receiver}$	$y, -x, -y, x$		
Delay	**Duration**		
$d1$	2 s		

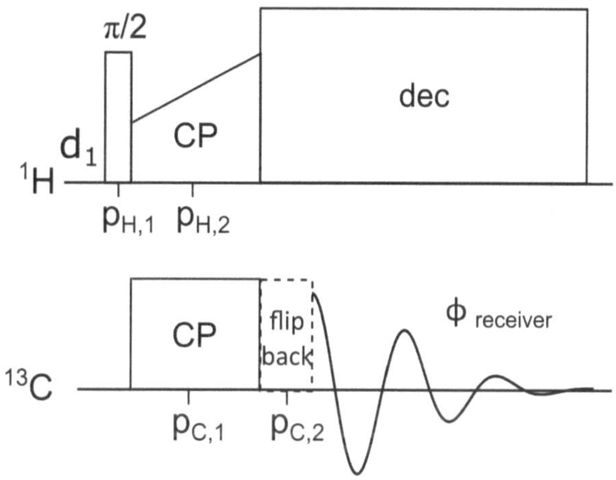

Fig. 8 The ^{1}H-^{13}C CPMAS pulse sequence

Spectra such as that shown in Fig. 4 would normally be obtained using this experiment because of the enhanced ^{13}C NMR sensitivity relative to single-pulse ^{13}C experiments. Cross polarization from ^{1}H to ^{13}C produces more ^{13}C NMR signal per scan because each ^{1}H atom has a larger magnetic moment than each ^{13}C, and ^{1}H atoms outnumber ^{13}C atoms within each amino acid. Furthermore, the use of the ^{1}H spin system as the source of detected magnetization allows for more rapid scanning, since T_1 is usually shorter for ^{1}H than ^{13}C in peptide samples.

More complex NMR experiments include 2D NMR experiments and measurements of decay under dipolar coupling. Simplified versions of these pulse sequences are shown in Fig. 9, with details provided when specific 2D and dipolar coupling experiments are discussed later. The 2D NMR experiment requires acquisition of a set of FIDs corresponding to series of delay values (t_1) in the pulse sequence (see Fig. 9a). The frequency indicated by the vertical axis in Fig. 5 corresponds to the Fourier transform of NMR peak intensity as a function of t_1, and is referred to as the "indirect

Table 2
Typical parameters for a ^{1}H-^{13}C cross-polarization experiment

MAS rate: 25 kHz (rotor period, $\tau_r = 40$ μs)			
Pulse	**Power (kHz)**	**Duration (μs)**	**Phase**
$p_{H,1}(\pi/2)$	110	2.27	$y, -y$
$p_{H,2}$	Linear ramp: 50–100	2000	x
$p_{C,1}$	50	5	$x, x, y, y, -x, -x, -y, -y$
$p_{C,2}$	0[a]	5	$y, y, -x, -x, -y, -y, x, x$
$\phi_{Receiver}$	$x, -x, y, -y, -x, x, -y, y$		
Delay	**Duration**		
$d1$	2 s		
Decoupling period	**Power (kHz)**	**Method**	
dec_1	100	TPPM	

[a]Comments: zero except during calibration

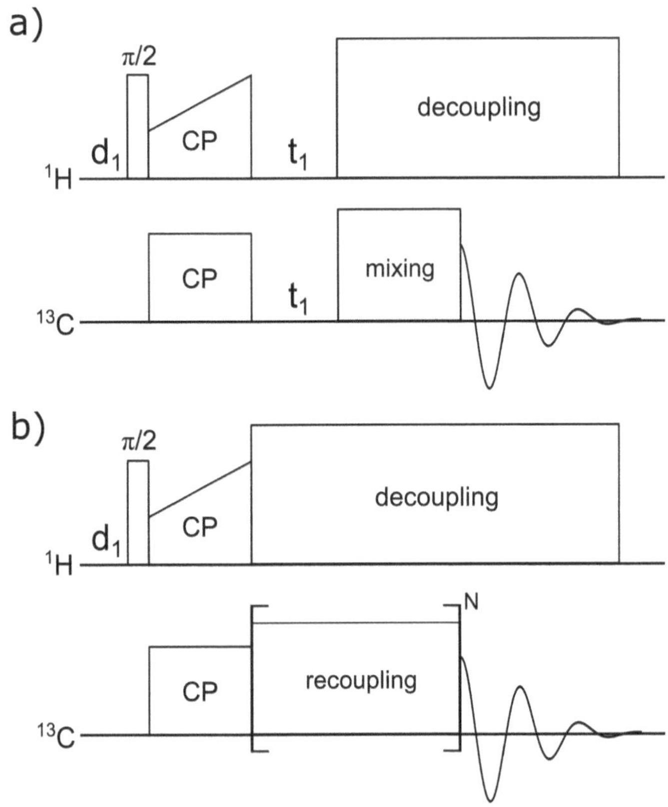

Fig. 9 General schemes for the 2D (**a**) and dipolar recoupling (**b**) NMR pulse sequences presented in this chapter

dimension." The frequency indicated by the horizontal axis of the 2D NMR spectrum in Fig. 5 (direct dimension) corresponds to the Fourier transform of each detected FID. For phase sensitivity in the indirect dimension, two FIDs must be collected for each value of t_1, each corresponding to a different phase for the pulse following the t_1 delay [100]. Dipolar recoupling pulses are used in "mixing" periods between the t_1 periods and the signal acquisition periods in order to introduce the effects of spin–spin couplings necessary to produce crosspeaks.[11] Figure 9b illustrates the general scheme for pulse sequences of experiments that measure decay of NMR peaks under the influence of dipolar couplings. The vertical axis of the simulated data in Fig. 6a corresponds to integrated peak intensities for acquired NMR signals. The horizontal axis corresponds to effective time for spin dynamics under the effects of dipolar couplings. Dipolar evolution time does not necessarily correspond directly with a delay in the recoupling pulse sequence: different evolution times could be achieved by applying a variable number (N) of repeated blocks of dipolar recoupling pulses, or by varying positions of pulses within the recoupling period. Dipolar recoupling experiments that involve variation of effective dipolar coupling times without changing the overall duration of applied pulses are called "constant time" pulse sequences. Constant time pulse sequences are designed to compensate for time-dependent loss of NMR coherences during application of the pulse sequence.

2.6 Strategy for Structural Characterization

The process of structural investigation generally follows an iterative procedure, such as that depicted in Fig. 10, in which NMR experiments and corresponding isotopic labels are each chosen to probe specific structural features. Formulation of NMR-constrained structural models for self-assembling peptides thus requires a series of experiments on samples with different isotopic labels. As sample preparation can be expensive and time consuming, it is desirable to obtain maximal structural information from each isotopically labeled sample. The largest contributors of sample costs include costs of isotopically labeled materials and peptide synthesis costs.[12] Choice of isotopic labeling requires a reasonable hypothesized structure (or set of candidate structures), which can be revised as more experimental data are obtained. For 2D NMR experiments, we desire minimal overlap of peaks and informative crosspeaks. Overlap of NMR peaks can lead to ambiguity in spectral assignments and interpretation of crosspeaks.[13] Informative crosspeaks

---

[11] It is possible to detect crosspeaks in 2D spectra even if dipolar recoupling is not employed during mixing periods, since MAS does not completely eliminate spin–spin couplings.

[12] The most expensive FMOC-protected amino acids are those with acidic or basic sidechains, because of the need for sidechain production during peptide synthesis.

[13] Isotopic labeling with ^{13}C or ^{15}N sites within sidechain aromatic or functional groups is often advantageous, because these sites often correspond to spectrally isolated NMR signals.

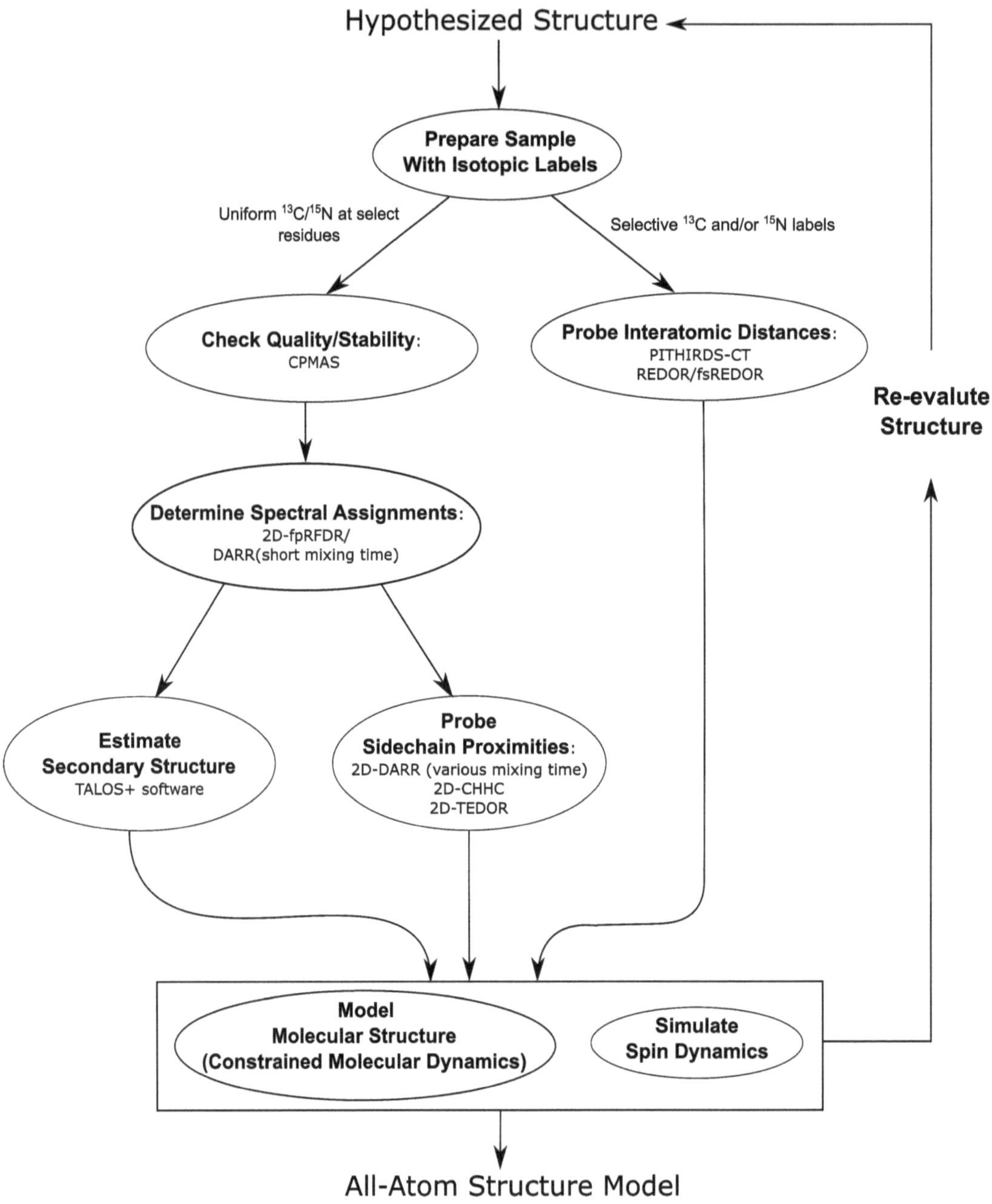

Fig. 10 Flowchart showing an overall approach for assignment and structure determination in peptide/protein structure characterization by solid-state NMR

tend to be those between amino acids that are nonadjacent in the primary structure but adjacent in the molecular structure. For dipolar recoupling measurements, isotopic labels are chosen so that decays in NMR signal intensity inform on unknown interatomic distances. Choice of labels (and interpretation of NMR data) can be facilitated significantly by computer modeling, which

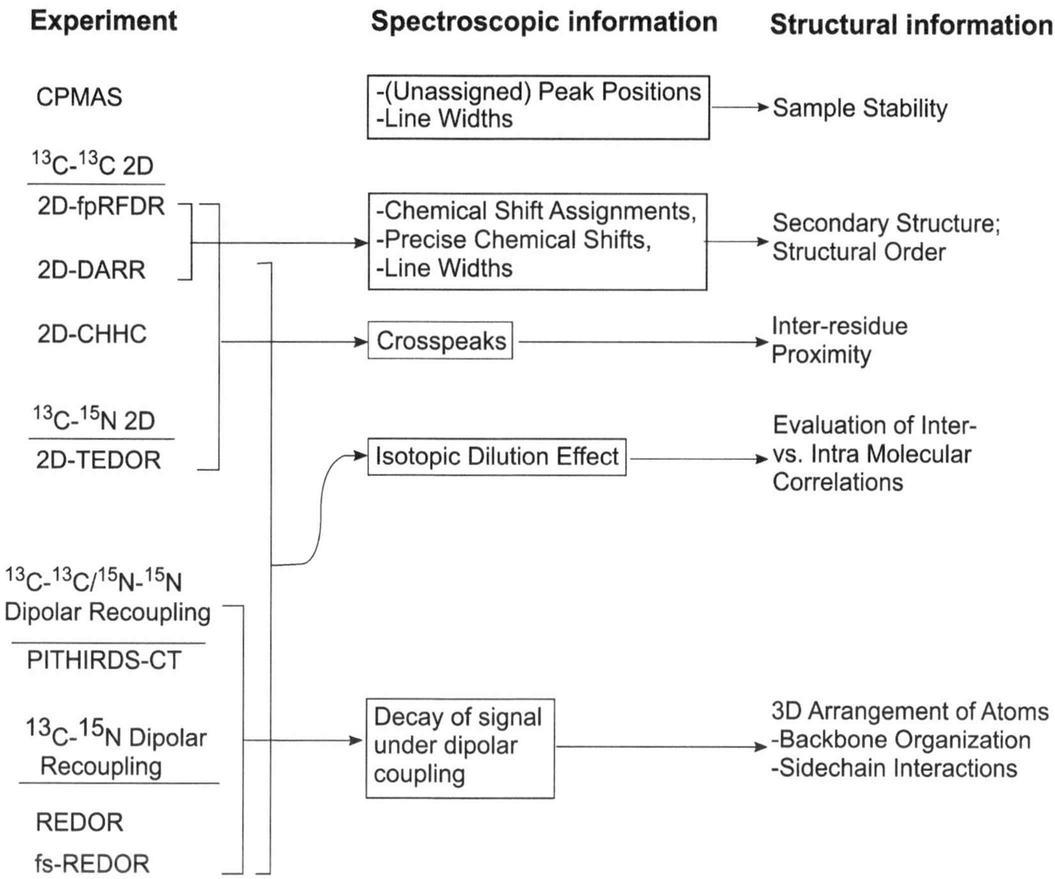

Fig. 11 An overall summary of spectroscopic and structural information that can be obtained from different solid-state NMR techniques

includes molecular modeling (constrained molecular dynamics simulations) and modeling of nuclear spin dynamics during dipolar recoupling experiments [97, 98].

NMR observables that can be interpreted in terms of structure include positions and line widths of NMR peaks, crosspeaks in 2D NMR spectra, and peak decay curves for dipolar recoupling experiments (see Fig. 11). Peak positions and line widths report on secondary structure (backbone conformation) and structural order, respectively. Crosspeaks indicate sets of spatially proximate atoms. Dipolar recoupling decay curves provide more quantitative information about 3D arrangements of atoms. Our nucleus of choice for detection of NMR signals is usually ^{13}C. Samples can be selectively labeled with ^{13}C, and spectral resolution (separation of different peaks) can be achieved with MAS and ^{1}H decoupling (see Fig. 12). Although ^{1}H NMR spectroscopy is more sensitive (larger magnetic moment, higher natural abundance, shorter T_1), ^{1}H NMR spectra of typical samples do not exhibit resolved peaks from different sites (Fig. 13). Poor ^{1}H resolution makes it difficult

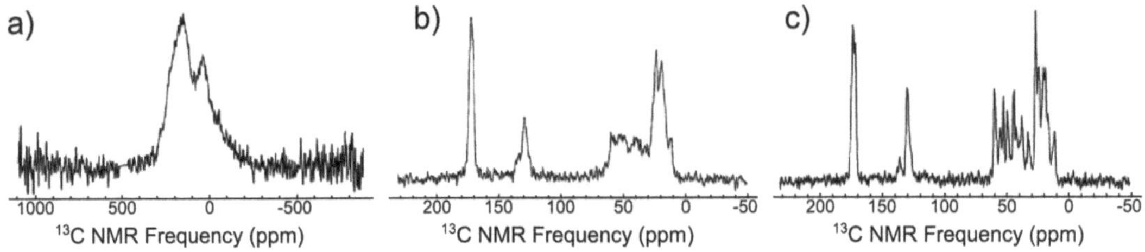

Fig. 12 ^{1}H-^{13}C cross-polarization NMR spectra of amyloid fibrils of the 40-residue isoform of the Alzheimer's amyloid β-peptide, acquired on a 400 MHz magnet. The sample was uniformly ^{13}C- and ^{15}N-labeled at seven amino acids: F19, V24, G25, A30, I31, L34, and M35. (**a**) A ^{13}C NMR spectrum acquired without MAS or ^{1}H decoupling. (**b**) A spectrum acquired with 20 kHz MAS and no ^{1}H decoupling. (**c**) A spectrum acquired with 20 kHz MAS and 110 kHz ^{1}H decoupling

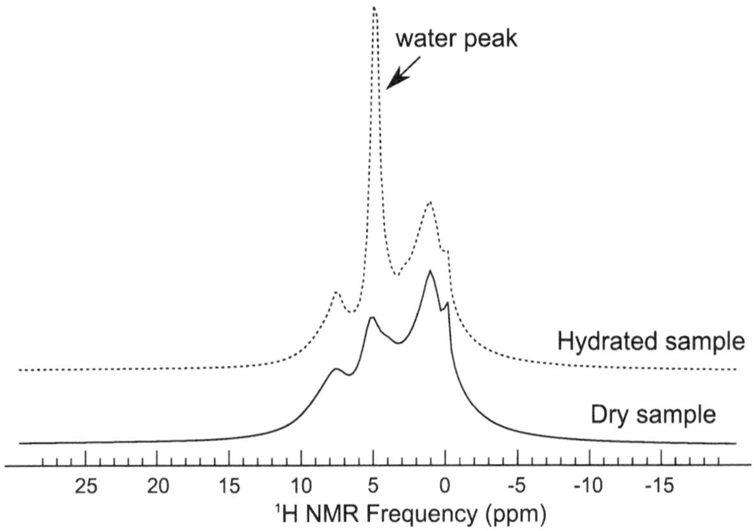

Fig. 13 ^{1}H NMR Spectra comparison between dry (solid) and hydrated (dashed) sample

to extract structural information from ^{1}H spectra, but we do collect ^{1}H spectra when it is important to monitor levels of sample hydration [101]. Signals from ^{15}N also contain important structural information, but lower NMR sensitivity for ^{15}N (smaller magnetic moment than ^{13}C) often makes it preferable to measure ^{15}N signals indirectly via interactions with ^{13}C. The choice of isotopic labeling scheme for a sample depends on the intended NMR experiment. Samples studied by 2D NMR tend to have uniform ^{13}C and ^{15}N isotopic labels at a small number (~7 or less) of amino acids (see Fig. 1b). Isotopic labeling schemes chosen for dipolar recoupling experiments tend to avoid more than one labeled site per amino acid (Fig. 1c). Multiple labeled sites within a single amino acid would be undesirable because inter-amino acid atomic distances tend to be longer, and decays in dipolar recoupling curves tend to be dominated by the shortest distances between labeled sites.

Commonly employed 2D NMR experiments include 2D finite pulse radio frequency driven recoupling (fpRFDR) [91, 102], 2D dipolar assisted rotational resonance (DARR) [103], 2D-CHHC [104], and 2D transferred echo double resonance (TEDOR) [105]. Spectral assignments for ^{13}C peaks are normally determined using 2D-fpRFDR (Fig. 5) or 2D-DARR with a short mixing period (~10 ms), for which crosspeaks would correspond mostly to bonded ^{13}C atoms.[14] An advantage of 2D-DARR over 2D-fpRFDR is that long (up to 1.5 s) mixing periods are possible, because high decoupling powers are not required during these periods. Longer mixing periods make it possible to observe inter-amino acid crosspeaks. The 2D-CHHC experiment is advantageous for the analysis of antiparallel β-sheets. These 2D spectra can exhibit specific C_α-C_α crosspeaks because of a unique structural feature: when distinct amino acids are brought in close proximity by hydrogen bonding within antiparallel β-sheets, their H_α atoms (H covalently linked to C_α) are separated by only ~0.2 nm. When interpreting inter-amino acid crosspeaks for any 2D NMR technique, it can be unclear whether these crosspeaks are between atoms within the same molecule, or atoms that reside on different molecules. This ambiguity can be resolved by preparing samples that contain mixtures of isotopically labeled and unlabeled peptide (e.g., where 30–50% of the molecules in a sample are labeled) and observing the effects of isotopic dilution on crosspeak intensity [4, 6, 106]. Attenuation of crosspeak intensity relative to diagonal signals and intra-amino acid crosspeaks indicates magnetic interactions between atoms on adjacent molecules.

Dipolar recoupling NMR experiments can be used to probe 3D arrangements of isotopically labeled sites more quantitatively. Samples labeled for these experiments tend to be more selectively labeled than samples prepared for 2D NMR experiments. Samples are usually characterized by a small number of NMR peaks, for which spectral assignments are unambiguous. We are interested in measuring peak intensity decay as a function of time under the influence of dipolar coupling. The scheme in Fig. 9b illustrates how this measurement could be accomplished. However, this figure does not show experimental features that are necessary to compensate for relaxation-induced loss of NMR coherence that would depend on the overall time of application of a dipolar recoupling sequence. Our method of choice for measuring ^{13}C-^{13}C dipolar couplings is PITHIRDS-CT (not an acronym) [57]. For probing ^{13}C-^{15}N dipolar coupling, we use rotational echo double resonance (REDOR) [107]. Frequency-selective REDOR

[14] In our experience, 2D-fpRFDR tends to more selectively produce crosspeaks between directly bonded ^{13}C atoms at MAS speeds above 20 kHz, when compared to 2D-DARR experiments with short (10 ms) mixing times. It is also possible to observe these crosspeaks at shorter mixing times (2 ms) with 2D-fpRFDR, such that overall signal is stronger due to less T_2 relaxation.

(fsREDOR) [108] has the added feature of frequency-selective pulses that are used to eliminate the effects of intra-amino acid dipolar couplings to sites with peaks that are isolated in frequency. This experiment can be used to measure dipolar couplings between sidechain carbonyl ^{13}C and sidechain ^{15}N atoms.

3 Solid-State NMR Experiments

3.1 Calibration

It is necessary to calibrate precise frequencies of NMR peaks, set the sample rotation angle to the magic angle, and establish the relationship between radio frequency pulse powers and $\vec{B_1}$ field strength. Our descriptions of calibration experiments assume that the reader has a basic understanding of NMR spectrometer operation and possesses the skills listed below.

1. Knowledge of how to tune an NMR probe.
2. Shim the $\vec{B_0}$ magnetic field so that it is homogeneous over the volume of the sample. Homogeneity of $\vec{B_0}$ could be assessed by measuring ^{13}C line widths for adamantane, which should be much less than the narrowest lines expected from the samples being studied [86].
3. The ability to measure the necessary duration for a pulse of a specific tip angle, or set the power of a pulse if a specific duration and tip angle are required. For a fixed tip angle, pulse duration would scale linearly with $|\vec{B_1}|$ and proportional to the square root of pulse power [83].
4. An understanding of the conventions for reporting NMR frequencies (referenced ppm scale) and setting spectrometer frequencies [83].
5. The ability to measure spin-lattice relaxation time T_1 and set the initial delay for each scan (d_1) appropriately [83, 85].

Magnet shimming, or the adjustment of current through electromagnets in order to tune the homogeneity of $\vec{B_0}$, is usually performed on liquid samples [86, 109]: ^2H NMR could be performed on D_2O, or ^{31}P NMR could be performed on phosphoric acid [110].

NMR peak frequency calibration of NMR signal from any nucleus (^1H, ^{13}C and ^{15}N) is equivalent to precise measurement of the $\vec{B_0}$ field strength. Once NMR frequency is calibrated for a single nucleus, frequency calibrations necessary for other nuclei studied on the same magnet can be determined by calculations [111]. Frequency calibrations are typically performed through measurements on standard samples that exhibit NMR signals with known and stable frequencies (chemical shifts in ppm) [95]. TMS is a common standard sample for ^{13}C NMR, with a single ^{13}C peak set at 0 ppm. A single-pulse experiment would be sufficient to measure the NMR spectrum of TMS. However, this liquid is not routinely used for frequency calibration with solid-state NMR spectrometers because of its volatility and the inconvenience of loading it into sample rotors. Instead, a more common solid-state ^{13}C standard is adamantane, which reliably exhibits peaks at 38.48 and 29.45 ppm. The NMR signal of adamantane can be measured using a single pulse or ^{1}H-^{13}C CPMAS experiment (Fig. 8). The ^{13}C NMR signal is strongest for a ^{1}H-^{13}C CPMAS experiment when the ^{1}H carrier is at the central frequency of the ^{1}H NMR spectrum; this spectrum can be measured using a single-pulse experiment.

Calibration of the sample rotation angle is readily performed using samples with NMR line shapes that are sensitive to this angle. The magnitude of quadrupolar spinning sidebands for ^{79}Br NMR KBr powder are maximized for sample rotation at the magic angle; this measurement can be performed with single-pulse NMR [86, 112]. The sample rotation angle can also be calibrated using ^{1}H-^{13}C CPMAS NMR of polycrystalline Gly selectively labeled with ^{13}C at the carbonyl site (^{13}CO Gly). The ^{13}C NMR line width of this sample is minimized at the magic angle [113]. The same sample could be used for peak frequency calibration, as long as the ^{13}CO peak frequency from this sample is calibrated against adamantane. When using an amino acid to calibrate spectrometer frequency, it is important to be aware that peak positions are sensitive to the possible coexistence of multiple crystal structures with sample-dependent abundances [113]. It is also common practice to use an NMR rotor containing a mixture of adamantane and KBr powers, so that the same sample could be used to calibrate both NMR peak frequency and the sample rotation angle.

The powers of radio frequency fields can be calibrated using CPMAS experiments. With the ^{1}H-^{13}C CPMAS experiment, the ^{1}H radio frequency field can be calibrated by varying the power of the $p_{H,\,1}$ pulse. Calibration of ^{1}H pulse power is most easily performed by doubling the duration of this pulse so that it corresponds to a π pulse at the desired power level; the desired input power would correspond to the lowest power that would yield no detected ^{13}C signal. The optional flip-back pulse could be used to calibrate the power of the ^{13}C pulses in a similar manner. When multiple ^{13}C peaks are observed in a spectrum, the ^{13}C carrier frequency should be placed on resonance with a peak that is strong and relatively isolated (such as a carbonyl ^{13}C peak); off-resonant effects could lead to inaccurate measurements of pulse tip angle. A $\frac{\pi}{2}$ flip-back pulse would eliminate observed signal for a peak that is on resonance. Analogously, the ^{1}H-^{15}N CPMAS experiment could be employed to calibrate ^{15}N pulse powers.

When working with a self-assembled peptide sample, our laboratory routinely performs the calibration experiment in the following order:

1. Insert a standard adamantane/KBr sample. Spin the sample at the speed that is expected for the experiment to be performed.
2. Tune the probe for ^{79}Br NMR. Employ single-pulse ^{79}Br NMR to measure the ^{79}Br NMR signal. Adjust the sample rotation angle to maximize the ^{79}Br quadrupolar sidebands.
3. Tune the probe for a double-resonance experiment (pulses on ^{1}H and ^{13}C, with detection of ^{13}C NMR).
4. Employ a single-pulse ^{1}H NMR experiment to determine the ^{1}H carrier frequency that is centered on the adamantane ^{1}H spectrum. Use a single pulse experiment to measure the ^{1}H NMR spectrum (Fig. 7). This experiment does not require precise calibration of the ^{1}H radio frequency field strength. We usually use a short pulse (e.g., 2 μs) and an input power within the limit specified by the probe manufacturer. We typically use a delay d_1 of 5 s. Use this spectrum to determine the ^{1}H carrier frequency, which should be at the center of the ^{1}H spectrum.
5. Employ CPMAS experiments to calibrate radio frequency pulse powers and NMR peak frequencies. Before the ^{1}H pulse powers are calibrated, use a conservative ^{1}H pulse power (particularly for decoupling) to protect the NMR probe electronics.
   (a) Calibrate the ^{1}H pulse power by varying the power (or duration) of the $p_{H,\ 1}$ pulse. Once the ^{1}H pulse power calibration is known, the ^{1}H pulse powers can be set. Our laboratory typically employs 110 kHz powers for ^{1}H pulses, including decoupling pulses. For cross polarization, polarization transfer efficiency is typically maximized when the average power of the simultaneous ^{1}H and ^{13}C pulses differ in power by the sample rotation rate [55, 114]. For example, if the sample is spinning at 25 kHz and a 50 kHz ^{13}C pulse is used for cross polarization, then the average power for the ^{1}H cross-polarization pulse should be 75 kHz. In our experience, the use of a ramped ^{1}H pulse during cross polarization ensures little sensitivity of the polarization transfer efficiency to the average ^{1}H pulse power.
   (b) Calibrate the ^{13}C pulse power using the ^{1}H-^{13}C CPMAS experiment. By setting the ^{13}C carrier frequency to the frequency of the left ^{13}C peak, it is possible to calibrate ^{13}C pulse power by adjusting the power or duration of $p_{C,\ 2}$ flip-back pulse. Use the obtained ^{13}C pulse calibration to set the powers of the ^{13}C pulses. Our laboratory

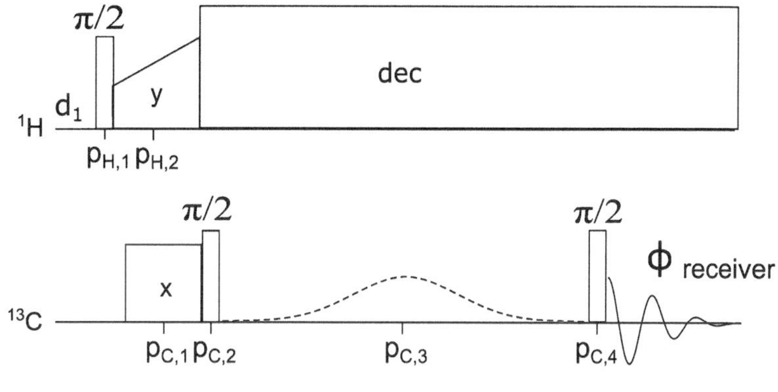

Fig. 14 Pulse sequence for calibration of a frequency-selective ^{13}C pulse

typically uses 50 kHz pulse power for ^{13}C pulses (5 μs $\frac{\pi}{2}$ pulses).

(c) Set the frequency calibration so that the adamantane ^{13}C NMR peaks are at their known NMR frequencies. Record the ^{13}C carrier frequency that corresponds to 100 ppm. This is the carrier frequency that we typically employ for ^{13}C NMR measurements because it is approximately at the center of the chemical shift range of ^{13}C signals.

(d) If the intended experiment involves ^{15}N NMR, calibrate the ^{15}N pulse power using a ^{1}H-^{15}N CPMAS experiment on a Gly or comparable sample (adamantane and KBr do not contain ^{15}N atoms).

6. Insert and spin the sample to be studied. Retune the probe. Recalibrate the radio frequency field strengths using CPMAS experiments. It will not be necessary to recalibrate NMR peak frequencies, since these depend on $\left|\vec{B}_0\right|$ and not the sample.

The pulse sequence in Fig. 14 (typical parameters in Table 3) shows how to calibrate a frequency-selective Gaussian ^{13}C π pulse. The Gaussian pulse is drawn with a dashed line because it appears in the pulse sequence every second scan. The ^{13}C carrier frequency of the experiment (or just the Gaussian pulse) should be on-resonance with the peak to be selectively inverted. Alternation of the receiver phase would ensure that any ^{13}C NMR peaks not inverted by the Gaussian pulse would be averaged away after even numbers of scans. The duration and power of the Gaussian pulse can be optimized to maximize the intensity of the selectively inverted peak and achieve the desired frequency selectivity. For the selective pulse, a trade-off should be expected between selectivity (increased for longer pulses) and sensitivity, which is reduced by relaxation effects for pulses that are too long.

3.2 2D Experiments

Multidimensional NMR methods are unified by a single concept: multidimensional spectra are correlation maps that indicate the presence of interactions between nuclear spins with distinct NMR

Table 3
Typical parameters for calibration of frequency-selective pulse

MAS rate: 10 kHz (rotor period, $\tau_r = 100$ μs)			
Pulse	**Power (kHz)**	**Duration (μs)**	**Phase**
$p_{H,1}(\pi/2)$	110	2.27	$(x)_8, (-x)_8$
$p_{H,2}$	Linear ramp between 40 and 80	2000	y
$p_{C,1}$	50	2000	y
$p_{C,2}\ (\pi/2)$	50	5	$x, -x$
$p_{C,3}\ (\pi$, Gaussian)	0.5[a]	1000	x
$p_{C,4}\ (\pi/2)$	50	5	$x, x, y, y, -x, -x, -y, -y$
$\phi_{Receiver}$	$x, -x, y, -y, -x, x, -y, y, -x, x, -y, y, x, -x, y, -y$		
Delay	**Duration**		
$d1$	2 s		
Decoupling period	**Power (kHz)**	**Method**	
dec$_1$	110	TPPM	

[a]Comments: The power designated for the Gaussian pulse is the average power

frequencies. In the solid-state NMR methods discussed here, these interactions are eliminated (or greatly attenuated) by MAS and ^{1}H decoupling. To produce informative crosspeaks, dipolar recoupling pulse sequences are applied during mixing periods to temporarily reintroduce spin–spin interactions. Spectral frequencies are encoded during time periods t_1 and t_2, during which nuclear magnetization precesses about the z-axis under minimal influence of orientation-dependent or spin–spin interactions. This scheme maximizes spectral resolution while providing a means of identifying couplings that inform on structure. The 2D methods described here all involve direct detection of ^{13}C NMR signals, so all the horizontal frequency axes correspond to ^{13}C NMR frequency. The ^{13}C-^{13}C 2D NMR experiments described here involve precession of ^{13}C magnetization during t_1 and report on couplings between ^{13}C nuclei. The ^{13}C-^{13}C 2D NMR experiments described here have peaks on the diagonal with intensities that depend on frequency in a way that approximately resembles ^{1}H-^{13}C CPMAS, such as that shown in Fig. 4. There are ^{13}C-^{13}C 2D methods that do not exhibit peaks on the diagonal, but we do not discuss them here [115–118]. Heteronuclear (^{15}N-^{13}C) 2D NMR spectra do not have diagonal signals.

Choice of MAS speed is important for 2D NMR experiments. Atomic sites with large anisotropies (orientation dependences) in chemical shift (large line widths without MAS) exhibit spinning sidebands, or satellite peaks, under MAS. Spinning sidebands will occur at frequencies of the primary NMR peak plus and minus the sample rotation rates. It is possible to detect crosspeaks between spinning sidebands and other coupled atoms. In practice, sidebands are most easily detected for carbonyl ^{13}C signals and are also readily detectable for aromatic ^{13}C peaks. Care must be taken to avoid overlap between spinning sidebands and crosspeaks of interest. If such overlap is unavoidable, desired information can be obtained by performing a 2D NMR experiment at multiple spinning speeds. Furthermore, polarization transfer efficiencies for recoupling methods used during mixing periods can be highly dependent on MAS speed.

4 2D Finite Pulse Radio Frequency Driven Recoupling (2D-fpRFDR)

Figure 15 and Table 4 specify the pulse sequence and typical experimental parameters for the 2D-fpRFDR experiment [91]. This experiment produces ^{13}C-^{13}C 2D exchange spectra with signals on the diagonal and crosspeaks that are approximately symmetric about the diagonal. This symmetry occurs because polarization transfers can occur in both directions between coupled pairs of spins: for example, Fig. 5 includes crosspeaks at coordinates (53, 32.2) and (32.2, 53) because polarization transfers from R9 C_α to R9 C_β occur at the same rate as transfers from R9 C_β to R9 C_α during the fpRFDR mixing period. At mixing times near 1.28 ms ($N = 32$), crosspeaks are observed primarily between directly bonded ^{13}C atoms (e.g., between ^{13}C-labeled C_α and C_β sites), as with the simulated spectrum depicted in Fig. 5. Weak crosspeaks could also be observed between ^{13}C atoms connected by two bonds (e.g., between ^{13}C-labeled C_α and C_γ sites). The 2D-fpRFDR technique has been used to obtain ^{13}C spectral assignments and precise NMR peak positions for each labeled site for self-assembled

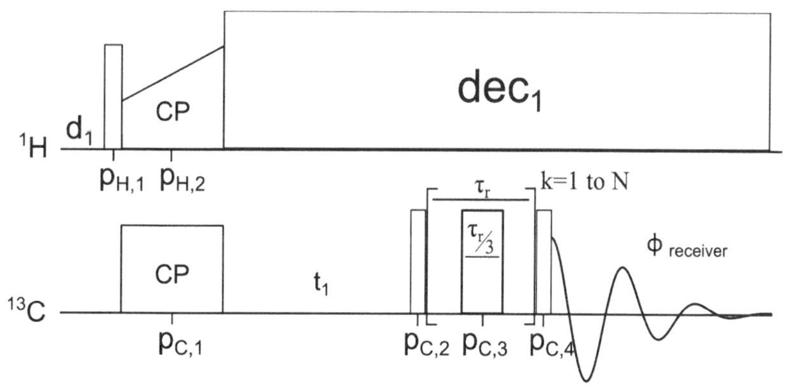

Fig. 15 2D Finite Pulse Radio Frequency Driven Recoupling (2D-fpRFDR) pulse sequence

Table 4
Typical parameters for 2D-fpRFDR experiment

MAS rate: 25 kHz (rotor period, $\tau_r = 40$ μs)			
Pulse	**Power (kHz)**	**Duration (μs)**	**Phase**
$p_{H,1}$	110	2.27	$(x)_8, (-x)_8$
$p_{H,2}$	Linear ramp between 50 and 100	2000	y
$p_{C,1}$	50	2000	y
$p_{C,2}$ $(\pi/2)$	50	5	$x, -x$ and $y, -y$ for each t_1
$p_{C,3}$ (π)[a]	37.5	$\frac{\tau_r}{3}$=13.33	$x, y, x, y, y, x, y, x, -x, -y, -x, -y, -y, -x, -y, -x$
$p_{C,4}$ $(\pi/2)$	50	5	$x, x, y, y, -x, -x, -y, -y$
$\phi_{Receiver}$	$x, -x, y, -y, -x, x, -y, y, -x, x, -y, y, x, -x, y, -y$		
Delay	**Duration (s)**		
$d1$	2		
t_1	Varied for 2D acquisition		
Decoupling period	**Power (kHz)**		**Method**
dec_1	110		Two-pulse phase modulation (TPPM)

[a]Comments: $N = 32$

peptides labeled with uniform ^{13}C at select amino acids. Analysis of secondary chemical shifts (values relative to the same atoms on the same amino acids within random coil peptides) for ^{13}C atoms near the peptide backbone (carbonyl, C_α, C_β) has been used to assess secondary structure [119–121]. Analysis of line widths is a basis for evaluating structural order, and this analysis has been used to identify unstructured regions of peptide assemblies [122, 123]. In addition, comparisons of crosspeak positions and line shapes have been used as a basis for evaluating structural variation between different samples [73, 106, 124].

5 2D Dipolar Assisted Rotational Resonance (2D-DARR)

The 2D-DARR technique produces ^{13}C-^{13}C 2D exchange spectra that are similar to those produced by the 2D-fpRFDR experiment, but without the need for high-power ^1H decoupling during the mixing period. Consequently, the experiment can be performed with a wider range of possible mixing times. At mixing times of ~10 ms or less, 2D-DARR spectra exhibit crosspeaks that

Table 5
Typical parameters for 2D-DARR experiments

MAS rate: 10 kHz (rotor period, $\tau_r = 100$ μs)			
Pulse	**Power (kHz)**	**Duration (μs)**	**Phase**
$p_{H,1}$	110	2.27	$(x)_8, (-x)_8$
$p_{H,2}$	Linear ramp between 40 and 80	2000	y
$p_{C,1}$	50	2000	y
$p_{C,2}$ $(\pi/2)$	50	5	$x, -x$ and $y, -y$ for each t_1
$p_{C,3}$	50	5	$x, x, y, y, -x, -x, -y, -y$
$\phi_{Receiver}$	$x, -x, y, -y, -x, x, -y, y, -x, x, -y, y, x, -x, y, -y$		
Delay	**Duration (s)**		
$d1$	2		
t_1	Varied for 2D acquisition		
Mixing time	**Duration (ms)**		
τ	10–1500		
Decoupling period	**Power (kHz)**		**Method**
dec_1	110		TPPM

[a]Comments: During mixing, ^1H continuous wave radio frequency should be applied with power equal to the MAS rotation rate (10 kHz)

correspond mostly to ^{13}C atoms within a single amino acid [125]. As mixing times are increased to 50 ms, additional crosspeaks can be observed that correspond to adjacent amino acids within the primary structure. With mixing times at 500 ms or above, crosspeaks include those that report on amino acids brought into close proximity by molecular folding. Typical parameters for setting up a 2D-DARR experiment are shown in Table 5 and the corresponding pulse sequence is shown in Fig. 16.

Figure 17a shows a simulated 2D-DARR spectrum (for ~500 ms mixing) detailing the expected crosspeaks ("+" markings) corresponding to contacts between I32 and V40 brought together by antiparallel arrangement of β-sheets (Fig. 17b) [6]. To further determine if the contact is intramolecular or intermolecular, an isotopically diluted sample could be probed using the same technique. If the inter-amino acid crosspeaks are between atoms on different molecules, the crosspeaks shown at positions marked by "+" will be attenuated in the isotopically diluted sample when the spectra are normalized to the intra-amino acid crosspeaks. At longer mixing times, crosspeak patterns in 2D-DARR spectra can become crowded. For this reason, isotopic labeling of aromatic

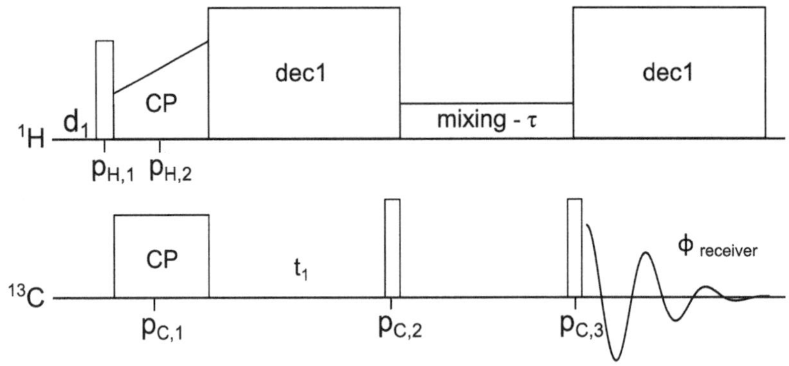

Fig. 16 2D Dipolar Assisted Rotational Resonance (2D-DARR) pulse sequence

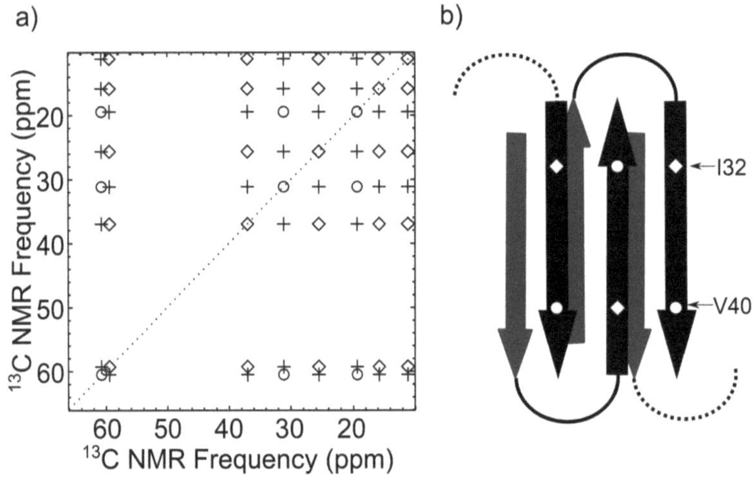

Fig. 17 (**a**) A schematic of the expected 2D-DARR spectrum (aliphatic region) for oligomers of the 42-residue isoform of the Alzheimer's amyloid-β peptide [6]. The sample is uniformly ^{13}C-labeled at the I32 and V40 residues. Empty diamonds and empty circles represent on-diagonal peak positions as well as intra-sidechain crosspeaks within I32 and V40, respectively. The cross marks represent expected inter-residue crosspeaks between the I32 and V40. b) Schematic picture showing I32 and V40 are brought into proximity by an antiparallel β-sheet arrangement

amino acids (e.g., F) is advantageous, because aromatic ^{13}C NMR peaks are spectrally isolated from carbonyl and aliphatic signals. Concerns about spectral overlap with spinning sidebands are also relevant to 2D-DARR spectra. In addition, magnetization transfer can be enhanced when the spinning sidebands are near crosspeaks of interest. Crosspeaks corresponding to longer distances (above ~0.5 nm) are also more difficult to detect when MAS speeds are too fast (above ~20 kHz).

6 2D-CHHC

An antiparallel β-sheet structure will place H_α atoms in close proximity (<0.3 nm) for specific pairs of amino acids [104]. This configuration predicts a strong ^1H-^1H dipolar coupling between these H_α atoms. The presence of this coupling can be indirectly detected using the 2D-CHHC measurement through the detection of C_α-C_α crosspeaks. Typical parameters for setting up a CHHC experiment are shown in Table 6 and the pulse sequence is shown

Table 6
Typical parameters for 2D-CHHC experiments

MAS rate: 10 kHz (rotor period, $\tau_r = 100$ μs)			
Pulse	**Power (kHz)**	**Duration (μs)**	**Phase**
$p_{H,1}$	110	2.27	$(y)_{16}, (-y)_{16}$
$p_{H,2}$	Linear ramp: 40–80	2000	x
$p_{H,3}$	Linear ramp between 66.67 and 53.33	150	$x, x, y, y, -x, -x, -y, -y$
$p_{H,4}$	110	2.27	$y, y, -x, -x, -y, -y, x, x$
$p_{H,5}$	110	2.27	$-y, -y, x, x, y, y, -x, -x, y, y, -x, -x, -y, -y, x, x$
$p_{H,6}$	Linear ramp between 66.67 and 53.33	150	$x, x, y, y, -x, -x, -y, -y$
$p_{C,1}$	50	2000	x
$p_{C,2}$ ($\pi/2$ pulse)	50	5	y
$p_{C,3}$	50	5	$-y, y$ and $-x, x$ for each t_1
$p_{C,4}$	50	5	x
$p_{C,5}$	50	5	$x, x, y, y, -x, -x, -y, -y$
$\phi_{Receiver}$	$x, -x, y, -y, -x, x, -y, y, -x, x, -y, y, x, -x, y, -y, -x, x, -y, y, x, -x, y, -y, x, -x, y, -y, -x, x, -y, y$		
Delay	**Duration (ms)**		
d_1	2		
t_d (z-filter)	3		
t_1	Varied during 2D acquisition		
Mixing time	**Duration (μs)**		
τ	$2\tau_r = 200$		
Decoupling period	**Power (kHz)**		**Method**
dec_1	110		TPPM

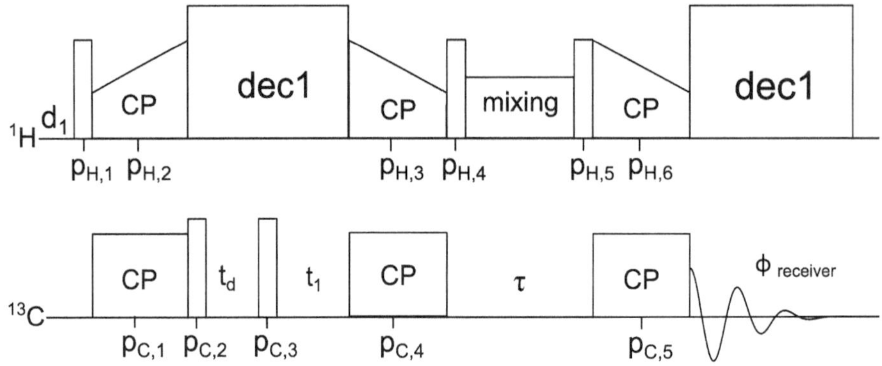

Fig. 18 2D-CHHC pulse sequence

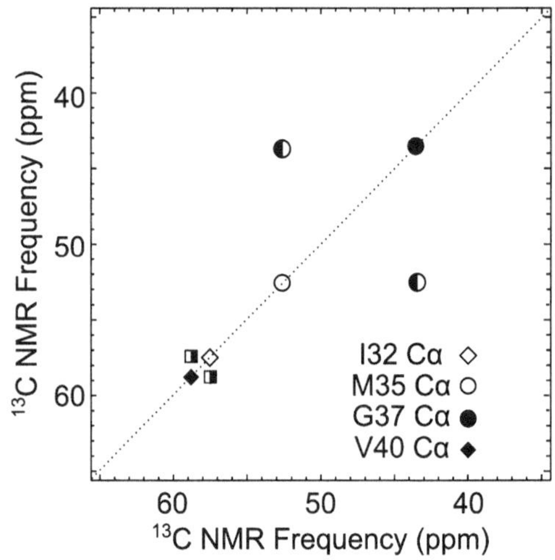

Fig. 19 Positions of diagonal peaks (solid/empty diamonds and circles) and crosspeaks (half-filled diamonds and circles) for α-carbons for the 2D-CHHC experiment of oligomers of the 42-residue isoform of the Alzheimer's amyloid-β peptide [6]. The crosspeaks between I32 and V40 and between M35 and G35 are anticipated because of antiparallel arrangements of β-strands; these residues were uniformly labeled with ^{13}C and ^{15}N. In the experimental data, we were unable to clearly resolve the I32 C_α-V40 C_α crosspeaks because of the close peak positions observed for the two sites

in Fig. 18. Figure 19 is a simulated spectrum from a 2D CHHC experiment corresponding to results obtained previously for an Aβ42 oligomer sample [6]. Expected C_α-C_α crosspeaks are shown in half-filled circles. The labeling scheme for 2D-CHHC experiment should be carefully chosen to avoid spectral overlap. While spectral overlap is a concern with all 2D NMR experiments, the 2D-CHHC experiment is particularly sensitive to this issue because all C_α sites have similar chemical shifts. As a result, expected 2D

CHHC crosspeaks may become undetectable due to overlap with the diagonal. For example, in Fig. 19 chemical shifts for C_α sites in M35 and G37 are better separated and crosspeaks show less overlap with the diagonal than those for I32 and V40. Also, it is important to use short cross-polarization pulses before and after the mixing period ($p_{C,\,3}$ and $p_{H,\,3}$) to achieve selective polarization transfers between C_α and H_α atoms. The efficiency of cross-polarization transfer tends to decrease as MAS rate is increased above 10 kHz.

7 2D Transferred Echo Double Resonance (2D-TEDOR)

The 2D-TEDOR [105] experiment produces a ^{15}N-^{13}C 2D exchange spectrum. For this experiment, ^{13}C is normally the directly detected nucleus (horizontal axis) because of higher sensitivity. Figure 20 shows a recent version of this experiment [105]. Typical parameters for setting up a 2D-TEDOR experiment are shown in Table 7. This experiment has been used to establish inter-sidechain proximity, providing information similar to that offered by 2D-DARR experiments (see Fig. 21) [4]. The technique can also be used for spectral assignments [126]. By varying the mixing time, it is possible to analyze heteronuclear dipolar interactions between ^{13}C and ^{15}N in order to assess ^{13}C-^{15}N distances [127, 128].

7.1 Dipolar Recoupling Experiments

Dipolar recoupling NMR measurements refer to the measurement of peak intensity as a function of evolution time under the influence of dipolar couplings. These measurements are sometimes called NMR distance measurements, but the shapes of measured decay curves are not determined solely by distance when interactions are more complex than what would be expected for coupled pairs of spins. Since magnetic dipolar couplings are strongly dependent on distance (r^{-3}), decays are influenced most significantly by the spacing between the most proximate pairs of spins. When groups of more than two spins are arranged at similar interatomic distances, the shapes of observed decay curves are influenced by the 3D arrangements of atoms [56, 129, 130]. Data from dipolar recoupling measurements are most readily interpreted when they are compared to predictions of computer simulations of nuclear spin dynamics. Commonly used spin nuclear simulation software packages include SIMPSON [98] and SPINEVOLUTION [97]. Since a dipolar recoupling measurement is usually only concerned with only one NMR peak, NMR sensitivity to this peak may be enhanced by pulsed spin-locking [131].

The time-dependent loss of coherences that underlie detectable signals is a commonly encountered challenge with dipolar recoupling experiments. The loss of signal could be caused by relaxation phenomena or imperfections in the pulse sequence. As a result, NMR signal could decay as dipolar recoupling pulses are applied for

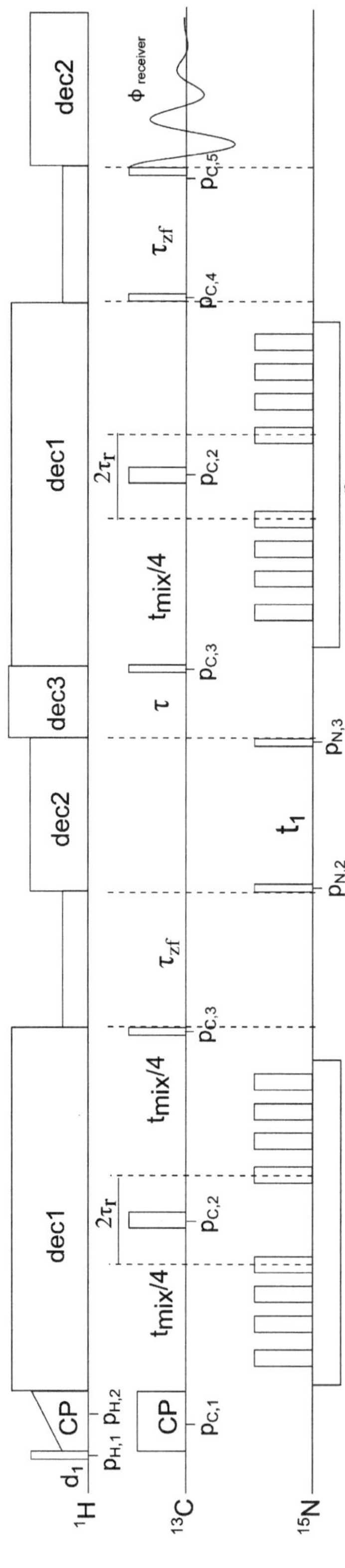

Fig. 20 2D Z-filtered Transferred echo double resonance (2D-TEDOR) pulse sequence

Table 7
Typical parameters for 2D ZF-TEDOR experiments

MAS rate: 9 kHz (rotor period, τ_r = ~110 μs)			
Pulse	**Power (kHz)**	**Duration (μs)**	**Phase**
$p_{H,1}$ ($\pi/2$)	110	2.27	$(x)_{16}, (-x)_{16}$
$p_{H,2}$	Linear ramp: 39.33–78.67	2000	x
$p_{C,1}$	50	2000	x
$p_{C,2}$ (π)	50	10	x, x, y, y
$p_{C,3}$ ($\pi/2$)	50	5	x
$p_{C,4}$ ($\pi/2$)	50	5	$y, y, -y, -y$
$p_{C,5}$ ($\pi/2$)			$(x)_4, (y)_4, (-x)_4, (-y)_4$
$p_{N,1}$ (recoupling)	50	10	$-y, y, y, -y, x, -x, -x, x, y, -y, -y, y,$ $-x, x, x, -x, y, -y, -y, y, -x, x, x, -x,$ $-y, y, y, -y, x, -x, -x, x$
$p_{N,2}$ ($\pi/2$)	50	5	$x, -x$
$p_{N,3}$ ($\pi/2$)	50	5	$x, -x$
$\phi_{Receiver}$	$-y, y, y, -y, x, -x, -x, x, y, -y, -y, y, -x, x, x, -x,$ $y, -y, -y, y, -x, x, x, -x, -y, y, y, -y, x, -x, -x, x$		
Delay	**Duration (s)**		
d_1	2		
t_1	Varied for 2D acquisition		
τ_{zf}	0.0002		
Decoupling period	**Power (kHz)**		**Method**
dec_1	110		TPPM
dec_2	110		PM
dec_3	110		

Comments: Two $p_{N,1}$ pulses are applied per rotor period. The number of $p_{N,1}$ applied at a time depends on the choice of t_{mix}, which is typically between 1 and 20 ms

varying length of time, even if the nuclear spins being probed are not coupled to one another. For the heteronuclear (^{15}N-^{13}C) dipolar recoupling techniques REDOR and fsREDOR, data are plotted as a ratio of measured ^{13}C intensity (S) scaled to the peak intensity measured for the same experiment conducted without the ^{15}N pulses (S_0). This method of data collection assumes that any

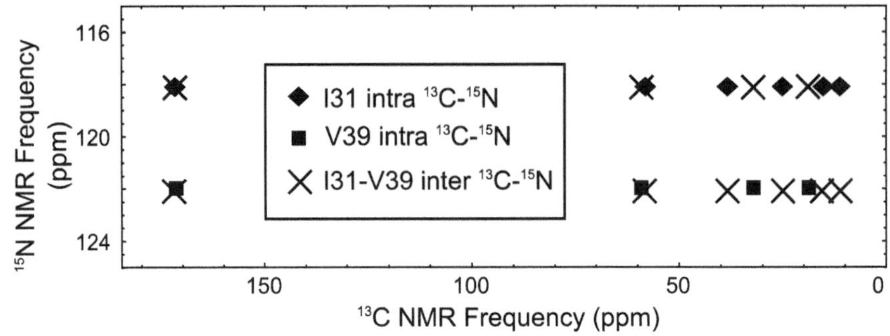

Fig. 21 Simulated 2D-TEDOR data for amyloid fibrils of the 40-residue isoform of the Alzheimer's amyloid-β peptide [4]. The peptide was uniformly ^{13}C- and ^{15}N-labeled at I31 and V39 and other amino acids for which signals are not shown. The solid diamonds and solid squares indicate positions of the intra-residue ^{13}C-^{15}N crosspeaks for I31 and V39, respectively. The cross marks correspond to the inter-residue ^{13}C-^{15}N crosspeaks

mechanism for loss of ^{13}C intensity that occurs during application of the pulse sequence would affect S and S_0 in the same way. For the homonuclear dipolar recoupling method PITHIRDS-CT, effective time for evolution of ^{13}C-^{13}C dipolar couplings is varied by changing placement of pulses within a pulse sequence that does not change in overall duration.

8 Rotational Echo Double Resonance (REDOR) and Frequency-Selective REDOR (fsREDOR)

Rotational Echo Double Resonance (REDOR) reports on the strength of the heteronuclear dipolar coupling and the dipolar evolution time to infer spatial proximities [107, 132]. This technique has been used to measure distances for samples that are selectively labeled with different nuclei (^{13}C and ^{15}N). The pulse sequence of REDOR is shown in Fig. 22 and typical parameters for setting up a REDOR experiment are shown in Table 8 [132]. Figure 23 shows simulated REDOR curves for different atomic distances. REDOR experiments have been used to determine the register of constituent β-strands within amyloid fibrils [133–135]. The technique has also been used to detect the presence β-hairpin conformation in the MAX1 nanofiber structure [7].

Frequency-selective REDOR (fsREDOR) [108] introduces a frequency-selective spin echo to selectively recouple the ^{13}C-^{15}N dipolar interaction samples with multiple ^{13}C and ^{15}N isotopic labels (Fig. 24 and Table 9). The frequency-selective pulses can be used to isolate pairs of spectrally resolved ^{13}C and ^{15}N atoms by eliminating the effects of interactions involving other nearby ^{13}C and ^{15}N atoms. This feature extends the utility of REDOR to samples with uniformly ^{13}C and ^{15}N and amino acids with functional sidechains. Sidechain carboxylic acid ^{13}C and ^{15}N atoms

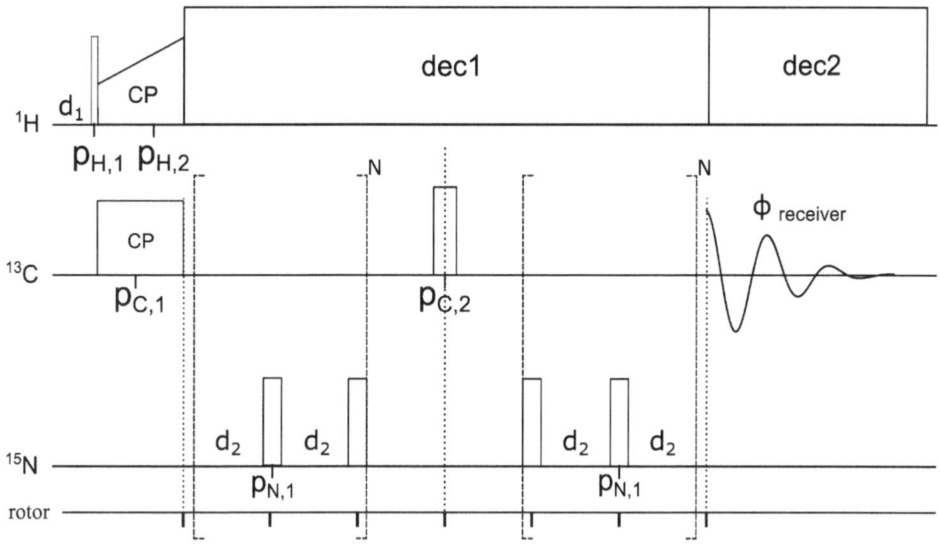

Fig. 22 Rotational Echo Double Resonance (REDOR) pulse sequence

Table 8
Typical parameters for REDOR experiment

MAS rate: 10 kHz (rotor period, $\tau_r = 100$ μs)			
Pulse	**Power (kHz)**	**Duration (μs)**	**Phase**
$p_{H,1}(\pi/2)$	110	2.27	$y, -y$
$p_{H,2}$	Linear ramp: 40–80	2000	x
$p_{C,1}$	50	2000	$x, x, y, y, -x, -x, -y, -y$
$p_{C,2}$ (π)	50	10	$x, x, y, y, -x, -x, -y, -y$
$p_{N,1}$ (π)	50	10	x, y
$\phi_{Receiver}$	$x, -x, y, -y, -x, x, -y, y$		
Delay	**Duration**		
$d1$	2 s		
Decoupling period	**Power (kHz)**	**Method**	
dec_1	110	CW	
dec_2	110	TPPM	

Comments: Loop parameter N: 1, 2, 3,...
The experiment is run as a pseudo-2D sequence, where instead of t_1 the loop parameter n is increased from 1 to 2, 3,...

often correspond to NMR peaks at frequencies that are distinct from signals from other labeled sites. The fsREDOR technique has been applied on uniformly $^{13}C/^{15}N$-labeled Aβ fibril samples to

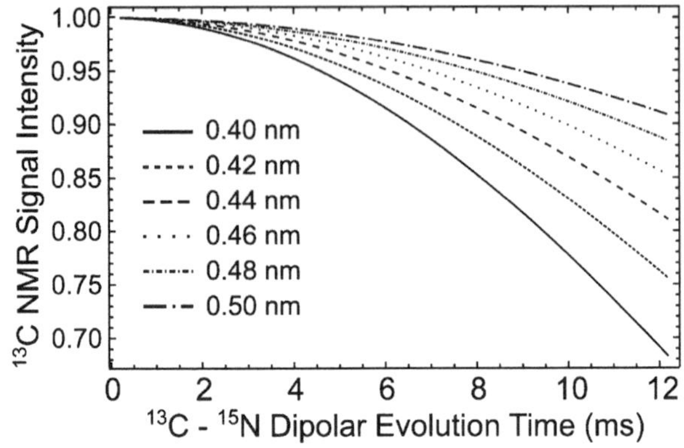

Fig. 23 Simulated REDOR dephasing curves indicating different atomic distances between a pair of ^{13}C and ^{15}N atoms, as shown by different solid and dashed lines

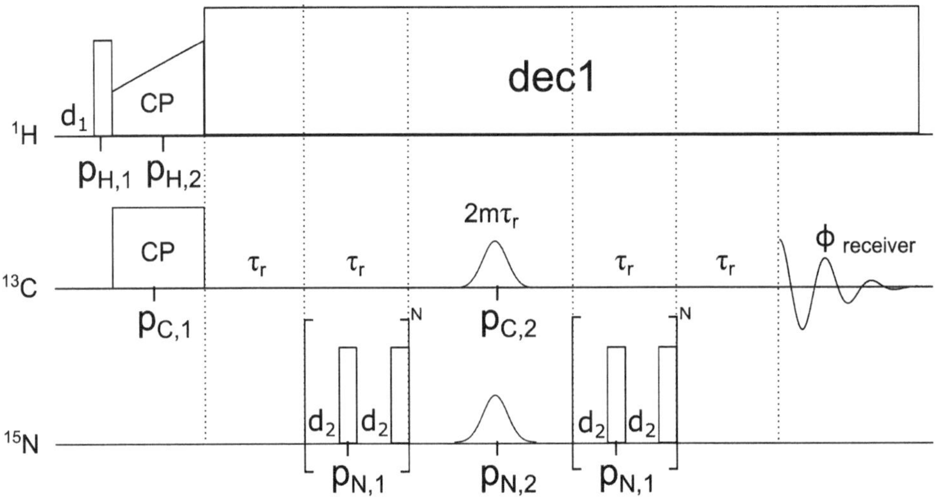

Fig. 24 Frequency-Selective Rotational Echo Double Resonance (fsREDOR) pulse sequence

probe the interaction between the positively charged sidechain ^{15}N (e.g., K28) and negatively charged sidechain carbonyl ^{13}C atoms (e.g., D23), which can form salt bridges in Aβ fibrils [4, 8, 136].

9 PITHIRDS-CT

PITHIRDS-CT [57] is a constant-time homonuclear (^{13}C-^{13}C or ^{15}N–^{15}N) dipolar recoupling technique that has been used to probe 3D arrangements of ^{13}C or ^{15}N atoms in selectively labeled samples (see Fig. 25 and Table 10). For PITHIRDS-CT experiments on ^{13}C, samples are usually labeled selectively at carbonyl or methyl sites. Since carbonyl and methyl sites have very different ^{13}C

Table 9
Typical parameters for fsREDOR experiments

MAS rate: 10 kHz (rotor period, $\tau_r = 100$ μs)			
Pulse	**Power (kHz)**	**Duration (μs)**	**Phase**
$p_{H,1}(\pi/2)$	110	2.27	y
$p_{H,2}$	Linear ramp: 40–80	2000	x
$p_{C,1}$	50	2000	$(x)_4, (y)_4, (-x)_4, (-y)_4$
$p_{C,2}$ (Gaussian, π)	0.5	1000	$x, y, -x, -y, y, -x, -y, x, -x, -y, x, y, -y, x, y, -x$
$p_{N,1}$ (π)	50	10	$x, y, x, y, y, x, y, x, -x, -y, -x, -y, -y, -x, -y, -x$
$p_{N,2}$(Gaussian, π)	0.5	1000	x
$\phi_{Receiver}$	$x, -x, x, -x, y, -y, y, -y, -x, x, -x, x, -y, y, -y, y$		
Delay	**Duration**		
$d1$	2 s		
Decoupling period	**Power (kHz)**	**Method**	
dec_1	110	TPPM	

Comments: Loop parameter N: 1, 2, 3...

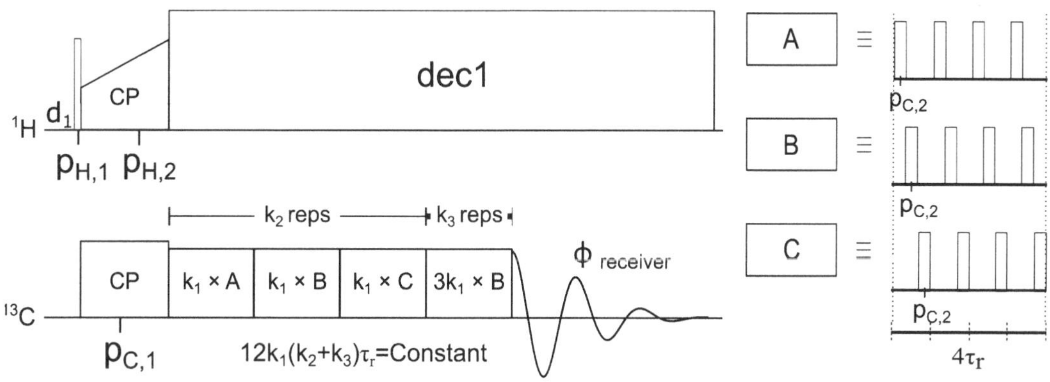

Fig. 25 PITHIRDS-CT pulse sequence

chemical shifts, it is possible to label a sample with ^{13}C at a carbonyl site and a methyl site [5, 99]. If such a labeling scheme is used, the labeled carbonyl site should be sufficiently far from the labeled methyl (>1 nm) so that it can be assumed that there is no dipolar interaction between carbonyl and methyl ^{13}C atoms.

Table 10
Typical parameters for PITHIRDS-CT experiments

MAS rate: 20 kHz (rotor period, $\tau_r = 50$ μs)			
Pulse	**Power (kHz)**	**Duration (μs)**	**Phase**
$p_{H,1}(\pi/2)$	110	2.27	y
$p_{H,2}$	Linear ramp: 46.67–93.34	2000	x
$p_{C,1}$	50	2000	x, x
$p_{C,2}(\pi)^a$	30	$\tau_r/3 = 16.67$	$x, y, x, y, y, x, y, x, -x, -y, -x, -y, -x, -y, -x, -y$
$\phi_{Receiver}$	x		
Delay	**Duration**		
$d1$	2 s		
Decoupling period	**Power (kHz)**	**Method**	
dec$_1$	110	TPPM	

[a]Comments: The xy-16 phase cycle is applied to $p_{C,2}$
The constants $k_1 = 4$, $k_2 + k_3 = 32$, for 38.4 ms total recoupling time [57]

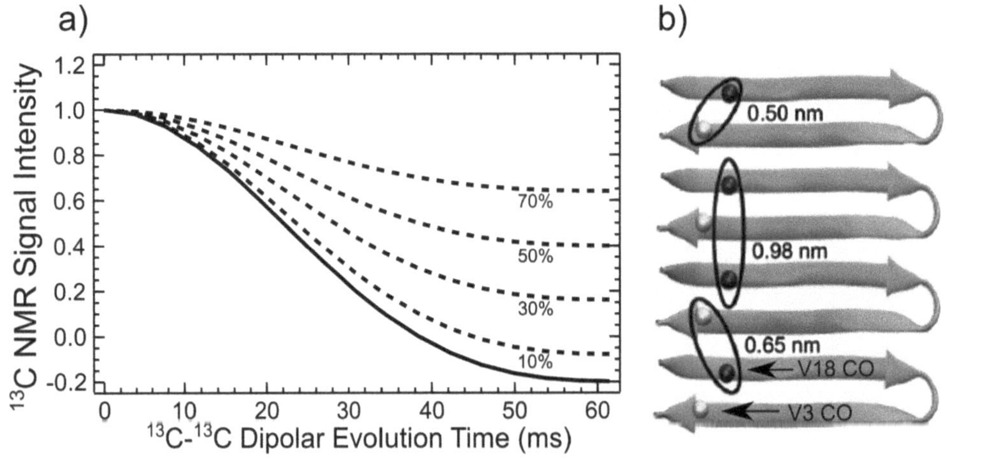

Fig. 26 (**a**) The solid curve shows PITHIRDS-CT simulated data for eight ^{13}C nuclei at the V3 CO and V18 CO positions within MAX8 β-sheets as shown in (**b**) [53]. Dashed curves show how the curve changes upon incorporation of the indicated percentages of MAX8 molecules in minor conformations that do not exhibit ^{13}C-^{13}C dipolar couplings

The PITHIRDS-CT experiment is particularly useful for identifying in-register parallel β-sheets. In such a molecular configuration, equivalent backbone sites on adjacent molecules are arranged within 0.5 nm of one another. PITHIRDS-CT data can also provide a basis for studying samples with structures that deviate from in-register parallel arrangements of β-strands, including those with registry shifts, antiparallel β-sheets, and structural heterogeneities [5, 7, 53, 99]. As an example, Fig. 26 shows simulated PITHIRDS-CT curves for a sample of MAX8 nanofibers with ^{13}C at two carbonyl sites. The two labeled carbonyl sites produce NMR peaks at the same frequency. The curves show the expected behavior for molecules in β-hairpin conformations (solid line) and the effect of the presence of molecules in conformations that do not correspond to measurable ^{13}C-^{13}C dipolar couplings (dashed curves). PITHIRDS-CT technique can also be used to measure backbone torsion angles [57].

Acknowledgements

This work was supported by the National Institute on Aging of the National Institutes of Health (award number R01AG045703). The content is solely the responsibility of the authors and does not necessarily represent the official views of the National Institutes of Health. A portion of the work is financially supported by the National Science Foundation (DMR-105221 to AKP) and the startup at Georgia Institute of Technology. The authors also gratefully acknowledge Terrone L. Rosenberry, Ankita Gupta, and Smaranda Birlea for the proofreading of this manuscript.

References

1. Nelson MT, Humphrey W, Gursoy A et al (1996) NAMD: a parallel, object oriented molecular dynamics program. Int J High Perform Comput Appl 10(4):251–268
2. Case DA, Cheatham TE, Darden T et al (2005) The amber biomolecular simulation programs. J Comput Chem 26 (16):1668–1688
3. Schwieters CD, Kuszewski JJ, Tjandra N et al (2003) The Xplor-NIH NMR molecular structure determination package. J Magn Reson 160(1):65–73
4. Paravastu AK, Leapman RD, Yau WM et al (2008) Molecular structural basis for polymorphism in Alzheimer's β-amyloid fibrils. Proc Natl Acad Sci U S A 105 (47):18349–18354
5. Cormier AR, Pang X, Zimmerman MI et al (2013) Molecular structure of RADA16-I designer self-assembling peptide nanofibers. ACS Nano 7(9):7562–7572
6. Huang D, Zimmerman MI, Martin PK et al (2015) Antiparallel beta-sheet structure within the C-terminal region of 42-residue Alzheimer's amyloid-beta peptides when they form 150-kDa oligomers. J Mol Biol 427(13):2319–2328
7. Nagy-Smith K, Moore E, Schneider J et al (2015) Molecular structure of monomorphic peptide fibrils within a kinetically trapped hydrogel network. Proc Natl Acad Sci U S A 112(32):9816–9821
8. Lu J-X, Qiang W, Yau W-M et al (2013) Molecular structure of β-amyloid fibrils in Alzheimer's disease brain tissue. Cell 154 (6):1257–1268
9. Arimon M, Diez-Perez I, Kogan MJ et al (2005) Fine structure study of Aβ1-42 fibrillogenesis with atomic force microscopy. FASEB J 19(10):1344–1346
10. Fändrich M, Meinhardt J, Grigorieff N (2009) Structural polymorphism of Alzheimer Aβ and other amyloid fibrils. Prion 3 (2):89–93
11. Petkova AT, Leapman RD, Guo ZH et al (2005) Self-propagating, molecular-level polymorphism in Alzheimer's β-amyloid fibrils. Science 307(5707):262–265
12. Cardoso I, Goldsbury CS, Muller SA et al (2002) Transthyretin fibrillogenesis entails the assembly of monomers: a molecular model for in vitro assembled transthyretin amyloid-like fibrils. J Mol Biol 317 (5):683–695

13. Schmidt M, Sachse C, Richter W et al (2009) Comparison of Alzheimer Abeta(1-40) and Abeta(1-42) amyloid fibrils reveals similar protofilament structures. Proc Natl Acad Sci U S A 106(47):19813–19818
14. Goldsbury C, Baxa U, Simon MN et al (2011) Amyloid structure and assembly: insights from scanning transmission electron microscopy. J Struct Biol 173(1):1–13
15. Chen B, Thurber KR, Shewmaker F et al (2009) Measurement of amyloid fibril mass-per-length by tilted-beam transmission electron microscopy. Proc Natl Acad Sci U S A 106(34):14339–14344
16. Wagner DE, Phillips CL, Ali WM et al (2005) Toward the development of peptide nanofilaments and nanoropes as smart materials. Proc Natl Acad Sci U S A 102(36):12656–12661
17. Woolfson DN, Gribbon C, Channon KJ et al (2008) MagicWand: a single, designed peptide that assembles to stable, ordered alpha-helical fibers. Biochemistry 47 (39):10365–10371
18. Cerf E, Sarroukh R, Tamamizu-Kato S et al (2009) Antiparallel β-sheet: a signature structure of the oligomeric amyloid β-peptide. Biochem J 421:415–423
19. Sarroukh R, Goormaghtigh E, Ruysschaert J-M et al (2013) ATR-FTIR: a "rejuvenated" tool to investigate amyloid proteins. BBA-Biomembranes 1828(10):2328–2338
20. Greenfield NJ (1999) Applications of circular dichroism in protein and peptide analysis. TrAC Trends Anal Chem 18(4):236–244
21. Kelly SM, Jess TJ, Price NC (2005) How to study proteins by circular dichroism. Biochim Biophys Acta 1751(2):119–139
22. Ono K, Condron MM, Teplow DB (2009) Structure-neurotoxicity relationships of amyloid beta-protein oligomers. Proc Natl Acad Sci U S A 106(35):14745–14750
23. Biancalana M, Koide S (2010) Molecular mechanism of Thioflavin-T binding to amyloid fibrils. Biochim Biophys Acta 1804 (7):1405–1412
24. Saeed SM, Fine G (1967) Thioflavin-T for amyloid detection. Am J Clin Pathol 47 (5):588–593
25. Wolfe LS, Calabrese MF, Nath A et al (2010) Protein-induced photophysical changes to the amyloid indicator dye thioflavin T. Proc Natl Acad Sci 107(39):16863–16868
26. Khurana R, Coleman C, Ionescu-Zanetti C et al (2005) Mechanism of thioflavin T binding to amyloid fibrils. J Struct Biol 151 (3):229–238
27. Linke RP (2006) Congo red staining of amyloid: improvements and practical guide for a more precise diagnosis of amyloid and the different amyloidoses. In: Uversky VN, Fink AL (eds) Protein misfolding, aggregation, and conformational diseases: part a: protein aggregation and conformational diseases. Springer US, Boston, pp 239–276
28. Schütz AK, Soragni A, Hornemann S et al (2011) The amyloid-Congo red interface at atomic resolution. Angew Chem Int Ed 50 (26):5956–5960
29. Klunk WE, Pettegrew JW, Abraham DJ (1989) Quantitative evaluation of congo red binding to amyloid-like proteins with a beta-pleated sheet conformation. J Histochem Cytochem 37(8):1273–1281
30. Jimenez JL, Guijarro JL, Orlova E et al (1999) Cryo-electron microscopy structure of an SH3 amyloid fibril and model of the molecular packing. EMBO J 18(4):815–821
31. Zhang R, Hu X, Khant H et al (2009) Interprotofilament interactions between Alzheimer's Abeta1-42 peptides in amyloid fibrils revealed by cryoEM. Proc Natl Acad Sci U S A 106(12):4653–4658
32. Yucel T, Micklitsch CM, Schneider JP et al (2008) Direct observation of early-time hydrogelation in β-hairpin peptide self-assembly. Macromolecules 41 (15):5763–5772
33. McDonald M, Box H, Bian W et al (2012) Fiber diffraction data indicate a hollow core for the Alzheimer's a beta 3-fold symmetric fibril. J Mol Biol 423(3):454–461
34. Morris K, Serpell L (2010) From natural to designer self-assembling biopolymers, the structural characterisation of fibrous proteins and peptides using fibre diffraction. Chem Soc Rev 39(9):3445–3453
35. Sunde M, Serpell LC, Bartlam M et al (1997) Common core structure of amyloid fibrils by synchrotron X ray diffraction. J Mol Biol 273 (3):729–739
36. Diaz-Avalos R, Long C, Fontano E et al (2003) Cross-beta order and diversity in nanocrystals of an amyloid-forming peptide. J Mol Biol 330(5):1165–1175
37. Stroud JC, Liu C, Teng PK et al (2012) Toxic fibrillar oligomers of amyloid-β have cross-β structure. Proc Natl Acad Sci U S A 109 (20):7717–7722
38. Whittemore NA, Mishra R, Kheterpal I et al (2005) Hydrogen-deuterium (H/D) exchange mapping of a ss(1-40) amyloid fibril secondary structure using nuclear magnetic

resonance spectroscopy. Biochemistry 44 (11):4434–4441
39. Vilar M, Wang L, Riek R (2012) Structural studies of amyloids by quenched hydrogen--deuterium exchange by NMR. Amyloid Proteins Methods Protoc 849:185–198
40. Gu L, Liu C, Stroud JC et al (2014) Antiparallel triple-strand architecture for prefibrillar Aβ42 oligomers. J Biol Chem 289 (39):27300–27313
41. Karyagina I, Becker S, Giller K et al (2011) Electron paramagnetic resonance spectroscopy measures the distance between the external β-strands of folded α-synuclein in amyloid fibrils. Biophys J 101(1):L1–L3
42. Gu L, Liu C, Guo Z (2013) Structural insights into Aβ 42 oligomers using site-directed spin labeling. J Biol Chem 288 (26):18673–18683
43. Margittai M, Langen R (2008) Fibrils with parallel in-register structure constitute a major class of amyloid fibrils: molecular insights from electron paramagnetic resonance spectroscopy. Q Rev. Biophys 41 (3–4):265–297
44. Torok M, Milton S, Kayed R et al (2002) Structural and dynamic features of Alzheimer's Aβ peptide in amyloid fibrils studied by site-directed spin labeling. J Biol Chem 277(43):40810–40815
45. Frare E, Mossuto MF, de Laureto PP et al (2006) Identification of the core structure of lysozyme amyloid fibrils by proteolysis. J Mol Biol 361(3):551–561
46. Kheterpal I, Williams A, Murphy C et al (2001) Structural features of the Aβ amyloid fibril elucidated by limited proteolysis. Biochemistry 40(39):11757–11767
47. Chan JCC (2012) Solid-state NMR techniques for the structural determination of amyloid fibrils. Top Curr Chem 306:47–88
48. Habenstein B, Loquet A (2016) Solid-state NMR: an emerging technique in structural biology of self-assemblies. Biophys Chem 210:14–26
49. Goldbourt A (2013) Biomolecular magic-angle spinning solid-state NMR: recent methods and applications. Curr Opin Biotechnol 24(4):705–715
50. Tycko R (2011) Solid-state NMR studies of amyloid fibril structure. Annu Rev. Phys Chem 62:279–299
51. Tycko R (2006) Solid-state NMR as a probe of amyloid structure. Protein Pept Lett 13 (3):229–234
52. Tycko R (2000) Solid-state NMR as a probe of amyloid fibril structure. Curr Opin Chem Biol 4(5):500–506
53. Leonard SR, Cormier AR, Pang X et al (2013) Solid-state NMR evidence for beta-hairpin structure within MAX8 designer peptide nanofibers. Biophys J 105(1):222–230
54. Andrew ER, Bradbury A, Eades RG (1959) Removal of dipolar broadening of nuclear magnetic resonance spectra of solids by specimen rotation. Nature 183(4678):1802–1803
55. Wu XL, Zilm KW (1993) Cross polarization with high-speed magic-angle spinning. J Magn Reson A 104(2):154–165
56. Bennett AE, Griffin RG, Ok JH et al (1992) Chemical shift correlation spectroscopy in rotating solids: radio frequency-driven dipolar recoupling and longitudinal exchange. J Chem Phys 96(11):8624
57. Tycko R (2007) Symmetry-based constant-time homonuclear dipolar recoupling in solid state NMR. J Chem Phys 126 (6):064506–064506
58. Taboada L, Nicolás E, Giralt E (2001) One-pot full peptide deprotection in Fmoc-based solid-phase peptide synthesis: methionine sulfoxide reduction with Bu4NBr. Tetrahedron Lett 42(10):1891–1893
59. Kates SA, Albericio F (2000) Solid-phase synthesis: a practical guide. Marcel Dekker, New York
60. Giano MC, Pochan DJ, Schneider JP (2011) Controlled biodegradation of self-assembling beta-hairpin peptide hydrogels by proteolysis with matrix metalloproteinase-13. Biomaterials 32(27):6471–6477
61. Wang SSS, Tobler SA, Good TA et al (2003) Hydrogen exchange-mass spectrometry analysis of beta-amyloid peptide structure. Biochemistry 42(31):9507–9514
62. Weinkauf R, Schanen P, Yang D et al (1995) Elementary processes in peptides—electron-mobility and dissociations in peptide cations in the gas-phase. J Phys Chem 99 (28):11255–11265
63. Tycko R (2014) Physical and structural basis for polymorphism in amyloid fibrils. Protein Sci 23(11):1528–1539
64. Rangachari V, Moore BD, Reed DK et al (2007) Amyloid-β(1-42) rapidly forms protofibrils and oligomers by distinct pathways in low concentrations of sodium dodecylsulfate. Biochemistry 46:12451–12462
65. Nichols MR, Moss MA, Reed DK et al (2002) Growth of β-amyloid(1-40) protofibrils by monomer elongation and lateral association. Characterization of distinct products by light scattering and atomic force microscopy. Biochemistry 41:6115–6127
66. Walsh DM, Hartley DM, Kusumoto Y et al (1999) Amyloid b-protein fibrillogenesis:

structure and biological activity of protofibrillar intermediates. J Biol Chem 274:25945–25952
67. Kodali R, Wetzel R (2007) Polymorphism in the intermediates and products of amyloid assembly. Curr Opin Struct Biol 17(1):48–57
68. Kodali R, Williams AD, Chemuru S et al (2010) Abeta(1-40) forms five distinct amyloid structures whose beta-sheet contents and fibril stabilities are correlated. J Mol Biol 401 (3):503–517
69. Gosal WS, Morten IJ, Hewitt EW et al (2005) Competing pathways determine fibril morphology in the self-assembly of beta(2)-microglobulin into amyloid. J Mol Biol 351 (4):850–864
70. Lewandowski JR, van der Wel PCA, Rigney M et al (2011) Structural complexity of a composite amyloid fibril. J Am Chem Soc 133 (37):14686–14698
71. Krysmann MJ, Castelletto V, Kelarakis A et al (2008) Self-assembly and hydrogelation of an amyloid peptide fragment. Biochemistry 47 (16):4597–4605
72. Paravastu AK, Qahwash I, Leapman RD et al (2009) Seeded growth of beta-amyloid fibrils from Alzheimer's brain-derived fibrils produces a distinct fibril structure. Proc Natl Acad Sci U S A 106(18):7443–7448
73. Petkova AT, Buntkowsky G, Dyda F et al (2004) Solid state NMR reveals a pH-dependent antiparallel beta-sheet registry in fibrils formed by a beta-amyloid peptide. J Mol Biol 335(1):247–260
74. Tycko R, Wickner RB (2013) Molecular structures of amyloid and prion fibrils: consensus versus controversy. Acc Chem Res 46 (7):1487–1496
75. Moore BD, Rangachari V, Tay WM et al (2009) Biophysical analyses of synthetic amyloid-β(1-42) aggregates before and after covalent cross-linking. Implications for deducing the structure of endogenous amyloid-β oligomers. Biochemistry 48:11796–11806
76. Cormier AR, Lopez-Majada JM, Alamo RG et al (2013) Distinct solid and solution state self-assembly pathways of RADA16-I designer peptide. J Pept Sci 19(8):477–484
77. Yokoi H, Kinoshita T, Zhang SG (2005) Dynamic reassembly of peptide RADA16 nanofiber scaffold. Proc Natl Acad Sci U S A 102(24):8414–8419
78. Eby DM, Johnson GR, Farmer BL et al (2011) Supramolecular assembly of a biomineralizing antimicrobial peptide in coarse-grained Monte Carlo simulations. Phys Chem Chem Phys 13(3):1123–1130
79. McNeill SA, Gor'kov PL, Shetty K et al (2009) A low-E magic angle spinning probe for biological solid state NMR at 750 MHz. J Magn Reson 197(2):135–144
80. Griffiths DJ (1986) Introduction to electrodynamics, 2nd edn. Prentice Hall, Englewood Cliffs
81. Slichter CP (1996) Principles of magnetic resonance. Solid-state sciences, vol 1, 3rd edn. Springer, Berlin
82. Abragam A (1961) Principles of nuclear magnetism, International series of monographs on physics, vol 32. Clarendon Press, New York
83. Levitt MH (2001) Spin dynamics: basics of nuclear magnetic resonance. Wiley, New York
84. Bloch F (1946) Nuclear induction. Phys Rev 70(7–8):460–474
85. Fukushima E, Roeder SBW (1981) Experimental pulse NMR: a nuts and bolts approach. Perseus Books, Reading
86. Berger S, Braun S (2004) 200 and more NMR experiments: a practical course, 3rd edn. Wiley-VCH, Weinheim
87. Grey CP, Tycko R (2009) Solid-state NMR in biological and materials physics. Phys Today 62(9):44–49
88. Bennett AE, Rienstra CM, Auger M et al (1995) Heteronuclear decoupling in rotating solids. J Chem Phys 103(16):6951–6958
89. Vanderhart DL, Earl WL, Garroway AN (1981) Resolution in C-13 NMR of organic-solids using high-power proton decoupling and magic-angle sample spinning. J Magn Reson 44(2):361–401
90. Chen LL, Kaiser JM, Lai JF et al (2007) J-based 2D homonuclear and heteronuclear correlation in solid-state proteins. Magn Reson Chem 45:S84–S92
91. Ishii Y, Balbach JJ, Tycko R (2001) Measurement of dipole-coupled lineshapes in a many-spin system by constant-time two-dimensional solid state NMR with high-speed magic-angle spinning. Chem Phys 266 (2–3):231–236
92. Pines A, Gibby MG, Waugh JS (1973) Proton-enhanced NMR of dilute spins in solids. J Chem Phys 59(2):569–590
93. Schaefer J, Stejskal EO (1976) C-13 nuclear magnetic-resonance of polymers spinning at magic angle. J Am Chem Soc 98 (4):1031–1032
94. Griffiths DJ (1995) Introduction to quantum mechanics. Prentice Hall, Upper Saddle River
95. Morcombe CR, Zilm KW (2003) Chemical shift referencing in MAS solid state NMR. J Magn Reson 162(2):479–486

96. Fritzsching KJ, Yang Y, Schmidt-Rohr K et al (2013) Practical use of chemical shift databases for protein solid-state NMR: 2D chemical shift maps and amino-acid assignment with secondary-structure information. J Biomol NMR 56(2):155–167
97. Veshtort M, Griffin RG (2006) SPINEVOLUTION: a powerful tool for the simulation of solid and liquid state NMR experiments. J Magn Reson 178(2):248–282
98. Bak M, Rasmussen JT, Nielsen NC (2000) SIMPSON: a general simulation program for solid-state NMR spectroscopy. J Magn Reson 147(2):296–330
99. Tay WM, Huang D, Rosenberry TL et al (2013) The Alzheimer's amyloid-β(1-42) peptide forms off-pathway oligomers and fibrils that are distinguished structurally by intermolecular organization. J Mol Biol 425:2494–2508
100. States DJ, Haberkorn RA, Ruben DJ (1982) A two-dimensional nuclear overhauser experiment with pure absorption phase in four quadrants. J Magn Reson (1969) 48 (2):286–292
101. Gueron M, Plateau P, Decorps M (1991) Solvent signal suppression in NMR. Prog Nucl Magn Reson Spectrosc 23:135–209
102. Ishii, Y (2001) 13C–13C dipolar recoupling under very fast magic angle spinning in solid-state nuclear magnetic resonance: Applications to distance measurements, spectral assignments, and high-throughput secondary-structure determination. J Chem Phys 114(19):8473–8483
103. Takegoshi K, Nakamura S, Terao T (2001) 13C-1H dipolar-assisted rotational resonance in magic-angle spinning NMR. Chem Phys Lett 344(5–6):631–637
104. Tycko R, Ishii Y (2003) Constraints on supramolecular structure in amyloid fibrils from two-dimensional solid-state NMR spectroscopy with uniform isotopic labeling. J Am Chem Soc 125(22):6606–6607
105. Jaroniec CP, Filip C, Griffin RG (2002) 3D TEDOR NMR experiments for the simultaneous measurement of multiple carbon–nitrogen distances in uniformly13C,15 N-labeled solids. J Am Chem Soc 124(36):10728–10742
106. Petkova AT, Yau WM, Tycko R (2006) Experimental constraints on quaternary structure in Alzheimer's β-amyloid fibrils. Biochemistry 45(2):498–512
107. Gullion T, Schaefer J (1989) Rotational-echo double-resonance NMR. J Magn Reson 81 (1):196–200
108. Jaroniec CP, Tounge BA, Herzfeld J et al (2001) Frequency selective heteronuclear dipolar recoupling in rotating solids: accurate13C – 15 N distance measurements in uniformly 13C,15 N-labeled peptides. J Am Chem Soc 123(15):3507–3519
109. Anderson WA (1961) Electrical current shims for correcting magnetic fields. Rev Sci Instrum 32(3):241
110. Harris RK, Becker ED, De Menezes SMC et al (2001) NMR nomenclature. Nuclear spin properties and conventions for chemical shifts—(IUPAC recommendations 2001). Pure Appl Chem 73(11):1795–1818
111. Cavanaugh J, Fairbrother WJ, Palmer AG et al (2006) Protein NMR spectroscopy, 2nd edn. Academic, Raleigh
112. Taylor RE (2004) Setting up 13C CP/MAS experiments. Concepts Magn Reson A 22 (1):37–49
113. Taylor RE (2004) C-13 CP/MAS: application to glycine. Concepts Magn Reson Part A 22a(2):79–89
114. Stejskal E, Schaefer J, Waugh J (1977) Magic-angle spinning and polarization transfer in proton-enhanced NMR. J Magn Reson (1969) 28(1):105–112
115. Hohwy M, Rienstra C, Jaroniec C et al (1999) Fivefold symmetric homonuclear dipolar recoupling in rotating solids: application to double quantum spectroscopy. J Chem Phys 110(16):7983–7992
116. Massiot D, Fayon F, Capron M et al (2002) Modelling one-and two-dimensional solid-state NMR spectra. Magn Reson Chem 40 (1):70–76
117. Schneider R, Seidel K, Etzkorn M et al (2009) Probing molecular motion by double-quantum (13C, 13C) solid-state NMR spectroscopy: application to ubiquitin. J Am Chem Soc 132(1):223–233
118. Johnson RL, Anderson JM, Shanks BH et al (2013) Spectrally edited 2D 13 C 13 C NMR spectra without diagonal ridge for characterizing 13 C-enriched low-temperature carbon materials. J Magn Reson 234:112–124
119. Wishart D, Sykes B (1994) The 13C chemical-shift index: a simple method for the identification of protein secondary structure using 13C chemical-shift data. J Biomol NMR 4(2):171–180
120. Wishart DS (2011) Interpreting protein chemical shift data. Prog Nucl Mag Res Sp 58(1–2):62–87
121. Shen Y, Delaglio F, Cornilescu G et al (2009) TALOS plus: a hybrid method for predicting protein backbone torsion angles from NMR

chemical shifts. J Biomol NMR 44 (4):213–223
122. Petkova AT, Ishii Y, Tycko R (2002) Probing the structure of Alzheimer's beta amyloid fibrils by two-dimensional C-13-C-13 and C-13-N-15 solid state NMR methods. Biophys J 82(1):320A–320A
123. Petkova AT, Ishii Y, Balbach JJ et al (2002) A structural model for Alzheimer's β-amyloid fibrils based on experimental constraints from solid state NMR. Proc Natl Acad Sci U S A 99(26):16742–16747
124. Paravastu AK, Petkova AT, Tycko R (2006) Polymorphic fibril formation by residues 10-40 of the Alzheimer's β-amyloid peptide. Biophys J 90(12):4618–4629
125. Ramamoorthy A (2005) NMR spectroscopy of biological solids. CRC Press, New York
126. Colvin MT, Silvers R, Frohm B et al (2015) High resolution structural characterization of Abeta42 amyloid fibrils by magic angle spinning NMR. J Am Chem Soc 137 (23):7509–7518
127. Jaroniec CP, MacPhee CE, Bajaj VS et al (2004) High-resolution molecular structure of a peptide in an amyloid fibril determined by magic angle spinning NMR spectroscopy. Proc Natl Acad Sci U S A 101(3):711–716
128. Debelouchina GT, Bayro MJ, Fitzpatrick AW et al (2013) Higher order amyloid fibril structure by MAS NMR and DNP spectroscopy. J Am Chem Soc 135(51):19237–19247
129. Bennett AE, Rienstra CM, Griffiths JM et al (1998) Homonuclear radio frequency-driven recoupling in rotating solids. J Chem Phys 108(22):9463–9479
130. Gullion T, Vega S (1992) A simple magic angle spinning NMR experiment for the dephasing of rotational echoes of dipolar coupled homonuclear spin pairs. Chem Phys Lett 194:424–428
131. Petkova AT, Tycko R (2002) Sensitivity enhancement in structural measurements by solid state NMR through pulsed spin locking. J Magn Reson 155(2):293–299
132. Jaroniec CP, Tounge BA, Rienstra CM et al (2000) Recoupling of heteronuclear dipolar interactions with rotational-echo double-resonance at high magic-angle spinning frequencies. J Magn Reson 146(1):132–139
133. Verel R, Tomka IT, Bertozzi C et al (2008) Polymorphism in an amyloid-like fibril-forming model peptide. Angew Chem Int Ed 47 (31):5842–5845
134. Balbach JJ, Ishii Y, Antzutkin ON et al (2000) Amyloid fibril formation by a beta(16-22), a seven-residue fragment of the Alzheimer's beta-amyloid peptide, and structural characterization by solid state NMR. Biochemistry 39(45):13748–13759
135. Xiao Y, Ma B, McElheny D et al (2015) Abeta (1-42) fibril structure illuminates self-recognition and replication of amyloid in Alzheimer's disease. Nat Struct Mol Biol 22 (6):499–505
136. Tycko R, Sciarretta KL, Orgel JPRO et al (2009) Evidence for novel β-sheet structures in iowa mutant β-amyloid fibrils. Biochemistry 48(26):6072–6084

Chapter 3

ATR-FTIR Analysis of Amyloid Proteins

Jean-Marie Ruysschaert and Vincent Raussens

Abstract

Attenuated total reflection FTIR (ATR-FTIR) has been used for decades to study protein secondary structures. More recently, it reveals also to be an exquisite and sensitive tool to study and discriminate amyloid aggregates. Based on the analysis of specific spectral features of β-sheet structures, we present here a detailed protocol to differentiate oligomers vs. fibrils. This protocol, applicable to all amyloid proteins, demonstrates the power of this inexpensive, rapid, and low protein material-demanding method.

Key words Amyloid, Aggregates, Oligomers, Fibrils, FTIR, ATR, Structure

Abbreviations

ATR	Attenuated total reflection
FTIR	Fourier-transform infrared
FWHH	Full width at half height
HFIP	Hexafluoro-isopropanol
IRE	Internal reflection element
MCT	Mercury-Cadmium-Telluride
TFE	Trifluoroethanol

1 Introduction

One of the major differences between the different aggregated forms of amyloid proteins is their secondary structure (antiparallel β-sheet for oligomers and parallel β-sheet for fibrils) and this is related to their respective cytotoxicity [1–6]. We, and others, have demonstrated that ATR-FTIR is an exquisite and versatile method in the study and distinction of the different amyloid aggregated forms that overcome some of the limitations of high-resolution techniques [4, 7–9]. ATR-FTIR can also provide an almost continuous structural view of protein conversion during the aggregation process [8]. Furthermore, we demonstrated that the presence of

Bradley L. Nilsson and Todd M. Doran (eds.), *Peptide Self-Assembly: Methods and Protocols*, Methods in Molecular Biology, vol. 1777, https://doi.org/10.1007/978-1-4939-7811-3_3, © Springer Science+Business Media, LLC, part of Springer Nature 2018

lipids does not prevent the analysis of the amyloid sample [10]. The interaction of these proteins with membranes is known to be a major cause of their toxicity [11, 12]. In addition, while the major component of amyloid deposits is the protein responsible for the disease, other components, such as other proteins, are also present in the deposits. ATR-FTIR allows studying the effects of these proteins on the amyloid aggregation process [13–15].

Briefly, these capabilities come from two main advantages of ATR-FTIR. First, infrared spectroscopy, through the analysis of the amide I (~1700–1600 cm^{-1}) and less informative amide II (~1600–1500 cm^{-1}) regions, is a sensitive and accurate method to address protein secondary structure [16, 17]. It is one of the few low-resolution structural methods that can distinguish between antiparallel and parallel β-sheet arrangements, which is of the utmost importance in the study of amyloid proteins [18–21]. A full analysis of the protein structure will require a complete study of the amide I band by either curve-fitting or other means [22, 23]. These have been described elsewhere, and thus will not be addressed here. The analysis that we present here will focus on the spectral signatures of β-sheet structures that define aggregates in a fibrillar or oligomeric state. In antiparallel β-sheet structures, the amide I region displays two stretching frequency absorptions: a major one located at $\sim$1630 cm^{-1}, and a minor component characterized by an average wavenumber around 1695 cm^{-1} caused by a strong dipolar coupling, approximately fivefold weaker than the absorption at 1630 cm^{-1}. Because the 1695/1630 intensity ratio has been suggested to be proportional to the percentage of antiparallel arrangement of the β-strands within a β-sheet, this ratio will be the most useful criterion for the characterization of oligomers. For parallel β-sheet structures, the amide I region features the major component only. Amyloid fibrils typically display absorbance between 1611 and 1630 cm^{-1}, while for native β-sheet proteins it extends from 1630 to 1643 cm^{-1}. These differences are attributed to more extended β-sheets and/or longer β-strands formation as well as more planar sheet formation for amyloid fibrils.

Finally, attenuated total reflection (ATR) [24] is a sampling method particularly suitable for infrared spectroscopy because of the reflection of the light into a high refractive index crystal. This crystal, called the internal reflection element (IRE, see Fig. 1), creates an evanescent wave perpendicular and outside the surface of the IRE at the point(s) of reflection. In the mid-infrared range, this evanescent wave extends ~1–2 μm from the surface and can interact with a sample in close contact with it. For biological and macromolecular samples ATR-FTIR has multiple advantages: (1) because the sample is in close contact with the surface and excess water can be removed, liquid water signal in the spectrum can be reduced or eliminated; (2) working directly at the surface of the IRE, there is no drawback to using aggregated or insoluble

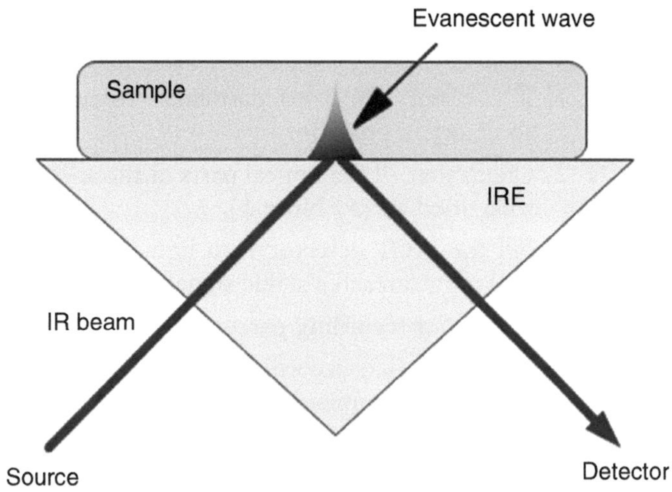

Fig. 1 Schematic representation of ATR principle. Upon reflection inside the IRE, an evanescent wave is produced outside the IRE crystal that can interact with the sample. The IRE can be a prism (as represented) or a hemisphere giving rise to a single reflection or longer bars with bevel-edges allowing multiple reflections. The crystal can be made of diamond, germanium, ZnSe, silicon, or KRS-5

samples neither lipids, as is the case with other biophysical techniques. These properties are exquisitely adapted to amyloid protein study and circumvent major problems encountered with most high-resolution methods (3).

In this protocol, we detail a procedure to analyze the parallel and antiparallel b-sheet content in amyloid protein aggregates. We also describe the steps necessary to determine the relative amounts of oligomer and fibril in the sample. While this analysis is general to almost all amyloid proteins, we recommend that other biophysical techniques be performed in-parallel to characterize the species formed during amyloid aggregate assembly.

2 Materials

1. Fourier-transform infrared (FTIR) spectrophotometer with mid-infrared capabilities and equipped either directly or through a special accessory for attenuated total reflection (ATR).
2. A liquid nitrogen-cooled MCT detector. If available, a thermostatized room will improve the spectral quality of the results (*see* **Note 1**).
3. An internal reflection element (IRE), such as either Ge or diamond (*see* **Note 2**).
4. Amyloid fibril sample containing a low salt concentration (*see* **Note 3**).

3 Methods

3.1 Preparation of the FTIR Spectrometer and the IRE

1. If necessary for your particular instrument, place the ATR accessory in position.
2. Check that all the optical parts of the spectrometer are purged with dried air (*see* **Note 1**).
3. Fill the MCT detector with liquid N_2 (*see* **Note 4**) and wait ~30 min to reach a stable signal.
4. Load your recording parameters (*see* **Note 5**).
5. Clean the IRE according to the manufacturer's notices. We strongly recommend cleaning the IRE crystal just before the experiment. Usually aggregated amyloid proteins do not require a specific cleaning procedure (*see* **Note 6**).
6. If needed, put the cleaned IRE back to its correct position in the IR beam.
7. Check if enough signal reaches the detector.
8. Record a background spectrum with the parameters used to record the sample spectrum (only the number of scans can be modified but there is usually no reason to do so). The background spectrum should be recorded just before sample recording (*see* **Note 7**).

3.2 Sample Spreading and Spectrum Recording

1. Prepare protein sample(s) according to your own experience or from protocols found in literature (*see* **Note 8**).
2. Spread the sample using a micropipet tip. Do not hesitate to touch the IRE with the tip. Volume of the sample to spread depends strongly on the IRE used (e.g., for a 2 mm × 2 mm single reflection prism IRE, 1 or 2 μL are sufficient, for longer (3–5 cm) trapezoidal IRE 10–20 μL might be spread at once, *see* **Note 9**).
3. Under a soft N_2 (or another nonreactive gas) flow (*see* **Note 10**) evaporate the liquid water (or solvent) excess in the sample. To control the evaporation process, start the spectrum "recording" in a previsualization, and not acquisition, mode. Display the full spectrum, watch the intensity of the major water band (~3000–3500 cm^{-1}) decreasing until it reaches a stable level (*see* **Note 11**). If your spectrometer does not allow a previsualization mode, the best way to know if the excess water is removed and your sample is correct for recording is to check the visual appearance of the sample film. With low volume and low salt concentration the film should be smooth and clear sometimes with iridescence (due to thin-film interference). For larger volumes and/or higher salts, the film usually looks smooth and matte (*see* **Note 12**).

4. When the excess liquid water is removed and the signal is stable (*see* **Note 13**), start recording the sample spectrum.
5. If not done automatically, store the spectrum file(s) on a computer in a format compatible with your analysis program.
6. Record a water vapor spectrum. To obtain a water vapor spectrum matching the water vapor features appearing in your sample, do not displace, remove, clean, or open anything on your spectrometer. Record a new background, then open slightly the sample compartment for a few seconds, and then record a new spectrum. Store the resulting water vapor spectrum (*see* **Note 14**).

3.3 Spectral Processing

All processing steps described here can be done on any spectral analysis software including those coming with recent spectrometers. Most of these steps can be automatized to some extent using macro recording. Thanks to Prof. Erik Goormaghtigh, our laboratory has developed a piece of software entirely dedicated to these kinds of analyses, and all the described steps can be done in just one step for multiple spectra. This software, called "Kinetics," runs under Matlab©, and is available upon request from Erik Goormaghtigh (egoor@ulb.ac.be) or the corresponding author.

1. Subtract water vapor spectrum from your sample spectrum (Fig. 2a, b). An easy and convenient way to find the subtraction coefficient is using the area of a specific water vapor peak. We use one of the most intense peaks in the amide I and II region, the 1559 cm^{-1} peak. This peak appears on the side of the amide II band and is usually quite distinct from other features (black peak in Fig. 2a). Using your favorite analysis software, integrate the peak between 1562 and 1555 cm^{-1} on both sample and water vapor spectra. The ratio of the two areas gives you the subtraction coefficient to be applied to correctly remove the water vapor contribution in your sample spectrum.
2. Apply a linear baseline subtraction on the entire spectrum (*Fig. 2c for the amide I and II region, see Fig. 2 legend for baseline selected points*). The points to apply for the baseline should be selected according to sample spectrum (*see* **Note 15**). The same baseline points should be kept throughout your analyses for consistency.
3. Smooth the spectrum to 4 cm^{-1} by convolution (Fig. 2c).
4. For better visualization, especially when the protein is not extremely rich in β-sheet (<50%), the resolution can be increased by Fourier self-deconvolution (using a Lorentzian line shape with a full width at half height (FWHH) of 20 cm^{-1} and a Gaussian line shape for apodization with a FWHH of 13.333 cm^{-1} resulting in an enhancement factor $K = 1.5$) [26] (Fig. 2e). This allows a visualization of bands otherwise difficult to assess, especially the 1695 cm^{-1} band

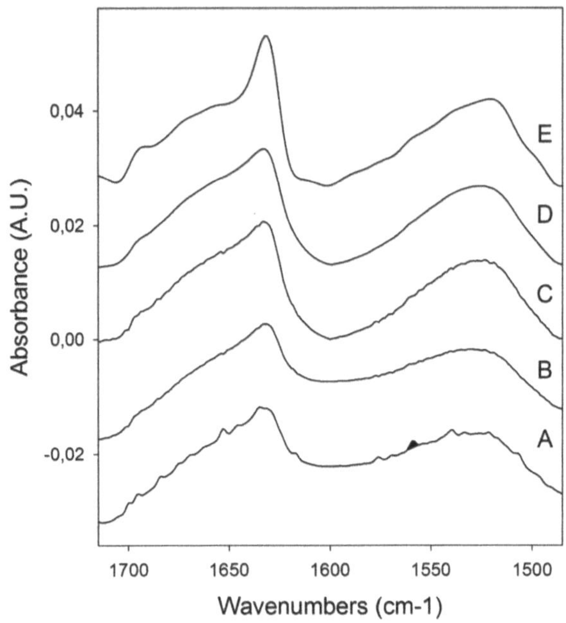

Fig. 2 Example of spectral processing in the amide I and II region. (**a**) original recorded spectrum of Aβ40 oligomers (the black area peak at 1559 cm^{-1} is the one used to compute the subtraction coefficient of the water vapor spectrum); (**b**) spectrum A after water vapor subtraction; (**c**) spectrum B after baseline subtraction (baseline points applied in the displayed windows are located, respectively, at 1711, 1599, and 1482 cm^{-1}; other baseline points used are positioned at 3990, 3700, 2450, 2245, 1744, 1430, 1160, and 810 cm^{-1} (not shown)); (**d**) spectrum C after smoothing by convolution at 4 cm^{-1}; (**e**) spectrum D after resolution enhancement by self-deconvolution. All spectra have been shifted for better visualization except for spectrum C

which, when concomitant with the intense ~1630 cm^{-1} band, indicates an antiparallel β-sheet and the presence of oligomers.

3.4 Spectral Analysis

3.4.1 Case 1: Samples Have High β-Sheet Content (≥55–60%) (Specific and Simplest Case, Fig. 3, Left)

1. Locate the two main components (*see* **Note 16**) of the antiparallel β-sheet on the self-deconvoluted spectrum (corresponding to **step 4** in Subheading 3.3).
2. Measure the spectral intensities at these two wavenumbers on the non-deconvoluted spectrum (corresponding to **step 3** in Subheading 3.3, see also legend of Fig. 3).
3. Calculate the 1695/1630 cm^{-1} intensity ratio (*see* **Note 16**).
4. Inspect the amide II region; if a shoulder or an increased intensity appears around 1530 cm^{-1}, it indicates the presence of oligomers (*black arrows in Fig. 3*).

3.4.2 Case 2: Samples Have Low β-Sheet Content (<50%) (General Case, Fig. 3 Right)

1. Locate the different amide I components using self-deconvolution with a higher enhancement factor (for example, using a Lorentzian line shape with a full width at half height (FWHH) of 30 cm^{-1} and a Gaussian line shape for apodization

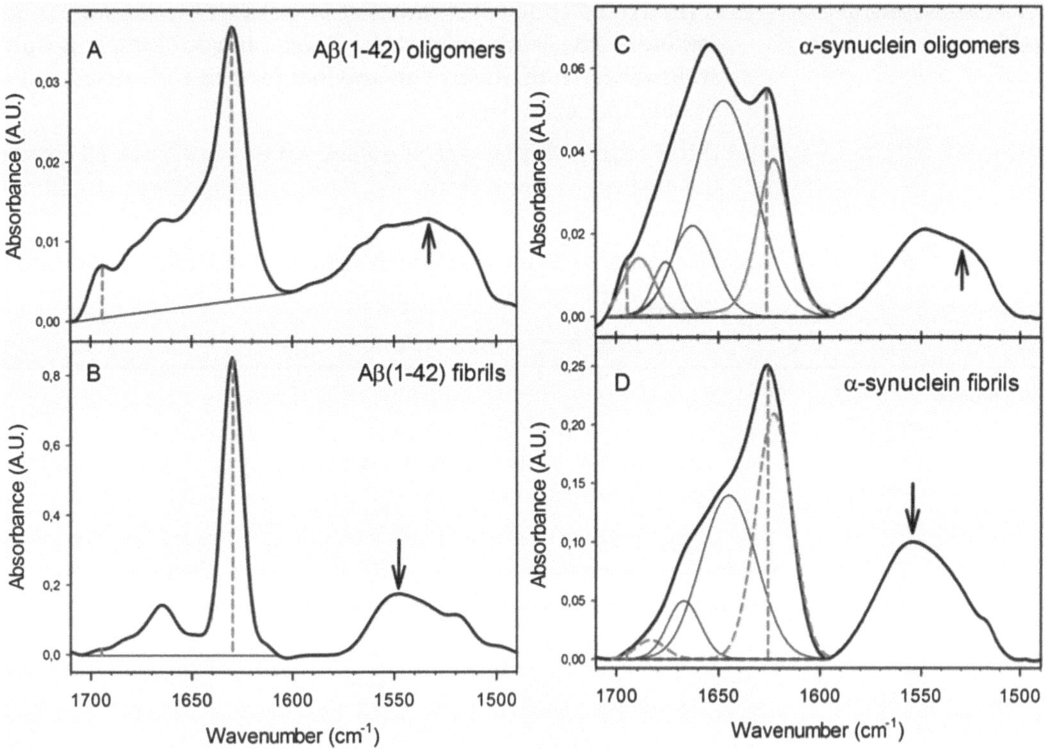

Fig. 3 Examples of oligomer vs. fibril quantifications for Aβ42 (left) and α-synuclein (right). Aβ42 (left) represents a case where the percentage of β-sheet is high (>60%). Therefore, a "simple" intensity (gray dashed vertical lines) ratio measurement at 1695 and 1630 cm^{-1} will conveniently differentiate oligomers (**a**) from fibrils (**b**). The black arrows in the amide II region show the characteristic shift when going from oligomers to fibrils with the disappearance of intensity around 1530 cm^{-1}. In the case of α-synuclein (right), the β-sheet percentage is lower (≤40%). A curve-fitting of the amide I regions shows that for both oligomers (**c**) and fibrils (**d**) overlapping structures (represented by curves in black under the spectrum amide I bands) have significant contributions in the recorded intensities (gray dashed vertical lines). The ratio should be measured by the area of the curves corresponding to the two contributions of β-sheet (gray dashed curves). In the case of fibrils (**d**), the curve supposedly corresponding to the 1695 cm^{-1} band is significantly displaced (gray dashed curve) and is probably not representative of β-sheet anymore. Here again, the black arrows show the characteristic change in the amide II region

with a FWHH of 16.666 cm^{-1} resulting in an enhancement factor $K = 1.8$) [26].

2. Using all the identified components, proceed to curve-fitting (*see* **Note 17**) of the non-deconvoluted spectrum (corresponding to **step 3** in Subheading 3.3, see also legend of Fig. 3).
3. Measure the area of the curves corresponding to the ~1695 and ~1630 cm^{-1} contributions and calculate the 1695/1630 band area ratio (*see* **Note 16**).
4. Inspect the amide II as in **step 4** in Subheading 3.4.1.

3.4.3 Interpretation of the Results

1. If the 1695/1630 cm^{-1} intensity or band area ratios are ≥ 0.25 (*see* **Note 18**), it means that the β-sheet in your sample is fully antiparallel. In the case of an amyloid protein this corresponds to oligomeric species.
2. If this ratio is close to zero (at least ≤ 0.05) (*see* **Note 18**), then the β-sheet is entirely parallel, and it corresponds to fibrils (*see* **Note 19**).
3. The inspection of the amide II region (if available) as indicated should confirm this interpretation.

4 Notes

1. Alternatively, with nitrogen, be extremely cautious with nitrogen in a closed room.
2. Many crystals might be used for ATR in infrared spectroscopy (e.g., Germanium, ZnSe, KRS-5, Silicon, Diamond) and all will give good results. For practical reasons, we will recommend to use either Ge or diamond for their higher refractive index and especially hardness allowing easier handling and cleaning procedures.
3. The rule of thumb is that the sample should contain less salt than protein (i.e., in a given volume, the total weight of all the salts or other components should be equal or inferior to the weight of the protein). For fibrils, simple washing after pelleting the fibrils by centrifugation, with few changes of the supernatant into a low salt buffer or water, is the easiest way to achieve this. For oligomers, it is more difficult to devise a general recipe. A tip to remove most of the salt from any sample quickly and easily is to use 0.025 μm pore size MF-Millipore© membrane (25 mm in diameter) composed of mixed cellulose (VSWP02500 MF-Millipore© Membrane Filter). To wash out salts, fill a beaker with distilled or deionized water, let one of the Millipore filters float at the surface with the glossy side facing up, and gently deposit a drop (10–20 μL) of your sample to be desalted on the filter (in the middle to avoid it falls in the beaker). Wait 30–45 min, recover you sample, and it should be desalted enough to record a good infrared spectrum. This "method" is neither analytical nor quantitative, but in our hands it produces good results. The drawbacks are the following: your sample will be slightly diluted at the end; if your sample contains small peptide(s), they might be lost. Conversely, the latter may be advantageous should one wish to remove short peptide monomers from your sample.
4. First, you must always apply security rules specific to liquid N_2 handling; liquid N_2 can be extremely harmful! Second, the

liquid N_2 reservoir of the MCT detector is not always easily accessible, and it may be difficult to follow the filling process. It is therefore easy to overfill the reservoir. If this happens, then the liquid N_2 will accumulate at the bottom of the detector compartment and quickly cause water to condense on the mirrors and detector window. This will cause the signal to drop dramatically and cause water to accumulate on the optic parts, which is detrimental to the instrument. If you are familiar with the spectrometer you can open the detector compartment and flow N_2 gas over the optic parts to remove all the condensation and bring the parts to room temperature. If you are not confident enough, you will have to wait until the excess liquid N_2 evaporates and the detector compartment warms up. This can be a long process.

5. These parameters should be for proteins or biological samples. Do not load specific ATR parameters provided by the manufacturer, this is not needed.

   Here are example spectral parameters used to obtain the spectra displayed in Figs. 2 and 3 and acquired using an Equinox 55 spectrometer (Bruker).

   (a) Recording:
   - Resolution: 2 cm^{-1}.
   - Number of scans: 256.
   - Data saved: absorbance, single channel, background.

   (b) Optic:
   - Source setting: MIR-source.
   - Beamsplitter: KBr.
   - Optical filter setting: OPEN.
   - Iris aperture: 3500 μm.
   - Detector setting: MCT.
   - Scanner velocity: 8, 40.0 kHz.
   - Sample signal gain: Automatic.
   - Background signal gain: Automatic.
   - Switch gain: ON.
   - Window in points: 350.
   - Delay before measurement: 1 s.

   (c) Acquisition:
   - Wanted high frequency limit: 5266 cm^{-1}
   - Wanted low frequency limit: 0 cm^{-1}
   - Low pass filter: 1, 16 kHz
   - Acquisition mode: Double sided; Forward-backward
   - Correlation mode: Around peak, Low.

(d) FT (Fourier Transform):
   - Phase resolution: 16.
   - Phase correction mode: Mertz.
   - Apodization function: Blackman-Harris 3-term.
   - Zerofilling factor: 1.

(e) Display:
   - Display single scan before measurement: ON.
   - Window: 4000–800 cm^{-1}.

6. If you notice any problems with specific proteins, use a mixture (50:50 vol:vol) of strong detergent and methanol. If still not working, then trifluoroethanol (TFE) or 1,1,1,3,3,3-hexafluoroisopropanol (HFIP) can be used with caution.
7. The question arises whether one should record a new background for each sample if using the same setting and the same IRE each time. The answer is that if you plan to record only a limited number of spectra on the same day, we will advise to record a new background. On the contrary, if you need to record a big number of spectra on the same day (e.g., when following a time-dependent aggregation process requiring an important number of time points), then just cleaning the IRE crystal well and checking that the baseline spectrum returns to zero after cleaning is sufficient.
8. Preparation is highly dependent on the type of amyloid protein considered and if both fibrils and oligomers are needed separately or as a mixture. Check the literature to find most appropriate preparation conditions. If none exist so far, you will have to develop your own protocol.

   If the preparation conditions of your amyloid protein have been previously fully optimized, then you can move to the ATR-FTIR experiment without any further requirements. Otherwise, a complete and complementary characterization (e.g., thioflavin T fluorescence, conformational antibodies recognition, electronic microscopy, AFM) [25] will be mandatory to be sure you are indeed working with the correct and expected sample(s).
9. If not enough protein is available in the volumes required to obtain a good quality spectrum, do not try to increase the volume to spread. After evaporation of the excess solvent, a new volume can be spread again on top of the previous one. However, be careful of the final salt content!
10. The N_2 flow should be strong enough to evaporate the excess water or solvent in a few minutes but soft enough to avoid displacing the sample.

11. It is better to monitor this intensity at ~3400 cm^{-1}, which is "on the side" of the band. This is because the maximum of the water band is close to the amide A band and may cause confusion.
12. If the film is not smooth, there is probably too many salts from the buffer left in your sample.
13. Due to buffer composition and IRE material, in some occurrences we note a significant increase in liquid water spectral intensity upon N_2 flow removal. In these cases, we suggest to leave a light N_2 flow on the sample during the acquisition of the spectrum.
14. You do not need to record a water vapor spectrum for each sample. However, you should know that vapor spectra are extremely influenced by atmospheric conditions, especially temperature. If between experiments the time is short and the general conditions are similar, you do not need to record each time a new vapor spectrum. Nevertheless, we will recommend recording at least one vapor spectrum for each day of experiments and more if, during the day, the atmospheric conditions drastically changed.
15. For example, in the presence of lipids, the point on the side of the amide I will probably be displaced. Select points close to the edge of the bands or at the lowest intensity between bands if the intensity does not return to zero or close (see legend of Fig. 2 for an example).
16. For simplicity reasons, we always use the ~1695 and ~1630 cm^{-1} wavenumbers to talk about the two main spectral components of β-sheet. Amyloids typically display absorbance between 1611 and 1630 cm^{-1}, and the 1695 cm^{-1} band might also be displaced due to differences in the dipolar coupling between the two bands.
17. A description of the entire curve-fitting procedure is beyond the scope of this protocol. Curve-fitting has been described in many good articles and reviews (e.g., [22]), and the reader is referred to these for further explanations.
18. These values have been experimentally assessed using well-defined soluble protein structure and different amyloid protein preparations [4, 7–9].
19. Recently, Wang et al. have demonstrated that it is possible to calculate the quantity of fibrils in a complex protein mix based on many different band area ratios [27].

Acknowledgements

The authors would like to thank all the talented Ph.D. students and postdoctoral fellows who help to establish and improve this protocol. V.R. is Senior Research Associate for the National Fund for Scientific Research (F.R.S.-FNRS, Belgium). V.R. acknowledges funding from the F.R.S.-FNRS (PDR grant #70214.12) and SAO-FRA (grant S#14025).

References

1. Kayed R, Head E, Thompson JL, McIntire TM, Milton SC, Cotman CW, Glabe CG (2003) Common structure of soluble amyloid oligomers implies common mechanism of pathogenesis. Science 300:486–489
2. Laganowsky A, Liu C, Sawaya MR, Whitelegge JP, Park J, Zhao M, Pensalfini A, Soriaga AB, Landau M, Teng PK, Cascio D, Glabe C, Eisenberg D (2012) Atomic view of a toxic amyloid small oligomer. Science 335:1228–1231
3. Breydo L, Uversky VN (2015) Structural, morphological, and functional diversity of amyloid oligomers. FEBS Lett 589:2640–2648
4. Sarroukh R, Goormaghtigh E, Ruysschaert J-M, Raussens V (2013) ATR-FTIR: a "rejuvenated" tool to investigate amyloid proteins. Biochim Biophys Acta 1828(10):2328–2338
5. Gustot A, Raussens V, Dehousse M, Dumoulin M, Bryant CE, Ruysschaert J-M, Lonez C (2013) Activation of innate immunity by lysozyme fibrils is critically dependent on cross-β sheet structure. Cell Mol Life Sci 70:2999–3012
6. Gustot A, Gallea JI, Sarroukh R, Celej MS, Ruysschaert JM, Raussens V (2015) Amyloid fibrils are the molecular trigger of inflammation in Parkinson's disease. Biochem J 471:323–333
7. Cerf E, Sarroukh R, Tamamizu-Kato S, Breydo L, Derclaye S, Dufrêne YF, Narayanaswami V, Goormaghtigh E, Ruysschaert J-M, Raussens V (2009) Antiparallel β-sheet: a signature structure of the oligomeric amyloid β-peptide. Biochem J 421:415–423
8. Sarroukh R, Cerf E, Derclaye S, Dufrêne YF, Goormaghtigh E, Ruysschaert J-M, Raussens V (2011) Transformation of amyloid β(1–40) oligomers into fibrils is characterized by a major change in secondary structure. Cell Mol Life Sci 68:1429–1438
9. Celej MS, Sarroukh R, Goormaghtigh E, Fidelio G, Ruysschaert JM, Raussens V (2012) Toxic prefibrillar alpha-synuclein amyloid oligomers adopt a distinctive antiparallel beta-sheet structure. Biochem J 443:719–726
10. de Planque MR, Raussens V, Contera SA, Rijkers DT, Liskamp RM, Ruysschaert JM, Ryan JF, Separovic F, Watts A (2007) β-Sheet structured beta-amyloid(1–40) perturbs phosphatidylcholine model membranes. J Mol Biol 368:982–997
11. Cecchi C, Stefani M (2013) The amyloid-cell membrane system. The interplay between the biophysical features of oligomers/fibrils and cell membrane defines amyloid toxicity. Biophys Chem 182:30–43
12. Jang H, Arce FT, Ramachandran S, Kagan BL, Lal R, Nussinov R (2014) Disordered amyloidogenic peptides may insert into the membrane and assemble into common cyclic structural motifs. *Chem Soc Rev* 43:6750–6764
13. Cerf E, Gustot A, Ruysschaert J-M, Goormaghtigh E, Raussens V (2011) Pathogenic E4 isoform of apolipoprotein E favors the oligomeric state of Aβ. FASEB J 25:1585–1595
14. Itkin A, Dupres V, Dufrêne YF, Bechinger B, Ruysschaert J-M, Raussens V (2011) Calcium ions promote formation of amyloid β-peptide (1–40) oligomers causally implicated in neuronal toxicity of Alzheimer's disease. PLoS One 6:e18250
15. Bharadwaj P, Head R, Martins R, Raussens V, Sarroukh R, Jegasothy H, Waddington L, Bennett L (2013) Modulation of amyloid-β 1–42 structure and toxicity by proline-rich whey peptides. Food Funct 4(1):92–103
16. Miyazawa T, Blout ER (1961) The infrared spectra of polypeptides in different conformations: amide I and amide II bands. J Am Chem Soc 83:712–719
17. Krimm S, Abe Y (1972) Intermolecular interaction effects in the amide I vibrations of polypeptides. Proc Natl Acad Sci U S A 69:2788–2792

18. Chirgadze YN, Nevskaya NA (1976) Infrared spectra and resonance interaction of amide-I vibration of the antiparallel-chain pleated sheet. Biopolymers 15:607–625
19. Chirgadze YN, Nevskaya NA (1976) Infrared spectra and resonance interaction of amide-I vibration of the parallel-chain pleated sheets. *Biopolymers* 15:627–636
20. Kubelka J, Keiderling TA (2001) Differentiation of beta-sheet-forming structures: ab initio-based simulations of IR absorption and vibrational CD for model peptide and protein beta-sheets. J Am Chem Soc 123:12048–12058
21. Karjalainen EL, Ravi HK, Barth A (2011) Simulation of the amide I absorption of stacked β-sheets. J Phys Chem B 115:749–757
22. Goormaghtigh E, Cabiaux V, Ruysschaert JM (1990) Secondary structure and dosage of soluble and membrane proteins by attenuated total reflection Fourier-transform infrared spectroscopy on hydrated films. Eur J Biochem 193:409–420
23. Goormaghtigh E, Ruysschaert JM, Raussens V (2006) Evaluation of the information content in infrared spectra for protein secondary structure determination. Biophys J 90:2946–2957
24. Harrick NJ (1967) Internal reflection spectroscopy. Wiley, New York
25. Bitan G, Fradinger EA, Spring SM, Teplow DB (2005) Neurotoxic protein oligomers—what you see is not always what you get. Amyloid 12:88–95
26. Kauppinen JK, Moffat DJ, Cameron DG, Mantsch HH (1981) Noise in Fourier self-deconvolution. Appl Optics 20:1866–1879
27. Wang P, Bohr W, Otto M, Danzer KM, Mizaikoff B (2015) Quantifying amyloid fibrils in protein mixtures via infrared attenuated-total-reflection spectroscopy. Anal Bioanal Chem 407:4015–4021

Chapter 4

Imaging Protein Fibers at the Nanoscale and In Situ

Angelo Bella, Michael Shaw, Emiliana De Santis, and Maxim G. Ryadnov

Abstract

Protein self-assembly offers a rich repertoire of tools and technologies. However, despite significant progress in this area, a deterministic measure of the phenomenon, which might lead to predictable relationships between protein components, assembly mechanisms, and ultimately function, is lacking. Often the challenge relates to the choice of the most informative and precise measurements that can link the chemistry of the building blocks with the resulting assembly, ideally in situ and in real time. Using the example of protein fibrillogenesis—a self-assembly process fundamental to nearly every aspect of biological organization, from viral assembly to tissue restoration—this chapter demonstrates how protein self-assembly can be visually and precisely measured while providing measurement protocols applicable to other self-assembly systems.

Key words Molecular self-assembly, Nanometrology, Real-time imaging, Super-resolution microscopy, Nanoscale biophysics, Protein fibrillogenesis

1 Introduction

The ability to construct biological matter from the molecule up holds promise for technologies that are set to shape our future [1]. Synthetic biology, regenerative medicine, and advanced materials are just a few areas where engineered self-assembling systems can contribute effective solutions [2]. Such solutions are not limited to commercial adaptations, and can be insightful to advance the mechanistic understanding of biomolecular assembly [3]. From infection to tissue restoration, self-assembly mechanisms underpin a variety of biological processes by encoding the hierarchical organization of predominantly protein molecules into the complex architectures of viruses, cells, and tissues.

There exist different forms and shapes of self-assembled systems. Most of them can be grouped into the generic units of particles (viruses) and tubes or fibers (microtubules, collagen fibers) [4]. These units alone can support specialist functions or can

Bradley L. Nilsson and Todd M. Doran (eds.), *Peptide Self-Assembly: Methods and Protocols*, Methods in Molecular Biology, vol. 1777, https://doi.org/10.1007/978-1-4939-7811-3_4,

arrange into higher-order functional systems (extracellular matrices, cells). Arguably, however, protein fibers present the richest repertoire of functional assemblies that are involved in both normal and pathogenic processes.

Indeed, diverse information including, but not limited to, cell proliferation and tissue development as well as neurodegeneration and viral assembly is realized through protein fibrillogenesis [5–9]. Resulting structures are near crystalline [10], which implies a nanoscale order, and grow hierarchically [11], which implies their assembly is polar and directional [12]. To understand the process and link it with the assembled structures requires suitable fibrillogenesis models that would be free of other events and modifications, which are not necessarily involved in fiber formation. De novo amino-acid sequences based on archetypal folding motifs provide a straightforward solution [13]. This so-called reductionist approach allows distinguishing folding-assembly pathways from unrelated events, thereby helping to reveal fiber formation as a cooperative process at the nanoscale. To achieve this, high-resolution and real-time imaging approaches are necessary.

Since this chapter is a practical guide to the imaging of protein fibers, it starts with basic design principles to enable protein fibrillogenesis, for which we focus on helical assemblies. It then describes microscopy methods, traditional and innovative, which can be used for the visual characterization of fiber formation in situ and in real time (Fig. 1).

1.1 Basic Principles for Designing Protein Fibers

Naturally occurring fibers use monomeric subunits whose amino-acid sequences support particular oligomerization states and propagation modes. Longitudinal assembly appears to dominate and often occurs at the expense of complementary subunits placed axially out of register. Examples may include α-helical intermediate filaments assembled from staggered monomers [20]. β-Pleated

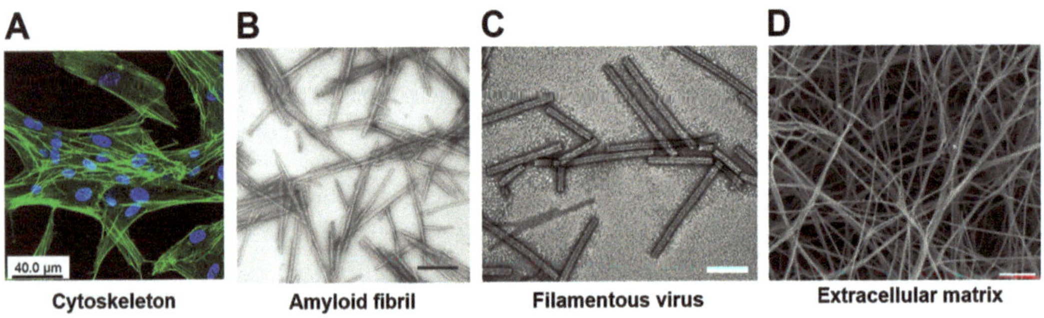

Fig. 1 Examples of naturally occurring protein fibers. (**a**) Actin fibers (green) forming cytoskeleton with nuclear DNA highlighted in blue [14]. (**b**) Transthyretin-derived amyloid fibrils (scale bar 200 nm) [15]. (**c**) Tobacco mosaic virus (scale bar 100 nm) [16]. (**d**) Type I collagen extracellular matrix (scale bar 1 μm) [17]. Reproduced: by permission of the Royal society of chemistry (**a**, **c**), from [15], copyright (2002) National Academy of Sciences, USA (**b**) and by permission of the Journal of Cell Science by Company of Biologists Ltd (**d**)

amyloid fibrils that grow from out-of-register β-sheets [21] or collagen fibers consisting of tropocollagen molecules held together side by side in an offset fashion [22].

Over the last decade a number of de novo designs based on α-helical coiled-coil motifs applied the same principles of staggered assembly. Coiled coils are rope-like bundles of two or more helices that interdigitate into a super-helix. The sequences of coiled coils show an amphipathic heptad repeat of polar (P) and hydrophobic (H) residues, namely PHPPHPP [25]. The repeat is often designated *gabcdef* and depicted schematically using coiled-coil helical wheels with 3.5 residues per turn (as opposed to 3.6 residues per turn for non-oligomerizing helices). A clear advantage of coiled coils is that the heptad repeat, which spans 1 nm, can be used as a construction module [26] and when combined with staggered assembly, the sequences comprising as few as 4 heptads, each giving two two-heptad overhangs, can assemble into micron-long fibers (Fig. 2) [13, 23].

To ensure complementary sequence offsets, designed coiled coils use charge-charge interactions at *g* and *e* sites of successive heptads (*g-e'* interactions) [27]. Specifically, each of two generic heptads that use cationic (lysines) and anionic (glutamates) residues can be paired sequentially to provide the oppositely charged sticky ends [10]. The choice of hydrophobic residues in *a* and *d* defines the oligomerization state of the self-assembling subunit (Fig. 2) including fiber-forming sticky-ended dimers [10, 14], trimers [23, 28], and pentamers [29].

The other sites of the heptad unit (*b*, *c* and *f*) are usually polar (glutamines) or small (alanines) amino acids having high helical propensities. Single aromatic residues (tryptophans, tyrosines) in *f* are often used as chromophores for accurate concentration measurements. Additionally, to fine-tune *inter*-stagger interactions tryptophans can be regularly spaced at super-helix pitch distances (1.63 nm), thereby promoting the zippering up of coiled-coil ropes into discrete filaments while arresting fiber thickening (Fig. 3) [30].

1.2 Morphology on Demand

In contrast to β-structured and collagen-like fibers, α-helical designs allow for a better control and substantial orthogonality over the assembly. This property has been successfully used to rationally engineer nanoscale order [10] and nano-to-microscale morphologies into resulting fibrous materials including kinked [31], branched [32], and network-like [24, 33] architectures (Fig. 2). The head-to-tail cyclization of coiled-coil sequences was shown to guide the arbitrary propagation of self-assembling subunits resulting in matrices and nets that can span unprecedented nano-to-sub-millimeter dimensions [24, 33]. Unlike fibers derived from other folding types, helical designs are strongly prone to thickening by maturation. Yet, by carefully tuning interfacial coiled-coil contacts, it is possible to limit the lateral association of

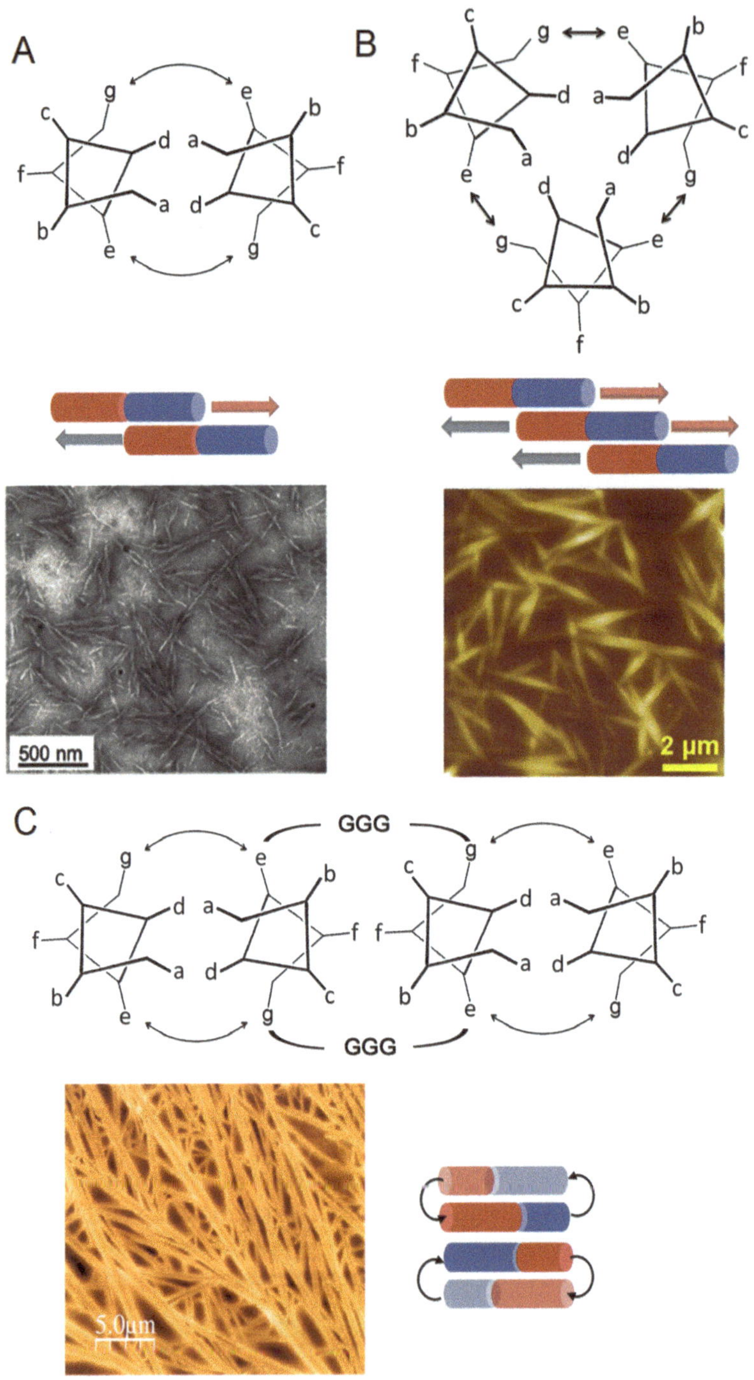

Fig. 2 Coiled-coil heptad repeats, *gabcdef,* configured into helical wheels for dimeric (T2, **a**) [14], trimeric (FiM, **b**) [23], and dimeric cyclized (Cycl_one, **c**) [24] self-assembling subunits (upper) together with the resulting fibrillar structures (lower). Reproduced by permission of the Royal society of chemistry (**a**), Adapted by permission from Macmillan Publishers Ltd.: Scientific Reports, [23] Copyright 2014 (**b**) and Angew Chem Intl Ed, John Wiley and Sons [24] (**c**)

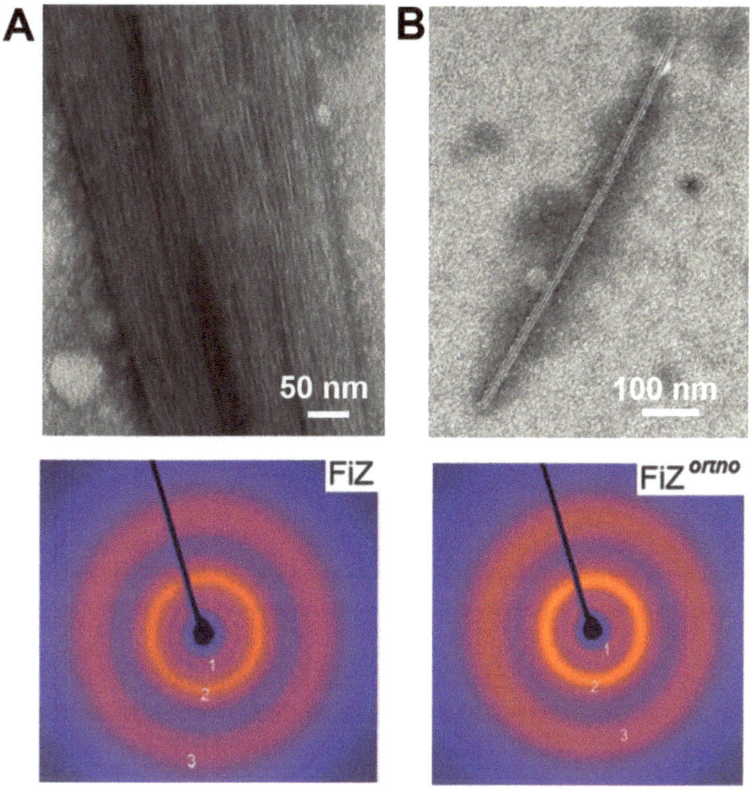

Fig. 3 Zippering up of coiled-coil ropes into individual filaments via interfacial aromatic interactions. (**a**) Transmission electron micrographs (top) and X-ray diffraction patterns (bottom) for a filamentous coiled-coil zipper (FiZ, **a**), which gives individual filaments bundled together without maturation, and its orthogonal version (FiZ^{ortho}, **b**), whose filaments are less prone to bundling up. Reproduced from [30] with permission from the PCCP Owner Societies

coiled-coil bundles [30]. Moreover, the length and thickness of resulting fibers can be empirically correlated with the length of designed peptide sequences [14].

1.3 Material Requirements for Imaging

Protein fibers exhibit nanoscale order which can be engineered rationally [10]. The accurate characterization of the assembled structures requires high-resolution methods. These provide end-point measurements that do not depend on the knowledge of intermediate steps in fiber formation or growth kinetics. X-ray fiber diffractions (XRD) from oriented and partly dried fiber assemblies are obtained to gain detailed insights into fiber ultrastructure [34]. XRD data is used to complement the visual evidence of fiber assembly providing structural correlations with fiber morphology. Transmission electron microscopy (TEM) is the most commonly used visualization technique in this regard.

Figure 3 gives an example of correlations between coiled-coil structure, packing, and assembly. The XRD patterns for the two designs of coiled-coil fibers show typical intra-helical distances for discrete coiled coils and close packing distances between coiled coils in assembled filaments. The pattern for the orthogonal design matches closely the size of 3D hexagonal lattices for axially packed coiled-coil staggers, which confirms super-helix contiguity despite the orthogonal (via a side chain) linkage of two two-heptad overhangs. TEM confirm the filamentous assembly of the coiled-coil staggers with consistent differences in the bundling of individual filaments [30].

TEM provides a nanometer spatial resolution, but does not allow for trade-offs with temporal measurements. Cryogenic TEM coupled with tomography offers a somewhat more advanced approach in gaining more precise and less biased structural information that is free of sample staining and drying. However, self-assembly is a living mechanistic process, the complete understanding of which demands imaging capabilities with sufficient temporal resolution to monitor fiber growth [12, 22]. Optical microscopy methods are minimally invasive and well suited for imaging live processes under ambient conditions, but are limited in spatial resolution (~200 nm), which often prevents from detecting small structural changes. Super-resolution fluorescence imaging improves spatial resolution beyond the diffraction limit (approximately half the wavelength of light used for imaging), with structured illumination microscopy (SIM) being of particular interest for the present application owing to its fast image acquisition rate [35], low sample light exposure, compatibility with conventional fluorescent labeling methods. SIM works by illuminating the sample with spatially patterned light, causing previously undetectable high spatial frequency image information to be visible as low frequency moiré fringes [36]. By capturing a series of images under different illumination patterns this high frequency information can be shifted back to its origin in Fourier space allowing reconstruction of a "super-resolved" image with lateral spatial resolution down to ~100 nm [37]. Such images can be used to track microscopic self-assembly in real time; however for further insights in the process which can be gained through even higher spatial resolution it is necessary to consider alternative methods. Atomic force microscopy (AFM), in particular, can be performed in water and does not require the use of labels to generate image contrast. However, AFM measurements are position- and surface-dependent and subject to sample drifting. A compromise therefore is often proposed in materials themselves.

For example, to delineate fibrillogenesis as a mechanistic process, homogeneous assembly models are crucial. Naturally occurring structures, taken out of their native context, are not suitable for the purpose. Large heterogeneities can be observed in reconstituted fiber morphologies (extracellular matrices) or in growth rates derived from inherent structural polymorphism (amyloids) [38]. In contrast, archetypal designs based on simplified sequence templates (coiled-coil heptads) and generic assembly modes (staggers) can give relatively uniform fiber structures of statistically significant densities to allow for the analysis of growth kinetics [23]. Peptide designs can readily accommodate orthogonal modifications to incorporate fluorophores (i.e., *f* positions in coiled coils), thereby supporting in situ and real-time imaging of fiber formation. More complex fibrous assemblies including branching matrices and knotted networks are far less predictable. Their designs can only be partly satisfied by deliberately introduced promiscuity in fiber formation, which provides a level of control over resulting structures (networked as opposed to straight fibers). All in all, the SIM imaging of an archetypal protein fiber appears as an obvious solution to the real-time measurements of fiber growth kinetics.

2 Materials

2.1 Peptide Synthesis

2.1.1 9-Fluorenylmethoxycarbonyl (Fmoc) Solid Phase Peptide Synthesis

1. Solvents: Dimethylformamide (DMF), *N,N*-diisopropylethylamine (DIPEA), dichloromethane (DCM), trifluoroacetic acid (TFA), tri-isopropyl silane (TIS).
2. Fmoc-amino acids: Fmoc-Ala-OH, Fmoc-Gln(Trt)-OH, Fmoc-Glu-(O^{t}Bu)-OH, Fmoc-Glu-OAll, Fmoc-Ile-OH, Fmoc-Lys(Boc)-OH, Fmoc-Leu-OH, Fmoc-Lys(Aloc)-OH, Fmoc-Tyr(^{t}Bu).
3. Activating agent: O-benzotriazole-*N,N,N′,N′*-tetramethyluronium hexafluorophosphate (HBTU).
4. Resins: (100–200 mesh 0.36–0.67 mmol/g) Fmoc-Gln(Trt)-Wang resin or Rink Amide MBHA.
5. Fmoc deprotection: 20% piperidine in DMF.
6. Allyl deprotection: Tetrakis(triphenylphosphine)palladium (0) (4 eq) in $CHCl_3$/AcOH/DIPEA 3.6/0.2/0.2 mL.
7. Coupling mixture: 0.5 M HBTU and 1 M DIPEA in DMF.
8. Cleavage cocktail: TFA/TIS/water (95:2.5:2.5).

2.1.2 Peptide Labeling

1. Alexa Fluor 488-carboxylic acid, succinimidyl ester.
2. DIPEA (2 eq) in DMF.

2.1.3 Reversed Phase High-Performance Liquid Chromatography

1. Mobile phases: 5% acetonitrile (AcN) + 95% water (buffer A), 5% Water + 95% AcN (buffer B) both containing 0.1% TFA. All solvents are HPLC grade.

Table 1
Sequence template with varied integral module numbers

Name[a]	Sequence
FiZ	KLAALKWKLAALKQKLAALKWELAALEQELAALEWELAALEQ
FiZ^{ortho}	KLAALKWKLAALKQKLAALKX(ELAALEQELAALEWELAALEQ)[b]
FiM	KLAALKQKLAALKKELAALEQELAALEQ
FiM(Alexa)	KLAALKQKLAALKK(Alexa488)ELAALEQELAALEQ
T2	KIAALKQKIAALKKEIAALEYEIAALEQ
Cycl_one	cyclo(KIAALKYEIAALEQKIAALKGGGEIAALEQKIAALKQEIAALEGGG)

[a]Original names
[b]X is a glutamate forming an isopeptide (orthogonal) bond with the first glutamate of the anionic domain

2. Stationary phase: Vydac C8 and C18 reverse-phase columns. Analytical (5 μm, 4.6 mm i.d. × 250 mm) and semi-preparative (5 μm, 10 mm i.d. × 250 mm).
3. High-Performance Liquid Chromatographic (HPLC) system (Jasco-2080 with UV 2072 detector).

2.1.4 Mass Spectrometry

1. MALDI ToF (Bruker Daltonics) Autoflex III smartbeam, with flex control as proprietary software.
2. 2,5-Dihydroxybenzoic acid or sinapinic acid as matrix.

2.2 Fiber Assembly

Peptides described in Table 1 below are incubated in MOPS (10 mM, pH 7.4) overnight prior to endpoint imaging (TEM, AFM). All solutions use filtered (0.22 μm) water with a resistivity of 18.2 MΩ.

2.3 Spectroscopy

2.3.1 Circular Dichroism

1. Quartz cuvettes (0.1 and 0.05 cm from Starna UK).
2. Chirascan Plus spectropolarimeter (Applied Photophysics Ltd.) equipped with a Peltier temperature controller.

2.3.2 Linear Dichroism

1. Micro-volume quartz cuvette flow cell with 0.5 mm annular gap and quartz capillaries (Kromatec Ltd., UK).
2. Jasco 810 spectropolarimeter using a photoelastic modulator ½ wave.

2.3.3 Fourier-Transform Infrared (FT-IR)

1. Tensor-37 series FTIR spectrophotometer with a BioATR II unit (Bruker Optics) with photovoltaic MCT detector and a Bruker optics workstation equipped with OPUS software.

2.4 Microscopy

2.4.1 Structured Illumination Microscopy

1. Custom-built structured illumination microscope [18, 19] based on an inverted microscope body (IX71, Olympus) with high numerical aperture (UAPON 100XOTIRF, Olympus) and silicone immersion (UPLSAPO 60XS, Olympus) objective

lenses for total internal reflection fluorescence (TIRF) and wide field epifluorescence imaging, respectively. Images are captured with a scientific CMOS camera (Flash 4.0 vl, Hamamatsu Photonics).

2. Time-lapse imaging: performed in TIRF mode using a ×100/1.49 objective lens with a typical exposure time of 100 ms. Super-resolution images of the assembled fibers are captured using a ×60/1.3 objective lens.
3. Sample mounting: samples are mounted in a high precision z-stage (NanoScan, Prior Scientific Ltd.) in closed loop with a focus stabilization system (CRISP, Applied Scientific Instrumentation). Illumination patterns generated using a spatial light modulator (SXGA-3DM, Forth Dimension Displays) illuminated using a 488 nm laser (Sapphire, Coherent Inc.). Additional optics and optomechanics are from Thorlabs and Qioptiq. All hardware is controlled using custom software that is written in LabVIEW 2013 (National Instruments). Software can be tailored to specific hardware requirements for bespoke microscopes or can be proprietary for commercial instruments.
4. Super-resolution image reconstruction: custom software written in Matlab (MathWorks) as detailed elsewhere [18, 19].
5. Filament segmentation: ImageJ software.
6. Sample chambers: chambered cover glass with 8 wells (NUNC Lab-TEK II, Thermo Scientific).

2.4.2 Transmission Electron Microscopy

1. Support grid: formvar/carbon coated copper/palladium support grids or copper grids coated with pioloform and carbon film. All subjected to plasma glow discharge before use.
2. Staining reagent: 2% phosphotungstic acid or 1% uranyl acetate in distilled water filtered through a 0.2 μm pore size Minisart unit.
3. Microscope: Philips BioTwin transmission electron microscopy with 80 kV accelerating voltage or FEI Tecnai 20 with 200 kV accelerating voltage.
4. Image analysis: ImageJ software.

2.4.3 Cryo-Transmission Electron Microscopy

1. Support grid: lacey carbon grids (EM Resolutions Ltd.) subjected to plasma glow discharge before use.
2. Sample freezing: plunge freezing into liquid ethane cooled with liquid N_2 and performed using a VITROBOT mark IV (FEI Company).
3. Microscope: FEI Tecnai 20 with 200 kV accelerating voltage and fitted with an Eagle 4k × 4k camera (FEI).
4. Image analysis: ImageJ software.

2.4.4 Atomic Force Microscopy

1. Clean silicon wafer 7.0 mm × 5.0 mm with 100 μm diameter and 525 μm thickness.
2. Cypher AFM instrument (Asylum research).
3. AFM tips: super sharp silicon probes (nanosensors) with resonant frequency of ~330 kHz. 5 nm radius of curvature, force constant 42 N/m.
4. Proprietary imaging software SPIP 6.0.2.

2.5 Fiber Diffraction of Partially Aligned Samples

2.5.1 Partial Alignment of Fibers: Stretched Frame Procedure

1. Borosilicate thin-walled capillaries (1.5-mm outer diameter, 1.17-mm inner diameter) (Harvard Apparatus).
2. Cutting stone (Hampton Research).
3. Standard beeswax: break into pieces, transfer the pieces into a glass beaker.
4. Mounting clay (Hampton research).
5. Petri dishes, 90-mm diameter.
6. Parafilm®.

2.5.2 Fiber Diffraction

1. Rigaku CuKα rotating anode X-ray source (wavelength 1.5418 Å) and R-AXIS IV++ diffractometer.
2. Instrument control and data collection through the Crystal-ClearTM software (Rigaku).
3. Data handled and converted to image files in Tiff formats.

3 Methods

3.1 Synthesis of Model Fiber-Forming Subunits

Model peptide sequences are relatively short and straightforward for chemical synthesis. These are assembled on a solid support (polymer resin beads) using fully automated peptide synthesizers. Site-specific modifications including fluorescent labeling, branching, and cyclization can be performed manually on the same substrate. After the modifications are complete, which can be confirmed by mass spectrometry, the assembly of the sequence can continue till completion, if necessary, and the synthesized product is then cleaved from the support for post-synthetic purification and analysis.

3.1.1 Peptide Synthesis

Solid-phase peptide synthesis is performed using standard Fmoc/^{t}Bu protocols that allow significant flexibility regarding coupling efficiencies, synthesis scale and time, solvent and reagent use and ultimately product yields. The same protocols were used for the incorporation of orthogonal protecting groups (allyl for Cycl_one, Mtt for FiZ^{ortho}) to allow site-specific side-chain modifications. Uronium salts (HBTU, HCTU) together with DIPEA were used as coupling reagents to assemble the sequences reported in Table 1.

1. Perform peptide synthesis at a 100 μM scale on a MBHA rink amide resin (100–200 Mesh)
2. Use an automated peptide synthesizer (CEM Liberty) according to the following synthesis cycle
   (a) Deprotect Fmoc-protected resin/amino acids by 20% piperidine in DMF (7 mL) for 5 min.
   (b) Wash (4×) the resin with DMF (7 mL).
   (c) Perform amino-acid coupling (10 min) using Fmoc-protected amino acid (2.5 mL, 0.2–0.4 M) with HBTU or HCTU (0.5 M, 1 mL) and DIPEA (1 mL, 1 M). All are in DMF.
   (d) Wash (4×) the resin with DMF (7 mL).
3. For Orthogonal Allyl/Alloc Deprotection
   (a) Add tetrakis (triphenylphosphine) palladium (0) (400 μM) in chloroform/DIPEA/AcOH (36:0.2:0.2) containing phenylsilane (4 eq.) as a scavenger.
   (b) Incubate for 2–16 h with gentle agitation in the dark.
   (c) Wash with 0.5% sodium diethyldithiocarbamate (w/v in DMF), 0.5% DIPEA (v/v in DMF) and DMF.
   (d) Repeat washing (3×).
4. For Post-synthesis Cleavage
   (a) Per-wash resin with DCM to remove residual DMF.
   (b) Keep the resin under cleavage cocktail (10 mL/g of resin) for 3–4 h.
   (c) Recover the liquid phase by filtration (main filtrate) and the resin rinsed with another 5 mL of the cocktail (second filtrate).
   (d) Add a threefold excess of diethyl ether to each filtrate.
   (e) Centrifuge the filtrates at 3500 rpm for 45 min at 4 °C.
   (f) Discharge the supernatant to give peptide precipitates.
5. Lyophilization
   (a) Dissolve the obtained precipitates in distilled water and freeze.
   (b) Freeze-dry the solidified sample under vacuum.

3.1.2 Labeling and Cyclization

Side chain modifications (fluorophores—FiM-Alexa), orthogonal conjugation (FiZ^{ortho}), and cyclization (Cycle_one) are carried out by applying Fmoc/^{t}Bu/Mtt or Fmoc/^{t}Bu/Allyl protocols (Table 1). Fmoc-Lys(Aloc)-OH, Fmoc-Lys(Mtt)-OH, and Fmoc-Glu(OAl)-OH derivatives are used to incorporate modification sites in fully protected sequences directly on resin. Allyl and Mtt chemistries are orthogonal to the Fmoc/^{t}Bu protocols and do not affect other residues.

Fluorescent Labeling of FiM-Alexa

1. Mix an equivalent of orthogonally deprotected resin (-Lys (Aloc)-) with Alexa Fluor 488 succinimidyl ester/DIPEA (1.1 eq. in DMF).
2. Perform the reaction at 60 °C for 10 min in a microwave unit (CEM Corp.).

Backbone Cyclization

1. Load Fmoc-Glu-OAll onto the MBHA resin as the first residue through the γ-carboxylate.
2. Assemble the sequence using the cycle described in **Step 2** of Subheading 3.1.1 and when complete remove the allyl group using the mixture in **Step 3** of Subheading 3.1.1 which will leave the other residues protected.
3. Once the Alloc group is removed cyclize the peptide directly on the resin using a fourfold excess of HBTU and DIPEA (total 4 mL) at 60 °C (10 min).
4. Repeat the cyclization step twice.
5. Cleave the resin and process the product using the protocol described in **Steps 4** and **5** of Subheading 3.1.1. Note: Upon cleavage the glutamate is converted into glutamine.

3.1.3 Reverse Phase-HPLC and Mass Spectrometry

1. Solubilize the crude peptide in distilled water.
2. Inject 5 mL of the obtained peptide solution on a semi-preparative RP-HPLC column (Vydac C_{18}, 5 μm).
3. Run a 10–70% gradient of 95% aq. CH_3CN containing 0.1% TFA over 50 min, with the flow rate of 4.7 mL/min.
4. Monitor the gradient at different wavelengths: 214 and 230 nm (peptide bonds), 280 nm (aromatic side chains), and 495 nm (Alexa Fluor 488).
5. Collect eluted fractions every 30 s.
6. Mix 1 μL of DHB matrix (10 mg/mL in CH_3CN/water 50:50 containing 0.1% TFA) with peptide (1 μL) and spot 1 μL of the obtained mixture on a MALDI-TOF plate to confirm peptide identity.
7. Combine pure fractions, freeze, and freeze-dry as per **Steps 4** and **5** of Subheading 3.1.1.
8. Solubilize an aliquot of the purified peptide to confirm and quantify its purity by analytical RP-HPLC (Vydac C_{18}, 5 μm) at the flow rate of 1 mL/min.

3.2 Fiber Assembly

3.2.1 Fiber Preparations for TEM Imaging

1. Prepare a stock solution (0.5–1 mL) of a pure peptide (2–3 mM).
2. Determinate peptide concentration by UV-Vis for FiM(Alexa) at 495 nm (ε of 71,000 M^{-1} cm^{-1}), for FiM at 214 nm calculating an extinction coefficient as described elsewhere [39] and at 280 nm for all other peptides (Table 1).

3. Prepare peptide samples (200 μL, 100 μM) in 10 mM MOPS or 10 mM phosphate buffer, pH 7.4.
4. Incubate at room temperature overnight.

3.2.2 Fiber Preparations for Optical Microscopy

1. Prepare a 10 μM solution of FIM (Alexa) in 10 mM MOPS, pH 7.4.
2. Prepare FiM(Alexa)/FiM samples at a 1:10^4 molar ratio using the obtained solution and the stock (FiM).
3. Place the resulting solution into a well of an 8-well chamber (Nunc™ Lab-Tek™).

3.2.3 Transmission Electron Microscopy (Staining)

1. Deposit 8 μL of a fiber preparation (as above) on a glow-discharged TEM grid.
2. Let the content of the droplet settle for 1 min.
3. Remove excess solution carefully by blotting paper from one side of the grid.
4. Alternatively, deposit a TEM grid upside down on a droplet placed on a piece of clean Parafilm®.
5. Remove excess water by blotting.
6. Stain the sample grid for 10 s with uranyl acetate, phosphotungstic acid, or ammonium molybdate (8 μL, 1% w/v).
7. Remove excess stain by blotting, and transfer the grid into a TEM microscope for imaging.

3.2.4 Cryo-transmission Electron Microscopy (No Staining)

1. Prepare samples in a climate chamber with 100% relative humidity.
2. Load a droplet (5 μL) of a peptide solution on a glow-discharged lacey carbon grid.
3. Let the content of the droplet settle for 2–10 s before blotting.
4. Plunge into liquid ethane cooled with liquid N_2 for freezing. Peptide nanostructures become embedded in a vitreous ice suspended inside the holes of the carbon grid.
5. Transfer the sample grid (without warming) into a Gatan 626 cryo-holder filled with liquid N_2 and image.

3.3 Spectroscopy

3.3.1 Circular Dichroism spectroscopy

CD Spectra

1. Prepare peptide sample (50–300 μM) in 10 mM MOPS or phosphate buffer, pH 7.4.
2. Place the sample in a quartz cuvette with 0.05 or 0.1 cm path length.
3. Record far-UV CD spectra in the 190–260 nm wavelength range collecting data points at 1 nm step, 1 nm bandwidth with 4 acquisitions.

4. Convert all the measurements, after baseline correction, into mean residue ellipticities by normalizing for the concentration of peptide bonds and cuvette path length (MRE, deg cm^2 $dmol^{-1}$ res^{-1}).

Thermal Denaturation Curves

1. Record unfolding curves at 222 nm in the range of 5–90 °C at 1 °C intervals with a ramp of 1 °C/min and 16 s equilibration time for each temperature point.
2. Record CD spectra in the 190–260 nm wavelength range after the melt and cooling.
3. Plot the thermal unfolding curves as a function of temperature in mean residue ellipticities.

3.3.2 Linear Dichroism Spectroscopy

1. Place a peptide aliquot (60 μL, 100 μM in 10 mM MOPS, pH 7.4) into a micro-volume quartz Couette flow cell.
2. Achieve molecular alignment by setting up a constant laminar flow of the sample solution between the two coaxial cylinders, a stationary quartz rod and a cylindrical capillary rotating at 3000 rpm.
3. Record all spectra in the 260–190 nm wavelength range at a 1 nm step size, 1 nm bandwidth, 50 nm/min scan speed, with a 4-s response time and three acquisitions.
4. Process the spectra by subtracting nonrotating baseline spectra and normalizing for the concentration of peptide bonds with conversion to a mean residue delta epsilon $\Delta\varepsilon$ (M^{-1} cm^{-1} res^{-1}).

3.3.3 Fourier-Transform Infrared Spectroscopy

1. Place a peptide aliquot (20 μL, 100 μM in 10 mM MOPS, pH 7.4) in a circular sampling area of a 2-mm radius with a path length of 6 μm. The multi-reflection ATR accessory used is based on a dual-crystal technology, which has an upper silicon crystal and a hemispherical zinc selenide (ZnSe) lower crystal that does not come into contact with the sample.
2. Maintain the temperature of the sample at 20 °C.
3. Collect FTIR spectra in the range of 4000 and 850 cm^{-1} with a resolution of 4 cm^{-1}, a scanner velocity 20 kHz, 128 scans, a phase resolution of 32, and a zero filling factor 4.

3.4 Fiber Diffraction of Partially Aligned Samples

3.4.1 X-Ray Fiber Diffraction

1. Prepare XRD specimens by suspending a droplet of a peptide solution between the ends of wax-coated capillaries and allow them to dry.
2. Examine the peptides dried on the wax head forming a protuberance.
3. Compare all XRD patterns versus the XRD of the wax heads supporting the samples.

4. Mount XRD specimens vertically onto the four-axis goniometer of a RAXIS IV++ X-ray diffractometer (Rigaku) equipped with a rotating anode generator.
5. XRD data (rings) for samples with no positional ratio between reflections correspond to average distances between objects. The average distance is the spacing calculated from the position of the ring: spacing is equal to $d = 2 \times 3.14/q$, where q is a scattering vector for the position of the ring.

3.5 Microscopy

3.5.1 Optical Microscopy

Wide-Field Fluorescence Microscopy

1. Deposit a peptide aliquot (6 μL) on a microscope slide and cover the slide with a glass slip.
2. Image fluorophore-labeled fibers using an ×40 lens.
3. Use the DIC imaging mode to optimize the focus and record fluorescence using relevant excitation wavelengths.
4. Process the images using Carl Zeiss Vision imaging software before overlaying.

Structured Illumination Microscopy

1. Perform fiber imaging using a custom-built (or commercial) SIM system as described in Subheading 2.4.1.
2. Immediately prior to imaging, transfer peptide solution to a well of an 8-well chambered coverslip mounted on the microscope sample stage.
3. Capture time-lapse images under illumination of the sample with a series of nine sinusoidal irradiance patterns, with the corresponding phase gratings displayed on the SLM set such that the first diffracted orders fell inside the TIRF ring of the objective lens [18, 19].
4. Acquire endpoint images with a coarser grating displayed on the SLM to allow illumination of the full depth of the sample.
5. Reconstruct the images of fibers formed at the endpoint of the self-assembly process as described in [18].

   Note: Fig. 4a, b shows a typical SIM image reconstruction of densely packed fibers, illustrating how the improved image contrast and spatial resolution of SIM enable a clearer visualization of individual fibers against the fluorescent background. For time-lapse imaging, owing to the relatively low signal-to-noise ratio in the raw images, each sequence of raw images is summed up to create a diffraction-limited TIRF image with improved illumination uniformity and signal-to-noise.
6. Analyze time-lapse TIRF images to determine fiber growth kinetics.
7. Prior to segmentation, correct lateral offsets between images in each image series using intensity-based registration (using fluorescent clusters as markers), Fig. 4c.

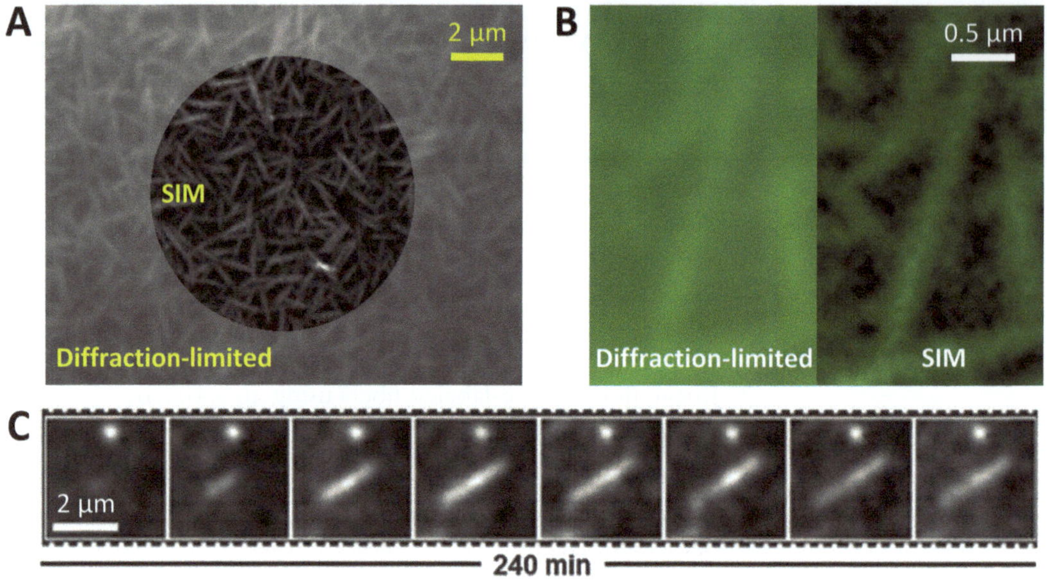

Fig. 4 In situ and live imaging. (**a**) Composite SIM (center circle) and diffraction-limited wide-field image (outside the circle) showing protein fibers assembled from staggered coiled-coil trimers labeled with Alexa488. (**b**) Montage showing diffraction-limited wide-field (left) and SIM (right) images of the same region, emphasizing how the removal of the out-of-focus light and improved spatial resolution in SIM allows clearer visualization of individual fibers against the fluorescent background. (**c**) Time-lapse TIRF images of a self-assembling protein fiber. The bright round feature in the upper part of each image is a fluorescent aggregate used for image registration. Adapted by permission from Macmillan Publishers Ltd: Scientific Reports, [23] Copyright 2014

8. Normalize the intensity of each image and apply a top-hat filter to increase image contrast.
9. Segment Individual fibers and track them through the time-lapse sequence using open active contours implemented in the ImageJ plugin JFilament [40].
   Note: Values for foreground and background are defined individually for each fiber tracked in the sequence. All other contour parameters are empirically optimized once and assigned the same values for all fibers in the sequence.

3.5.2 High-Resolution Microscopy

Transmission Electron Microscopy

1. Record micrographs using an FEI Tecnai 20 twin lens scanning transmission electron microscope, operated at 200 kV and equipped with an Eagle 4k × 4k CCD camera.

Atomic Force Microscopy

1. Clean a silicon wafer using ethanol and dry the substrate under nitrogen.
2. Immerge the clean silicon wafer in 100–300 µM of peptide solution in MOPS at pH 7.4 and incubate overnight at room temperature.

3. Record topographic, amplitude, and phase AFM images using tapping mode AFM. All images are flattened with a first-order line-wise correction fit.
4. Process Images using proprietary software (SRIP 6.0.2).

References

1. Whitesides GM, Boncheva M (2002) Beyond molecules: self-assembly of mesoscopic and macroscopic components. Proc Natl Acad Sci U S A 99:4769–4774
2. Adler-Abramovich L, Gazit E (2014) The physical properties of supramolecular peptide assemblies: from building block association to technological applications. Chem Soc Rev 43:6881–9683
3. Whitesides G, Mathias J, Seto C (1991) Molecular self-assembly and nanochemistry: a chemical strategy for the synthesis of nanostructures. Science 80:1312–1319
4. De Santis E, Ryadnov MG (2015) Peptide self-assembly for nanomaterials: the old new kid on the block. Chem Soc Rev 44:8288–8300
5. Hay EH (2013) Cell biology of extracellular matrix, 2nd edn. Springer Science & Business Media, Dordrecht
6. Grosse R, Vartiainen MK (2013) To be or not to be assembled: progressing into nuclear actin filaments. Nat Rev Mol Cell Biol 14:693–697
7. Mienaltowski MJ, Birk DE (2014) Structure, physiology, and biochemistry of collagens. Adv Exp Med Biol 802:5–29
8. Ross CA, Poirier MA (2004) Protein aggregation and neurodegenerative disease. Nat Med 10:S10–S17
9. Van Raaij MJ, Mitraki A, Lavigne G, Cusack S (1999) A triple beta-spiral in the adenovirus fibre shaft reveals a new structural motif for a fibrous protein. Nature 401:935–938
10. Papapostolou D, Smith AM, Atkins EDT, Oliver SJ, Ryadnov MG, Serpell LC, Woolfson DN (2007) Engineering nanoscale order into a designed protein fiber. Proc Natl Acad Sci U S A 104:10853–10858
11. Smith AM, Acquah SFA, Bone NK, Harold W, Ryadnov MG, Stevens MSP, Walton DRM, Woolfson DN (2004) Polar assembly in a designed protein fiber. Angew Chem Int Ed Engl 44:325–328
12. Scheibel T, Kowal AS, Bloom JD, Lindquist SL (2001) Bidirectional amyloid fiber growth for a yeast prion determinant. Curr Biol 11:366–369
13. Woolfson DN, Ryadnov MG (2006) Peptide-based fibrous biomaterials: some things old, new and borrowed. Curr Opin Chem Biol 10:559–567
14. De Santis E, Faruqui N, Noble JE, Ryadnov MG (2014) Exploitable length correlations in peptide nanofibres. Nanoscale 6:11425–11430
15. Jaroniec CP, MacPhee CE, Astrof NS, Dobson CM, Griffin RG (2002) Molecular conformation of a peptide fragment of transthyretin in an amyloid fibril. Proc Natl Acad Sci U S A 99:16748–16753
16. Rong J, Oberbeck F, Wang X, Li X, Oxsher J, Niu Z, Wang Q (2009) Tobacco mosaic virus templated synthesis of one dimensional inorganic–polymer hybrid fibres. J Mater Chem 19:2841
17. Poincloux R, Lizárraga F, Chavrier P (2009) Matrix invasion by tumour cells: a focus on MT1-MMP trafficking to invadopodia. J Cell Sci 122:3015–3024
18. O'Holleran K, Shaw M (2014) Optimized approaches for optical sectioning and resolution enhancement in 2D structured illumination microscopy. Biomed Opt Express 5:2580–2590
19. Shaw M, Zajiczek L, O'Holleran K (2015) High speed structured illumination microscopy in optically thick samples. Methods 88:11–19
20. Chernyatina AA, Hess JF, Guzenko D, Voss JC, Strelkov SV (2016) How to study intermediate filaments in atomic detail. Methods Enzymol 568:3–33
21. Liu C, Zhao M, Jiang L, Cheng PN, Park J, Sawaya MR, Pensalfini A, Gou D, Berk AJ, Glabe CG, Nowick J, Eisenberg D (2012) Out-of-register β-sheets suggest a pathway to toxic amyloid aggregates. Proc Natl Acad Sci U S A 109:20913–20918
22. Starborg T, Lu Y, Meadows RS, Kadler KE, Holmes DF (2008) Electron microscopy in cell-matrix research. Methods 45:53–64
23. Bella A, Shaw M, Ray S, Ryadnov MG (2014) Filming protein fibrillogenesis in real time. Sci Rep 4:7529
24. Bella A, Ray S, Shaw M, Ryadnov MG (2012) Arbitrary self-assembly of peptide extracellular microscopic matrices. Angew Chem Int Ed Engl 51:428–431

25. Hartmann MD, Mendler CT, Bassler J, Karamichali I, Ridderbusch O, Lupas AN, Avarez BH (2016) α/β coiled coils. elife 5:11861
26. Ryadnov MG (2007) Peptide alpha-helices for synthetic nanostructures. Biochem Soc Trans 35:487–491
27. Ryadnov MG, Bella A, Timson S, Woolfson DN (2009) Modular design of peptide fibrillar nano- to microstructures. J Am Chem Soc 131:13240–13241
28. Anzini P, Xu C, Hughes S, Magnotti E, Jiang T, Hemmingsen L, Demeler B, Conticello VP (2013) Controlling self-assembly of a peptide-based material via metal-ion induced registry shift. J Am Chem Soc 135:10278–10281
29. Potekhin SA, Melnik TN, Popov V, Lanina NF, Vazina AA, Rigler P, Verdini AS, Corradin G, Kajava AV (2001) De novo design of fibrils made of short α-helical coiled coil peptides. Chem Biol 8:1025–1032
30. De Santis E, Castelletto V, Ryadnov MG (2015) Interfacial zippering-up of coiled-coil protein filaments. Phys Chem Chem Phys 17:31055–31060
31. Ryadnov MG, Woolfson DN (2003) Engineering the morphology of a self-assembling protein fibre. Nat Mater 2:329–332
32. Ryadnov MG, Woolfson DN (2005) MaP peptides: programming the self-assembly of peptide-based mesoscopic matrices. J Am Chem Soc 127:12407–12415
33. Faruqui N, Bella A, Ravi J, Ray S, Lamarre B, Ryadnov MG (2014) Differentially instructive extracellular protein micro-nets. J Am Chem Soc 136:7889–7898
34. Makin OS, Serpell LC (2005) X-ray diffraction studies of amyloid structure. Methods Mol Biol 299:67–80
35. Kner P, Chhun BB, Griffis ER, Winoto L, Gustafsson MG (2009) Super-resolution video microscopy of live cells by structured illumination. Nat Methods 6:339–342
36. Gustafsson MG (2000) Surpassing the lateral resolution limit by a factor of two using structured illumination microscopy. J Microsc 198:82–87
37. Gustafsson MG, Shao L, Carlton PM, Wang CJ, Golubovskaya IN, Cande WZ, Agard DA, Sedat JW (2008) Three-dimensional resolution doubling in wide-field fluorescence microscopy by structured illumination. Biophys J 94:4957–4970
38. Pinotsi D, Buell AK, Galvagnion C, Dobson CM, Kaminski-Schierle GS, Kaminski CF (2014) Direct observation of heterogeneous amyloid fibril growth kinetics via two-color super-resolution microscopy. Nano Lett 14:339–345
39. Kuipers BJH, Gruppen H (2007) Prediction of molar extinction coefficients of proteins and peptides using UV absorption of the constituent amino acids at 214 nm to enable quantitative reverse phase high-performance liquid chromatography-mass spectrometry analysis. J Agric Food Chem 55:5445–5451
40. Smith MB, Li H, Shen T, Huang X, Yusuf E, Vavylonis D (2010) Segmentation and tracking of cytoskeletal filaments using open active contours. Cytoskeleton 67:693–705

Chapter 5

Replica Exchange Molecular Dynamics: A Practical Application Protocol with Solutions to Common Problems and a Peptide Aggregation and Self-Assembly Example

Ruxi Qi, Guanghong Wei, Buyong Ma, and Ruth Nussinov

Abstract

Protein aggregation is associated with many human diseases such as Alzheimer's disease (AD), Parkinson's disease (PD), and type II diabetes (T2D). Understanding the molecular mechanism of protein aggregation is essential for therapy development. Molecular dynamics (MD) simulations have been shown as powerful tools to study protein aggregation. However, conventional MD simulations can hardly sample the whole conformational space of complex protein systems within acceptable simulation time as it can be easily trapped in local minimum-energy states. Many enhanced sampling methods have been developed. Among these, the replica exchange molecular dynamics (REMD) method has gained great popularity. By combining MD simulation with the Monte Carlo algorithm, the REMD method is capable of overcoming high energy-barriers easily and of sampling sufficiently the conformational space of proteins. In this chapter, we present a brief introduction to REMD method and a practical application protocol with a case study of the dimerization of the 11–25 fragment of human islet amyloid polypeptide (hIAPP(11–25)), using the GROMACS software. We also provide solutions to problems that are often encountered in practical use, and provide some useful scripts/commands from our research that can be easily adapted to other systems.

Key words Replica exchange method, Molecular dynamics simulations, Free energy landscape, Protein aggregation, Human islet amyloid polypeptide, GROMACS, REMD

1 Introduction

1.1 Replica Exchange Molecular Dynamics Study of Peptide Aggregation and Self-Assembly

Proteins and peptides can self-assemble into β-sheet rich filaments that form naturally in bacteria and abnormally in tissues of patients with neurodegenerative diseases. Such assemblies also have remarkable mechanical properties with potential nanotechnological applications [1, 2]. The pathological self-assembly of amyloidogenic proteins is related to many human diseases such as Alzheimer's disease (AD), Parkinson's disease (PD), and type II diabetes (T2D) [3–5]. Revealing the mechanism of protein self-assembly is essential for the development of effective inhibitors against amyloidosis and designing new biological materials. Computational

Bradley L. Nilsson and Todd M. Doran (eds.), *Peptide Self-Assembly: Methods and Protocols*, Methods in Molecular Biology, vol. 1777, https://doi.org/10.1007/978-1-4939-7811-3_5,

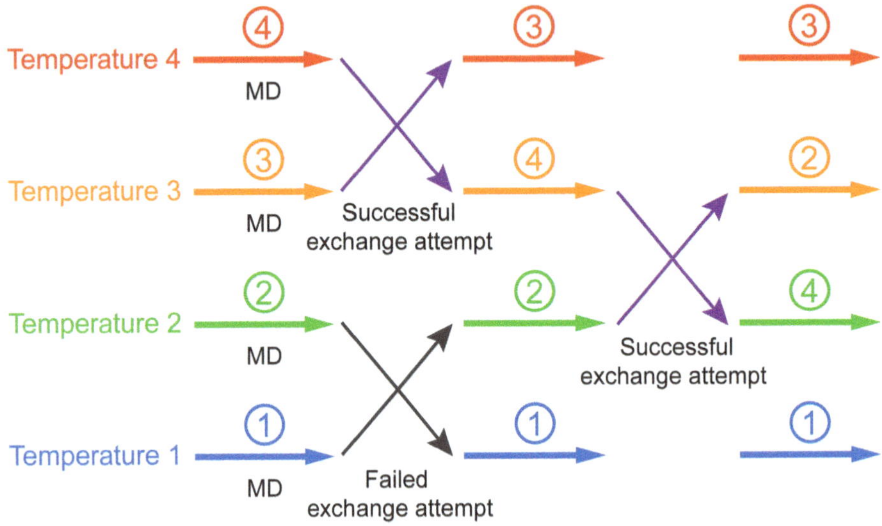

Fig. 1 Illustration of replica exchange molecular dynamics (REMD) method

approaches are powerful tools to study the early self-assembly step of protein aggregation, which is challenging to characterize experimentally due to the fast aggregation process and the transient nature of oligomers. Conventional molecular dynamics (MD) simulations can hardly explore the complete free energy surface of biomolecules as the simulated system could be easily trapped in local free-energy-minimum conformations. Thus, enhanced sampling methods have been developed in recent years [6, 7], such as metadynamics [8] and replica exchange molecular dynamics (REMD) [9]. Here we focus on the REMD method, a highly practical and efficient algorithm to study peptide self-assembly.

The REMD method was first introduced in the study of biomolecules by Okamoto and coworkers [9]. It is a hybrid method by combining MD simulations with the Monte Carlo algorithm. In REMD simulations, several copies (replicas) of the same system are simulated in parallel using MD simulations at different temperatures or at the same temperature but using different Hamiltonians. Swapping between neighboring replicas is periodically attempted with a probability given by the Metropolis criterion (Fig. 1). This method generates a generalized ensemble of the simulated system. In this way, REMD is capable of overcoming high-energy barriers easily and sampling conformational space sufficiently, which allows exploration of the free energy landscape of protein aggregates. Furthermore, the parallel feature of the REMD method makes it suitable for the highly parallel computer cluster resources that are often available nowadays.

In this chapter, we present a brief introduction to REMD method and a practical application to study the initial dimerization step of the 11–25 fragment of human islet amyloid polypeptide (hIAPP(11–25)) aggregation [10]. We describe the detailed protocol for performing a REMD simulation and data analysis, and discuss the problems that are often encountered in REMD simulations. We also provide some useful scripts/commands from our research experience that can be easily adapted to other biomolecular systems.

1.2 REMD Method

The methodology of REMD was presented in detail in a previous study [9]. Here, we provide a brief description. Considering a system with N particles of mass $m_k (k = 1, \ldots, N)$, with coordinates $q \equiv (q_1, \ldots, q_N)$ and momenta $p \equiv (p_1, \ldots, p_N)$, the Hamiltonian of the system is

$$H(q, p) = K(p) + V(q)$$

where $K(p)$ is the kinetic energy $\left(K(p) = \sum_{k=1}^{N} \frac{p_k^2}{2m_k} \right)$, and $V(q)$ is the potential energy.

In the canonical ensemble at temperature T, the probability of finding the system in state $x \equiv (q, p)$ is

$$\rho_B(x, T) = \exp[-\beta H(q, p)]$$

where $\beta = 1/k_B T$, and k_B is Boltzmann constant.

The simulation system has the generalized ensemble, where M non-interacting copies (replicas) of the above system are distributed at M different unique temperatures $T_m (m = 1, \ldots, M)$. The replica $i (i = 1, \ldots, M)$ corresponds to the temperature $m\ (m = 1, \ldots, M)$.

Let $X = \left(x_i^{[i(1)]}, \ldots, x_i^{[i(M)]} \right) = \left(x_{m(1)}^{[i]}, \ldots, x_{m(M)}^{[M]} \right)$ denotes the state of the generalized ensemble, and X is specified by the M sets of coordinates $q^{[i]}$ and momenta $p^{[i]}$ of N particles in replica i at temperature T_m:

$$x_m^i \equiv \left(q^{[i]}, p^{[i]} \right)_m$$

Since each replica is independent from others, the probability of the generalized ensemble in state X is

$$\rho_{\mathrm{REM}}(X) = \exp\left\{ -\sum_{i=1}^{M} \beta_{m(i)} H\left(q^{[i]}, p^{[i]} \right) \right\}$$
$$= \exp\left\{ -\sum_{m=1}^{M} \beta_m H\left(q^{[i(m)]}, p^{[i(m)]} \right) \right\}$$

where $i(m)$ and $m(i)$ are connected by

$$\begin{cases} i = i(m) \\ m = m(i) \end{cases}$$

The simulation tends to exchange replica i (at temperature T_m) and j (at temperature T_n):

$$X = \left(\ldots, x_m^{[i]}, \ldots, x_n^{[j]}, \ldots\right) \to X' = \left(\ldots, x_m^{[j]'}, \ldots, x_n^{[i]'}, \ldots\right)$$

or in detail,

$$\begin{cases} x_m^i \equiv \left(q^{[i]}, p^{[i]}\right)_m \to x_m^{[j]'} \equiv \left(q^{[j]}, p^{[j]'}\right)_m \\ x_n^j \equiv \left(q^{[j]}, p^{[j]}\right)_n \to x_n^{[i]'} \equiv \left(q^{[i]}, p^{[i]'}\right)_n \end{cases}$$

Note that exchanging temperature between two replicas is equivalent to exchanging two configurations at adjacent temperatures. In this REMD protocol, the momenta should also be rescaled,

$$\left\{ p^{[i]'} \equiv \sqrt{\frac{T_n}{T_m}} p^{[i]} \quad p^{[j]'} \equiv \sqrt{\frac{T_m}{T_n}} p^{[j]} \right.$$

This scheme rescales uniformly the momenta of all particles to keep at a proper temperature, and furthermore, provides a neat way to satisfy the detailed balance condition by canceling out the kinetic energy terms, as shown below.

For transition probability $w(X \to X')$, the detailed balance condition requires

$$\rho_{\mathrm{REM}}(X) w(X \to X') = \rho_{\mathrm{REM}}(X') w(X' \to X)$$

Thus, we have

$$\frac{w(X \to X')}{w(X' \to X)} = \exp\left\{-\beta_m\left[K\left(p^{[j]'}\right) + V\left(q^{[j]}\right)\right] - \beta_n\left[K\left(p^{[i]'}\right) + V\left(q^{[i]}\right)\right]\right.$$

$$\left. +\beta_m\left[K\left(p^{[i]}\right) + V\left(q^{[i]}\right)\right] + \beta_n\left[K\left(p^{[j]}\right) + V\left(q^{[j]}\right)\right]\right\}$$

$$= \exp\left\{-\beta_m \frac{T_m}{T_n} K\left(p^{[j]}\right) - \beta_n \frac{T_n}{T_m} K\left(p^{[i]}\right) + \beta_m K\left(p^{[i]}\right) + \beta_n K\left(p^{[j]}\right)\right.$$

$$\left. -\beta_m\left[V\left(q^{[j]}\right) - V\left(q^{[i]}\right)\right] - \beta_n\left[V\left(q^{[i]}\right) - V\left(q^{[j]}\right)\right]\right\} = \exp(-\Delta)$$

where $\Delta \equiv [\beta_n - \beta_m](V(q^{[i]}) - V(q^{[j]}))$.

This can be satisfied by introducing the Metropolis criterion [11]:

$$w(X \to X') \equiv w\left(x_m^{[i]} | x_n^{[j]}\right) = \min(1, \exp(-\Delta))$$

2 Materials

2.1 GROMACS-4.5.3 [12]

The REMD simulation can be performed using popular MD simulation packages including GROMACS [12], AMBER [13], CHARMM [14], NAMD [15], etc. For this case study we use GROMACS-4.5.3. The implementation in other GROMACS versions is straightforward. When using other simulation packages, we recommend referring to the documents or tutorials provided by the official sites of these packages.

2.2 High Performance Computing (HPC) Cluster

REMD simulations are highly parallel and resource demanding. HPC cluster installed with the GROMACS-4.5.3 and a standard message passing interface (MPI) library are required. Typically two cores per replica can give nice productivity in a HPC cluster equipped with Intel Xeon X5650 CPUs or above.

2.3 Visual Molecular Dynamics (VMD)

In this protocol, the versatile package VMD [16] is used for molecular modeling and structure visualization. VMD supports computers running MacOS X, Unix, or Windows systems, and is distributed free of charge.

2.4 Linux Shell

Some shell scripts coded in Bash are provided for data preparation and file processing. These scripts can run on a normal Linux distribution or a virtual Linux environment such as Cygwin on Windows.

3 Methods

Here we illustrate how to perform a REMD study from scratch using the GROMACS-4.5.3 and VMD packages via a case study of hIAPP(11–25) dimer system at a constant pressure and temperature (*NPT*) ensemble. The peptide with amino acid sequence of RLANFLVHSSNNFGA is capped by an acetyl (CH_3CO) group at the N-terminus and a NH_2 group at the C-terminus, as done experimentally [17]. Note that the REMD algorithm was initially developed for canonical ensemble (*NVT*, constant volume and constant temperature). It can be easily adapted to the *NPT* ensemble with the Hamiltonian of the system as

$$H(q,p) = K(p) + V(q) + PV$$

where P and V are the pressure and volume of the system, respectively. The contribution of volume fluctuations to the total energy is negligible [18].

The REMD method combines the replica exchange algorithm with multiple MD simulations. Thus, the initial step of the REMD simulation is identical to that of the conventional MD simulation. In the final step, a series of parallel MD replicas at different tem peratures are created. Here we illustrate the major processes of carrying out a REMD study without too much details, unless there are some key issues needing further clarification.

3.1 Construct an Initial Configuration of hIAPP(11–25) Dimer

The first step in performing a REMD study is to construct a starting configuration (in some cases, more than one initial configurations are needed) (*see* **Note 1**).

1. Use VMD TCL console to construct a configuration in which the two hIAPP(11–25) molecules have a vertical chain distance of 1.4 nm, i.e., with no atomic contacts (*see* **Note 2**).

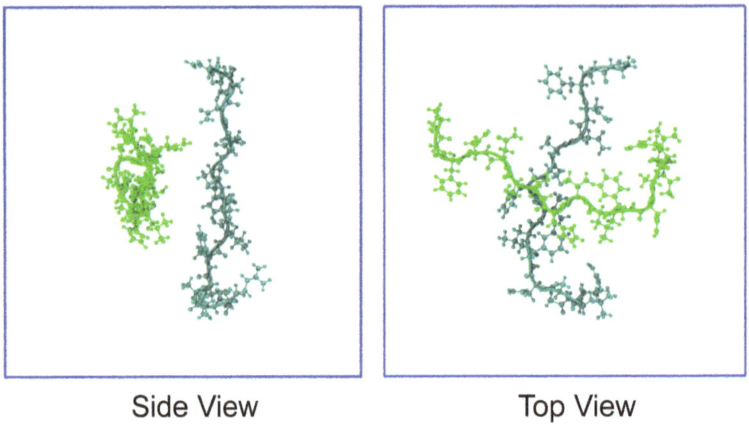

Fig. 2 The initial conformation of hIAPP(11–25) dimer

To construct the desired configuration, one can construct a configuration by manipulating the molecules via modifying and combining the following TCL commands (# comments):

```
set chainA [atomselect 0 "all"] # Select chain A
set chainB [atomselect 1 "all"] # Select chain B
measure center $chainA # Measure the center of mass of chain A
measure center $chainB
$chainA moveby {0 0 5} # Move chain A by 5 Angstrom in Z direction
$chainA move [transaxis z 30 deg] # Rotate chain A by 30 degree around Z axis
$chainA writepdb chainA.pdb
$chainB writepdb chainB.pdb
```

2. Merge the two chains into one pdb file, e.g., dimer.pdb, by Linux Shell commands, such as cat, and then use the Gromacs command *editconf* to convert it to proper pdb format (Fig. 2).
3. Use *pdb2gmx* command tool to create the coordinate and topology files.

```
pdb2gmx -f dimer.pdb -o dimer.gro -p dimer.top -inter –ignh
```

3.2 Choose a Simulation Box

1. After we created an initial configuration, now we need to solvate the molecule in a water box by the following command.

```
editconf -f dimer.gro -o dimer_box.gro –c -box 5 5 5
```

Note that the box should be big enough so that the molecule will not interact with its neighboring mirror (i.e., within the potential distance cutoff) under periodic boundary condition (PBC). Beware that the protein molecule rotates in solution. As a general rule of thumb, the rotational correlation time τ_c (time taken to rotate by one radian) of a monomeric protein in solution in nanoseconds is approximately 0.6 multiplying its molecular weight in kDa [19], which means that a 90° rotation of a protein takes time ~[0.94 × molecular weight (in kDa)] ns. For a globular protein system a more spherical box shape such as dodecahedron is highly recommended, which saves as much as 29% water molecules (*see* **Note 3**).

3.3 Do Energy Minimization and Equilibrium

1. After creating a simulation box, we need to do the energy minimization and equilibrium in both vacuum and solvent. The following commands do the job.

   grompp -f em.mdp -c dimer_box.gro -p dimer.top -o dimer_em_vac.tpr

   mdrun -v -deffnm dimer_em_vac

   genbox -cp dimer_em_vac.gro -cs spc216.gro -p dimer.top -o dimer_sol.gro grompp -f em.mdp -c dimer_sol.gro -p dimer.top -o dimer_em_sol.tpr

   mdrun -v -deffnm dimer_em_sol

2. Now add counter ions and extra salt ions to get a neutral system usually with a final salt concentration of 0.1 M.

   grompp -f em.mdp -c dimer_em_sol.gro -p dimer.top -o dimer_ions.tpr

   genion -s dimer_ions.tpr -p dimer.top -o dimer_ions.gro -neutral -conc 0.1

   Prompt selection: SOL

   grompp -f em.mdp -c dimer_ions.gro -p dimer.top -o dimer_em_ions.tpr

   mdrun -v -deffnm dimer_em_ions

3. Equilibrate the system in NVT ensemble first and then in NPT ensemble to further relax the system and reduce the possibility of system exploding in production run afterwards.

   grompp -f NVT.mdp -c dimer_em_ions.gro -p dimer.top -o dimer_nvt.tpr

   mdrun -v -deffnm dimer_nvt

   grompp -f NPT.mdp -c dimer_nvt.gro -p dimer.top -o dimer_npt.tpr

   mdrun -v -deffnm dimer_npt

3.4 Set Temperature Distribution and Create Replicas

Now that we have a well-equilibrated system, it's time to construct replicas for the REMD simulation. A key step in the REMD method is to set a proper temperature distribution, which should give a sufficiently high temperature and similar acceptance ratio between all adjacent replica pairs over the entire temperature range [9, 20]. In addition, for efficiency each replica should spend equal amounts of time at each temperature [21]. The upper limit of the temperature distribution should be high enough so that no trapping in a local minimum-energy state occurs. However, this is not straightforward to verify even though random walks among replicas are observed. Typically one can examine the selection of the highest temperature by comparing the energy distribution with a classical MD run at an expected temperature, which should not produce a significant deviation. Previous studies showed that the maximum temperature should be just above the temperature where the activation free energy vanishes and any replicas higher than that would decrease the efficiency due to non-Arrhenius behavior [22, 23]. In the REMD study of hIAPP(11–25) dimer, 453 K was chosen and verified as a sufficient temperature upper limit. The lower limit temperature is usually near the temperature of highest interest (mostly at the physiological temperature of 310 K for the study of peptide aggregation).

Temperatures that are distributed exponentially usually provide a uniform acceptance ratio in the limit of constant heat capacity [24]. Different algorithms have been developed to further optimize this distribution. Here we recommend an algorithm derived by Patriksson and van der Spoel [20] where the energy distributions are assumed to be Gaussian. A temperature distribution can be provided once the number of atoms in the system, upper and lower temperature limits, and a desired exchange probability are given. No further prior knowledge about energies or temperatures is needed.

1. Generate the temperature distribution ranging from 310 to 453 K using the web server (http://folding.bmc.uu.se/remd).

   It should be noted that the temperature differences between adjacent replicas should not be too large in order to obtain sufficient potential energy overlaps between the two replicas, thus the acceptance ratio between them is not too small (as a rule of thumb, 0.2~0.3 is good). A short (e.g., 1 ns) REMD test is practicable to check the acceptance ratio.
2. Generate the replicas of the hIAPP(11–25) dimer system.
   With a selected temperature distribution, the replicas can be generated using the following Bash script create_replicas.sh (with instructions on how to prepare files in # comments):

```
#---------Begin---------
#! /bin/bash
# First prepare the file temperature.dat with your temperature
  distribution arranged in two columns as:
# 0 310.0
# 1 312.84
# 2 315.7
```

```
# ...

# Then in remd.mdp file, specify any temperature value with the
  "TEMP" string, e.g.
# ref_t = TEMP TEMP
# gen_temp = TEMP
# and specify the remd.mdp file directory here.
mdpFile=/DIRECTORY/TO/YOUR/REMD.MDP

# Execute this script by:
# ./create_replicas.sh < temperature.dat

while read replica temp
do
sed -e 's/TEMP/'$temp'/g' $mdpFile > rep_$replica.mdp
# Option 1. Generate *.tpr with single starting structure, mod-
  ify it if needed
grompp -f rep_$replica.mdp -c dimer_npt.gro -p dimer.top -o
    dimer_$replica.tpr
# Option 2. Generate *.tpr with 4 (or any other number) start-
  ing structures (should be named as dimer_npt_0.gro
  dimer_npt_1.gro dimer_npt_2.gro dimer_npt_3.gro etc),
  modify it if needed
# grompp -f rep_$replica.mdp -c dimer_npt_$(($replica%4)).
  gro -p dimer.top -o dimer_$replica.tpr
done
rm \#* mdout.mdp
#---------End---------
```

3. Run a 1-ns REMD simulation to check whether the acceptance ratio between two adjacent replicas is within 0.2~0.3. Adjust the temperature limits and/or the exchange probability to regenerate the desired temperature distribution.

3.5 Running the Simulation

1. Now, run the REMD simulation by the following command (suppose we have 44 replicas):

```
mdrun -s dimer_.tpr -multi 44 -replex 1000 -maxh 240 -noappend
```

For data safety and subsequent ease-of-coding, we recommend using the parameters –noappend to save the output data separately, and –maxh to gracefully terminate REMD jobs before reaching the HPC cluster wall time limit (e.g., 10 days) that can easily corrupt data. To extend or restart a normally terminated/crashed job, checkpoint (.cpt) files are needed. However, data outputs including .cpt files are not always saved at the same pace among all replicas (*see* **Note 4**). A practical trick to avoid corrupting a .cpt file is to use the parameter –cpt, e.g., 60 (min) and set long enough time intervals of writing out .cpt file.

The -replex 1000 was used in our REMD simulation, which means we make the exchange trial between a pair of replicas every 1000 integration steps (with an integration time step of 2 fs). Different swap time settings can affect the sampling efficiency of REMD simulation. Details on this issue are discussed in **Note 5**.

3.6 Data Analysis

3.6.1 Convergence Check of REMD Simulation

The first step to analyze the data after finishing REMD run is to check the convergence of the simulation, as physical variables can only be correctly calculated based on a converged sampling. There are several ways to assess the convergence (Figs. 3, 4, and 5), including checking physical properties between two independent time intervals and checking whether a replica spans full temperature space multiple times by tracing a selected replica with simulation time.

1. Monitor the time evolution of hIAPP(11–25) dimer replicas by tracing one replica (replica 1) in temperature space.

   The time evolution of replica 1 shows that the peptide visited sufficiently the entire temperature space (Fig. 3).

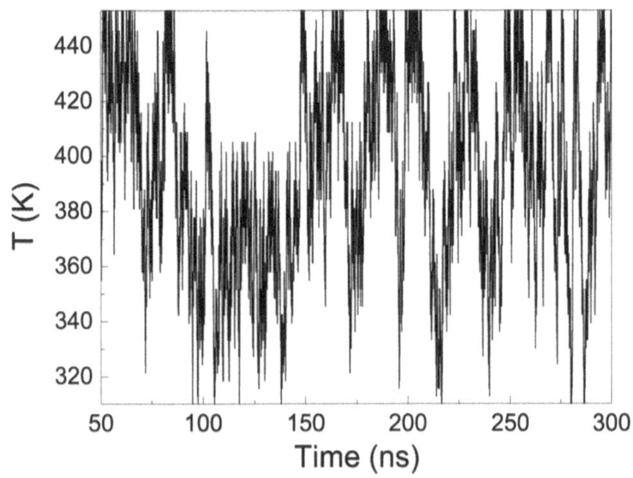

Fig. 3 Convergence check of the REMD run of hIAPP(11–25) dimer. The time evolution of replica-1 in temperature space

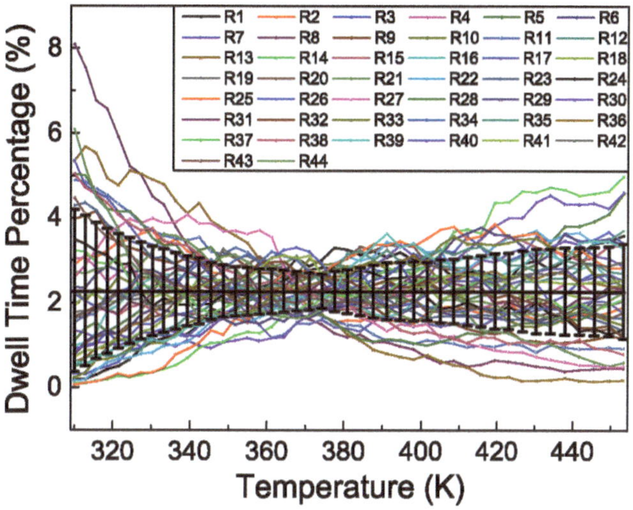

Fig. 4 The percentage of dwell time at each temperature for all of the 44 replicas. The overall standard deviations are shown in bars. With a certain exchange probability, a good sampling efficiency requires that the time a replica spends at every temperature site should be similar. So that the unbiased random walk of a replica through the whole temperature space is guaranteed

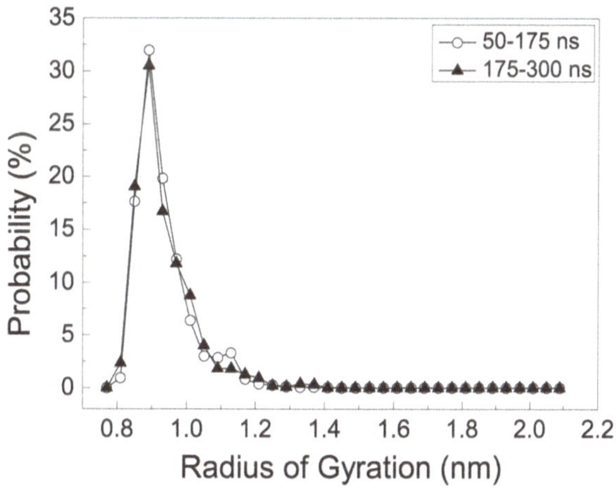

Fig. 5 Convergence check of the REMD run of hIAPP(11–25) dimer. Probability distributions of Rg within two different time intervals of 50–175 ns and 175–300 ns are overlapped very well

2. Calculate the percentage of dwell time at each temperature for each of the 44 replicas.

   Figure 4 demonstrates each replica samples all of the 44 temperatures through the 300-ns simulation time, indicating sufficient sampling. The overall standard deviations also show an appropriate sampling of the REMD simulation. For rigorous results, we need to check the distributions of some physical variables within two or more independent time intervals. For

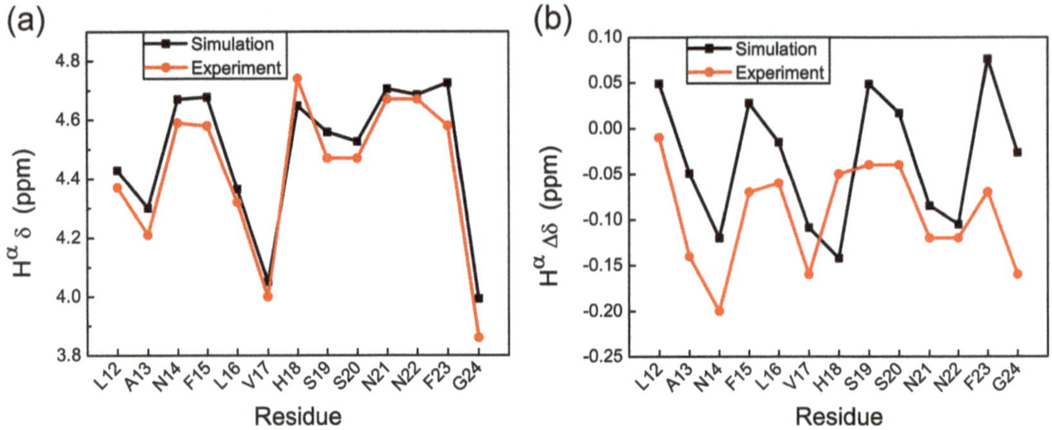

Fig. 6 Comparison between calculated and experimental chemical shifts of hIAPP(11–25) monomer along with representative REMD-generated α-helical and β-structures at 310 K. (**a**) H^{α} chemical shifts (δ), (**b**) H^{α} secondary chemical shifts ($\Delta\delta$)

example, in our hIAPP(11–25) dimer system, the convergence of the REMD run was checked using the probability distributions of Rg generated within two independent time intervals: 50–175 ns and 175–300 ns. As seen from Fig. 5, the two probability distribution curves overlap very well, indicating a good convergence of the REMD simulations.

3. Compare simulation results with experimental data if available. A comparison of simulation with experimental results is necessary if experimental data are available. A good agreement between them can validate the reliability of the simulation results. As in our hIAPP(11–25) monomer REMD simulation, we calculated the H^{α} chemical shifts of each amino acid residue using SPARTA+ [25], and compared it with solution NMR data [17]. As shown in Fig. 6a, b, our simulations essentially reproduced the experimental H^{α} chemical shifts. The calculated H^{α} secondary chemical shifts were also in agreement with solution NMR experimental observations, with a Pearson correlation coefficient of 0.57. These results demonstrate that our REMD simulations reasonably converged.

3.6.2 Fix the Periodic Boundary Condition (PBC) Problem

Simulations are normally performed under PBC. Molecules are often seen in a bond-broken state if they cross the PBC boundaries. We need to restore all atoms into the same unit cell to avoid bond-broken state. Generally the following steps can do the job.

1. Make a group in .ndx file using make_ndx to select a group of atoms or molecules as center, which will make the whole system gathered into one unit cell. Then do the following command:

```
trjconv -f traj.xtc -o center.xtc -n index.ndx -center -boxcenter
    rect
```

2. trjconv -f center.xtc -o inbox.xtc -ur compact -pbc atom.
3. trjconv -f inbox.xtc -s topol.tpr -o final.xtc -pbc mol.

Then we need to check the final.xtc with VMD. If there are still molecules that diffuse away (often encountered with multi-peptides system), repeat the **steps 1~3** and use a different center group in the **step 1**.

3.6.3 Calculate Chain-Independent RMSD for Multi-Peptide System

It is often useful to cluster all the conformations sampled in the REMD simulation, either for charactering the conformational ensemble or for further use in other analyses (such as Markov State Model (MSM)) [26, 27]. However, a common issue frequently encountered with the RMSD-based clustering is the exchange effect of topologically identical peptide chains in the coordinate files. As more typically shown in our previous REMD study on Aβ(16–22) octamer system [28], the eight chains are not covalently connected, and their relative spatial positions in one conformation may be different from those of another. The RMSD value calculated directly using the coordinate file from the REMD run may not be the smallest and cannot be used for structure similarity comparison. In this situation, the permutation of all chains should be considered, although this may drastically increase the computational cost. While there is no such issue with single peptide/protein, ignoring the chain exchange effect in multi-peptide/protein system will cause errors in RMSD calculation.

3.6.4 Calculate the Connectivity Length to Characterize the Degree of Peptide Aggregation

Oligomerization is the initial process of protein aggregation and fibrillation, in which a quantity capable of capturing the degree of aggregation would be very useful. We have introduced a mathematical parameter—connectivity length (CL) to characterize the ordering of the β-sheet-rich oligomers in our previous study [29]. This method can also be used to characterize the extent of aggregation. CL is defined as the sum of the square root of the sizes of all oligomers (or β-sheet sizes in the characterization of oligomer ordering). For example, for an octa-peptide system, the resulting two-tetramer aggregate (4,4) will give a $CL = \sqrt{4} + \sqrt{4} = 4$; while another arrangement (1,3,2,2) (i.e., a system consisting of one monomer, one trimer, and two dimers) will give a $CL = \sqrt{1} + \sqrt{3} + \sqrt{2} + \sqrt{2} = 5.56$. Therefore, CL reflects the extent of oligomerization (or the ordering the oligomers). The smaller the CL is, the stronger the peptide oligomerization becomes (or the more ordered the oligomer is).

3.6.5 Construct the Free Energy Landscape

REMD simulations sample the conformational space more efficiently than conventional MD simulations, making it quite suitable to investigate peptide aggregation. Here, taking the hIAPP(11–25) dimer system as an example, we illustrate the procedure of free energy landscape construction.

1. Choose two variables that can reflect the essential information of the system as order parameters.
2. Calculate the free energy of different states based on binned-state-statistics using $-RT \ln P(x, y)$, where R, T, P are, respectively, the gas constant, temperature, and probability of specific state.

   For the hIAPP(11–25) dimer (Fig. 5), the two-dimensional free energy surface is constructed using $-RT \ln [P(\text{Rg, RMSD})]$, where $P(\text{Rg, RMSD})$ is the probability of a conformation having a certain value of Rg and RMSD. Here, Rg and RMSD denote the radius of gyration of the dimer and the root mean square of deviation of the configuration referenced to the structure E in Fig. 5 of [10]. Five minimum energy basins are observed in free energy landscape. The corresponding representative snapshots show hIAPP(11–25) dimers mostly adopt disordered conformations and to a lesser extent antiparallel β-sheet conformations. However, caution should be taken when explaining the result of a free energy landscape, as the information reflected by just two reaction coordinates could be incomplete.
3. Use different reaction coordinates to construct the free energy landscape.
   To optimize the free energy analysis, the landscape based on different reaction coordinates should be plotted. For the hIAPP (11–25) dimer study, the free energy landscape based on Rg and Hydrogen bond was also constructed and are in good agreement with the one described here (Fig. 7).

3.6.6 Tips on REMD Data Manipulation

REMD simulation generates large data set, often in hundreds of gigabytes. Data transfer and manipulation should be taken carefully. We recommend a Linux tool *rsync* for data upload/download, which is better than the typical *scp* tool (*see* **Note 6**). In addition, a practical trick to save space in GROMACS REMD simulation is to convert the .trr format into the .xtc format, which significantly reduces the space occupation by removing the velocities and compressing the data, but still keeps a sufficient resolution (*see* **Note 6**).

3.6.7 Trajectory Visualization with VMD

There are versatile applications in VMD for MD trajectory visualization and analysis. When using VMD, a novice user may encounter the issue that the secondary structure doesn't change properly with trajectory animation. A TCL script can solve the problem (*see* **Note 7**).

All-atom water REMD is one of the most rigorous simulation methods currently used to explore the free energy surface of a protein system. With the proper system size and physical conditions, REMD can provide structure insights into various biomolecules. As an example here we illustrate the simulation of the peptide oligomer system, hIAPP(11–25), to obtain clues of peptide assemblies and aggregation mechanism. Our extensive REMD

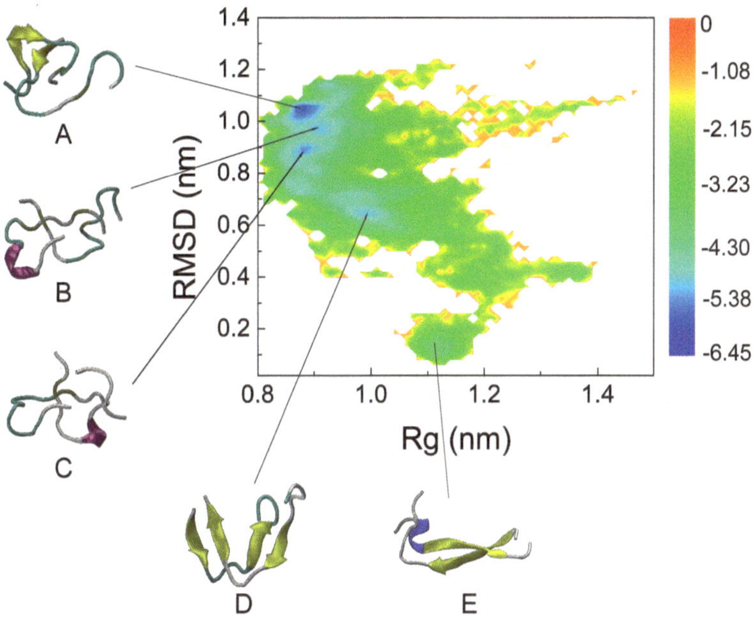

Fig. 7 Free energy surface (in kcal/mol) of hIAPP(11–25) dimer at 310 K as a function of radius of gyration (Rg) and backbone RMSD. The reference structure for RMSD calculation is structure E given in Fig. 5 of [10]. The free energy surface contains five minimum energy basins, centered at (Rg, RMSD) values of (0.88 nm, 1.06 nm), (0.91 nm, 0.96 nm), (0.88 nm, 0.9 nm), (0.99 nm, 0.64 nm), and (1.12 nm, 0.12 nm). Representative structures in each minimum-energy basin along with their probabilities: 14.98% (A), 6.24% (B), 6.24% (C), 14.41% (D), and 2.25% (E)

simulations revealed the conformational distribution feature and the α-Helix to β-Sheet transition mechanism of the hIAPP(11–25) system.

4 Notes

1. In REMD simulations, multiple starting configurations are recommended.
2. Although the *editconf* tool implemented in GROMACS package can also do the same thing, the VMD TCL commands are more powerful.
3. For a dodecahedron simulation box, the system visualization in VMD will be shown as a triclinic box, which can be transferred to a normal dodecahedron view by the following GROMACS command:

   trjconv -f tri.xtc -s md.tpr –pbc mol –ur compact -o dodeca.xtc

4. Due to the parallel feature of REMD simulation jobs, the .cpt files can be easily corrupted if all jobs are not terminated properly (e.g., by an abrupt job killing), so that the .cpt files of all replicas

don't match when attempting a job continuation or restart. To check whether a .cpt file is damaged or not, use the following command:

gmxcheck -f state.cpt

A batch checking in advance for all checkpoint files can be performed by the following Shell script, which can avoid queuing-time waste in HPC cluster when .cpt files are actually corrupted.

```
#!/bin/bash
for i in $(seq 0 43) #Replica 0~43
do
  if gmxcheck -f state$i\_prev.cpt &>/dev/null
  then
    echo $i.ok
  else
    echo $i.no
  fi
done
```

5. Generally it has been assumed that for sampling efficiency a long enough relaxation time is needed for each replica's MD run after a swap. There has been a long debate on the exchange attempt frequency in REMD. Some argue that the time interval for exchange of conformations between adjacent replicas should not be shorter than the potential energy autocorrelation time [30, 31]. Others argue that the sampling efficiency increases with the increasing exchange-attempt frequency [32, 33]. Combining computational efficiency considerations and REMD sampling, a rule-of-thumb attempt exchange time is 500~1500 integration steps (with a typical integration time step of 2 fs for biomolecules).
6. The Linux tool *rsync* is powerful for data backup and/or synchronization with the feature of breakpoint-resume, which can be conveniently used for transferring large REMD data. Simply use the following command to download data:
rsync -vaPz user@server-address:/Directory/of/data ./Local/directory

Note: caution with the parameter -u, which can improperly skip a partly transferred file.

We provide the following Shell script for converting all .trr files in a current directory to .xtc files, which can then be transferred by *rsync*. A secure deletion of the converted .trr files is executed to avoid the surge of disk occupation.

```
#----Begin----
#!/bin/bash
echo -e "\e[32mListing trr files...\e[0m"
echo ""
if ls *.trr
then
  echo ""
  echo -e "\e[32mNow converting all .trr to .trr.xtc\e[0m"
echo -e "\e[33mPlease wait\e[0m"
  echo -e "\e[33m...\e[0m"
  for i in 'ls *.trr'
  do
    if
      trjconv -f $i -o $i.xtc &>trr2xtc.out
    then
      rm $i
    else
      echo -e '\e[31m***Fatal error: Oh Gosh! Something bad
 happens! Please check the output file trr2xtc.out\e[0m'
      break
    fi
  done
  echo ""
  echo -e "\e[5m\e[32mComplete.\e[0m"
else
  echo ""
  echo -e "\e[31m***Error: Failed listing trr file. Please check
 the directory.\e[0m"
fi
#----End----
```

7. Type the following command in VMD TCL console to get a dynamic secondary structure visualization.

```
source ss.tcl
```

The content of ss.tcl is:

```
#----Begin----
  trace variable vmd_frame(0) w pvar
```

```
proc pvar {name element op} {
   mol ssrecalc top
}
#----End----
```

Acknowledgments

This project has been funded in whole or in part with Federal funds from the National Cancer Institute, National Institutes of Health, under contract number HHSN261200800001E. The content of this publication does not necessarily reflect the views or policies of the Department of Health and Human Services, nor does mention of trade names, commercial products, or organizations imply endorsement by the U.S. Government. This research was supported (in part) by the Intramural Research Program of the NIH, National Cancer Institute, Center for Cancer Research.

References

1. Paparcone R, Cranford SW, Buehler MJ (2011) Self-folding and aggregation of amyloid nanofibrils. Nanoscale 3:1748–1755
2. Knowles TP, Buehler MJ (2011) Nanomechanics of functional and pathological amyloid materials. Nat Nanotechnol 6:469–479
3. Kelly JW (1998) The alternative conformations of amyloidogenic proteins and their multi-step assembly pathways. Curr Opin Struct Biol 8:101–106
4. Rochet J-C, Lansbury PT Jr (2000) Amyloid fibrillogenesis: themes and variations. Curr Opin Struct Biol 10:60–68
5. Clark A, Lewis CE, Willis AC, Cooper GJS, Morris JF, Reid KBM, Turner RC (1987) Islet amyloid formed from diabetes-associated peptide may be pathogenic in type-2 diabetes. Lancet 330:231–234
6. Mitsutake A, Mori Y, Okamoto Y (2013) Enhanced sampling algorithms. Methods Mol Biol (Clifton, NJ) 924:153–195
7. Barducci A, Pfaendtner J, Bonomi M (2015) Tackling sampling challenges in biomolecular simulations. Methods Mol Biol (Clifton, NJ) 1215:151–171
8. Laio A, Parrinello M (2002) Escaping free-energy minima. Proc Natl Acad Sci U S A 99:12562–12566
9. Sugita Y, Okamoto Y (1999) Replica-exchange molecular dynamics method for protein folding. Chem Phys Lett 314:141–151
10. Qi R, Luo Y, Ma B, Nussinov R, Wei G (2014) Conformational distribution and alpha-helix to beta-sheet transition of human amylin fragment dimer. Biomacromolecules 15:122–131
11. Metropolis N, Rosenbluth AW, Rosenbluth MN, Teller AH, Teller E (1953) Equation of state calculations by fast computing machines. J Chem Phys 21:1087–1092
12. Berendsen HJC, van der Spoel D, van Drunen R (1995) GROMACS: a message-passing parallel molecular dynamics implementation. Comput Phys Commun 91:43–56
13. Cornell WD, Cieplak P, Bayly CI, Gould IR, Merz KM, Ferguson DM, Spellmeyer DC, Fox T, Caldwell JW, Kollman PA (1995) A second generation force field for the simulation of proteins, nucleic acids, and organic molecules. J Am Chem Soc 117:5179–5197
14. Brooks B, Bruccoleri R, Olafson B, States D, Swaminathan S (1983) Karplus M: CHARMM: a program for macromolecular energy minimization and dynamics calculations. J Comput Chem 4:187–217
15. Phillips JC, Braun R, Wang W, Gumbart J, Tajkhorshid E, Villa E, Chipot C, Skeel RD, Kalé LV, Schulten K (2005) Scalable molecular dynamics with NAMD. J Comput Chem 26:1781–1802
16. Humphrey W, Dalke A, Schulten K (1996) VMD: visual molecular dynamics. J Mol Graph 14:33–38

17. Liu G, Prabhakar A, Aucoin D, Simon M, Sparks S, Robbins KJ, Sheen A, Petty SA, Lazo ND (2010) Mechanistic studies of peptide self-assembly: transient α-helices to stable β-sheets. J Am Chem Soc 132:18223–18232
18. Seibert MM, Patriksson A, Hess B, van der Spoel D (2005) Reproducible polypeptide folding and structure prediction using molecular dynamics simulations. J Mol Biol 354: 173–183
19. Cavanagh J, Fairbrother WJ, Palmer AG, Skelton NJ (1996) Protein NMR spectroscopy: principles and practice. Psychnol J
20. Patriksson A, van der Spoel D (2008) A temperature predictor for parallel tempering simulations. Phys Chem Chem Phys 10:2073–2077
21. Rathore N, Chopra M, de Pablo JJ (2005) Optimal allocation of replicas in parallel tempering simulations. J Chem Phys 122:024111
22. Zheng W, Andrec M, Gallicchio E, Levy RM (2007) Simulating replica exchange simulations of protein folding with a kinetic network model. Proc Natl Acad Sci 104: 15340–15345
23. Nymeyer H (2008) How efficient is replica exchange molecular dynamics? An analytic approach. J Chem Theory Comput 4:626–636
24. Kofke DA (2002) On the acceptance probability of replica-exchange Monte Carlo trials. J Chem Phys 117:6911–6914
25. Shen Y, Bax A (2010) SPARTA+: a modest improvement in empirical NMR chemical shift prediction by means of an artificial neural network. J Biomol NMR 48:13–22
26. Bowman GR, Huang X, Pande VS (2009) Using generalized ensemble simulations and Markov state models to identify conformational states. Methods (San Diego, Calif) 49:197–201
27. Qiao Q, Qi R, Wei G, Huang X (2016) Dynamics of the conformational transitions during the dimerization of an intrinsically disordered peptide: a case study on the human islet amyloid polypeptide fragment. Phys Chem Chem Phys 18:29892–29904
28. Li H, Luo Y, Derreumaux P, Wei G (2011) Carbon nanotube inhibits the formation of β-sheet-rich oligomers of the Alzheimer's amyloid-β(16-22) peptide. Biophys J 101: 2267–2276
29. Lu Y, Derreumaux P, Guo Z, Mousseau N, Wei G (2009) Thermodynamics and dynamics of amyloid peptide oligomerization are sequence dependent. Proteins 75:954–963
30. Periole X, Mark AE (2007) Convergence and sampling efficiency in replica exchange simulations of peptide folding in explicit solvent. J Chem Phys 126:014903
31. Abraham MJ, Gready JE (2008) Ensuring mixing efficiency of replica-exchange molecular dynamics simulations. J Chem Theory Comput 4:1119–1128
32. Sindhikara DJ, Emerson DJ, Roitberg AE (2010) Exchange often and properly in replica exchange molecular dynamics. J Chem Theory Comput 6:2804–2808
33. Sindhikara D, Meng Y, Roitberg AE (2008) Exchange frequency in replica exchange molecular dynamics. J Chem Phys 128:024103

Chapter 6

An Aggregate Weight-Normalized Thioflavin-T Measurement Scale for Characterizing Polymorphic Amyloids and Assembly Intermediates

Ronald Wetzel, Saketh Chemuru, Pinaki Misra, Ravi Kodali, Smita Mukherjee, and Karunakar Kar

Abstract

The red shift in the fluorescence excitation spectra of thioflavin dyes upon binding to fibrils has been a boon to the amyloid field, offering simple and effective methods for the qualitative detection of amyloid in tissue samples and for quantitation of particular fibril preparations with gravimetric linearity. The quantitative aspect of the thioflavin T (ThT) response, however, comes with an important caveat that bestows both significant limitations and great untapped power. It is now well established that amyloid fibrils of different proteins, as well as polymorphic fibrils of the same protein, can exhibit vastly different ThT fluorescence intensities for the same weight concentration of aggregates. Furthermore, the aggregated intermediates commonly observed in amyloid assembly reactions can exhibit aggregate weight-normalized (AWN) ThT fluorescence intensities that vary from essentially zero through a wide range of intermediate values before reaching the intensity of homogeneous, mature amyloid. These features make it very difficult to quantitatively interpret, without additional data, the time-dependent development of ThT fluorescence intensity in an assembly reaction. In this chapter, we describe a method for coupling ex situ ThT fluorescence determinations with an analytical HPLC supported sedimentation assay (also described in detail) that can provide significant new insights into amyloid assembly reactions. The time dependent aggregation data provided by the sedimentation assay reveals a time course of aggregation that is largely independent of aggregate properties. In addition, the combination of these data with ThT measurements of the same reaction time points reveals important aspects of average aggregate structure at each time point. Examples of the use and potential value of AWN-ThT measurements during amyloid assembly Aβ and polyglutamine peptides are provided.

Key words Amyloid, Oligomer, Thioflavin T, Centrifugation, HPLC, Assembly intermediates, Polymorphic, Fibrils, Nucleation, Elongation

1 Introduction

Aggregation reactions are a ubiquitous feature of the evolution, biological life cycles, and physical behavior of proteins [1], playing important roles both in the normal generation and flux of polypeptides in the cell and in a large and growing number of diseases

Bradley L. Nilsson and Todd M. Doran (eds.), *Peptide Self-Assembly: Methods and Protocols*, Methods in Molecular Biology, vol. 1777, https://doi.org/10.1007/978-1-4939-7811-3_6,

[2, 3]. In addition, amyloid fibrils and related protein aggregates comprise the starting point for the design of new, biology-inspired materials [4]. The earliest quantitative treatments of thermally induced aggregation of proteins deduced both the generic two-step nature of globular protein aggregation and the unusually high apparent activation energies of the reaction [5], two insights that contributed to our understanding of both the aggregation process and the nature of protein stability [6]. With the general acceptance of a model for proteins as covalent linear polymers, aggregation came to be viewed as an intertwining of disordered polypeptide chains producing multimeric species with no regular 3D structure [6]. This view became more nuanced in the 1960s and 1970s, as one kind of protein aggregated structure, the amyloid fibril [3], was recognized as containing one dominant type of protein molecule in each fibril type [7] while exhibiting paracrystallinity in its X-ray diffraction patterns [8] and ordered morphology in electron microscopy [9]. In this same time frame, ideas began to emerge supporting a more regular, ordered mechanism even for some "amorphous" protein aggregation reactions [10]. Work over the past 20 years has further supported the importance of regular atomic-level structure in amyloid and related protein aggregates, while at the same time demonstrating that amyloid self-assembly is far more complex than the relatively simple crystallization of rigid molecules. We now appreciate that there can be any number of alternative aggregate structures of the same protein [11, 12], some of which may be intermediates populated during assembly pathways [13–15] and others of which are polymorphic forms of mature amyloid [16–18]. The current challenge is to understand these structures and assembly mechanisms at the atomic level, and these goals require a continued effort to identify new tools as well as refine old ones.

Aromatic dyes such as thioflavin T (ThT) emerged in the early phase of the protein aggregation renaissance as a highly convenient method to qualitatively identify amyloid deposits in autopsy tissue [3] and to quantify the formation of fibrillary mass in vitro [19]. More recently, derivatives of these dyes were developed for use in quantifying amyloid deposits in living patients by noninvasive imaging methods [20]. ThT is very widely used in in vitro studies of amyloid assembly [21], including, most popularly, as an in situ measure of amyloid formation that can be monitored on small volumes in real time to obtain information on the kinetics of assembly [22]. While this method can be very useful in providing qualitative data, it suffers from at least two potential complications. First, at least in some cases, the presence of ThT in the reaction mixture appears to substantially alter the assembly mechanism [23]. Second, it is by now very well established that different aggregated forms of amyloidogenic proteins, including different polymorphic mature fibrils, can exhibit quite different ThT fluorescent intensities [18]. This means that although the development of ThT fluorescence in a final assembly reaction mixture may well be indicative of the presence of an amyloid-like structure, the intensity of that fluorescence will not necessarily characterize the identity of the aggregate or the degree of completeness of the reaction. More importantly, perhaps, it also means that the development of raw ThT fluorescence intensity over time in an assembly reaction cannot reliably be interpreted in quantitative, mechanistic terms. These limitations exist in spite of the fact that, for any one assembly state, ThT binding and its fluorescence intensity is highly reproducible and sensitive to structural features such as the state of the H-bonding network [18]. Thus, while there is a great potential in the ThT method to provide insights into structure and mechanism, it is almost always largely wasted and/or abused, through a lack of appreciation of its limitations and how to overcome them.

The problem is that the ThT intensity determined for an aggregate-containing sample is actually a combination of three different qualities: the total percentage of starting monomer that has been converted to aggregates, the distribution of different aggregate types within the sample, and the absolute ThT intensities, per mass, that are characteristic features of each of these aggregate types. Knowing only the measured ThT intensity of this sample is clearly not enough to distinguish, for example, between a suspension containing a small amount of an aggregate giving an intense ThT response, and a reaction mixture containing 100% of an aggregate exhibiting a low ThT response. The missing link to a fuller understanding of such suspensions is to know the percentage of starting monomer that has been converted to aggregates at each time point. There are ways to estimate this value, one of which relies on coupling a centrifugation step that separates aggregates from low molecular weight species with an analytical method to quantify the amount of low molecular weight material in the centrifugation supernatant [18]. One particularly sensitive, accurate, and flexible analytical method for quantifying such samples involves determining the A_{215} peak area from a reverse phase high performance liquid chromatography (RP-HPLC) run on an aliquot of the supernatant, interpreted based on a standard curve generated from the same peptide [24, 25].

Here we describe a simple extension of the classical ThT method in which we combine ex situ ThT fluorescence measurements with this analytical HPLC method for determining the mass of starting monomer that has been converted to aggregates, either in an ongoing self-assembly reaction or in a final reaction mixture. We show how this method provides information that neither measurement alone can provide, giving insights into structures of final products and structural transitions during self-assembly. In this application, the sedimentation/HPLC assay effectively replaces the ThT assay as the measure of the global extent of aggregation, and the new aggregate-weight normalized ThT (AWN-ThT) values provide insights into the structures of the aggregates that are populated along the reaction coordinate.

1.1 ThT Analysis of Aggregate Suspensions

The basis of the sensitivity of ThT to amyloid-like structures appears to be the ability of this molecule to bind to specific sites within the β-sheets of the aggregate [26], leading to a shift in the excitation spectrum of the bound ThT to a higher wavelength [27]. By exciting within this red-shifted wavelength region, most or all of the emitted fluorescence is generated by the aggregate-bound dye, not the free dye. Under appropriate conditions of measurement, this fluorescence intensity is linear with the mass concentration of aggregates in suspension [19], allowing quantitation of homogeneous aggregate suspensions as well as assembly kinetics studies [21]. Since even the least stable amyloid fibril tends to dissociate very slowly in native buffers [23], it is possible to analyze amyloid suspensions by transferring aliquots into a cuvette containing a ThT solution followed by a reading of fluorescence intensity [27]. Alternatively, for reactions where it can be established that the presence of ThT does not influence the assembly reaction [23], kinetics might conveniently be assessed using in situ ThT, for example, in a fluorescence microtiter plate reader in time-drive mode [22]. It is important to note, however, that the final ThT fluorescence of a particular weight of amyloid can be significantly higher, under standard measurement conditions, if the amyloid was assembled in the presence of ThT [23]. This suggests that one factor responsible for polymorph-dependent ThT signal intensities may be the degree of access of the ThT to relatively buried sites within the fibrils. In any case, from a practical point of view, the amplified ThT fluorescence exhibited by fibrils grown in the presence of ThT can be expected to compromise any attempt to use ThT intensities as a universal characteristic feature of a particular amyloid fibril. As such, the method described here focuses on fibrils that have been grown without ThT in the growth buffer.

In this chapter we describe methods that can be used routinely to determine ThT fluorescence of kinetics time points and final amyloid products, and important considerations for how to make these measurements robust and reproducible. These can be applied in isolation, in the traditional manner, or, preferably, in combination with a knowledge of the weight of aggregate present in each sample.

1.2 Sedimentation/HPLC Analysis of Aggregate Suspensions

The aggregate-weight normalization of ThT values described here depends on the availability of a convenient, sensitive, and reproducible method for gauging the progress of an aggregation reaction as monomers are converted to multimers. Analytical RP-HPLC determination of residual monomer after ultracentrifugation of reaction mixture aliquots fulfills these requirements [24, 25]. Thus, for a simple amyloid assembly reaction involving no intermediate aggregated states, the conversion of monomer to amyloid over time, as tracked by the HPLC sedimentation assay, parallels aggregation time courses measured by circular dichroism (CD), ThT fluorescence, and light scattering [28] techniques. For a number of kinetics data applications, such as analysis of nucleation mechanisms and monitoring of intermediate states, it would seem to be critical to quantify all types of aggregates as well as monomer. Thus, the availability of a mass-based method (rather than, for example, a beta-centric method like ThT intensity) is very important. Another effective method for determining the mass concentration of monomers, not described here, involves the quite analogous integration of the monomer peak in size exclusion chromatography (SEC) runs on reaction aliquots (K.W. Drombosky, R. Kodali and R. Wetzel, Ms. submitted). Although this SEC method gave similar monomer-loss curves to the centrifugation/RP-HPLC method in peptide aggregation studies, it cannot be used with mixtures of peptides or to identify chemical damage to peptides during aggregation.

The sedimentation/HPLC assay developed in our laboratory has a number of important features, and many considerations must be evaluated during experimental design. Since the A_{215} extinction coefficients of peptides can differ appreciably due to different residue compositions, it is important to determine an individual standard curve for each peptide being studied. Applying an independently measured extinction coefficient directly to HPLC peak areas is not good enough, however, since this does not correct for sequence-dependent, systematic losses of material in the HPLC analysis, which are especially important when small amounts of peptide are injected. Counter-intuitively, as explained later, determination of the concentration of the stock solution required for preparation of the standard curve by amino acid composition analysis may not be optimal. Using a tabletop ultracentrifuge, rather than a standard Eppendorf-style centrifuge, is sometimes important [28], since oligomers, protofibrils, and small fibrils may not pellet quantitatively at relatively low *g* forces. If this is a concern, and ultracentrifugation is not an option, the above-mentioned SEC method can be used.

As described here, the sedimentation/HPLC analysis can be used in isolation, as a preferred way to monitor aggregation kinetics in such a way as to capture all types of aggregates, not only amyloid-like aggregates. Or, preferably, the data can be used in parallel to normalize independently obtained ThT fluorescence values to gain insight into aggregate structures.

2 Materials, Equipment, Instrumentation, and Data Presentation

2.1 Equipment and Instrumentation

1. Gradient HPLC system with analytical capability such as Agilent 1100 and 1200 systems equipped with a solvent degasser, multi-wavelength detector, autosampler, and ChemStation software (*see* **Note 1**).
2. Analytical and preparative HPLC columns. For both purification and analysis of β-amyloid (Aβ) and polyglutamine (polyQ) peptides, RP-HPLC column supports are suitable. The preparative column may be a 9.4 × 250 mm Agilent Zorbax SB-C3 [5 μm particle size]) and the analytical column may be a 4.6 × 50 mm Agilent XDB RP-HPLC column, C3 or C8 for Aβ peptides and C8 or C18 for polyQ peptides (*see* **Note 2**).
3. Single pump LC system, such as an Agilent 1100 or 1200 system with detector and fraction collector (*see* **Note 3**).
4. HPLC size exclusion chromatography column, such as a Superdex 200 GE 10/300 column (*see* **Note 4**).
5. A high speed ultracentrifuge, such as a Beckman Optima TLX tabletop ultracentrifuge equipped with a TLA 100 fixed angle rotor (maximum speed 100,000 rpm corresponding to 436,000 × *g*).
6. A spectrofluorometer and quartz cuvette, such as a Horiba Fluoromax-4 spectrofluorometer and 4 mm square fluorescence cuvettes.
7. UV–vis spectrophotometer.

2.2 Materials

1. Thioflavin T (ThT) stock solution: 1.5 mM ThT in deionized water and filtered through a 0.22 μM filter. The solution can be stored at room temperature (RT) in 15 mL Falcon tubes or glass vials protected from light by wrapping in aluminum foil. Typically, such stock solutions are stable for several months. ThT stock solutions should be checked periodically for clarity and absence of particles. If these are present a new stock solution must be made. The solid ThT is stable for years at RT.
2. HPLC-grade water.
3. HPLC-grade acetonitrile.
4. Hexafluoro-2-propanol (HFIP).
5. Sequanal grade trifluoroacetic acid (TFA).
6. Guanidinium chloride (Gdn.HCl, for full-length Aβ peptides).
7. HPLC Buffer A: 0.05% TFA in HPLC-grade water.
8. HPLC Buffer B: 0.05% TFA in HPLC-grade acetonitrile.
9. Purified amyloid-forming peptides, such as $A\beta_{42}$, $A\beta_{43}$, and $K_2Q_{23}K_2$. The peptides used to produce the data shown in

the example figures were obtained by custom synthesis from the Keck Biotechnology Center at Yale University (http://info.med.yale.edu/wmkeck/), and, when necessary, purified and disaggregated as described in the source publications for the figures [15, 29].

10. Phosphate-buffered saline (PBS).

2.3 Data Analysis and Presentation

1. Curve-fitting software, such as Origin.
2. Plotted data can be fit to equations to test experimental models or simply to guide the eye in order to allow comparisons and analysis of trends. In the three figures reproduced here, data are fit in Origin 2015 to guide the eye only, as follows: Boltzman (Fig. 1, □); BiDoseRes (Fig. 1, ●, ■); B-spline (Fig. 2, ●, ■; Fig. 3, ●); double Boltzman (Fig. 3, ■).

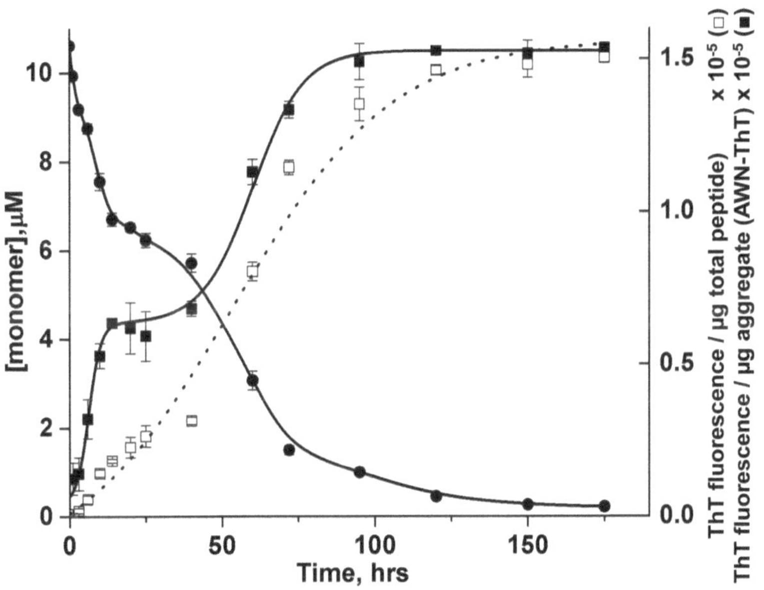

Fig. 1 Self-assembly of $A\beta_{43}$ monitored by sedimentation and ThT assays. Freshly disaggregated $A\beta_{43}$ was incubated without agitation at 10 μM in PBS at 37 °C and monitored by the sedimentation/HPLC assay (filled circle) and standard ThT assay (open square). The latter values were normalized to aggregate weight to yield AWN-ThT values (closed square) as described in this chapter. If only ThT were used to monitor self-assembly, the irregular but systematic increase in ThT intensity would be assumed to reflect some features of nucleated growth of the ThT-positive amyloid fibrils. However, the AWN-ThT time course suggests a more complex self-assembly reaction involving at least three aggregated species: a ThT-negative early intermediate, a later intermediate, populated between 10 and 40 h, exhibiting intermediate AWN-ThT values, and a mature amyloid fibril that dominates the reaction after 100 h. Importantly, upward trends in the AWN-ThT value, as seen between 0 and 10 h, and again between 40 and 100 h, indicate time periods of shifting balance between at least two different aggregate types of different AWN-ThT values. Thus, the general upward trend in the crude ThT signal between 40 and 100 h is seen to be due to two overlapping phenomena: (1) the increased proportion of monomeric $A\beta_{43}$ that is being incorporated into aggregates (as seen by the sedimentation assay), and (2) the steadily increasing proportion of total aggregates that exhibit the maximal AWN-ThT value associated with mature amyloid. Figure adapted, with additional data, from [15]

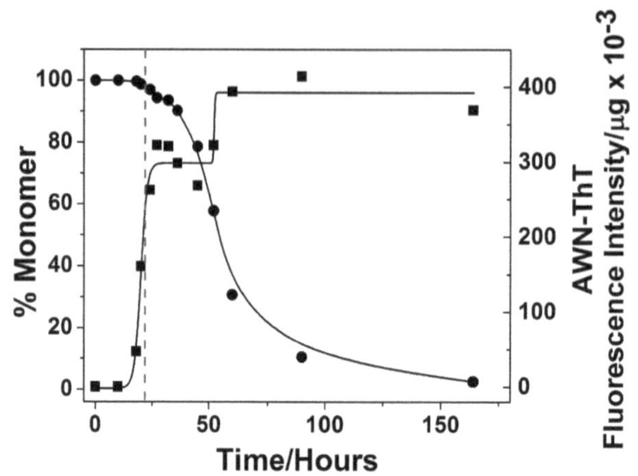

Fig. 2 Self-assembly of 215 μM $K_2Q_{23}K_2$ at 37 °C in PBS. This peptide was previously shown to undergo amyloid nucleation via a classical nucleated growth mechanism without the involvement of non-β oligomeric intermediates and with a critical nucleus (n*) of 4 [38]. Under these conditions, as measured by the sedimentation assay (filled circle), no detectible aggregation occurs within the first 30–35 h of incubation. In the analogous reaction shown here, at around 45 h (vertical dashed line) the monomer concentration after sedimentation begins to drop, indicating the onset of nucleated growth. Immediately upon initiation of aggregation, the AWN-ThT values (filled square) dramatically increase to a high, almost maximal AWN-ThT value, indicating that the first aggregates formed are already amyloid-like in structure. Reaction mixtures homogeneous with fibrils of maximal AWN-ThT values appear by about 60 h, at a point where about 70% of monomer has been converted. While these data are largely consistent with the proposed mechanism, they also suggest some previously unappreciated subtleties. In particular, while the conventional view of nucleated growth processes is that the bulk of the aggregate mass is generated through elongation seeded by rapidly formed mature fibrils, the AWN-ThT data suggest that for $K_2Q_{23}K_2$ this phase only obtains for the last 30% of monomer converted. Before that, the aggregate pool consists of at least some fibrils that are not yet completely mature, as characterized by AWN-ThT. One interpretation of these data is that the 25–60 h period is dominated by seeded elongation of an amyloid polymorph, which undergoes a further transformation to a structure exhibiting a somewhat higher AWN-ThT value late in the reaction. These data illustrate the power of the AWN-ThT scale to confirm mechanistic details while also revealing some previously unappreciated aspects of fibril self-assembly and structure. Figure reproduced from [29]

3 Methods

In general, disaggregation of polyglutamine peptides was by the mixed TFA/HFIP method, as described in detail previously [25], while the long Aβ molecules $A\beta_{42}$ and $A\beta_{43}$ were disaggregated by the Gdn.HCl/SEC method [29, 30]. Peptides should be disaggregated freshly, just before use. Stored disaggregated samples cannot be trusted to remain disaggregated.

3.1 Sedimentation/HPLC Analysis

1. Prepare stock solution aliquots of the peptide of interest at accurate concentrations. Stocks can be stably stored for later use to generate standard curves for analysis of peak areas. This is

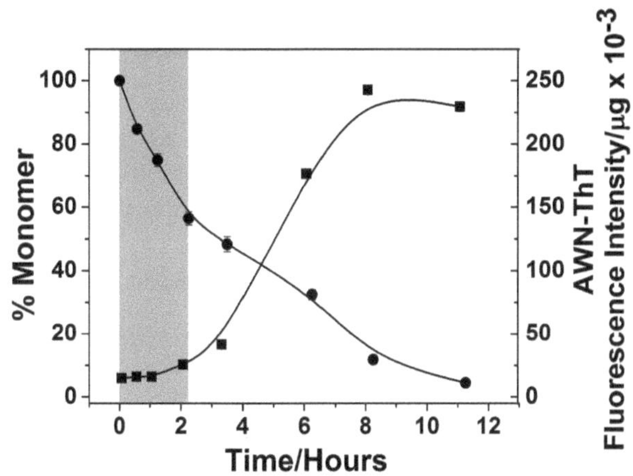

Fig. 3 Self-assembly of 2.7 μM Aβ_{42} in PBS at 37 °C without stirring, as monitored by sedimentation (filled circle) and AWN-ThT (filled square) assays. Aggregates exhibiting no AWN-ThT signal develop immediately from $t = 0$ to $t = 120$ mins (shaded region), after which continued aggregation is accompanied by ever-increasing AWN-ThT values until 8 h, when the loss of monomer from solution is ~85% complete. Thus, in contrast with $K_2Q_{23}K_2$ (Fig. 2), for which the vast bulk of the AWN-ThT signal amplitude develops immediately upon aggregation onset, for Aβ_{42} the bulk of the AWN-ThT signal develops quite gradually as monomer is being lost from solution. According to these data, amyloid mass derived from classical seeded elongation of mature fibrils may only be derived from the last 15% of the aggregation reaction. Up to that point, amyloid-like aggregates appear to develop via a combination of monomer recruitment and transformations of non-β aggregates through a series of aggregates with ever-increasing levels of ThT-sensing β-structure [15, 29]. Figure reproduced from [29]

conveniently accomplished in combination with HPLC purification of the peptide.

2. Purify peptides using a preparative HPLC and a linear gradient of HPLC Buffer A and HPLC Buffer B.
3. Collect chromatography fractions and analyze by analytical HPLC and/or mass spectrometry to determine fractions of acceptable purity.
4. Combine fractions and mix.
5. Determine the A_{215} of the chromatographic pool, adjusting, if necessary, by dilution with an appropriate mixture of HPLC buffers A and B in order to obtain an absorbance in the linear range of the UV–Vis instrument.
6. Dilute further, if necessary, to obtain stock solution aliquots of desired concentration for convenience of later setting up standard curves (*see* **Note 5**).
7. Determine stock solution concentration from the A_{215} using a calculated extinction coefficient as the sum of individual amino acid extinction coefficients [31, 32] (*see* **Note 6**). Store the stock solution aliquots, which should be in approximately 0.05% TFA in a mixture of CH_3CN and water, by flash-freezing

in liquid nitrogen or dry-ice/acetone, then storing in a –80 °C freezer.

8. Determine a standard curve by multiple analytical HPLC runs and analyses of appropriate weights of the peptide of interest (*see* **Note 7**). Analytical HPLC is run using an appropriate RP-HPLC column and a gradient of acetonitrile in water with Buffer A and B identical to the purification system described in **step 1** of Subheading 3.1.

9. After blank gradient subtraction, determine HPLC peak areas and plot vs. peptide mass loaded. A straight line should be obtained, with the possible exception of curvature in the region of low peptide loads due to larger fractional absorption by the column and hence low recovery of peptides. In the analysis of peptide aggregation kinetics, it is preferable to choose sample sizes expected to deliver peptide amounts in the linear range of the standard curve. At the same time, if analyzing very low concentrations—for example, at the end of the reaction when fibril growth has come to equilibrium, it is acceptable to work in a nonlinear part of the standard curve, so long as reproducibility has been demonstrated and precautions are taken against run-to-run carryover (*see* **Note 8**).

10. To set up an aggregation kinetics run, disaggregate (*see* **Notes 9** and **10**) an appropriate amount of purified peptide and use the analytical RP-HPLC assay to determine the concentration of soluble monomeric peptide. From this stable stock of monomeric peptide, adjust the concentration by dilution to obtain a solution that will generate the desired starting concentration after buffer components are added to initiate aggregation.

11. Just before adding buffer components and adjusting to temperature, process an aliquot of this peptide solution in exactly the same way that later reaction time points will be processed, and determine the peptide concentration of the centrifugation supernatant. Adjust this value for the volume change brought about when buffer components are added. Do this analysis in replicates to get the most reliable value possible, since all other values will later be related to this critical "starting concentration" value to determine percent aggregation. Obtaining a $t = 0$ value in this way is preferable to simply taking an aliquot immediately after aggregation is initiated, since aggregation—such as initial oligomer formation—can be very fast, and since—as stated above—the starting concentration value is critical.

12. Optimize analytical HPLC gradient if necessary to allow separation of the peptide from any chemical degradation products generated during the aggregation reaction, and from any additional peptides purposefully added to the incubation, such as

potential inhibitors or co-aggregation peptides (*see* **Note 11**). Make sure to include a column wash step as part of the programed gradient for each analytical run. Including this however does not obviate the need to also run blank runs between each analytical run to minimize sample carryover errors.

13. Using the optimized gradient, analyze time points as desired, by gently mixing the reaction, removing an aliquot, exposing to high speed centrifugation (*see* **Note 12**), carefully removing an aliquot of the resulting supernatant (*see* **Note 13**), and diluting it into a measured volume of HPLC Buffer A of the HPLC solvent system for injection into the analytical HPLC. If there will be frequent time points taken in the early stage of the reaction, it is important to stabilize the resulting centrifugation supernatants against further aggregation while the time points wait in the queue for analysis (*see* **Note 14**). It is imperative to measure all volumes as accurately as possible (*see* **Note 13**) and to keep track of all dilutions in order later to back-calculate the corresponding concentrations of monomer in the reaction mixture at different time points. Plan at the outset what sort of replicates will be needed (*see* **Note 15**).
14. As the aggregation reaction proceeds, the concentration of monomer left in solution grows smaller and smaller. For accurate determination of the full time course of the reaction, as well as for determination of the critical concentration C_r (*see* **Note 13**), it is important to be prepared to inject larger volumes of reaction supernatants as the reaction proceeds toward equilibrium. This requires some foresight in ensuring that the starting volume of the reaction is large enough that sufficient volume will still be available in the last phase of the reaction to allow taking forward time points as well as setting up reverse, fibril dissociation reactions (*see* **Note 13**).
15. As the amyloid assembly reaction draws to an end, probe for a robust endpoint concentration of monomer, a measure of fibril stability [23], by checking for fibril disassembly. Just before reaching the projected equilibrium position in the fibril growth direction, gently dilute an aliquot of the reaction by 10-fold to 100-fold into a new reaction tube and continue the incubation (*see* **Note 16**). Immediately analyze the monomer concentration in this newly generated dissociation reaction, then continue to analyze reaction time points for monomer concentration until it no longer changes with time. Depending on the peptide and conditions, this might take from a day to several months (*see* **Note 16**). It is expected that the monomer concentration will rise to a plateau value that is essentially equivalent to the final concentration of monomer obtained in the fibril growth direction.

3.2 Standard Ex Situ ThT Kinetics

1. Identify appropriate slit widths as described here in **steps 1–5**. In general, it is a mistake to simply choose slit width values from the literature on the peptide of interest, since linearity will depend on instrument design. To ensure that slit widths achieve the required linearity of fluorescence signal with peptide concentration, determine acceptable slit widths for ThT analysis with your instrument by following **steps 2–5**.
2. From a complete amyloid assembly reaction mixture, as prepared above, transfer an aliquot of the reaction, equivalent to 1 μg of peptide, into a fluorescence cuvette containing a 10 μM solution of ThT in PBS (*see* **Note 17**). Using this amyloid suspension and working at an excitation wavelength (λ_{ex}) of 445 nm and an emission wavelength (λ_{em}) of 489 nm (*see* **Note 18**), optimize the slit width settings, as described next.
3. For the excitation slit width, the goal is to maximize the elicited fluorescence while avoiding photobleaching of the ThT that leads to signal degradation. Choose an interim value for the emission slit width (perhaps 2 nm) and a starting excitation slit width of 1 nm.
4. Gently mix (gentle vortexing or inversion) the fibril-ThT mixture just before taking each reading. Collect fluorescence over a 1 min time period. Fluorescence values may fluctuate due to migration of large fibril particles through the beam, but the average signal over repeated 5 or 10 s increments should essentially remain constant. If this is confirmed, increase the excitation slit width to 2 nm and repeat. Continue this protocol, expanding slit width in small increments, until there is a clear time dependent decay in the short-increment signal average. This indicates that the incident light is causing ThT photobleaching, leading to signal loss and therefore errors in the ThT fluorescence. Drop back from this slit width to one that gives a high, but stable, fluorescence signal.
5. For the emission slit width, the goal is to maximize the elicited fluorescence while avoiding saturation of the fluorescence detector, which will dampen the signal strength at high intensity and compromise linearity. Set the excitation slit width to the optimized value obtained in **step 4** of Subheading 3.2, and set the initial emission slit width to 1 nm. Gently mix the fibril-ThT mixture just before taking each reading. Iteratively determine fluorescence signals at increasing emission slit widths until fluorescence yield no longer increases linearly with slit width increases. Drop back in slit width to the top of the linear range.
6. You are now ready to monitor ThT fluorescence by the standard ex situ approach on an ongoing amyloid assembly reaction. First, determine the fluorescence intensity of 10 μM ThT alone in the ThT buffer of choice.

7. Determine the volume of aliquots of the ongoing reaction mixture that will be required to deliver 1 μg of total peptide for ThT readings, and adjust the required ThT concentration and volume accordingly (*see* **Note 19**).
8. Follow the aggregation reaction by taking aliquots of the reaction mixture at defined reaction times (the same time points as are being analyzed by the sedimentation/HPLC assay) and determining the ThT fluorescence under standard conditions.
9. Record the fluorescence intensity at each reaction time point (*see* **Note 21**).

3.3 Weight-Normalized ThT Values

1. From the monomer concentrations determined for each time point obtained by the sedimentation / HPLC assay (Subheading 3.1), calculate the weight fraction of peptide aggregated (AgF_{wt}), which is [weight concentration of monomer at time t_0—weight concentration of monomer at time t_n]/weight concentration of monomer at time t_0. Where M = monomer, this is equivalent to:

$$\mathrm{AgFwt}_{t=n} = \left([\mathrm{M}]_{t=0} - [\mathrm{M}]_{t=n}\right)/[\mathrm{M}]_{t=0}$$

2. For each time point, obtain ThT_{net} by subtracting the blank fluorescence value (ThT_{bl}) obtained for ThT plus buffer from the sample ThT fluorescence intensity (ThT_{spl}) obtained for amyloid plus ThT plus buffer:

$$\mathrm{ThT}_{\mathrm{net}} = \mathrm{ThT}_{\mathrm{spl}} - \mathrm{ThT}_{\mathrm{bl}}$$

3. For each time point, calculate the aggregate weight-normalized ThT (AWN-ThT) value by dividing ThT_{net} by AgFwt:

$$\text{AWN-ThT} = \mathrm{ThT}_{\mathrm{net}}/\mathrm{AgF}_{\mathrm{wt}}$$

These values can be plotted with respect to aggregation time to get a picture of how aggregate morphology changes during aggregate assembly (*see* Figs. 1, 2, and 3). Formally these values can be expressed as arbitrary fluorescence units per aggregate concentration. Alternatively, if one defines one's standard assay conditions to include a specific ThT solution volume in the cuvette, the unit is arbitrary fluorescence units per weight of aggregate.

3.4 Interpretation of AWN-ThT Values and Time Courses

1. AWN-ThT values and time courses often provide significantly more detail on the nature of an aggregate population compared with traditional ThT intensity data (for a comparison, see the two ThT curves in Fig. 1). Thus, for a reaction end-point that

Table 1
Properties of $A\beta_{40}$ polymorphic fibrils[a,b]

Polymorphic form	A	B	C	D	E	Z
AWN-ThT	55 ± 10	35 ± 4	242 ± 25	17 ± 3	190 ± 25	12 ± 2
C_r, μM[c]	1.06 ± 0.18	20.3 ± 2.5	0.22 ± 0.06	1.35 ± 0.07	0.13 ± 0.03	1.36 ± 0.15
N-H exchanged[d]	16.7 ± 1.3	24.2 ± 0.2	10.5 ± 0.11	13.8 ± 0.32	14.1 ± 0.44	24.2 ± 0.26

[a]Data from [18]
[b]Monomeric Aβ(1–40) incubated as follows: PBS, 37 °C, no agitation (A); phosphate, 24 °C, no agitation (B); PBS, 24 °C, with agitation (C); PBS, 4 °C, with agitation (D); phosphate, NaCl, SDS, 37 °C, no agitation (E); Tris–HCl, NaCl, $ZnCl_2$, 37 °C, no agitation (Z)
[c]Monomer concentration of amyloid assembly reaction at equilibrium
[d]Hydrogen-deuterium exchange of isolated aggregate, corrected for artifactual exchange

has reached equilibrium (no further change in AWN-ThT with time) and which appears to be a uniform amyloid product by EM, FTIR, X-ray fiber diffraction, etc., the AWN-ThT value can be taken to be a characteristic value for the particular amyloid polymorph formed in the reaction (Table 1). In addition, a spontaneous amyloid formation reaction profile that features a sudden, dramatic rise of the AWN-ThT value, just as aggregates start to form, to a value at or near that of the final product can be taken as an indication of the absence of any oligomeric intermediates on the self-assembly pathway (Fig. 2). In contrast, a spontaneous amyloid formation reaction profile exhibiting an early stage during which sedimentable aggregates form that exhibit no ThT response (Fig. 3) can be taken to indicate the initial formation of oligomeric aggregates with no amyloid-like β-structure. Further, a time course featuring a very gradual increase in AWN-ThT values that parallels the increase in sedimentable aggregates might suggest the presence along the time course of one or more aggregates with structures that may be amyloid-like to some degree, but which are not identical to mature fibrils. Whether or not these implied intermediates (which should be detectible by other means, such as light scattering and electron microscopy) are on- or off-pathway cannot be deduced solely from the AWN-ThT time course.

2. While providing a significant improvement over a raw ThT fluorescence time course (e.g., Fig. 1, □), and while providing important new insights into self-assembly, some AWN-ThT profiles remain difficult to fully interpret without additional work. Thus, a time course showing a gradual rise in AWN-ThT (e.g., Figs. 1 (■) and 3 (■)) over time must represent either (a) a series of homogeneous aggregate populations with steadily increasing AWN-ThT values, (b) a series of mixtures of aggregate types that shift over time in the relative

proportions of the different aggregates represented, or (c) a combination of a and b. While difficult to interpret in terms of the assembly mechanism, such profiles can be seen as an indication that the reaction is not a simple case of an early nucleation event that generates a mature amyloid fibril, followed by the elongation (with perhaps aggregate-dependent secondary nucleation [33, 34]) of that fibril. If this were the case, in spite of the small mass of aggregate in the early stages of the reaction, the AWN-ThT response would already be maximized or nearly maximized (Fig. 2, ■), in that it represents the structure of the mature fibril.

3. By adopting the procedures described in this chapter laboratories working on protein aggregation will obtain new tools for better understanding the nature of their self-assembly reactions and aggregated products and intermediates. The ratios of these AWN-ThT values for different polymorphic amyloids and of intermediates to final products should be universal. The absolute values of AWN-ThT values, however, will not be quantitatively comparable between laboratories or instruments, since ThT fluorescence is normally reported in terms of arbitrary intensity units. This use of arbitrary fluorescence units is necessitated not only because of lack of standardization of parameters such as λ_{ex}, λ_{em}, and slit widths, but equally importantly because of great variability among fluorescence spectrometers. The ability to quantitatively compare ThT results between instruments and laboratories is an important goal, and is achievable. One method for accomplishing this will be described in a subsequent publication (S. Mukherjee, R. Kodali and R. Wetzel, Ms. in preparation).

4 Notes

1. The analytical HPLC system should include the following features: (a) two-pump system capable of delivering variations of the standard gradients used for peptide chromatography with stable, low-noise UV–Vis baselines; (b) sensitive UV–Vis detector capable of obtaining readings in the absorbance range of the peptide bond, i.e., ~A_{215}; (c) system software capable of programmable control of automatic sample injection and gradient control, as well as data collection and both automatic and manual data analysis (i.e., peak integration). In addition, it is preferable, but not required, that the system includes (a) an autosampler (preferably with a cooling function) for programmed analysis of multiple kinetics time points; (b) pumps and peripherals capable of driving analysis in microbore columns, which greatly speeds analysis time; (c) capability to vary injection loop size in order to analyze very low concentration samples by injecting relatively large volumes.

2. In general, reverse phase HPLC media should be of small, spherical, uniformly sized particles for high resolution separations, with chemistries chosen depending on the peptide or protein being investigated.
3. Minimum requirements are a medium pressure pump for driving the size exclusion column, a UV detector for monitoring peptide elution, and pump components in contact with solvent that are stable to corrosive salts, such as titanium or PEEK.
4. SEC column requirements for use in disaggregation include: (a) a size fractionation range capable of baseline separation of small oligomers, monomer, and the low-MW salt fraction; (b) stability to transient exposure to high concentrations of denaturants like Gdn.HCl; (c) low nonspecific adsorption of peptide monomer by the column support; and (d) ability to be cleaned and regenerated if aggregated protein builds up within the column after disaggregation of peptides. It is advisable that good recovery of monomer from the chosen column (c) be confirmed by using the sensitive concentration determination procedures described here.
5. Concentration should be low enough to obtain accurate dispensation by the autosampler when different volumes are delivered to supply different masses of peptide injected. Typically, our stock solutions for generating standard curves are in the 50–250 μg/mL range, for convenient delivery of samples for HPLC analysis ranging from 0.5 to 20 μg, which is the linear range for several peptides studied in this way.
6. Classically, techniques such as amino acid composition analysis (which depends on the modification and quantitation of amino acids released from a protein sample by hydrolysis), are considered the gold standard for determining protein concentrations. However, we have struggled to obtain consistent and realistic results with this method when applied to amyloidogenic peptides, possibly because of a tendency of the peptides to adsorb to surfaces during preparation of the highly diluted samples required for the technique. The alternative recommended method here is to use independently determined extinction coefficients for each amino acid type to calculate a coefficient for the entire peptide, at a time just after purification when it is denatured and well solvated in the HPLC elution buffer.
7. Do not use the calculated extinction coefficient to directly interpret peak areas from HPLC runs, as this would fail to account for nonquantitative and variable recovery of peptide mass in the chromatography. This can be substantial, especially for hydrophobic peptides, and especially at low loads of peptide mass. Recovery will vary with the peptide and with the

chromatography conditions. Using a standard curve generated from the peak areas recovered from the chromatography corrects for the variable recovery. A typical weight range for generation of a standard curve is 1–10 μg. The low end of the useful range is limited by the increased percentage of peptide adsorbed nonspecifically on the resin as well as by detector sensitivity. The high end of the range is limited by column capacity and potentially by the limits of detector linearity.

8. When constructing a standard curve, start by analyzing the smallest amount of peptide and work toward the largest, to avoid errors due to peptide carryover between runs. This is not practical when following a reaction time course, however, since monomer concentrations start high and tail off. However, under such conditions, especially when higher masses of peptide are being analyzed, one can and should run blank gradients between each peptide analysis to suppress carryover.

9. Different peptides will require different disaggregation protocols, depending on the peptide's character and the investigator's priorities (*see* **Note 10**). The pI of the peptide is an important variable. During the disaggregation and workup, it is critical to avoid even transient exposure to pH values similar to the pI. Thus, for Aβ peptides it is important to avoid mild acid, while mild acid is fine for polyglutamine peptides flanked with lysine residues [25]. One typical disaggregation protocol [25] involves sequential exposure to a mixture of TFA and HFIP, gentle removal of solvent through a nitrogen stream, exposure to vacuum, resuspension of the dried film in water of a pH distant from the pI, centrifugation and/or membrane filtration to remove fine particulates (potential seeds). After quick determination of the concentration of the resulting supernatant/filtrate, the aggregation reaction is initiated by addition of buffer components and adjustment to the desired incubation temperature.

10. While there are a number of effective disaggregation protocols in the literature, we feel it is critical to avoid a situation where there is residual organic solvent in the final "disaggregated" peptide solution. This includes protocols involving dilution into buffer from a DMSO or HFIP solution. It is also important to avoid protocols using evaporation from organic solvents that do not include final exposure to strong (and dry; i.e., not a water aspirator) vacuum, since even undetectable levels of organic solvent, tightly bound to the peptide, can influence aggregation reactions, for example, by inducing oligomer formation. An excellent disaggregation method involves dissolution of the peptide in a high, denaturing concentration of Gdn. HCl followed directly by preparative size exclusion chromatography with isolation of the monomer peak [30]. One

disadvantage of this method is that it is not easily downward scalable, making it difficult to carry out reactions where only a small amount of peptide is available or required. Our published protocol [25] is more downward scalable, and should be effective so long as there is strict adherence to the procedure. With the SEC disaggregation method, it is important to avoid prolonged incubation of the disaggregated monomer before starting the aggregation reaction, since oligomerization can begin as soon as the monomer emerges from the SEC column.

11. In the absence of separation challenges, the ideal gradient will be optimized to minimize turnaround time between injections. Microbore HPLC systems can greatly shorten such turnaround times. But if there are chemical decomposition products, such as from Met oxidation, generated during aggregation reactions, it is important to separate these, since they may well have their own, less favorable, K_a for fibril growth, and if not analytically separated from the intact peptide, their contribution to the monomer peak will distort aggregation kinetics and the position of equilibrium at the end of the reaction. A test incubation should be run and a sample of the monomer peak from late in the aggregation reaction should be analyzed by mass spectrometry and compared to starting peptide to ensure there are no unseparated decomposition products. Similarly, if there are plans to co-incubate the peptide of interest with other molecules with A_{215} absorption, the separation gradient should be challenged for its ability to separate these molecules. Inject each molecule separately, then together, to probe the quality of the separation. If co-elution of peptides is observed, change the separation protocol (different gradient, different solvents, different column material) until (preferably) baseline separation is achieved. It should be noted that a great strength of using analytical HPLC to monitor aggregation reactions is this separation ability, which is not possible if centrifugation supernatants are simply assayed for total protein or total UV absorbance, or if monomers are quantified using SEC peaks.

12. The effectiveness of centrifugation conditions to remove aggregates depends on the sedimentation parameters of the aggregates [35]. Thus, inclusions and clusters of amyloid can effectively be removed by a relatively low speed centrifugation [36], and most fibrils of significant size can be removed in an Eppendorf centrifugation [28]. However, isolated small amyloid fibrils [28] and non-β oligomers [37] likely require more stringent conditions, and we recommend routine centrifugation in a tabletop ultracentrifuge to optimize removal. It is possible that even these conditions may be insufficient to remove small multimers of small peptides [35]. One test for the efficacy of centrifugation conditions is to expose a

supernatant to filtration with a very small pore membrane, then to use the highly accurate analytical HPLC analysis to compare the peptide concentration before and after filtration [38]. If there is no significant difference, one can exclude the possibility that oligomers larger than the nominal pore size of the membrane are being counted as part of the low MW or monomer fraction. However, some peptide monomers, such as various huntingtin exon1-related peptides, appear to bind tightly to some membranes (M. Jayaraman and B. Sahoo, unpublished observations), which would invalidate this test. This artifact can be tested by analyzing freshly disaggregated monomer before and after membrane filtration. Another approach to optimizing the accuracy of estimating aggregate formation is to determine monomer concentrations of time points by integrating monomer peaks in SEC (K.W. Drombosky, R. Kodali and R. Wetzel, Ms. submitted) with the caveat that degraded peptides and added peptide inhibitors will compromise the measurement.

13. Some spontaneous amyloid formation reactions proceed very slowly (i.e., months or longer) at convenient and accessible concentrations [38]. While exposing such peptide solutions to stresses such as high heat or agitation may overcome the normal barriers to nucleation and allow data to be obtained, this is not very helpful if one is focused on understanding nucleation under more physiological conditions. If the concentration dependence of nucleation is steep (reflecting a multimeric nucleus in a classical nucleated growth process), it is possible to speed things up by greatly increasing concentration. This, however, may introduce other problems if the peptide is expensive and/or difficult to come by. It is possible to minimize material requirements in such high concentration reactions, while maintaining accuracy and reproducibility, by incubating relatively small volumes of highly concentrated peptide and monitoring loss of monomer in the sedimentation/HPLC assay by a modified workup protocol involving gravimetric determination of small aliquot volumes using an analytical balance [38]. Especially for long incubation of small volumes, it is also obviously important to have complete control over evaporation losses during the long reaction time. Volume losses through evaporation can be minimized by sealing the reaction tube, incubating in a water vapor equilibrated environment, and using a pulse spin in an Eppendorf followed by gentle swirling to recover condensed water from the inner lid. Volume losses that happen in spite of these precautions can be monitored by tracking the weight loss (analytical balance) of the reaction tube and comparing that to the total aliquot volumes removed over time.

14. If there are many samples of sedimentation supernatants to be analyzed, it is very convenient to queue up the samples in an autosampler tray for programmed injections. To minimize loss of monomer through aggregation during the wait time in the autosampler, the autosampler tray should be cooled; this is an option with some commercial autosamplers. It may also be necessary to adjust the pH of the samples to further minimize degradation. Strong acid or base will normally discourage aggregation, but may also lead to chemical changes to the peptide which will complicate interpretation of the resulting HPLC elution profile. Weak acid, on the other hand, may actually encourage further aggregation by taking some peptides closer to their pI value, as in the case of Aβ peptides [12]. It is important to optimize and customize autosampler stabilization conditions for the subject peptide. We have found that addition of formic acid to 20% stabilizes aggregation reaction time points for at least 20 h [25].
15. Variability in the time lags before aggregation onset in spontaneous amyloid formation kinetics are often ascribed to the stochastic nature of the nucleation process. However, stochastic variability is only to be expected in cases where only a few nuclei are projected to exist at any one time—which is actually pretty unlikely under most in vitro experimental conditions [39]. Thus, a solution of 10 μM peptide with (monomeric) nucleation equilibrium constant K_{n*} of 10^{-10} [40] is projected to populate in pseudo-steady-state about 600,000 independent nuclei in a 1 mL reaction volume. In this hypothetical example, only if the K_{n*} were in the range of 10^{-15} to 10^{-16}, yielding an average of 0.6–6 nuclei projected per 1 mL of a 10 μM sample, would nucleation kinetics be susceptible to great variability in nucleus formation kinetics. This means that spontaneous aggregation kinetics under most in vitro circumstances normally can be expected to be reproducible, and that an observed lack of reproducibility must therefore be attributed to lapses in experimental protocols and technique, such as incomplete disaggregation leading to presence of seeds in some samples. To confirm reproducibility of nucleation, one should set up complete replicates of reactions. To confirm reproducibility of sample workup, it is sufficient to take multiple aliquots of a single aggregation reaction at each time point.
16. Not all amyloid formation reactions are expected to be demonstrably reversible [4], either for those cases in which amyloid formation involves covalent bond formation between peptide subunits, or because the rate of fibril dissociation is prohibitively slow, or because the concentration of monomer at equilibrium is too low to reliably measure. However, for those systems featuring an accessible equilibrium point, the

determination of the equilibrium concentration of monomer provides valuable information on fibril stability [23]. Excessive manipulation of amyloid fibrils (freeze/thaw, sedimentation/resuspension, etc.) can compromise their seeding/dissociation properties. To characterize fibril dissociation, dilute an aliquot of an ongoing assembly reaction as it nears completion. The dilution should be enough to reduce the monomer concentration to 5–10 fold below the expected equilibrium concentration, but at the same time maintain the *total* peptide concentration (i.e., monomer plus fibril) well above the expected equilibrium concentration to ensure that fibrils will still be present at equilibrium. Fibril dissociation can take days to months, so it is important to appropriately plan for the length of the experiment, in terms of ensuring sufficient material for many time points, and including a bacteriostatic such as sodium azide.

17. Although the values of 1 μg peptide and 10 μM ThT will work for many amyloid reactions analyzed in many fluorescence formats, these quantities may have to be adjusted for some systems. Some fibrils give relatively poor ThT responses, necessitating that larger mass concentrations of fibrils be used in order to obtain good signal to noise. Some fibrils may require higher concentrations of ThT to saturate the response. To minimize sample requirements, choose a micro cuvette with a minimum necessary volume in the 100 μL range, if available. Amyloid fibrils of different peptides can exhibit mass-normalized ThT signals that vary by at least a factor of 20 [18]. Using intermediate slit width settings find a level of amyloid concentration giving adequate signal to noise. ThT binding/fluorescence normally saturates amyloid fibrils at 10 mM ThT or below. Once amyloid mass has been set, one can explore ThT concentrations to find the point of saturation. Avoid excessive concentrations of ThT, since this could increase buffer blank noise without significantly increasing amyloid signal.
18. The reader will find a variety of λ_{ex} and λ_{em} values listed in the literature, and many combinations giving similar performance are possible. Of course, once a pair of wavelengths is selected, it will be important to use these consistently within the lab to allow comparisons to past work. The absence from the literature of a standardized set of wavelengths is somewhat unfortunate. It has happened because there is no great incentive to standardization, since even if everyone used the same wavelengths it would not be possible to compare fluorescence intensities between laboratories and instruments (*see* **step 3** in Subheading 3.4).

19. For example, for a routine final volume for the ThT measurement of 100 μL: if 10 μL of the aggregation reaction is required to deliver a total of 1 μg peptide, then prepare a stock solution of 11.1 μM ThT and deliver 90 μL of that stock into the cuvette, followed by 10 μL of the reaction mixture (*see* **Note 20**) to yield a final [ThT] of 10 μM, gently mix, and read the fluorescence. To maximize correctness of the blank, construct the ThT blank in the same way, mixing 90 μL of the 11.1 μM ThT solution with 10 μL of the aggregation reaction buffer.
20. Calibrated micropipettes are not always accurate at low volumes, especially if delivering aggregate suspensions. The lower the aliquot volume, the more serious the challenge to reproducible delivery of correct volumes. To maximize accuracy if low volume aliquots are required, pipet the desired volume into a tared Eppendorf tube and reweigh to get the weight, and hence volume, of aliquot. Then, add ThT in buffer, adjusting the volume of ThT solution to achieve the required weight concentration of aggregates. It is also possible to improve accuracy by taking, and appropriately diluting, a larger volume of reaction mixture than is required to fill the cuvette. Maximizing the accuracy of each step in the protocol is important if optimal results are to be obtained.
21. The time course of fluorescence intensity obtained will be the standard type of ThT time course, which is an integral of the weight concentrations of each type of aggregate present at a particular time point, and the weight-normalized ThT fluorescence intensity of that aggregate type. The time course will trend upward, except that at very long time points, after amyloid assembly is complete, it may begin to trend downward. This effect, often observed during in vitro assembly studies (*see*, for example, [41]), presumably reflects some further structural change in the aggregate mass, possibly a "super-aggregation" of fibrils that may block some ThT binding sites or lead to uneven distributions of aggregates in the cuvette. Since this is not well understood, the assembly reaction is generally considered complete when the ThT signal is maximized, and the fibril product is correspondingly more well-behaved.

Acknowledgment

We acknowledge financial support from N.I.H. grant R01 GM099718.

References

1. Wetzel R (1994) Mutations and off-pathway aggregation of proteins. Trends Biotechnol 12(5):193–198. https://doi.org/10.1016/0167-7799(94)90082-5
2. Tyedmers J, Mogk A, Bukau B (2010) Cellular strategies for controlling protein aggregation. Nat Rev Mol Cell Biol 11(11):777–788. https://doi.org/10.1038/nrm2993
3. Buxbaum JN, Linke RP (2012) A molecular history of the amyloidoses. J Mol Biol 421 (2–3):142–159. https://doi.org/10.1016/j.jmb.2012.01.024
4. Krieg E, Bastings MMC, Besenius P, Rybtchinski B (2016) Supramolecular polymers in aqueous media. Chem Rev 116 (4):2414–2477. https://doi.org/10.1021/acs.chemrev.5b00369
5. Chick H, Martin CJ (1912) On the "heat coagulation" of proteins. Part IV. The conditions controlling the agglutination of proteins already acted on by hot water. J Physiol 45:261–295
6. Mirsky AE, Pauling L (1936) On the structure of native, denatured and coagulated protein. Proc Natl Acad Sci U S A 22:439–447
7. Glenner GG, Terry W, Harada M, Isersky C, Page D (1971) Amyloid fibril proteins: proof of homology with immunoglobulin light chains by sequence analyses. Science 172 (988):1150–1151
8. Eanes ED, Glenner GG (1968) X-ray diffraction studies on amyloid filaments. J Histochem Cytochem 16(11):673–677
9. Cohen AS, Calkins E (1959) Electron microscopic observations on a fibrous component in amyloid of diverse origins. Nature 183:1202–1203
10. London J, Skrzynia C, Goldberg ME (1974) Renaturation of Escherichia coli tryptophanase after exposure to 8 M urea. Evidence for the existence of nucleation centers. Eur J Biochem 47(2):409–415
11. Helms LR, Wetzel R (1996) Specificity of abnormal assembly in immunoglobulin light chain deposition disease and amyloidosis. J Mol Biol 257(1):77–86. https://doi.org/10.1006/jmbi.1996.0148
12. Wood SJ, Maleeff B, Hart T, Wetzel R (1996) Physical, morphological and functional differences between ph 5.8 and 7.4 aggregates of the Alzheimer's amyloid peptide Abeta. J Mol Biol 256(5):870–877. https://doi.org/10.1006/jmbi.1996.0133
13. Walsh DM, Lomakin A, Benedek GB, Condron MM, Teplow DB (1997) Amyloid beta-protein fibrillogenesis. Detection of a protofibrillar intermediate. J Biol Chem 272 (35):22364–22372
14. Harper JD, Wong SS, Lieber CM, Lansbury PT (1997) Observation of metastable Abeta amyloid protofibrils by atomic force microscopy. Chem Biol 4(2):119–125
15. Chemuru S, Kodali R, Wetzel R (2016) C-terminal threonine reduces Ab43 amyloidogenicity compared with Ab42. J Mol Biol 428 (2 Pt A):274–291. https://doi.org/10.1016/j.jmb.2015.06.008
16. Goldsbury CS, Cooper GJ, Goldie KN, Muller SA, Saafi EL, Gruijters WT, Misur MP, Engel A, Aebi U, Kistler J (1997) Polymorphic fibrillar assembly of human amylin. J Struct Biol 119(1):17–27
17. Petkova AT, Leapman RD, Guo Z, Yau WM, Mattson MP, Tycko R (2005) Self-propagating, molecular-level polymorphism in Alzheimer's b-amyloid fibrils. Science 307 (5707):262–265
18. Kodali R, Williams AD, Chemuru S, Wetzel R (2010) Abeta(1-40) forms five distinct amyloid structures whose beta-sheet contents and fibril stabilities are correlated. J Mol Biol 401 (3):503–517. https://doi.org/10.1016/j.jmb.2010.06.023
19. Naiki H, Higuchi K, Hosokawa M, Takeda T (1989) Fluorometric determination of amyloid fibrils *in vitro* using the fluorscent dye, thioflavine T. Anal Biochem 177:244–249
20. Klunk WE, Engler H, Nordberg A, Wang Y, Blomqvist G, Holt DP, Bergstrom M, Savitcheva I, Huang GF, Estrada S, Ausen B, Debnath ML, Barletta J, Price JC, Sandell J, Lopresti BJ, Wall A, Koivisto P, Antoni G, Mathis CA, Langstrom B (2004) Imaging brain amyloid in Alzheimer's disease with Pittsburgh compound-B. Ann Neurol 55 (3):306–319
21. Naiki H, Higuchi K, Nakakuki K, Takeda T (1991) Kinetic analysis of amyloid fibril polymerization in vitro. Lab Investig 65 (1):104–110
22. Collins SR, Douglass A, Vale RD, Weissman JS (2004) Mechanism of prion propagation: amyloid growth occurs by monomer addition. PLoS Biol 2(10):e321
23. O'Nuallain B, Shivaprasad S, Kheterpal I, Wetzel R (2005) Thermodynamics of abeta(1-40) amyloid fibril elongation. Biochemist 44 (38):12709–12718
24. Chen S, Berthelier V, Hamilton JB, O'Nuallain B, Wetzel R (2002) Amyloid-like

features of polyglutamine aggregates and their assembly kinetics. Biochemist 41 (23):7391–7399

25. O'Nuallain B, Thakur AK, Williams AD, Bhattacharyya AM, Chen S, Thiagarajan G, Wetzel R (2006) Kinetics and thermodynamics of amyloid assembly using a high-performance liquid chromatography-based sedimentation assay. Methods Enzymol 413:34–74. https://doi.org/10.1016/S0076-6879(06)13003-7
26. Biancalana M, Koide S (2010) Molecular mechanism of Thioflavin-T binding to amyloid fibrils. Biochim Biophys Acta 1804 (7):1405–1412
27. LeVine IIIH (1993) Thioflavine T interaction with synthetic Alzheimer's disease b-amyloid peptides: detection of amyloid aggregation in solution. Protein Sci 2:404–410
28. Chen S, Ferrone F, Wetzel R (2002) Huntington's disease age-of-onset linked to polyglutamine aggregation nucleation. Proc Natl Acad Sci U S A 99:11884–11889
29. Misra P, Kodali R, Chemuru S, Kar K, Wetzel R (2016) Rapid α-oligomer formation mediated by the Aβ C-terminus initiates an amyloid assembly pathway. Nat Commun 7:12419
30. Jan A, Hartley DM, Lashuel HA (2010) Preparation and characterization of toxic Abeta aggregates for structural and functional studies in Alzheimer's disease research. Nat Protoc 5 (6):1186–1209. https://doi.org/10.1038/nprot.2010.72
31. Kuipers BJ, Gruppen H (2007) Prediction of molar extinction coefficients of proteins and peptides using UV absorption of the constituent amino acids at 214 nm to enable quantitative reverse phase high-performance liquid chromatography-mass spectrometry analysis. J Agric Food Chem 55(14):545–551
32. Jayaraman M, Kodali R, Sahoo B, Thakur AK, Mayasundari A, Mishra R, Peterson CB, Wetzel R (2012) Slow amyloid nucleation via alpha-helix-rich oligomeric intermediates in short polyglutamine-containing huntingtin fragments. J Mol Biol 415(5):881–899. https://doi.org/10.1016/j.jmb.2011.12.010
33. Ferrone F (1999) Analysis of protein aggregation kinetics. Methods Enzymol 309:256–274
34. Knowles TP, Waudby CA, Devlin GL, Cohen SI, Aguzzi A, Vendruscolo M, Terentjev EM, Welland ME, Dobson CM (2009) An analytical solution to the kinetics of breakable filament assembly. Science 326(5959):1533–1537. https://doi.org/10.1126/science.1178250
35. Mok YF, Howlett GJ (2006) Sedimentation velocity analysis of amyloid oligomers and fibrils. Methods Enzymol 413:199–217. https://doi.org/10.1016/S0076-6879(06)13011-6
36. Sahoo B, Arduini I, Drombosky KW, Kodali R, Sanders LH, Greenamyre JT, Wetzel R (2016) Folding landscape of mutant Huntingtin exon1:diffusible multimers, oligomers and fibrils, and no detectable monomer. PLoS One. https://doi.org/10.1371/journal.pone.0155747
37. Williams AD, Sega M, Chen M, Kheterpal I, Geva M, Berthelier V, Kaleta DT, Cook KD, Wetzel R (2005) Structural properties of Abeta protofibrils stabilized by a small molecule. Proc Natl Acad Sci U S A 102(20):7115–7120
38. Kar K, Jayaraman M, Sahoo B, Kodali R, Wetzel R (2011) Critical nucleus size for disease-related polyglutamine aggregation is repeat-length dependent. Nat Struct Mol Biol 18 (3):328–336. https://doi.org/10.1038/nsmb.1992
39. Wetzel R (2006) Nucleation of huntingtin aggregation in cells. Nat Chem Biol 2 (6):297–298
40. Bhattacharyya AM, Thakur AK, Wetzel R (2005) Polyglutamine aggregation nucleation: thermodynamics of a highly unfavorable protein folding reaction. Proc Natl Acad Sci U S A 102(43):15400–15405
41. Morimoto A, Irie K, Murakami K, Masuda Y, Ohigashi H, Nagao M, Fukuda H, Shimizu T, Shirasawa T (2004) Analysis of the secondary structure of beta-amyloid (Abeta42) fibrils by systematic proline replacement. J Biol Chem 279(50):52781–52788

Chapter 7

Nanoparticle Tracking for Protein Aggregation Research

Xiaomeng Lu and Regina M. Murphy

Abstract

Protein aggregates are the pathological agents in several neurodegenerative disorders such as Alzheimer's and Huntington's disease. In the pharmaceutical industry, protein aggregation poses significant challenges to the manufacture of biologics. Nanoparticle tracking is an emerging technology that allows particle-by-particle measurement of aggregate size and concentration. The technique is solution based, and requires no labeling. Here we describe protocols for using nanoparticle tracking in protein aggregation research, and provide a few examples for illustrative purposes.

Key words Nanoparticles, Protein aggregation, Amyloid, Light scattering, Diffusion

1 Introduction

Aggregation of proteins, particularly into amyloid fibrils, is of considerable interest from a clinical perspective because of its role in causing neurodegenerative disorders such as Alzheimer's and Parkinson's. On a different front, manufacture of protein biologics for pharmaceutical applications is a growing business, where unwanted protein aggregation raises concerns about low yield, reduced efficacy, and risk of immunogenicity. Researchers are motivated to investigate diverse aspects of protein aggregation including: solution conditions that cause aggregation, formulations that prevent aggregation, the role of mutations in aggregation propensity, the size and morphology of aggregates, and the kinetics and mechanisms of growth. Experimental investigation of protein aggregation is challenging. Aggregates are polydisperse in size, with a broad and sometimes multimodal distribution. They can range in morphology from fibrillar to amorphous. Preferably, techniques allow characterization of aggregates without labels, because labels can interfere with or redirect aggregation, and in solution, because surface interactions can affect aggregation. Most desirable are methods that can be carried out noninvasively and nondestructively, that can be used with a variety of solvents, and that allow

Bradley L. Nilsson and Todd M. Doran (eds.), *Peptide Self-Assembly: Methods and Protocols*, Methods in Molecular Biology, vol. 1777, https://doi.org/10.1007/978-1-4939-7811-3_7,

measurements over time and over a range of concentrations. Experimental techniques for measuring aggregate concentration, size and morphology include size exclusion chromatography (SEC), dye-binding assays, electron and atomic force microscopy, and dynamic light scattering (DLS). Each technique provides distinctly different, and complementary, views of the "elephant," and each has advantages and limitations [1].

In this chapter we focus on nanoparticle tracking (NTA), a newer technique that holds much promise to fill a unique niche. In NTA, particles in the sample scatter light from an incident laser, and the scattered light is tracked using a CCD camera (Fig. 1a).

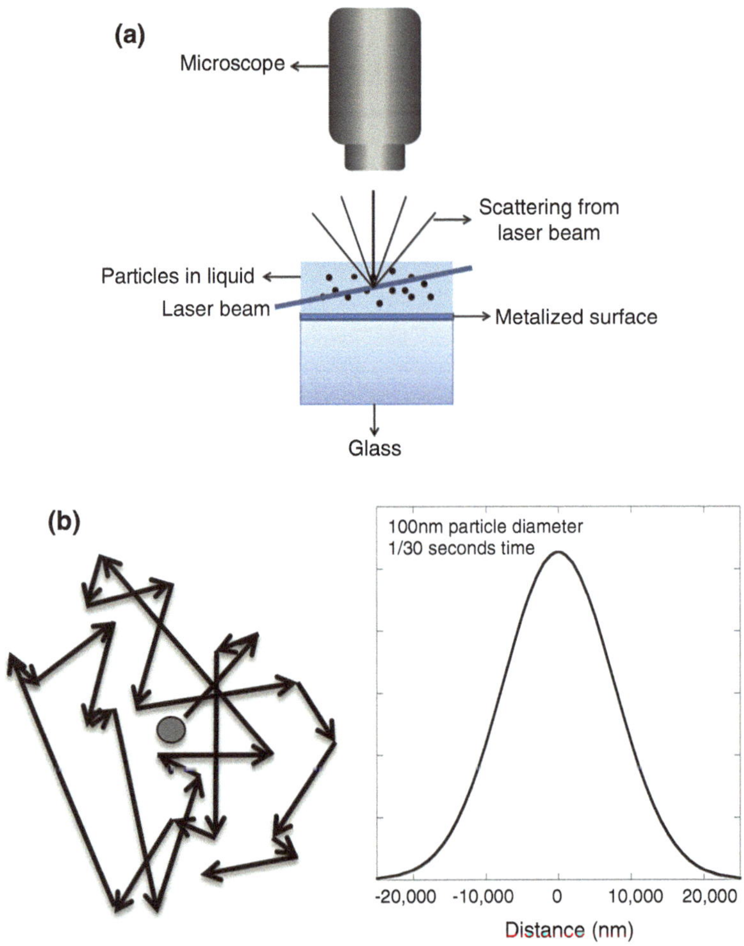

Fig. 1 (**a**) A schematic illustrating the principle of NTA operation. Light from the incident laser beam is scattered by the particles in solution, and the scattered light is captured by a camera perpendicular to the fluid plane. (**b**) Snapshots of a particle undergoing Brownian motion create a trajectory; the displacement *r* can be measured at each time interval *t*. The theoretical distribution of displacements follows a bell-shaped curve, which broadens with longer time intervals. The mean distance is zero, but the root-mean squared distance is not. For a 100 nm diameter particle diffusing in two dimensions in water at 298 K, with 1/30 s time intervals, $\sqrt{\overline{r^2}} = 750$ nm

Submicron particles move in random directions due to Brownian motion. By measuring the two-dimensional trajectory of the diffusing particle, the diffusion coefficient D is obtained:

$$\overline{r^2} = 4Dt \qquad (1)$$

where $\overline{r^2}$ is the mean squared displacement of the particle over a tracking time t. In NTA, a particle is detected because it scatters light, and its trajectory is tracked for a few seconds and video captured. The video is then evaluated frame by frame (typically at 30 frames per second), to measure the distance moved by the particle in each step, thereby building up a history of the particle motion. In the absence of any force fields or concentration gradients, the mean displacement $\bar{r} = 0$ (as long as sufficient statistics are built up) but the mean-squared displacement $\overline{r^2}$ is not (Fig. 1b). With a greater number of trajectories analyzed, the measured $\overline{r^2}$ approaches the true (statistical) value and therefore a more accurate measurement of D is obtained.

D is related to the hydrodynamic radius R_h of the particles by the Stokes-Einstein equation:

$$R_\mathrm{h} = \frac{kT}{6\pi\eta D} \qquad (2)$$

where k is Boltzmann's constant, T is the temperature, and η is the solvent viscosity. R_h of a spherical particle is simply the particle radius. For nonspherical particles, R_h is the radius of an equivalent spherical particle that has the same diffusivity as the particle of interest. In Eq. 2, it is assumed that particle motion is due to only translational diffusion. For high-aspect particles (such as needles or tubes, where the length-to-diameter ratio is large), rotational diffusion becomes a contributing factor. In NTA, rotational motion can cause "blinking" and distort the interpretation. Other factors that might complicate interpretation of the measured trajectory as simply R_h include flexing or bending motion of long flexible fibrils, or entanglement effects in highly concentrated solutions.

NTA is not an imaging technique. Rather, it detects scattered light from each particle. The intensity of the scattered light depends on the scattering cross section of the particle as well as the intensity of the incident beam. The scattering cross section depends on the particle size and its refractive index increment—a measure of the extent to which the particle scatters light in excess of the solvent. Larger particles will scatter light more intensely, as will particles with greater refractive index increment. There is some threshold of scattered intensity under which particles will no longer be detected, which depends on the beam intensity, the sensitivity of the camera, the particle size, and the particle chemistry. In practice, for protein aggregates the minimum size detectable by current NTA technology is ~30 nm. The minimum size is smaller for gold or other strongly scattering materials. On the other side of the size spectrum, aggregates larger than 1000–2000 nm (1–2 μm) diffuse too slowly for reliable measurement; these particles will scatter light strongly but one will not have an accurate measure of their size.

Since NTA is a particle-by-particle counting technique, one can directly determine the number (molar) concentration of particle aggregates. In contrast, SEC or dye-binding assays measure mass concentrations, and DLS provides no information about concentration. Thus, NTA measurements can be particularly advantageous for building kinetic models, because rates of reaction depend on molar and not mass concentration.

NTA and DLS are similar in that both are scattering techniques that measure particle diffusion. In DLS, one obtains an autocorrelation function that is the sum of contributions from all scattering particles in the sample volume. Obtaining size distributions from DLS measurements requires the use of constrained regularization or other statistical techniques, which are not always entirely reliable. Because individual particles are counted in NTA, the size distribution is directly obtained, as long as a sufficient number of particles are counted. The fact that larger particles scatter more strongly is a particular problem with DLS, where the signal from large particles can totally swamp out that from smaller particles. With NTA, this is less of a problem but is still not completely negligible. Because the laser used as the incident beam has a Gaussian distribution of intensity, smaller particles near the edge of the beam may miss detection whereas larger particles will be detected. Or, small particles may be hidden in the "halo" of strongly scattering particles. These factors can affect the accurate measurement of particle number concentrations, and are of particular concern when the sample size distribution is very broad.

One of the unique, but underutilized, advantages of NTA is that individual particles are measured for both size *and* intensity. One can thereby evaluate mixtures of particles with different material properties, as long as they have different refractive indices.

For more detailed information on the theory of nanoparticle tracking and analysis of data, readers are referred to [2] or [3]. Some additional useful references on applications of NTA include [4–10].

2 Materials

1. Buffers

   Water and most aqueous buffers can be used. For example, we have used ultrapure water (distilled, deionized, and filtered water), Tris buffer: 200 mM Tris–HCl, pH 8.4, phosphate-buffered saline PBS: 100 mM NaCl, 10 mM phosphate, pH

7.4, CHES: 10 mM *N*-cyclohexyl-2-aminoethanesulfonic acid, 140 mM NaCl, pH adjusted to 9 with 1 M NaOH, or 8 M urea/0.01 M glycine-NaOH, pH 10. Buffers are filtered through a 0.02 μm Whatman Anotop 10 filter (*see* **Note 1**).

2. Standards

   Nanosphere size standard (NIST Traceable Mean Diameter) polystyrene latex beads in several different sizes (e.g., 60, 100, and 240 nm) are available from Thermo Scientific (Fremont, CA) and can be used without further purification. Gold nanoparticles are available from BBI solutions (Cardiff, UK) and, because of their stronger scattering, are useful at smaller sizes (e.g., 20 nm).

3. Proteins

   Purity should be established by SDS-PAGE or other appropriate techniques. Samples should appear clear and colorless by eye, with no turbidity or sediment. Samples should be filtered through an appropriate size filter to remove contaminants (*see* **Note 2**).

4. Nanosight

   The Nanosight LM10 (Nanosight, Amesbury, UK) equipped with a 405 nm laser was used in all examples described here. Other laser wavelengths including 488 nm (blue), 532 nm (green), or 642 nm (red) are available, which are particularly useful for fluorescence-based methods (which are not discussed here). Two different cameras can be installed on the Nanosight: the standard CCD camera or a high-sensitivity CMOS camera. The higher sensitivity is better for very dilute solutions or smaller particles.

3 Methods

3.1 Instrument Setup, Sample Preparation and Data Collection

1. Assemble viewing unit. If necessary, switch off laser and remove power lead, and clean all parts (*see* **step 6**). Embed O-ring in its channel in the top plate. Lay the top plate onto the embedded optical flat gently. Tighten four metal screws alternately in opposite diagonal pairs. Do not overtighten. Attach the power lead to the viewing unit.
2. Check cleanliness. First check the cleanliness of the sample chamber and buffer preparation with ultrapure water and filtered buffer. Load water or buffer into the sample chamber using a syringe, allowing some excess to spill out of the outlet port. Advance sample slowly and avoid air bubbles (*see* **Note 3**). The observation of a completely blank background is required and represents the absence of scattering particles. If particles are detected in water or buffer, repeat cleaning of the instrument (*see* **step 6**) and filtering of the buffer until no particles are detected.

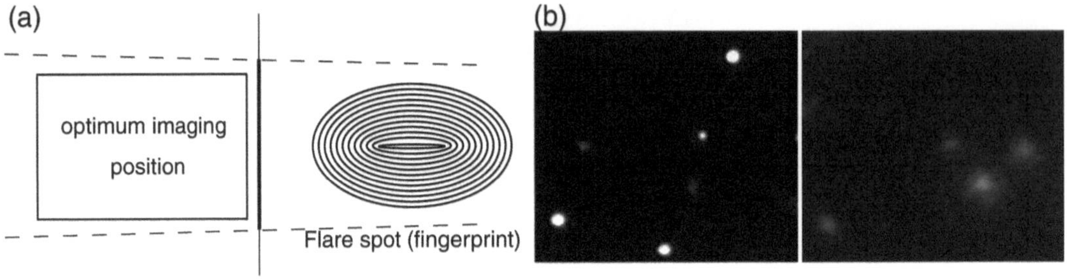

Fig. 2 Illustration of NTA data collection method. (**a**) Find the viewing field. Adjust the position of the microscope objective to find the flare spot (fingerprint) first, then move the objective toward left. The optimum imaging position is the left side of the vertical line. (**b**) Adjust the height of the microscope objective to obtain a clearly visible image of particles present in focus within the beam, as shown in the left panel. Avoid blurry particles shown in the right panel

3. Load sample. Sample volume should be greater than 0.5 mL. Load the sample using the same technique as for water or buffer. Turn on power. The laser beam should be a thin line passing through the sample chamber if the sample is properly prepared and loaded with a proper concentration. Samples may need to be diluted to reach the target particle number concentration (*see* **Notes 4** and **5**).
4. Adjust the microscope and find the viewing field. Place viewing unit onto microscope stage. Adjust the position of microscope objective to find the optimum imaging position. This is illustrated in Fig. 2a. The laser beam path is indicated by the dashed lines. Adjust the *x*- and *y*-axis positions, and find the fingerprint (flare spot). Adjust the *x*-axis position by moving to the left and find a line to the left of the flare spot. Continue moving to the left, and the optimum imaging position is on the left of the line. Then adjust the height of the microscope objective to obtain a clearly visible image of particles present within the beam, in focus. Particles should appear bright and sharp; avoid blurry or indistinct particles (Fig. 2b). It is easier to use the microscope to first find the approximate viewing field and then use the camera to find the optimal viewing field. You will need to adjust focus when switching from microscope to camera view.
5. Choose instrument settings and capture videos. Adjust the camera level so that all particles are visible but the background is dark (not milky). In SOP tab, use the standard measurement script function (five repeats of 60-s capture time, *see* **Note 6**), or load a script of your own (*see* **Note 7**). Run the script to initiate video capture. Be sure to keep the instrument steady and avoid any vibration or jarring. The instrument sits on vibration reduction pads and should be placed on a sturdy table, isolated from walls or any outside source of vibration such as traffic or trains. Check that the temperature in the cell is

recorded automatically, or measured and recorded (depending on instrument model).

6. Switch off the laser and disconnect power. Empty the sample chamber of all residual fluid using a syringe via the Luer ports. Rinse the sample chamber with filtered buffer or water a few times and expel from chamber before disassembling. Unscrew the screws securing the top plate and take out the optical flat. Rinse top plate and optical flat twice with water and ethanol, and then dry with an air can, paying special attention to the input and output Luer ports. Dampen a lens paper with water and carefully wipe once over the surface of the optical flat in situ and top plate. Repeat with a lens paper dampened with ethanol (*see* **Note 8**).

3.2 Video Analysis

1. Start with the recommended default screen gain value at 10.0 and select an appropriate detection threshold. Evaluate the video and decide "by eye" what is a particle. Adjust the detection threshold so that every particle that you identify visually is highlighted with a red cross (+) located at the center of the particle (which appears as a white dot). Particles are selected for identification based on the scattered intensity as well as some other factors. The red cross indicates that the particle has been recognized and its trajectory will be analyzed. If the detection threshold is too low, multiple red crosses might be erroneously assigned to a single particle, whereas a high detection threshold might miss less intensely scattering particles (*see* **Note 9**).
2. Click OK in processing setting message and videos will be analyzed, displayed, and exported. Analysis includes size, size distribution, number concentration, and intensity.

3.3 System Performance Check

Although NTA does not require calibration, it is recommended that the performance be checked every 6 months, or with every new operator. For this purpose, standardized polystyrene latex spheres or gold nanoparticles are useful (*see* **Note 10**).

1. To collect data on standardized latex beads: Prepare dilutions of 60, 100, or 240 nm Nanosphere size standard in deionized water. Adjust concentration if necessary. Inject into the sample chamber, capture videos, and analyze. The mean particle size should be very close to the standard. For example, in one experiment we obtained: 59 ± 3 nm for 60 nm beads, 99 ± 2 nm for 100 nm beads, and 239 ± 5 nm for 240 nm beads (mean $\pm$ SD from multiple repeat measurements).
2. Prepare a mixture of 60 and 100 nm latex beads at approximately 1:1 number ratio, and a mixture of 100 nm and 240 nm beads at 1:1 ratio. Inject into sample chamber, capture videos, and analyze. A bimodal distribution should be obtained. In one

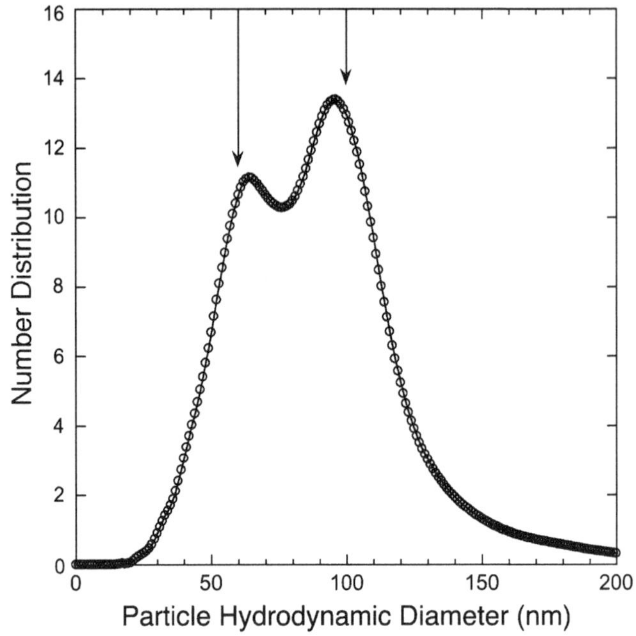

Fig. 3 Measured size distribution of a mixture of 60 and 100 nm particles at a 1.2:1 ratio. Arrows show true size of latex particles

experiment, we used a 1.2:1 ratio of 60 and 100 nm particles, and observed a bimodal distribution, with the peak sizes shifted slightly toward the center (64 and 96 nm, respectively). The measured number concentration of 60 nm beads was about 15% lower than the number of 100 nm beads, even though the actual number of 60 nm beads was 20% higher (Fig. 3). Similar phenomena were observed for 100 nm and 240 nm mixture. This demonstrates that the detection threshold varies with particle size, and that some smaller particles may escape detection in the presence of larger, more strongly scattering, particles.

3. To collect data on gold nanoparticles: Prepare several dilutions of 20 nm gold nanoparticles (BBI solutions, Cardiff, UK). Inject into sample chamber, capture, and analyze. The number concentration should be close to expected, (within 5–10%) and there should be a linear response with dilution as long as the sample was within the target concentration range of 10^7–10^9 particles/mL. For example, in one experiment, we obtained a mean size of 20.3 ± 0.5 nm; plotting the measured versus calculated number concentration yielded a straight line with a slope of 1.04 (compared to ideal slope of 1.0) and R^2 of 0.997.

3.4 Examples

In the following section, we provide details of three different applications that illustrate use of NTA in protein/peptide aggregation.

1. Example 1: Aggregate size distribution. Transthyretin (TTR) is a homotetrameric transport protein (56 kDa, 7 nm hydrodynamic diameter) that self-associates into fibrillar (amyloid) aggregates at acidic pH. Recombinant wt TTR was diluted to 0.35 mg/mL in acetate buffer (pH 4.4), quickly filtered, and immediately placed into the sample chamber. The size distribution is shown in Fig. 4a, and the result of binning the data is shown in Fig. 4b. The size distribution is unimodal and fairly broad. Virtually no particles of less than 40 nm diameter were detected. The lack of detection does not mean the lack of existence, rather it demonstrates the lower size limit of detection and that small particles are more difficult to detect in a heterogeneous sample that also contains larger particles. Samples were also analyzed by dynamic light scattering (DLS), and the mean hydrodynamic diameter is shown on the figure (*see* **Note 11**).
2. Example 2: Aggregation kinetics (*see* **Note 12**). Peptides containing repeat runs of 20 or 24 asparagines (N20 or N24, respectively), along with flanking lysines for solubility, were synthesized using standard solid-phase peptide synthesis and purified by reverse-phase HPLC [11]. Peptides were disaggregated, filtered through a 0.22 μm filter, and diluted into Tris buffer to a final concentration of 30 μM. Samples were

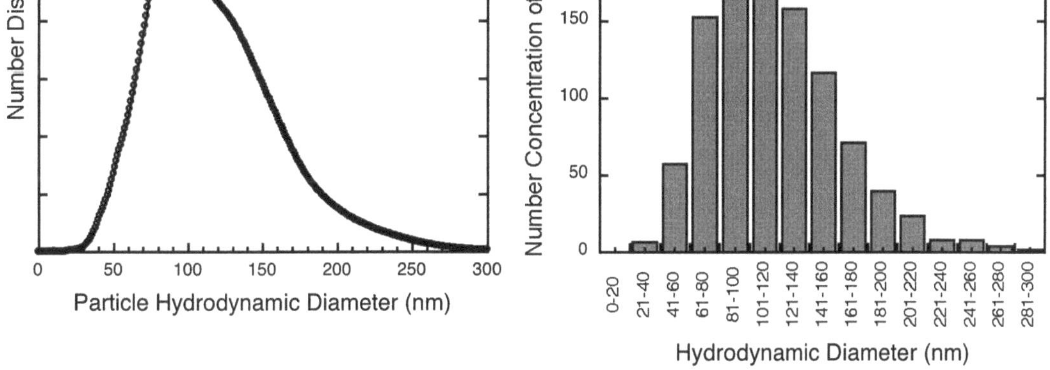

Fig. 4 Size distribution of TTR sample. The overall distribution is shown, as well as the option for binning

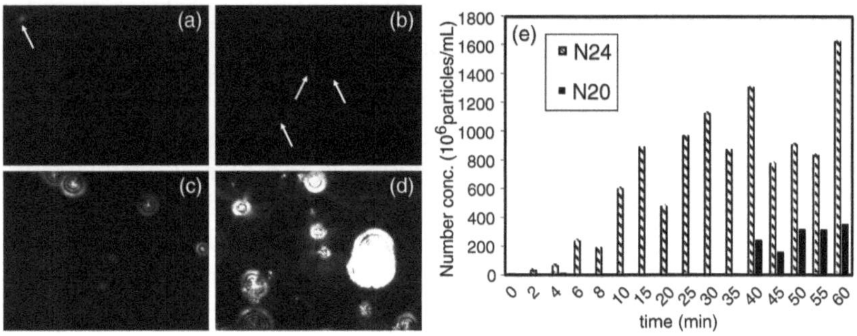


```
CAMERASETTINGSMSG
REPEATSTART
MESSAGE Please advance
sample, then press OK
CAPTURE 30
DELAY 85
REPEAT 4
CAPTURE 30
DELAY 265
REPEAT 9
CAPTURE 30
PROCESSSINGLESETTING
EXPORTRESULTS
```

Fig. 5 NTA analysis of N20 and N24 aggregation kinetics. Peptides were prepared at 30 μM in Tris buffer from monomeric stocks. (**a**–**d**) Still images taken from videos collected for (**a**) N20 at 35 min, (**b**) N24 at 2 min, (**c**) N20 at 40 min, and (**d**) N24 at 40 min. (**e**) Particle number concentration as a function of time for N20 or N24, and the script for 1 h measurement with 2-min intervals for the first 10 min and 5-min intervals for the last 50 min

immediately filtered through a 0.02 μm filter, and loaded into the chamber. A custom script was written to collect data with 2-min intervals for the first 10 min and 5-min intervals for the last 50 min, for a total of 1 h (Fig. 5). For N20, almost no particles (one or none in the viewing field) were observed up to 35 min. At 35 min, weakly scattering, rapidly diffusing particles started to appear; these grew in both size and number concentration over the next half hour. With N24, a few small particles appeared at 2 min and the particle number concentration increased steadily over the next 15 min (*see* **Note 13**). After 25 min, very large, strongly scattering aggregates appeared suddenly; the diffraction rings attest to the micron (or greater) size of the aggregates.

3. Example 3: Intensity versus size analysis. Peptides containing a repeat run of 24 glutamines (Q24) or the Alzheimer-related peptide beta-amyloid (Aβ) were allowed to aggregate. Samples were loaded into the chamber and data was collected. By checking Particle Data in the export data window, information on every tracked particle is obtained, including its hydrodynamic diameter and its scattered intensity. The "True" particles were selected, and the normalized scattered intensity was plotted versus particle size for every counted particle (Fig. 6). We observed segregation into two separate clusters, with aggregates of Q24 scattering less light than Aβ aggregates of equal size. This indicates that, for a given aggregate size, Q24 aggregates contain more water and are less "folded" than Aβ aggregates. Considered another way, for a given scattered intensity, Q24 aggregates diffuse more slowly and are therefore less compact than Aβ aggregates. Although both Q24 and Aβ assemble into fibrillar aggregates, these data indicate that there are structural differences between these aggregates.

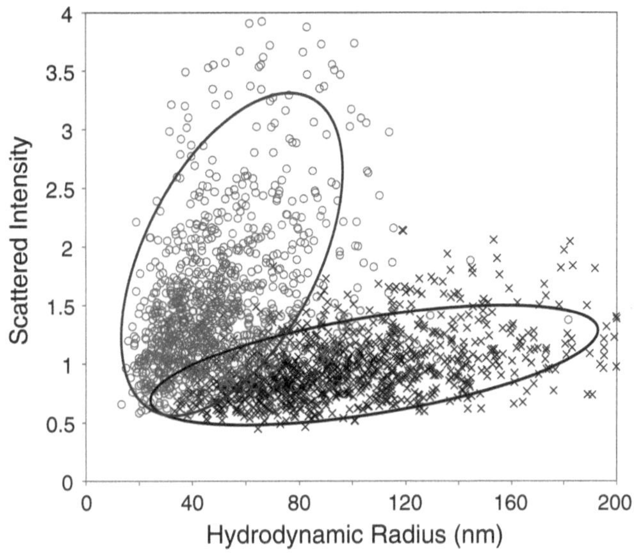

Fig. 6 Relationship between hydrodynamic radius and scattering intensity of individual particles. Comparison of Q24 (cross mark symbol) with Aβ (open circle)

4 Notes

1. Some liquids must not be used in NTA, including (a) solvents with significant optical absorption, (b) any glass etching solutions (hydrogen fluoride, potassium fluoride, etc.), or (c) O-ring incompatible solvents (DMSO, DMF, acetone, acetic acid, etc.). For aqueous buffers that contain high concentrations of solutes, or for nonaqueous solvents, the viscosity must be adjusted in SOP-Advance. Also, viscosity is a strong function of temperature and the operator should ensure that the correct viscosity is used. This is a critical step, because the relationship between diffusion coefficient and hydrodynamic radius is a function of solvent viscosity (Eq. 2).
2. Membrane filter sizes are given as a mean pore size, and there is a size distribution. Larger particles can initially pass through the filter, so the early few drops of filtrate should be discarded. Sedimenting particles will interfere with NTA measurements and must be avoided. These large particles can be removed by centrifugation or filtration.
3. After pulling the sample into the syringe, any air bubbles should be removed by inverting and tapping the syringe lightly, to force bubbles to float up, and then pushing gently till the liquid reaches the tip of the syringe. Having the sample chamber completely dry prior to sample loading helps to avoid air bubbles. The cell unit can be tilted (with the syringe injecting

vertically upward) to avoid trapping air bubbles. The sample chamber should be filled slowly until the liquid reaches the tip of the nozzle. It is important that sample solutions do not leak under the O-ring and around the optical flat as it could cause irreparable damage to the laser.

4. The optimum concentration range is 10^7–10^9 particles/mL. Samples containing fewer than 10^7 particles/mL have an insufficient number of particles in the field of view for statistically valid sampling. Samples containing more than 10^{10} particles/mL are likely to have particle trajectories that "cross over" one another. In a typical sample of an aggregation-prone protein, the protein mass concentration is determined by absorbance. However, neither the aggregate size nor the fraction of protein that is aggregated is known prior to NTA measurement, so one generally cannot a priori prepare a sample at the desired particle concentration range. The optimum concentration usually needs to be found by serial dilution, with no step greater than 100-fold dilution. A simple way to determine if a sample is too concentrated is to load the sample in the sample chamber and turn on the power. If the beam appears very bright and fuzzy, and "blooms" within the sample (looking at the sample cell chamber), or if the background appears like static on a television and particles are not distinguishable from the background (looking at the camera capture screen), then the sample is much too concentrated and needs further dilution. One word of caution: dilution may in some cases interfere with accurate characterization of the sample, for example, if aggregation is concentration-dependent.

5. A syringe pump increases the number of sample particles captured and tracked during a run. This provides a more representative sample population, thus yielding a more reproducible size distribution and more accurate concentration measurement. A syringe pump is particularly useful for very dilute samples where insufficient particles are tracked to provide a statistically robust measurement. A syringe pump can also be extremely useful to prevent settling of the sample during data collection. However, using a syringe pump requires higher sample volumes. It should also be noted that the shear stress produced by the syringe pump might induce aggregation of samples sensitive to these types of stresses.

6. The duration of each capture depends on the number of particles in the field of view; typically 60 s are sufficient if 20–60 particles are in the field of view in any single frame. (An individual particle is tracked for only a few seconds; during a 60-s capture particles will move in and out of the field of view.) A minimum of 500 total particle tracks per measurement should be collected. Longer measurement time is needed for

samples containing fewer particles, with the maximum measurement time about 3–4 min. At least three repeat measurements should be taken. Samples with high polydispersity and/or low particle concentrations require more repeats to reduce sampling biases. For kinetic studies where smaller time intervals are needed, shorter capture time (30-s) can be used.

7. As an example, the following script will collect five repeats of 60-s captures.

   CAMERASETTINGSMSG

   REPEATSTART

   MESSAGE Please advance sample, then press OK

   CAPTURE 60

   DELAY 1

   REPEAT 4

   PROCESSSINGLESETTING

   EXPORTRESULTS

8. To remove any adsorbed protein in the sample chamber, fill the chamber with 100 mM NaOH solution and hold for 30 min. Remove NaOH solution with syringe and follow the normal cleaning procedure.
9. It is a good idea to vary the detection threshold and compare distributions and number concentrations to see if there are any biases or other unwanted effects. Additional security against unwanted artifacts is provided by the software, as it will evaluate each identified particle and decide if particles are "true" or "false" particles. In general, it is better to go too low on detection threshold (false positives) than too high (false negatives), because the software will delete marginal ("false") particles. On the other hand, choosing a very low detection threshold means that a lot of particles are analyzed and determined to be false, which increases analysis time.
10. For materials with high refractive index (e.g., gold or silver colloidal), the lower size detection limit is 10–15 nm diameter. For particles with moderate refractive index (e.g., condensed polymers, protein aggregates), the lower size limit is around 25–35 nm. For weakly scattering materials (e.g., polymers, liposomes), detection limit is approximately 40 nm.
11. There are two important differences between NTA and DLS. First, the mean size determined by DLS is *z*-(intensity) averaged, which weights larger particles much more heavily. On the other hand, unaggregated protein and small (<40 nm) aggregates contribute to the DLS signal, but are too small to contribute to the distribution as measured by NTA.

12. Starting from time zero is important in kinetics study. The instrument setting should be adjusted well before sample injection. After loading the sample in chamber, adjust microscope quickly to get optimum view and start to capture immediately.
13. The particle number concentration (million particles per mL) is calculated by NTA. In our experience, the reported number is reasonably accurate for polystyrene spheres and gold nanoparticles, but large protein aggregates are under-counted [10].

Acknowledgements

The authors acknowledge helpful discussions with Dr. Dennis T. Yang. Financial support was provided by NSF CBET-1262729 and NIH R01AG033493.

References

1. Murphy RM, Roberts CJ (2013) Protein misfolding and aggregation research: some thoughts on improving quality and utility. Biotechnol Prog 29:1109–1115
2. Malloy A, Carr B (2006) Nanoparticle tracking analysis—the Halo (TM) system. Part Part Syst Charact 23:197–204
3. Gallego-Urrea JA, Tuoriniemi J, Hassellov M (2011) Applications of particle-tracking analysis to the determination of size distributions and concentrations of nanoparticles in environmental, biological and food samples. TrAC Trends Anal Chem 30:473–483
4. Filipe V, Hawe A, Jiskoot W (2010) Critical evaluation of nanoparticle tracking analysis (NTA) by NanoSight for the measurement of nanoparticles and protein aggregates. Pharm Res 27:796–810
5. Boyd RD, Pichaimuthu SK, Cuenat A (2011) New approach to inter-technique comparisons for nanoparticle size measurements; using atomic force microscopy, nanoparticle tracking analysis and dynamic light scattering. Colloids Surf A 387:35–42
6. Filipe V, Poole R, Kutscher M, Forier K, Braeckmans K, Jiskoot W (2011) Fluorescence single particle tracking for the characterization of submicron protein aggregates in biological fluids and complex formulations. Pharm Res 28:1112–1120
7. Gercel-Taylor C, Atay S, Tullis RH, Kesimer M, Taylor DD (2012) Nanoparticle analysis of circulating cell-derived vesicles in ovarian cancer patients. Anal Biochem 428:44–53
8. Kramberger P, Ciringer M, Strancar A, Peterka M (2012) Evaluation of nanoparticle tracking analysis for total virus particle determination. Virol J 9:265–274
9. Anderson W, Kozak D, Coleman VA, Jamting AK, Trau M (2013) A comparative study of submicron particle sizing platforms: accuracy, precision and resolution analysis of polydisperse particle size distributions. J Colloid Interface Sci 405:322–330
10. Yang DT, Lu XM, Fan YM, Murphy RM (2014) Evaluation of nanoparticle tracking for characterization of fibrillar protein aggregates. AICHE J 60:1236–1234
11. Lu XM, Murphy RM (2014) Synthesis and disaggregation of asparagine repeat-containing peptides. J Pept Sci 20:860–867

Chapter 8

Peptide Self-Assembly Measured Using Fluorescence Correlation Spectroscopy

Judith J. Mittag, Joachim O. Rädler, and Jennifer J. McManus

Abstract

Fluorescence correlation spectroscopy (FCS) is a flexible and powerful technique to measure the diffusion of fluorescently labeled particles. It has been important in examining a range of biological processes, from intracellular transport, to DNA hybridization. It is particularly suited to measuring the assembly of peptides, since peptides are often too small to be detected by standard light scattering methods, or may not contain aromatic amino acid residues, which limits the use of other spectroscopic techniques. In this protocol, we describe state-of-the-art sample preparation for $A\beta_{1-42}$ peptide solutions and the measurement and analysis of the self-assembly of the peptide to form fibrils via a number of intermediate states using FCS.

Key words Fluorescence correlation spectroscopy, Peptide, Self-assembly, Size distribution, Complex fluids, Polydispersity, Gaussian distribution model

1 Introduction

Fluorescence correlation spectroscopy (FCS) is a sensitive and versatile technique allowing the diffusion of fluorescently labeled nano- and meso-scale objects in solution to be measured down to picomolar concentrations [1]. The technique was first described in the early 1970s [2–4]. FCS monitors the intensity fluctuations of a fluorescently labeled particle as it passes through a fixed volume (Fig. 1a, b). The use of a confocal microscope with FCS reduces the measurement volume to less than 1 fL, allowing for single molecule detection [5, 6]. The time averaged fluorescence intensity fluctuations were used to determine an autocorrelation function (Fig. 1c), which contains information about the time taken for the labeled particle to pass through the measurement volume and the concentration of labeled particles in solution. This information in turn can be used to determine the particle size. One (a single label), two, or more color FCS can be performed when the wavelengths of emission maxima of the different dyes are sufficiently separated (or that filters are used) to eliminate cross-talk. If two labeled components

Bradley L. Nilsson and Todd M. Doran (eds.), *Peptide Self-Assembly: Methods and Protocols*, Methods in Molecular Biology, vol. 1777, https://doi.org/10.1007/978-1-4939-7811-3_8,

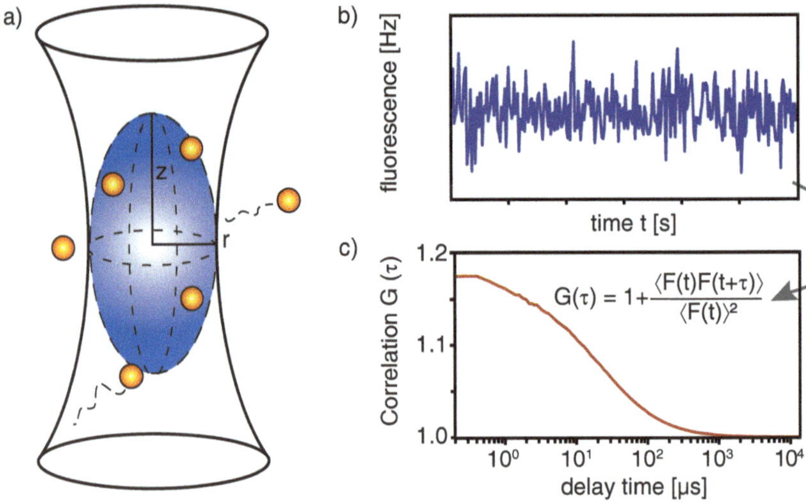

Fig. 1 (**a**) Confocal volume for a FCS experiment: fluorescently labeled particles diffuse through the confocal volume leading to fluctuations in the fluorescence intensity, (**b**) the fluctuations are recorded over the time of the experiment, and (**c**) the time averaged intensity fluctuations are recorded as an autocorrelation function, which can be used to determine the concentration of diffusing particles and the diffusion time and therefore their size

(of different colors) bind together, this can be detected by measuring a cross-correlation [7, 8]. Hence, the variety of ways in which measurements can be performed has allowed the technique to be used for a diverse range of measurements including the hybridization of DNA [6], the binding of peptides to lipid membranes [9], and the assembly of virus capsids [10].

One of the limitations of FCS is that it works best for monodisperse solutions, or solutions in which there are large differences in sizes between the diffusing species. Recent developments in data analysis techniques however have made FCS better suited for measurements of assembly processes where heterogeneous particles sizes are obtained [11–15]. The formation of a range of particle sizes is a typical occurrence in protein and peptide assembly and these more recent analysis techniques therefore offer new avenues to monitor the process. For these types of measurements, careful and consistent sample preparation is essential to ensure that the fitting methods are producing results based on robust data. We have measured the assembly process (fibrillogenesis) of amyloid-beta (Aβ), a peptide associated with the pathogenesis of Alzheimer's disease by FCS [15].

Here we describe a protocol to prepare peptide solutions containing $A\beta_{1-42}$ and the methods and procedures to acquire robust and reliable data during aggregation and fibril formation of the peptide using FCS. Furthermore, we describe in some detail the types of data analysis that can be performed on these heterogeneous solutions, including maximum entropy fitting (MEMFCS) and multi-Gaussian fitting (GDM), which recognize that the self-assembling peptide contains a number of different sized species that are not captured well using single component fits.

2 Materials

Prepare all aqueous solutions using ultrapure water (resistivity >18.1 MΩ), for example, by purifying deionized water through a Milli-Q water purification system.

1. Unlabeled $A\beta_{1-42}$ (*see* **Note 1**).
2. Labeled $A\beta_{1-42}$ (or labeling kit) (*see* **Note 2**).
3. Fluorescent dye with a known diffusion constant (for calibration), e.g., Alexa488 (*see* **Note 3** and Table 1).
4. Measurement chambers, e.g., LabTek, Nunc 8-well-plates, borosilicate bottom, 200–400 μL sample volume per well (Thermo Scientific) or Sensoplate plus, 20–100 μL sample volume per well (384-well, black, 175 μm glass bottom, Greiner Bio one) (*see* **Note 4**).
5. Poly-L-lysine coating solution: 100 μg/ml poly-L-lysine in Milli-Q water (*see* **Note 5**).
6. Immersion oil (*see* **Note 6**).
7. UV Quartz cuvette.
8. UV-vis spectrophotometer.
9. Ice.
10. 1% NH_4OH solution: 1 g of NH_4OH in 100 mL water.
11. Tris–HCl buffer: 100 mM Tris, pH 7.4 (*see* **Note 7**).

Table 1
Diffusion coefficients for a number of fluorescent small molecules that can be used for calibrating the FCS instrument

Dye	Diffusion constant [μm²/s]	Laser wavelength [nm]	Temperature [°C]	Reference
Alexa488	435	488	22.5 + 2%/°C	[19]
Alexa546	341	543	22.5 + 2%/°C	[19]
eGFP	95	488	22.5 + 2%/°C	[19]
Fluorescein	436	488	22.5 + 2%/°C	[19]
Rhodamine 6G	426	543	22.5 + 2%/°C	[19]
Cy5	370 ± 15	633	25	[20]
Alexa633	340 ± 10	633	25	[21]
Atto655	390	633	25	[22]

12. Ultracentrifuge.
13. Benchtop centrifuge.
14. Ultra-filtration unit with a 10 kDa molecular weight cutoff (e.g., Amicon ultra-filtration devices), or syringe drive filters with a 0.02 μm pore size (Whatman Anotop).
15. Glass vial (4 mL volume).
16. Eppendorf tubes.
17. Aluminum foil.
18. Kimwipes.
19. Nitrogen gas.
20. Data Analysis Software, e.g., Origin, Matlab, Igor.
21. Parafilm or adhesive film.

3 Methods

3.1 Sample Preparation

3.1.1 Dissolution of Peptide

1. To prepare unlabeled $A\beta_{1-42}$, add 1.8 mL of a 1% NH_4OH solution to 16 mg of the $A\beta_{1-42}$ peptide and add 1.8 mL of Tris–HCl (pH 7.4), which gives a 1 mM peptide solution [15]. Ensure that all the solid material is dissolved and avoid the introduction of air bubbles where possible (*see* **Note 8**).
2. Dissolve the labeled peptide (e.g., Aβ Hilyte-488, Anaspec Inc. Fremont, CA) per the supplier's instructions (*see* **Note 2**).

3.1.2 Removal of Preexisting Peptide Assemblies (e.g., Oligomers or Larger Aggregates) (See **Note 9**)

Preexisting assemblies can be present as artifacts of peptide lyophilization, sample preparation, or concentration steps. These should be removed prior to measurements. One (or more) of the following procedures can be used:

1. Ultracentrifugation: For Aβ, precool the rotor overnight at 4 °C and set the centrifuge temperature to 4 °C about 1 h before use (cooling may take some time). Use centrifuge tubes designed for ultracentrifugation (not the standard lab variety). Ensure an appropriate counterweight is present in the rotor. Centrifuge for 1 h at 100,000*g*. Remove only the upper 75% of the supernatant for further use and place on ice.
2. Ultra-filtration using, e.g., Amicon ultra-filtration devices to exclude components above a molecular weight cutoff (e.g., 10 kDa).
3. Filtration of samples through syringe driven filters, e.g., Millipore Durapore 0.22 μm filters or Whatman Anotop 0.02 μm filters.

3.1.3 Concentration Determination

1. After removing preexisting aggregates, determine the precise peptide concentration by measuring the absorbance at 280 nm for the unlabeled peptide and at 497 nm for Hilyte 488-labeled $A\beta_{1-42}$ on a UV-vis spectrophotometer.
2. Measure the concentration of unlabeled peptide using the formula $c = A/\varepsilon l$, where A is absorbance, ε is the molar extinction coefficient (equal to 1400 M^{-1} cm^{-1} (at 280 nm) for $A\beta_{1-42}$), and l is the path length in cm. For labeled protein, calculate the molar concentration using the formula

$$c = \left(\frac{A_{280} - (A_{\max,\text{dye}} - \text{CF})}{\varepsilon}\right) \cdot \text{dilution factor}$$

where $A_{\max,\ \text{dye}}$ is the absorbance maximum of dye, and CF is the correction factor for the dye (supplied by the manufacturer; for Hilyte 488, CF = 0.19).

3. Determine the labeling ratio using the formula:

$$\frac{\text{dyes [moles]}}{\text{protein [moles]}} = \left(\frac{A_{\max,\text{labeled}}}{\varepsilon' \cdot c \cdot \text{dilution factor}}\right)$$

where ε' is the extinction coefficient of the labeled peptide (e.g., For Hilyte 488 = 70,000 M^{-1} cm^{-1} at 497 nm).

3.2 Procedure for Coating Measurement Chambers with Poly-L-Lysine [16]—(See Note 5)

1. Dilute 0.5 mL poly-L-lysine stock solution into 2.83 mL water (or dilute to 15 μg/mL, total volume 3.33 mL).
2. Fill each chamber with 400 μL of the coating solution.
3. Incubate at room temperature for 1 h in the dark.
4. Aspirate the solution completely and store it in a glass vial in the fridge. The coating solution can be reused for ~3 months. Wrap aluminum foil around the vial (or use an amber vial) to protect the solution from light.
5. Rinse the chambers carefully with 400 μL of Milli-Q water 20 times to remove all unbound poly-L-lysine.
6. Dry the exterior glass bottom of the chamber slide carefully with a Kimwipe to avoid water marks.
7. Allow the chamber to dry in air at room temperature (or use a gentle stream of filtered nitrogen gas).

3.3 FCS Setup and Optimization

3.3.1 Preparation of FCS Instrument

Specific instructions for performing **steps 1–5** below will depend on the type of instrument used (see manufacturer instructions for assistance in performing these steps).

1. Switch on the laser 1 h prior to setup.

2. Place a droplet of water or oil on the immersion objective of the microscope.
3. Place a chamber on the stage above the objective (if using an inverted microscope).
4. Search for the second reflection, optimize the quality of your signal with the objective collar ring and adjust the laser power to a suitable level.
5. Perform the pinhole alignment.
6. Calibrate the FCS setup to determine the size of the confocal volume using a dye with a known diffusion constant (*see* Table 1). A concentration of 10–50 nM for the dye solution is typical (depending on the properties of the dye). Use the dilution buffer to dissolve the dye (*see* **Note 7**).
7. Add 200 μL of the dye solution to one of the chambers (or a suitable volume if different chambers are used).
8. Take 10 × 30 second measurements of the dye solution and determine the average.
9. Perform a one-component fit of the autocorrelation function using the instrumentation software (*see* Subheading 3.4.2) to determine the diffusion time (τ_D), and structure parameter (usually called S). Note these values.
10. Using D, the diffusion constant for the dye (from Table 1), first calculate r, using $\tau_D = r^2/4D$ and then z, using $S = z/r$. At this point do not change any of the components in the optical pathway or the calibration will need to be performed again.
11. Fix structure parameter (S) in the FCS software for the measurement of all other samples (*see* **Note 10**).

3.3.2 Optimizing the Labeled Protein Concentration (See **Note 7**)

1. Prepare your peptide samples (Subheadings 3.1.1–3.1.3). The labeled/unlabeled peptide ratio should be adjusted to achieve a labeled peptide concentration of ~10 nM (depending on the quality of the dye and the laser power used) to ensure a good signal-to-noise ratio.
2. If the measured signal is too weak or the signal-to-noise ratio is not good, increase the concentration of labeled peptide.

3.4 Sample Measurement and Analysis of FCS Data

3.4.1 Measuring an Autocorrelation Function for a Peptide Solution

1. Add the pre-prepared peptide solution to a sample chamber and seal with Parafilm or adhesive film to reduce evaporation.
2. Take a measurement (*see* **Notes 11, 12, 13** and **14**).

3.4.2 One-Component Fitting

To extract physically relevant information from the autocorrelation curve, fit an appropriate model function to the experimental data. The simplest fitting formula for a single component 3D freely diffusing species is:

$$G(\tau) = \frac{1}{N}\left(\frac{1}{1+\frac{\tau}{\tau_D}}\right)\left(\frac{1}{1+\frac{\tau}{S^2\tau_D}}\right)^{\frac{1}{2}}$$

where N is the average number of particles inside the confocal volume, τ is the correlation time, S is the structure parameter, and τ_D is the translational diffusion time of the molecule. An example of a "good" one-component fit is shown in Fig. 2a.

3.4.3 Two-Component Fitting

For solutions containing two species, fit using a two-component method, which is performed using the following expression:

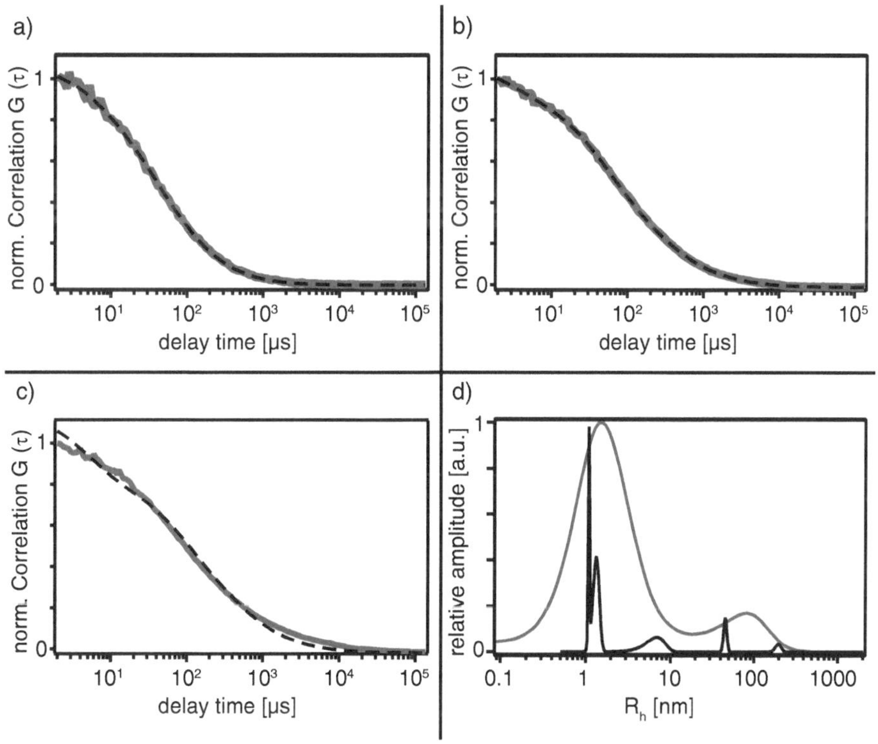

Fig. 2 Examples for data fitting. (**a**) An example of a "good" one-component fit. (**b**) A successful two-component fit, (**c**) a "bad" two-component fit that deviates from the measured data. This may indicate that a higher number of components are required to obtain a satisfactory fit of the data (data in gray, fit as black dashed line). (**d**) Amplitude distribution of a polydisperse $A\beta_{1-42}$ solution after 16 h analyzed using MEMFCS (gray) and GDM (black). While both results show similar overall trends, the GDM analysis gives more defined peaks than MEMFCS

$$G(\tau) = \frac{1}{N}\left[(1-y)\left(\frac{1}{1+\frac{\tau}{\tau_{D1}}}\right)\left(\frac{1}{1+\frac{\tau}{S^2\tau_{D1}}}\right)^{\frac{1}{2}} + y\left(\frac{1}{1+\frac{\tau}{\tau_{D2}}}\right)\left(\frac{1}{1+\frac{\tau}{S^2\tau_{D2}}}\right)^{\frac{1}{2}}\right] + 1$$

where τ_1 and τ_2 describe the diffusion times of the first and second diffusing species, y is the fraction of the second component in solution. This works particularly well if there is a significant difference in size between the two species. Figure 2b shows a satisfactory two-component fit. An example of a fit that does not accurately reflect the number of diffusing species in the solution is shown in Fig. 2c (*see* **Note 15**).

3.4.4 Higher Order Fitting: MEMFCS and GDM (Multicomponent Fitting) (See **Note 16**)

When more than two components are required to achieve a satisfactory fit, the analysis becomes more complicated. Three main approaches for multicomponent fits are used: CONTIN [11, 17], MEMFCS [12, 13], and GDM [14, 15]. The basis of each analysis method is the assumption of a quasi-continuous distribution of a large number of diffusing components. The major advantage of MEMFCS is that it provides a safe limit for interpretation of the data without any a priori assumptions and thereby reduces the risk of over-interpreting the data for highly polydisperse systems. GDM works in a similar way, but requires an assumption of the form of the amplitude distribution (i.e., the size ranges of the distributions used for the fit). While care must be exercised, it is faster and allows for better resolution of particle sizes than MEMFCS (Fig. 2d) (*see* **Note 17**).

3.4.5 Triplet Decay

The fluorescence of a molecule can switch on and off several times while diffusing through the confocal volume by decay to the triplet (dark) state. These fluctuations can provide an additional contribution to the autocorrelation curve. To account for this, a function that describes the triplet decay can be integrated into the fitting equation for the model of 3D freely diffusing particles as follows:

$$G_{\text{Triplet}}(\tau) = \left(1 + \frac{T}{1-T}\exp\left(-\frac{\tau}{\tau_{\text{T}}}\right)\right)$$

where τ_{T} is the triplet state relaxation time and T is the fraction of fluorophores in the dark state. The total autocorrelation curve then becomes a product of the triplet function and the model $G(\tau)$ as follows:

$$G_{\text{total}}(\tau) = G_{\text{Triplet}}(\tau) \cdot G(\tau)$$

3.4.6 Determination of the Hydrodynamic Radius

Once a suitable fitting procedure has been established, the particle size can be derived from the fitting parameters. For spherical molecules the hydrodynamic radius R_{h} can be calculated using the Stokes-Einstein-equation:

$$R_h = \frac{k_B T}{6\pi\eta D}$$

where k_B is the Boltzmann constant (1.38×10^{-23} J K^{-1}), T is temperature in Kelvin, η is the solvent viscosity (at temperature, T). It is also possible to insert correction factors to account for particles with nonspherical geometries [18].

4 Notes

1. $A\beta_{1-42}$ is available commercially (e.g., Anaspec, Fremont, CA, USA) from a number of different sources or can be produced recombinantly. Other peptides can also be prepared using the same protocol.
2. FCS measurements are usually performed when only a small portion of the peptide used is labeled (and hence a mixture of labeled and unlabeled peptide is used). The lowest possible amount of labeled peptide should be used. If a labeling kit is used to tag the peptide of interest, there are two options; covalent attachment of a fluorescent molecule, usually by conjugation to either a primary amine or a free cysteine; or the use of a binding dye, such as Thioflavin T, for which the fluorescence intensity increases significantly upon binding to amyloid-type assemblies. The choice will depend on the amino acid sequence of the peptide being examined and the types of assembly it will form. For methods that require covalent attachment of a dye, it is important to remove all excess (nonconjugated) dye. Exhaustive washing is required. The washings should be tested by measuring fluorescence intensity at the emission maximum of the dye in a fluorimeter to ensure that all excess dye has been removed. For both pre-labeled peptides and peptides labeled using a kit, the characteristics of the fluorescent label chosen should also be carefully evaluated for brightness, photostability (to ensure little or no bleaching), quantum efficiency, the size of dye (~1 nm) vs. the size of peptide (often less than 1 nm), the ability to determine the concentration of the labeled peptide and dye/protein ratio.
3. To effectively calibrate the FCS instrument, the confocal volume is determined by measuring the diffusion of a fluorophore for which the diffusion coefficient is known. Choose a dye best suited to the excitation lasers available and the filter set installed for emission. The excitation maximum for the calibration dye should match the excitation maximum for the fluorescent label used with the peptide. A list of commonly used calibration dyes and their corresponding diffusion constants are listed in Table 1.

4. There are a number of possible sample chambers (or micro fluidic devices) that can be used for FCS measurements. It is possible to measure FCS in very small volumes and the choice of sample chamber may depend on the availability of material. At a minimum, the chamber should preferably have a borosilicate glass base (but not too thick, since this will prohibit the adjustment of the collar ring of the objective lens).
5. FCS measurements are often performed in very low concentration samples. If the peptide binds to the sample chamber walls, this can reduce the bulk concentration significantly or indeed be responsible for the assembly of the peptide (via surface nucleation). The concentration of peptide (in a non-aggregating solution) can be monitored over time from the particle concentration data gathered in the autocorrelation function (as $1/N$). If the intercept of the autocorrelation function increases over time (without a concurrent increase in the diffusion time), peptide binding to the chamber walls should be suspected. If this occurs, surface binding can be pacified by coating the wells with a variety of other reagents. A procedure using poly-L-lysine has been described, but PEG, lipids, or BSA can also be used (using the same procedure).
6. The water used for the objective should be dust free. If present, the measurements may be affected. Filter the water through 0.22 μm filters prior to use.
7. It is important to ensure compatibility of the fluorescent dye and the buffer used. The dye brightness can be pH-dependent or the dye may need specific ions to be present (e.g., calcium). The solution conditions required for the fluorescent dye may not be compatible with the peptide (e.g., close to the isoelectric point, or wrong ionic strength), which could affect the assembly process. The selection of an appropriate buffer at a suitable pH and ionic strength should be given careful consideration.
8. Prepare a suitable buffer using analytical grade reagents. Filter through 0.22 μm filters (e.g., Millipore Durapore). It is also possible to perform measurements in complex fluids, such as blood plasma or cerebrospinal fluid. These are used instead of buffer. It is important that complex fluids are cell free, with no large particles (i.e., with sizes comparable to the confocal volume), as this will lead to light scattering.
9. The presence of dimers/oligomers and/or larger aggregates in the stock solution can make analysis of the data difficult. If possible, non-native higher order assemblies should be removed prior to the beginning of the measurement. If present, it will be difficult to accurately determine the size of the tagged peptide monomer and preexisting assemblies can alter

the kinetics of the assembly process. Prepare all samples consistently and just before measurements are performed.

10. If the laser power is increased (after the initial calibration), the calibration must be performed again. Increasing the laser power may lead to bleaching effects.
11. For measurements performed over several hours (typical for peptide assembly), sealing the sample chamber is important since evaporation will lead to an increase in concentration of the solutes and hence an overestimation of the number of particles. This may also mask a self-assembly process, which will decrease the particle number. The initial sample volume should also be large enough to ensure that evaporation doesn't dramatically alter the sample concentration.
12. For longer measurements, use oil immersion rather than water immersion for the objective. If one uses water, evaporation of the water droplet on the lens will occur. This needs to be monitored and re-applied if necessary. If the sample chamber is moved regularly throughout the experiment (e.g., if multiple wells are being measured simultaneously), re-application of the immersion fluid may also be required.
13. If large aggregates and/or dust are present in the sample during measurement, this will result in significant spikes in the intensity. Most software packages have integrated "dust filters." These tools exclude count rates that are above a threshold level (that can be specified) and data above this threshold is not used in calculating the autocorrelation function.
14. A large decrease in the number of particles may be due to the formation of very large aggregates (greater than 1 μm). If these particles are very large, sedimentation will occur and the formation will not be recorded in the autocorrelation function. Always visually inspect the sample chamber for large sedimented particles.
15. It can be very difficult to distinguish between a monomer and a dimer for the following reason. The autocorrelation function is an average of the time correlated intensity fluctuations of the diffusing species, where the intensity is proportional to the sixth power of the hydrodynamic radius. To distinguish clearly between two species in solution, one component should have twice the hydrodynamic radius of the other component. However, as a broad rule of thumb, a doubling in the hydrodynamic radius is equivalent to 8× increase in molecular weight. Therefore, a large increase in molecular weight is required to clearly distinguish different diffusing species using standard fitting procedures.
16. A multicomponent fit should be attempted when it is clear that either a one- or two-component fit do not accurately describe

your system (i.e., the fit is bad, *see* Fig. 2). Sample polydispersity is indicated when there are significant deviations between the data and the one- or two-component fits (*see* Fig. 2c). Even if the initial fit seems good (as indicated by the residuals), a multicomponent fit may still be warranted if the sizes produced by the one- or two-component fits are physically unrealistic (for example, smaller than either the peptide or the dye molecule), or if there is difficulty distinguishing between larger particles (those comparable in size to the confocal volume).

17. Fitting procedures for multicomponent, polydisperse solutions, e.g., MEMFCS and multi-Gaussian models require adequate computer power and fitting software (e.g., Igor or Matlab).

Acknowledgements

This work was made possible by funding from Science Foundation Ireland Stokes Lectureship (to J. J.McM); European Science Foundation networking programme "epitopeMap" (grant to J.O. R. and J. J. McM); EU FP7 (NanoTransKinetics grant to JJM, JOR), Deutsche Forschungsgemeinschaft (JJM, travel grant).

References

1. Ries J, Schwille P (2012) Fluorescence correlation spectroscopy. BioEssays 34:361–368
2. Magde D, Elson EL, Webb WW (1972) Thermodynamic fluctuations in a reacting system—measurement by fluorescence correlation spectroscopy. Phys Rev Lett 29(11):705–708
3. Elson EL, Magde D (1974) Fluorescence correlation spectroscopy. I. Conceptual basis and theory. Biopolymers 13:1–27
4. Magde D, Elson EL, Webb WW (1974) Fluorescence correlation spectroscopy. II. An experimental realization. Biopolymers 13:29–61
5. Rigler R, Mets Ű, Widengren J, Kask P (1993) Fluorescence correlation spectroscopy with high count rate and low background: analysis of translational diffusion. Eur Biophys J 22:169–175
6. Kinjo M, Rigler R (1995) Ultrasensitive hybridization analysis using fluorescence correlation spectroscopy. Nucl Acids Res 23 (10):1795–1799
7. Fitzpatrick J, Lillemeier B (2011) Fluorescence correlation spectroscopy: linking molecular dynamics to biological function *in vitro* and *in situ*. Curr Opin Struct Biol 21(5):650–660
8. Schwille P, Haustein E (2001) Fluorescence correlation spectroscopy—an introduction to its concepts and applications. Biophys Textbook Online 1(3):1–33
9. Rusu L, Gambhir A, McLaughlin S, Rädler JO (2004) Fluorescence correlation spectroscopy studies of peptide and protein binding phospholipid vesicles. Biophys J 87(2):1044–1053
10. Comas-Garcia M, Garmann RF, Singaram SW, Ben-Shaul A, Knobler CM, Gelbart WM (2014) Characterisation of viral capsid protein self-assembly around short single-stranded RNA. J Phys Chem B 118(27):7510–7519
11. Tjernberg LO, Pramanik A, Björling S, Thyberg P, Thyberg J, Nordstedt C, Berndt KD, Terenius L, Rigler R (1999) Amyloid β-peptide polymerization studied using fluorescence correlation spectroscopy. Chem Biol 6(1):53–62
12. Sengupta P, Garai K, Balaji J, Periasamy N, Maiti S (2003) Measuring size distribution in highly heterogeneous systems with fluorescence correlation spectroscopy. Biophys J 84 (3):1977–1984
13. Garai K, Sahoo B, Sengupta P, Maiti S (2008) Quasihomogeneous nucleation of amyloid beta yields numerical bounds for the critical radius, the surface tension, and the free energy barrier for nucleus formation. J Chem Phys 128 (4):045102-1–045102-7

14. Pal N, Verma SD, Singh MK, Singh MK, Sobhan S (2011) Fluorescence correlation spectroscopy: an efficient tool for measuring size, size-distribution and polydispersity of microemulsion droplets in solution. Anal Chem 83 (20):7736–7744
15. Mittag JJ, Milani S, Walsh DM, Rädler JO, McManus JJ (2014) Simultaneous measurement of a range of particle sizes during $A\beta_{1-42}$ fibrillogenesis quantified using fluorescence correlation spectroscopy. Biochem Biophys Res Commun 448(2):195–199
16. www.ibidi.com/service/application_notes/AN08_Coating.pdf
17. Provencher SW (1982) Contin: a general purpose constrained regularization program for inverting noisy linear algebraic and integral equations. Comput Phys Commun 27:229–242
18. Tirado M, Lopez Martinez C, Garcia de la Torre J (1984) Comparison of theories for the translational and rotational diffusion coefficients for rod-like macromolecules. Application to short DNA fragments. J Chem Phys 81(4):2047–2052
19. Petrášek Z, Schwille P (2008) Precise measurement of diffusion coefficients using scanning fluorescence correlation spectroscopy. Biophys J 94:1437–1448
20. Dertinger T, Loman A, Ewers B, Müller CB, Krämer B, Enderlein J (2008) The optics and performance of dual-focus fluorescence correlation spectroscopy. Opt Express 16 (19):14353–14368
21. Kapusta P (2010) Absolute diffusion coefficients: compilation of reference data for FCS calibration. PicoQuant GmbH Application Note
22. Dertinger T, Pacheco V, von der Hocht I, Hartman RH, Gregor I, Enderlein J (2007) Two-focus fluorescence correlation spectroscopy: a new tool for accurate and absolute diffusion measurement. Chem Phys Chem 8 (3):433–443

Part II

Methods for the Preparation of Functional Self-Assembled Peptide Biomaterials

Chapter 9

A General Method to Prepare Peptide-Based Supramolecular Hydrogels

Dan Yuan, Junfeng Shi, Ning Zhou, and Bing Xu

Abstract

As a type of versatile soft materials, supramolecular hydrogels made of peptides have received increasing research attention in past decade and have found useful applications in many areas. Meanwhile, numerous methods have been reported to initiate hydrogelation via noncovalent intermolecular interactions. Generally, most hydrogelation starts from a homogeneous solution and reaches a balance of water hydration and hydrophobic interactions, thereby resulting in hydrogelation. Here, we describe a general method to prepare supramolecular hydrogels of small peptides, and describe two examples to demonstrate hydrogel preparation.

Key words Peptide, Hydrogelation, pH, Enzyme

1 Introduction

Supramolecular hydrogels that result from the self-assembly of certain small organic molecules, including small peptides or peptide derivatives, are driven by noncovalent interactions and have emerged as versatile soft materials. These materials hold promise in many areas [1, 2]. For example, due to their inherent and excellent biocompatibility and biodegradability, supramolecular hydrogels are being explored to serve as scaffolds for tissue engineering [3–7], wound healing [8–10], and biomineralization [11, 12], vehicles for controlled drug release [13–16], matrices for protein microarray [17–19], a low cost platform for screening enzyme inhibitors or detecting enzymes [20–22], components for enzyme mimetics [23, 24], and anticancer nanomedicines [25, 26].

Supramolecular hydrogelators share common features with certain self-assembled structures in biology, such as amphiphilicity and supramolecular interactions (e.g., hydrogen bonding, aromatic–aromatic interaction, and electrostatic interaction). These noncovalent interactions lead to the formation of three-dimensional nanostructures. These networks also comprise the network

Bradley L. Nilsson and Todd M. Doran (eds.), *Peptide Self-Assembly: Methods and Protocols*, Methods in Molecular Biology, vol. 1777, https://doi.org/10.1007/978-1-4939-7811-3_9, © Springer Science+Business Media, LLC, part of Springer Nature 2018

matrices of hydrogels. The self-assembly of predictable noncovalent interactions enable supramolecular hydrogels to show rapid responses to a variety of external stimuli, including pH [27–29], temperature [30, 28], ionic strength [31, 32], irradiation of light [33, 34], ultrasound [35, 36], enzymatic conversion [37–45], and ligand–receptor interactions [33, 46–48].

Generally, a physical or chemical perturbation can initiate the hydrogelation that yields supramolecular hydrogels. For example, if a hydrogelator dissolves in an aqueous solution at a certain pH or temperature, changing pH or temperature alters the solubility of the hydrogelator to form a supersaturated solution and triggers the formation of networks of nanofibers, which results in a supramolecular hydrogel. Here, we choose a dipeptide of Fmoc-Ala-Ala-OH as a model peptide derivative and demonstrate the general procedure to trigger the hydrogelation via tuning the pH. We also describe an example wherein alkaline phosphatase is used to trigger molecular self-assembly of Fmoc-Tyr-OH in phosphate buffered saline (PBS) to result in hydrogelation.

2 Materials

Fmoc-Ala-Ala-OH, Phosphate-buffered saline (PBS, 1×), sodium hydroxide (NaOH), and hydrochloric acid solution (1 N, HCl) were from Fisher Scientific, Fmoc-Tyr(PO_3H_2)-OH was purchased from Sigma-Aldrich, and alkaline phosphatase (ALP, 1 KU, supplied with dilution buffer) was obtained from Biomatik USA. All solutions were prepared using deionized water (DI H_2O) unless indicated otherwise.

1. Fmoc-Ala-Ala-OH (such as from Santa Cruz Biotechnology).
2. PBS buffer (1×): 137 mM NaCl, 2.7 mM KCl, 8 mM Na_2HPO_4, and 2 mM KH_2PO_4, pH 7.4.
3. Fmoc-Tyr(PO_3H_2) (Such as from Novabiochem).
4. ALP buffer: 3 μL ALP stock solution (30 U/μL), 87 μL enzyme (1 U/μL ALP solution). Aliquot the 90 μL ALP buffer solution into several 0.3 mL microcentrifuge tubes, each containing 5 μL ALP buffer solution. Label them and store at 4 °C. ALP can be kept at 4 °C for at least 6 months.
5. 1 N NaOH: 4.0 g NaOH in 100 mL DI H_2O.
6. 0.1 N NaOH: 1 mL 1 N NaOH, 9 mL DI H_2O.
7. 0.1 N HCl: 1 mL 1 N HCl, 9 mL DI H_2O.
8. Transmission electron microscope (TEM).
9. Glow discharge thin carbon-coated copper TEM grid, such as 400 mesh from Pacific Grid-Tech.

10. 2% Uranyl acetate: weigh 1.0 g Uranyl Acetate Powder into a 100 mL flat flask, add 40 mL DI water and mix it until fully dissolved. Adjust the pH to 3.5 by using 1 N HCl, add DI H_2O to a final volume of 50 mL and filter the solution through a 0.45-μm filter. The solution should be kept at 4 °C in the dark.
11. Filter paper.
12. Circular dichroism (CD) spectropolarimeter.
13. 1 mm quartz cuvette.

3 Methods

Carry out all procedures at room temperature unless indicated otherwise.

3.1 Preparation of Fmoc-Ala-Ala-OH Hydrogel

1. Weigh out 1 mg of Fmoc-Ala-Ala-OH (2.6 μmol) into a 4 mL vial with a cap.
2. Add 0.6 mL DI H_2O to the above vial.
3. Turn the above suspension into a clear solution by adding 2 μL of 1 N NaOH solution.
4. Adjust pH of the solution carefully with the addition of 0.1 N HCl and 0.1 N NaOH. Sonicate the solution for 5 s with each addition of 2 μL 0.1 N HCl (*see* **Note 1**).
5. Close the cap and invert the vial to check whether the solution flows. If it flows, repeat **step 4**. If it doesn't flow when the vial is inverted, the hydrogel of Fmoc-Ala-Ala-OH has formed (see also the inverting-vial method described in [49].
6. Continue to Subheading 3.3 for biophysical characterization of Fmoc-Ala-Ala-OH hydrogel.

3.2 Prepare Hydrogel of Fmoc-Tyr-OH

1. Weigh 6 mg Fmoc-Tyr(PO_3H_2)-OH (12.4 μmol) into a 4 mL vial with a cap.
2. Add 0.6 mL PBS buffer to the above vial.
3. Sonicate the above mixture to obtain a clear solution.
4. Add 5 μL ALP solution (1 U/μL) to the above clear solution. Stir the solution slowly with the pipette tip to mix the ALP solution evenly (*see* **Note 2**).
5. Close the cap and store the above solution at room temperature.
6. The hydrogel of Fmoc-Tyr-OH should be formed within 1 h and can be verified by performing the inverted vial method as in **step 5** in Subheading 3.1.

3.3 Biophysical Characterizations of Peptide Hydrogels

1. Prepare samples for transmission electron microscope (TEM) sample preparation by adding 3–5 μL of the supramolecular hydrogel on glow discharge thin carbon-coated copper grid and incubate it for 30 s at room temperature.
2. Remove excess sample solution from the grid by blotting it with filter paper that is gently touched to the edge of the grid.
3. Wash the grid with three successive 3–5 μL drops of DI H_2O
4. Place a drop of 2% uranyl acetate on the grid and allow it to stand for 30 s, then blot the excess liquid with a piece of filter paper. Repeat the above procedure three times and let the grid air dry.
5. Visualize the deposited hydrogel structures on a TEM microscope at a high magnification (e.g., ×20,000, *see* **Note 3**).
6. Prepare a sample of the hydrogel for circular dichroism (CD) by placing 200 μL supramolecular hydrogel evenly without bubbles in a 1 mm thick quartz cuvette. If bubbles form, try to break the bubble using micropipette tips.
7. Measure the absorbance of circularly plane polarized light using a CD spectropolarimeter (*see* **Note 4**).

4 Notes

1. To avoid the formation of precipitate because of a sudden drop in pH, it is better to use a lower concentration of HCl or NaOH (e.g., 0.1 N) to adjust the pH of the solution.
2. To avoid the formation of precipitate because of a high cleavage rate by the enzyme, it is better to use a lower concentration of ALP (e.g., 1 U/μL or 0.5 U/μL). Adding a lower amount of ALP helps the formation of a clear supramolecular hydrogel. The formation time is varied according to the substrate.
3. Generally, we can visualize distinctive nanostructures, such as nanofibers, by TEM. This confirms that the hydrogels comprise well-ordered structures.
4. Using CD spectroscopy, higher-order structures derived by these hydrogels will give a minimum molar ellipticity at 216 nm. This is typical of β-sheet structures, and corresponds to secondary structure present in the nanostructures observed by TEM.

Acknowledgments

This work was partially supported by grant from NIH (CA142746) and Keck Foundation.

References

1. Du X, Zhou J, Shi J, Xu B (2015) Supramolecular Hydrogelators and hydrogels: from soft matter to molecular biomaterials. Chem Rev 115:13165–13307
2. Zhou J, Xu B (2015) Enzyme-instructed self-assembly: a multistep process for potential cancer therapy. Bioconjug Chem 26:987–999
3. Altunbas A, Pochan DJ (2012) Peptide-based and polypeptide-based hydrogels for drug delivery and tissue engineering. In: Deming T (ed) Peptide-based materials, pp 135–167
4. Ryan DM, Nilsson BL (2012) Self-assembled amino acids and dipeptides as noncovalent hydrogels for tissue engineering. Polym Chem 3:18–33
5. Jung H, Park JS, Yeom J, Selvapalam N, Park KM, Oh K et al (2014) 3d tissue engineered supramolecular hydrogels for controlled chondrogenesis of human mesenchymal stem cells. Biomacromolecules 15:707–714
6. He B, Yuan X, Wu J, Bai Y, Jiang DM (2015) Self-assembling peptide nanofiber scaffolds for bone tissue engineering. Sci Adv Mater 7:1221–1232
7. Latxague L, Ramin MA, Appavoo A, Berto P, Maisani M, Ehret C et al (2015) Control of stem-cell behavior by fine tuning the supramolecular assemblies of low-molecular-weight gelators. Angew Chem Int Ed 54:4517–4521
8. Grinstaff MW (2007) Designing hydrogel adhesives for corneal wound repair. Biomaterials 28:5205–5214
9. Yang, Z. M., Liang, G. L., Ma, M. L., Abbah, A. S., Lu, W. W. and Xu, B. (2007) D-glucosamine-based supramolecular hydrogels to improve wound healing. Chem Commun 843–845
10. Pinho E, Grootveld M, Soares G, Henriques M (2014) Cyclodextrin-based hydrogels toward improved wound dressings. Crit Rev Biotechnol 34:328–337
11. Schnepp ZAC, Gonzalez-McQuire R, Mann S (2006) Hybrid biocomposites based on calcium phosphate mineralization of self-assembled supramolecular hydrogels. Adv Mater 18:1869–1872
12. Shi N, Yin G, Han M, Xu Z (2008) Anions bonded on the supramolecular hydrogel surface as the growth center of biominerals. Colloids Surf B Biointerfaces 66:84–89
13. Sutton S, Campbell NL, Cooper AI, Kirkland M, Frith WJ, Adams DJ (2009) Controlled release from modified amino acid hydrogels governed by molecular size or network dynamics. Langmuir 25:10285–10291
14. Wang J, Wang Z, Gao J, Wang L, Yang Z, Kong D et al (2009) Incorporation of supramolecular hydrogels into agarose hydrogels-a potential drug delivery carrier. J Mater Chem 19:7892–7896
15. Mandal D, Mandal SK, Ghosh M, Das PK (2015) Phenylboronic acid appended pyrene-based low-molecular-weight injectable hydrogel: glucose-stimulated insulin release. Chem Eur J 21:12042–12052
16. Saboktakin MR, Tabatabaei RM (2015) Supramolecular hydrogels as drug delivery systems. Int J Biol Macromol 75:426–436
17. Kiyonaka S, Sada K, Yoshimura I, Shinkai S, Kato N, Hamachi I (2004) Semi-wet peptide/protein Array using supramolecular hydrogel. Nat Mater 3:58–64
18. Ikeda M, Ochi R, Hamachi I (2010) Supramolecular hydrogel-based protein and chemosensor array. Lab Chip 10:3325–3334
19. Yoshii T, Onogi S, Shigemitsu H, Hamachi I (2015) Chemically reactive supramolecular hydrogel coupled with a signal amplification system for enhanced analyte sensitivity. J Am Chem Soc 137:3360–3365
20. Ikeda M, Ochi R, Wada A, Hamachi I (2010) Supramolecular hydrogel capsule showing prostate specific antigen-responsive function for sensing and targeting prostate cancer cells. Chem Sci 1:491–498
21. Ren C, Zhang J, Chen M, Yang Z (2014) Self-assembling small molecules for the detection of important analytes. Chem Soc Rev 43:7257–7266
22. Huang P, Gao Y, Lin J, Hu H, Liao H-S, Yan X et al (2015) Tumor-specific formation of enzyme-instructed supramolecular self-assemblies as cancer theranostics. ACS Nano 9:9517–9527
23. Wang QG, Yang ZM, Zhang XQ, Xiao XD, Chang CK, Xu B (2007) A supramolecular-hydrogel-encapsulated hemin as an artificial enzyme to mimic peroxidase. Angew Chem Int Ed 46:4285–4289
24. Duncan KL, Ulijn RV (2015) Short peptides in minimalistic biocatalyst design. Biocatal Biotransformation 1:67–81
25. Su H, Koo JM, Cui H (2015) One-component nanomedicine. J Control Release 219:383–395
26. Tanaka A, Fukuoka Y, Morimoto Y, Honjo T, Koda D, Goto M et al (2015) Cancer cell death induced by the intracellular self-assembly of an enzyme-responsive supramolecular gelator. J Am Chem Soc 137:770–775

27. Roy S, Maiti DK, Panigrahi S, Basak D, Banerjee A (2012) A new hydrogel from an amino acid-based perylene bisimide and its semiconducting, photo-switching behaviour. RSC Adv 2:11053–11060
28. Kuang Y, Yuan D, Zhang Y, Kao A, Du X, Xu B (2013) Interactions between cellular proteins and morphologically different nanoscale aggregates of small molecules. RSC Adv 3:7704–7707
29. Bastings MMC, Koudstaal S, Kieltyka RE, Nakano Y, Pape ACH, Feyen DAM et al (2014) A fast Ph-switchable and self-healing supramolecular hydrogel carrier for guided, local catheter injection in the infarcted myocardium. Adv Healthcare Mater 3:70–78
30. Qiao Y, Lin YY, Yang ZY, Chen HF, Zhang SF, Yan Y et al (2010) Unique temperature-dependent supramolecular self-assembly: from hierarchical 1d nanostructures to super hydrogel. J Phys Chem B 114:11725–11730
31. Rao KV, Jayaramulu K, Maji TK, George SJ (2010) Supramolecular hydrogels and high-aspect-ratio nanofibers through charge-transfer-induced alternate coassembly. Angew Chem Int Ed 49:4218–4222
32. Bhattacharjee S, Bhattacharya S (2015) Charge transfer induces formation of stimuli-responsive, chiral, cohesive vesicles-on-a-string that eventually turn into a hydrogel. Chem Asian J 10:572–580
33. Zhang Y, Gu H, Yang Z, Xu B (2003) Supramolecular hydrogels respond to ligand–receptor interaction. J Am Chem Soc 125:13680–13681
34. Cao W, Zhang XL, Miao XM, Yang ZM, Xu HP (2013) Gamma-ray-responsive supramolecular hydrogel based on a diselenide-containing polymer and a peptide. Angew Chem Int Ed 52:6233–6237
35. Kuang Y, Gao Y, Shi J, Li J, Xu B (2014) The first supramolecular peptidic hydrogelator containing taurine. Chem Commun 50:2772–2774
36. Pappas CG, Mutasa T, Frederix P, Fleming S, Bai S, Debnath S et al (2015) Transient supramolecular reconfiguration of peptide nanostructures using ultrasound. Mater Horiz 2:198–202
37. Toledano S, Williams RJ, Jayawarna V, Ulijn RV (2006) Enzyme-triggered self-assembly of peptide hydrogels via reversed hydrolysis. J Am Chem Soc 128:1070–1071
38. Yang Z, Liang G, Wang L, Xu B (2006) Using a kinase/phosphatase switch to regulate a supramolecular hydrogel and forming the supramolecular hydrogel in vivo. J Am Chem Soc 128:3038–3043
39. Li J, Gao Y, Kuang Y, Shi J, Du X, Zhou J et al (2013) Dephosphorylation of D-peptide derivatives to form biofunctional, supramolecular nanofibers/hydrogels and their potential applications for intracellular imaging and intratumoral chemotherapy. J Am Chem Soc 135:9907–9914
40. Kuang Y, Shi J, Li J, Yuan D, Alberti KA, Xu Q et al (2014) Pericellular hydrogel/nanonets inhibit cancer cells. Angew Chem Int Ed 53:8104–8107
41. Shi J, Du X, Yuan D, Zhou J, Zhou N, Huang Y et al (2014) D-amino acids modulate the cellular response of enzymatic-instructed supramolecular nanofibers of small peptides. Biomacromolecules 15:3559–3568
42. Yuan D, Zhou R, Shi J, Du X, Li X, Xu B (2014) Enzyme-instructed self-assembly of hydrogelators consisting of nucleobases, amino acids, and saccharide. RSC Adv 4:26487–26490
43. Li J, Kuang Y, Shi J, Zhou J, Medina JE, Zhou R et al (2015) Enzyme-instructed intracellular molecular self-assembly to boost activity of cisplatin against drug-resistant ovarian cancer cells. Angew Chem Int Ed 54:13307–13311
44. Wu D, Du X, Shi J, Zhou J, Zhou N, Xu B (2015) The first Cd73-instructed supramolecular hydrogel. J Colloid Interface Sci 447:269–272
45. Zhou J, Du XW, Li J, Yamagata N, Xu B (2015) Taurine boosts cellular uptake of small D-peptides for enzyme-instructed intracellular molecular self-assembly. J Am Chem Soc 137:10040–10043
46. Shi J, Du X, Huang Y, Zhou J, Yuan D, Wu D et al (2015) Ligand–receptor interaction catalyzes the aggregation of small molecules to induce cell necroptosis. J Am Chem Soc 137:26–29
47. Shi JF, Du XW, Yuan D, Haburcak R, Zhou N, Xu B (2015) Supramolecular detoxification of neurotoxic nanofibrils of small molecules via morphological switch. Bioconjug Chem 26:1879–1883
48. Thompson MS, Tsurkan MV, Chwalek K, Bornhauser M, Schlierf M, Werner C et al (2015) Self-assembling hydrogels crosslinked solely by receptor-ligand interactions: Tunability, rationalization of physical properties, and 3d cell culture. Chem Eur J 21:3178–3182
49. Sreenivasachary N, Lehn JM (2005) Gelation-driven component selection in the generation of constitutional dynamic hydrogels based on guanine-quartet formation. Proc Natl Acad Sci U S A 102:5938–5943

Chapter 10

Recursive Directional Ligation Approach for Cloning Recombinant Spider Silks

Nina Dinjaski, Wenwen Huang, and David L. Kaplan

Abstract

Recent advances in genetic engineering have provided a route to produce various types of recombinant spider silks. Different cloning strategies have been applied to achieve this goal (e.g., concatemerization, step-by-step ligation, recursive directional ligation). Here we describe recursive directional ligation as an approach that allows for facile modularity and control over the size of the genetic cassettes. This approach is based on sequential ligation of genetic cassettes (monomers) where the junctions between them are formed without interrupting key gene sequences with additional base pairs.

Key words Cloning, Recombinant DNA techniques, Recursive ligation, Spider silk

1 Introduction

Recombinant DNA technology has had significant impact on biomaterial designs and production. One class of the biomaterials that benefited from advances in genetic engineering is spider silk [1–5]. Spider silks are remarkable natural polymers that consist of three domains: a repetitive middle core domain that dominates the protein chain, and non-repetitive N-terminal and C-terminal domains (Fig. 1) [6].

The large core domain is organized in a block copolymer-like arrangement, in which two sequences, poly-alanine (poly(A/GA)) and glycine-rich (GGX or GPGXX) polypeptides alternate (Fig. 1) [7]. Due to the repetitive nature of the core sequence, specific genetic engineering strategies had to be undertaken to meet the challenges of recombinant spider silk production (e.g., gene instability, a high frequency of homologous recombination, deletions, transcription errors) [1]. The most commonly used approaches toward this need are: (1) concatemerization, (2) step-by-step ligation, and (3) recursive directional ligation [8]. Concatemerization is a useful method when a library of genes of different sizes is desired, but has limitations in the preparation of genes with specific

Bradley L. Nilsson and Todd M. Doran (eds.), *Peptide Self-Assembly: Methods and Protocols*, Methods in Molecular Biology, vol. 1777, https://doi.org/10.1007/978-1-4939-7811-3_10, © Springer Science+Business Media, LLC, part of Springer Nature 2018

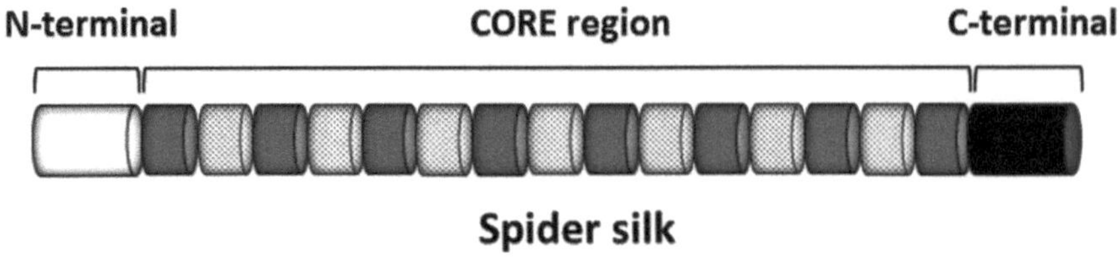

Fig. 1 Schematic representation of spider silk structure. Spider silks consist of repetitive middle core domain (gray and dotted), and non-repetitive N-terminal (white) and C-terminal domains (black). The core domain consists of: A block (gray), polyalanine/polyglycine rich domain that is responsible for β-sheet formation and is considered hydrophobic due to the high number of alanine residues; B block (dotted), composed of GGX repeats, thought to adopt a helical conformation and serve as a more hydrophilic link between crystalline β-sheet regions, as well as neighboring GGX helices in adjacent protein molecules that can help reinforce fiber alignment

sizes [9, 10]. Step-by-step ligation allows for control over the genetic sequence and the size of the construct [9, 11].

Here, we describe the application of recursive directional ligation approach for cloning of recombinant dragline silk, major ampullate protein 1 (MaSp I) from golden orb-weaver spider, *Nephila clavipes.* For silk cloning, this approach was first reported by Prince et al. in 1995 and used in the construction of MaSp1 and MaSp2, 24mer and 16mer, respectively [12]. This approach provides control over the sequence and the size of genetic construct, while also eliminating the restriction sites at the junctions between monomeric genetic cassettes without interrupting key gene sequences with additional base pairs [13]. The strategy for the cloning of recombinant spider silk-like sequences consists of three main steps: design of spider silk-like sequence, preparation of cloning vector, and recursive directional ligation (Fig. 2). Using this approach we were able to design a variety of silk-like constructs with different material properties [13–16], as well as to produce native-sized recombinant spider silk [7, 17]. Different host organisms have been applied to produce recombinant spider silk (e.g., *Escherichia coli*, *Pichia pastoris*, silkworm larvae, plants, mammalian cell lines, and transgenic animals) [8]. Here we use *E. coli* as the host, due to its well-characterized genetics.

2 Materials

Prepare all solutions using deionized water at room temperature unless otherwise specified. Store all reagents at room temperature unless indicated otherwise. Sterile conditions are not needed if not specified. Diligently follow all waste disposal regulations when disposing waste materials.

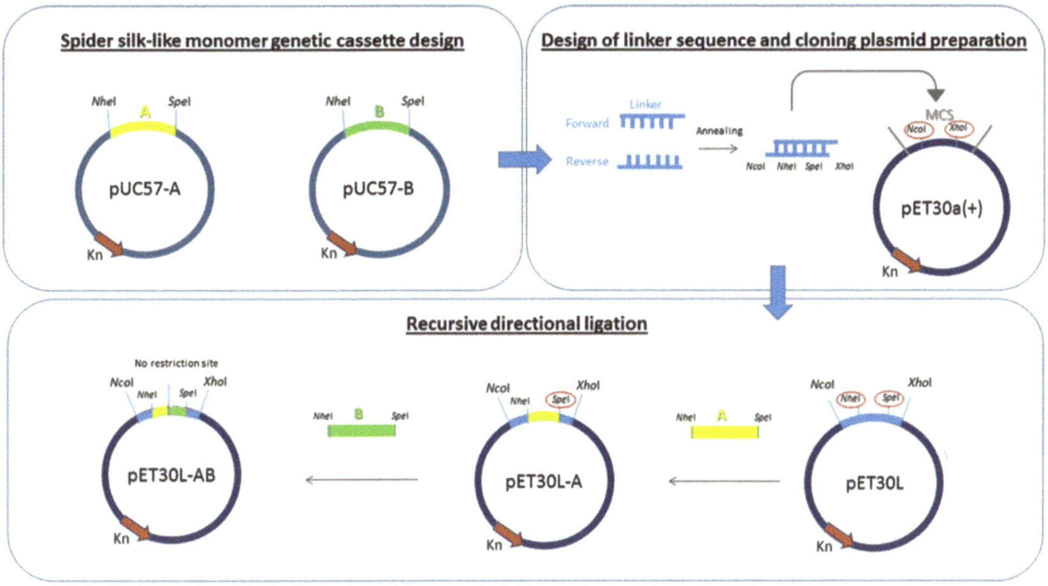

Fig. 2 Schematic representation of the strategy applied for spider silk cloning. The strategy consists of three main steps: design of spider silk-like sequence, preparation of cloning vector, and recursive directional ligation

2.1 Recombinant Spider Silk-Like Monomer Design

1. Spider silk genes: custom design MaSp 1 sequences (Figs. 1 and 3a), a hydrophobic poly-alanine rich block (A block) and a hydrophilic glycine-rich block (B block). The A block consists of one poly(GA) repeat (GAGAAAAGGAG) responsible for β-sheet formation, whereas the B block is composed of four GGX repeats, separated by the GSQGSGR sequence (*see* example Fig. 3a, **Note 1**). The A and B block sequences are synthesized as the *Nhe*I/*Spe*I insets (GenScript, Piscataway, NJ, USA) in pUC57 (Kn). Liberate the spider silk gene domain from pUC57 vector using restriction enzyme, *Nhe*I and *Spe*I (New England Biolabs, Ipswich, MA, USA).
2. Chemically competent cells: *E. coli* DH5α chemically competent cells for colony selection and plasmid isolation.
3. Antibiotic stock solution (1000× concentrate): 50 mg/mL kanamycin (Kn) prepared by dissolving 0.5 g of kanamycin sulfate powder into 10 mL deionized water in a 15 mL falcon tube. Filter with sterilized 0.22 μm PVDF filters in biohood to sterilize antibiotics solution (*see* **Note 2**). Store at −20 °C.
4. Culture plates for colony selection: dissolve 7.5 g of lysogeny broth (LB) powder in 300 mL deionized water in a 500 mL flask. Add 4.5 g agar powder into the LB solution. Stir well on a stir plate, and autoclave at 121 °C 15 psi for 20 min. Add 300 μL kanamycin stock solution at about 50 °C in biohood (*see* **Note 2**). Mix well on a stir plate. Pour LB/agar/Kn

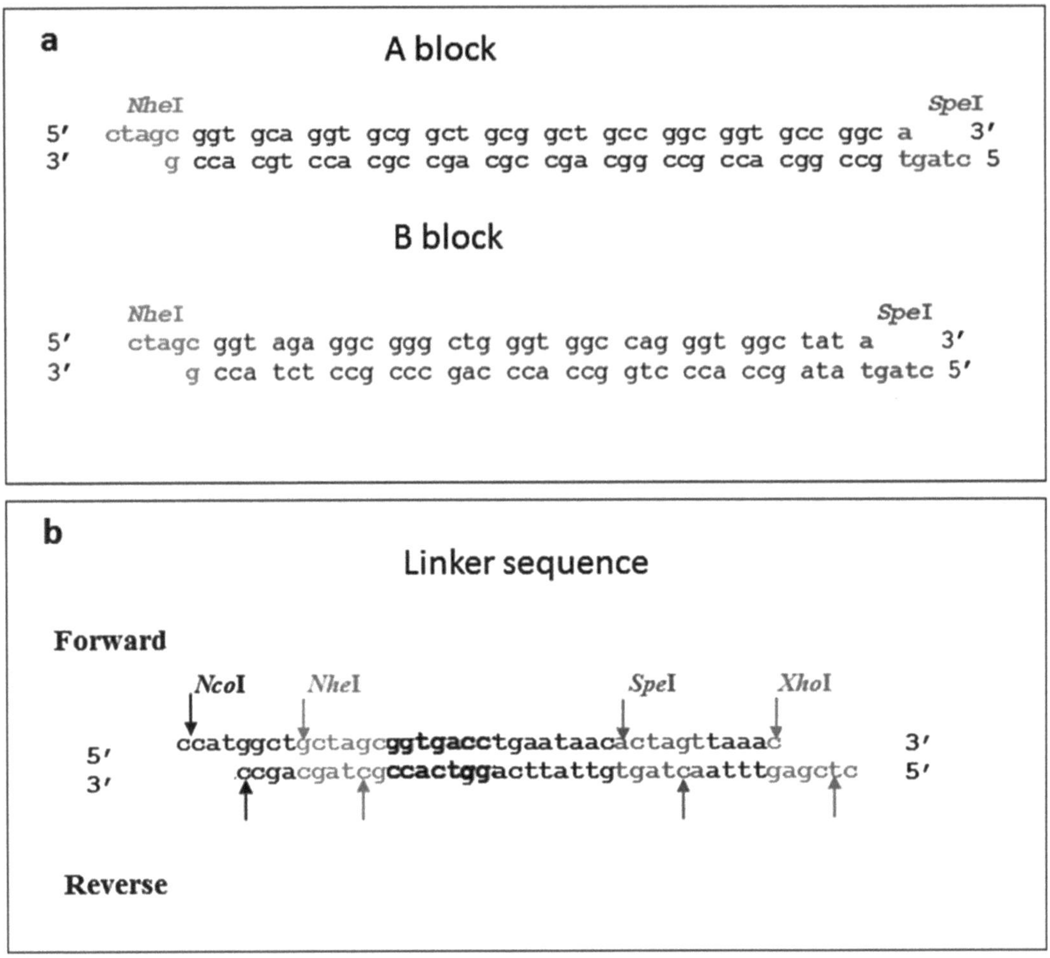

Fig. 3 Spider silk block sequences hydrophobic, A block and hydrophilic, B block (**a**) and pET30L linker sequence (**b**). The A block consists of one poly(GA) repeat (GAGAAAAAGGAG) responsible for β-sheet formation, whereas the B block is composed of four GGX repeats, separated by the GSQGSGR sequence. A cloning cassette linker was designed as *XhoI* and *NcoI* insert carrying *NheI* and *SpeI* restriction sites to modify commercially available pET30a(+) vector and generate pET30L

solution in sterilized Petri dishes, wait until solidified and put the lid back on for each plate in biohood. Store at 4 °C, in sterilized plastic bags.

5. Culture medium: LB containing 50 μg/mL Kn prepared by dissolving 25 g of LB powder in 1 L deionized water in a 4 L flask. Stir well on a stir plate, and autoclave at 121 °C 15 psi for 20 min. Add 1 mL kanamycin stock solution at 37 °C in biohood. Mix well on a stir plate. Store at room temperature.

6. 1% agarose gel: prepare 1×TAE (Tris-acetate-EDTA) buffer from 50×TAE buffer stock solution. To make 40 mL of a 1% agarose gel, dissolve 0.4 g of agarose powder in 40 mL of 1×TAE buffer by heating in a microwave 30s twice.

Let agarose solution cool down for 5 min. Add ethidium bromide (EtBr) (Fisher Scientific, Pittsburgh, PA, USA) to a final concentration of approximately 0.2–0.5 μg/mL (usually about 2–3 μL of lab stock solution per 100 mL gel). Seal the open ends of the plastic tray with the rubber stoppers. Pour the 1% agarose gel solution into the sealed plastic gel tray and position the 13-well comb at one end of the tray so that it is 0.5–1.0 mm above the bottom of the tray and not touching the rubber stopper. Allow the gel to harden and use it for visualization of spider silk monomer.

7. DNA isolation: QIAprep Spin Miniprep Kit (Qiagen, Valencia, CA, USA).
8. Gel extraction: QIAquick Gel Extraction Kit (Qiagen, Valencia, CA, USA).
9. Nanodrop: measurement of DNA concentration (*see* **Note 3**).
10. −20 °C freezer: for DNA storage.

2.2 Preparation of Plasmid for Spider Silk Cloning

1. Cloning cassette linker: the linker sequence should be designed with flanking *Xho*I and *Nco*I restriction sites for incorporation into pET30a(+) (Novagen, San Diego, CA, USA), and it should carry internal *Nhe*I and *Spe*I restriction sites (Fig. 3b). The linker is used to generate modified pET30a(+) vector, pET30L (*see* **Note 4**).
2. Annealed forward and reverse cloning cassette linker (1:1 v/v) (GenScript, Piscataway, NJ, USA).
3. Digested pET30a(+): digest pET30a(+) plasmid with *Nco*I and *Xho*I restriction enzymes (New England Biolabs, Ipswich, MA, USA).
4. pET30L vector: Ligate the double-stranded linker and digested pET30a(+) in a 20 μL ligation reaction by adding 2 μL of 10× T4 DNA Ligase Buffer (New England Biolabs, Ipswich, MA, USA), 0.02 pmol of pET30a(+) vector DNA, 0.06 pmol insert linker DNA, and 1 μL of T4 DNA Ligase (New England Biolabs, Ipswich, MA, USA). Add nuclease-free water to a total of 20 μL. The resulting cloning vector is referred to as pET30L.
5. Chemically competent cells: *E. coli* DH5α chemically competent cells for colony selection and plasmid isolation.
6. Culture plates: LB (Kn) plates for colony selection.
7. DNA isolation: QIAprep Spin Miniprep Kit (Qiagen, Valencia, CA, USA).
8. Digested pET30L: Digest pET30L with *Nco*I and *Xho*I restriction enzymes (New England Biolabs, Ipswich, MA, USA), in NEBuffer (New England Biolabs, Ipswich, MA, USA).

9. DNA visualization system: confirm incorporation of the insert by separating digested pET30a(+) and pET30L on 0.9% and 1% agarose gels and visualizing the bands using DNA imaging system (G:Box XR5, Syngene, Cambridge, UK) (*see* **Note 5**).
10. Sterile glycerol: store selected clones at −80 °C with glycerol (10% final concentration).

2.3 Cloning of Recombinant Spider Silk

1. Digested pET30L and pUC57-A: digest plasmids with *Nhe*I and *Spe*I restriction enzymes (New England Biolabs, Ipswich, MA, USA).
2. Ligated pET30L-A: Ligate pET30L-A with A block, and pET30L-A with B block by adding 2 μL of 10× T4 DNA Ligase Buffer (New England Biolabs, Ipswich, MA, USA), 0.02 pmol of vector DNA (pET30L or pET30L-A), 0.06 pmol insert DNA (A block or B block), and 1 μL of T4 DNA Ligase (New England Biolabs, Ipswich, MA, USA). Add nuclease-free water to a total of 20 μL.
3. Dephosphorylated and digested pET30L-A vector: Add 1/10 volume of 10× Antarctic Phosphatase Reaction Buffer (New England Biolabs, Ipswich, MA, USA) to 1–5 μg of digested pET30-A. Add 1 μL of Antarctic Phosphatase (5 units) and mix. Incubate for 15 min at 37 °C. Heat inactivate at 70 °C for 5 min.
4. Chemically competent cells: *E. coli* DH5α chemically competent cells for colony selection and plasmid isolation, and *E. coli* BL21 (DE3) for expression and protein production.
5. Culture plates: LB (Kn) plates for colony selection.
6. Agarose gel 0.9% and 1%: using gel electrophoresis separate linker, digest pET30L and pET30L-A on 0.9% and 1% agarose gel.
7. DNA isolation: QIAprep Spin Miniprep Kit (Qiagen, Valencia, CA, USA).
8. Gel extraction: QIAquick Gel Extraction Kit (Qiagen, Valencia, CA, USA) (*see* **Note 6**).
9. Nanodrop or similar UV-visible spectrophotometer: for DNA quantification.
10. Sterile glycerol: store selected clones at −80 °C with glycerol (10% final concentration) (*see* **Note 7**).

3 Methods

Carry out all the procedures at room temperature unless otherwise specified. Sterile conditions are not needed if not specified.

3.1 Recombinant Spider Silk-Like Monomer Design

1. Design the genes encoding recombinant silk-like proteins based on the sequence of the MaSp I protein of *Nephila clavipes* (Accession Numbers: P19837 and ACF19411.1) (*see* **Note 1**). Design two individual blocks, a hydrophobic poly-alanine rich block (A block) and a hydrophilic glycine-rich block (B block) (*see* example Fig. 3a). The A block consists of one poly (GA) repeat (GAGAAAAAGGAG) responsible for β-sheet formation, whereas the B block is composed of four GGX repeats, separated by the GSQGSGR sequence.
2. Optimize the codons for the production in *E. coli* and commercially obtain the sequence as the *Nhe*I/*Spe*I inset in pUC57 (Kn).
3. Transform high efficiency *E. coli* DH5α chemically competent cells following the protocol recommended by the manufacturer. Select the positive clones on LB plates supplemented with kanamycin (50 μg/mL) (*see* **Note 2**).
4. Inoculate selected clones in 4 mL of LB (Kn) medium (*see* **Note 2**). Culture overnight at 37 °C at 250 rpm. Isolate the plasmids using QIAprep Spin Miniprep Kit (Qiagen, Valencia, CA, USA). Check the concentration of isolated plasmid by nanodrop (*see* **Note 3**).
5. Perform the restriction reaction using *Nhe*I and *Spe*I endonucleases to isolate silk-like monomers. Prepare the reaction (50 μL total) by adding 1 μg of plasmid DNA, 5 μL of 10× NEBuffer, 1 μL of each restriction enzyme (*Nhe*I and *Spe*I), and add ultrapure water to 50 μL. Incubate 1 h at 37 °C (*see* **Note 4**).
6. Stop the restriction reaction by incubating the mix at 65 °C for 20 min.
7. Run 1% agarose gel to visualize released monomer (*see* **Note 5**). Perform the gel extraction using QIAquick Gel Extraction Kit (Qiagen, Valencia, CA, USA) following the procedure recommended by manufacturer (*see* **Note 6**). Analyze the concentration of DNA fragment by nanodrop and store at −20 °C.
8. Prepare LB (Kn) pre-cultures of selected clones carrying pUC57-AB, incubate overnight at 37 °C at 250 rpm. Prepare 10% glycerol stock and freeze at −80 °C (*see* **Note 7**).

3.2 Preparation of Plasmid for Spider Silk Cloning

1. Design a cloning cassette linker to modify commercially available pET30a(+) vector (Novagen, San Diego, CA, USA) (*see* example Fig. 3b). The sequence should be designed with *Xho*I and *Nco*I restriction sites for incorporation into pET30a(+) (Novagen, San Diego, CA, USA), and it should carry internal *Nhe*I and *Spe*I restriction sites internally to later generate pET30L. Commercially obtain the oligonucleotide linker

sequence (*see* **Note 8**). Prepare the oligonucleotide linker sequences as recommended by the manufacturer.

2. To generate the cloning cassette linker anneal two synthetic nucleotides, forward and reverse (1:1 v/v) by decreasing the temperature of a 20 pmol/μL oligonucleotide solution from 95 to 20 °C at a gradient of 0.1 °C/s. Mismatched double strands are denatured at 70 °C, followed by another temperature decrease to 20 °C. Repeat this cycle three times.
3. Digest pET30a(+) (Novagen, San Diego, CA, USA) with *Nco*I and *Xho*I by performing the restriction reaction (*see* **step 5** in Subheading 3.1). Run 0.9% agarose gel and gel extract the vector (*see* **Note 9**).
4. Ligate double-stranded linker and pET30a(+) (Novagen, San Diego, CA, USA) previously digested with *Nco*I and *Xho*I. Note that both restriction sites are preserved after ligation. The resulting cloning vector is referred to as pET30L. Perform ligation reaction (20 μL total) by adding 2 μL of 10× T4 DNA Ligase Buffer (New England Biolabs, Ipswich, MA, USA), 0.02 pmol of vector DNA (pET30a(+)) and 0.06 pmol insert DNA (double-stranded linker DNA) and 1 μL of T4 DNA Ligase (New England Biolabs, Ipswich, MA, USA). Add nuclease-free water to a total of 20 μL. Gently mix the reaction by pipetting up and down and microfuge briefly. Incubate the reaction at 16 °C overnight or at room temperature for 10 min. Heat inactivate at 65 °C for 10 min.
5. Chill the ligation mix on ice and transform 1–5 μL of the reaction into 50 μL high efficiency *E. coli* DH5α chemically competent cell (New England Biolabs, Ipswich, MA, USA) following the protocol recommended by the manufacturer. Select the positive clones on LB plates supplemented with kanamycin (50 μg/mL).
6. Isolate the plasmid using QIAprep Spin Miniprep Kit (*see* **step 4** in Subheading 3.1). Perform the restriction reaction using *Nco*I and *Xho*I enzymes (*see* **step 5** in Subheading 3.1). Run 1% agarose gel to confirm the successful ligation of pET30a(+) and linker sequence (*see* **Note 10**).
7. Store selected clones at −80 °C (*see* **step 8** in Subheading 3.1).

3.3 Cloning of Recombinant Spider Silk

1. Digest pET30L with *Nhe*I and *Spe*I by performing the restriction reaction (*see* **step 5** in Subheading 3.1). Run 0.9% agarose gel and gel extract the vector (*see* **step 7** in Subheading 3.1). Determine the plasmid concentration by nanodrop.
2. Ligate pET30L and monomer sequence corresponding to A block (*see* **step 7** in Subheading 3.1), both previously digested with *Nhe*I and *Spe*I (*see* **step 4** in Subheading 3.2). The resulting cloning vector is referred to as pET30L-A.

3. Transform high efficiency *E. coli* DH5α chemically competent cell (*see* **step 4** in Subheading 3.2) and select the cells on LB (Kn) plates.
4. Isolate the pET30L-A plasmid (*see* **step 4** in Subheading 3.1) and send selected clones for sequencing to confirm the incorporation of sequence A. Store successful clones at −80 °C (**Note** 7).
5. Use the successful pET30L-A clones to perform digestion with *Spe*I. Stop the reaction by incubation at 65 °C for 15 min.
6. To avoid the self-ligation of the pET30L-A vector digested with *Spe*I, remove 5′ phosphate from the DNA using Antarctic Phosphatase (New England Biolabs, Ipswich, MA, USA). To set up the reaction add 1/10 volume of 10× Antarctic Phosphatase Reaction Buffer to 1–5 μg of pET30-A cut with *Spe*I. Add 1 μL of Antarctic Phosphatase (5 units) and mix. Incubate for 15 min at 37 °C. Heat inactivate at 70 °C for 5 min.
7. Perform the ligation of pET30-A cut with *Spe*I and sequence corresponding to B block cut with *Spe*I and *Nhe*I (*see* **step** 7 in Subheading 3.1) (**Note 11**). Transform high efficiency *E. coli* DH5α chemically competent cell (*see* **step 4** in Subheading 3.2) and select the cells on LB (Kn) plates.
8. Isolate the plasmid (*see* **step 4** in Subheading 3.1) and send selected clones for sequencing to confirm the incorporation of sequence A. Store successful clones at −80 °C (**Note** 7).
9. For the expression and protein production transform the successful clones into *E. coli* BL(DE3) (New England Biolabs, Ipswich, MA, USA).

4 Notes

1. Use appropriate software (e.g., Bioedit) to design sequences. General recommendations for sequence design are: absence of dimerization capability, absence of significant hairpin formation (usually >3 bp), and low specific binding at the 3′ end.
2. Prepare the 50 °C water bath in advance to allow the autoclaved LB agar to cool down, but not start solidifying. After autoclaving leave the bottle in the bath until the temperature of the media reaches approximately 50 °C. Once the bottle can be held in your hands add 1 mL of 50 mg/mL Kn stock under sterile conditions. Do not autoclave kanamycin as it is thermosensitive. Kanamycin should be filter sterilized (pore size 0.22 μm). When pouring the plates, take care to avoid bubbles. If bubbles form, use pipetman to draw it back into the pipet while the agar is still liquid. In the case of LB broth, incubation

of the media in a water bath is not necessary, as the antibiotic can be added when the media cools down.

3. Concentration of the DNA should be determined on an agarose gel and by nanodrop. The concentration of DNA can be estimated on agarose gel when compared to the standards used for DNA quantification (e.g., New England Biolabs DNA ladder) [18]. In addition, agarose gel provides information on possible contamination and/or degradation of the DNA. Nanodrop allows for determination of DNA concentration by reading absorbance at 260 nm, where an OD (or absorbance) of 1.0 in a 1 cm path length = 50 μg/mL for double-stranded DNA, 40 μg/mL for single-stranded DNA and RNA, and 20–33 μg/mL for oligonucleotides. An absorbance ratio of 260 nm and 280 nm gives an estimate of the purity of the solution. Pure DNA and RNA solutions have OD_{260}/OD_{280} values of 1.8 and 2.0, respectively. Contamination of a sample with protein or phenol will result in a value significantly less than the above numbers, typically ≤ 1.6. This method is not useful for very low concentrations of DNA or RNA (<1 μg/mL).
4. All restriction enzymes should be maintained on ice while in use. Temperature change might cause enzyme degradation and activity loss. The enzymes should be added last to the reaction mix to allow for uniform distribution.
5. The percentage of agarose gel should be adjusted to the fragment size (*see* Table 1). If using ethidium bromide for DNA staining allow the agarose to cool down before adding ethidium bromide as the fumes of ethidium bromide are carcinogenic. When visualizing and cutting the bands use the minimal exposure to UV light as it might damage the DNA.
6. Cut the band corresponding to the monomer sequence and not the upper band that corresponds to plasmid backbone. Always use appropriate ladder when running the gel to confirm the size of the insert.
7. Due to the high density of 100% glycerol it is difficult to pipette, therefore 80% glycerol solution should be prepared and autoclaved. This will allow accurate pipetting. In addition, it is recommended to cut the pipette tip to allow for easier pipetting. The sterility of all solutions and equipment should be maintained. As glycerol is toxic to bacteria at high concentrations, once added to bacterial culture, the culture should be immediately frozen. The glycerol stock of all strains should be made and stored at −80 °C in the case there is the need to return back to any of the previous cloning steps.
8. Both forward and reverse linker oligonucleotides should be designed.

Table 1
Optimal agarose concentration for efficient separation of linear double-stranded DNA molecules by electrophoresis

% Agarose concentration in gel (w/v)	Efficient separation range for linear double-stranded DNA molecules (kb)
0.3	5–60
0.6	1–20
0.7	0.8–10
0.9	0.5–7
1.2	0.4–6
1.5	0.2–3
2.0	0.1–2

Dinjaski et al [18]

9. Even though there is no separation of two fragments of different size, gel extraction is needed in order to eliminate undigested plasmid that might negatively impact the cloning process. The concentration of undigested plasmid is usually low and cannot be detected on DNA agarose gel.
10. As the insertion of the linker does not change the antibiotic resistance it is difficult to select positive clones where the ligation was successful, as both positive and negative clones grow on LB (Kn) plates. Therefore, after plasmid isolation, digestion should be performed and positive clones sent for sequencing. In the case of pET30L, standard T7 and T7term primers can be used for sequencing.
11. *Nhe*I and *Spe*I are restriction endonucleases that produce compatible cohesive ends, therefore the plasmid pET30L-A cut only with *Spe*I, can be ligated to the B fragment cut with *Nhe*I and *Spe*I. Note that after the ligation of the cohesive ends formed after restriction with *Nhe*I and *Spe*I, the restriction site cannot be recognized by the two enzymes.

Acknowledgments

We thank the NIH (P41 EB002520), the AFOSR, and the NSF for support of this work.

References

1. Yigit S, Dinjaski N, Kaplan DL (2016) Fibrous proteins: at the crossroads of genetic engineering and biotechnological applications. Biotechnol Bioeng 113:913–929
2. Lin S, Ryu S, Tokareva O, Gronau G, Jacobsen MM, Huang W, Rizzo DJ, Li D, Staii C, Pugno NM, Wong JY, Kaplan DL, Buehler MJ (2015) Predictive modelling-based design and experiments for synthesis and spinning of bioinspired silk fibres. Nat Commun 6:6892
3. Huang W, Krishnaji ST, Hu X, Kaplan D, Cebe P (2011) Heat capacity of spider silk-like block copolymers. Macromolecules 44:5299–5309
4. Krishnaji ST, Bratzel G, Kinahan ME, Kluge JA, Staii C, Wong JY, Buehler MJ, Kaplan DL (2013) Sequence-structure-property relationships of recombinant spider silk proteins: integration of biopolymer design, processing, and modeling. Adv Funct Mater 23:241–253
5. Huang W, Krishnaji S, Kaplan D, Cebe P (2012) Thermal analysis of spider silk inspired di-block copolymers in the glass transition region by TMDSC. J Therm Anal Calorim 109:1193–1201
6. Ittah S, Cohen S, Garty S, Cohn D, Gat U (2006) An essential role for the C-terminal domain of a dragline spider silk protein in directing fiber formation. Biomacromolecules 7:1790–1795
7. Rabotyagova OS, Cebe P, Kaplan DL (2009) Self-assembly of genetically engineered spider silk block copolymers. Biomacromolecules 10:229–236
8. Tokareva O, Michalczechen-Lacerda VA, Rech EL, Kaplan DL (2013) Recombinant DNA production of spider silk proteins. Microb Biotechnol 6:651–663
9. Meyer DE, Chilkoti A (2002) Genetically encoded synthesis of protein-based polymers with precisely specified molecular weight and sequence by recursive directional ligation: examples from the elastin-like polypeptide system. Biomacromolecules 3:357–367
10. Wang Q, Xia X, Huang W, Lin Y, Xu Q, Kaplan DL (2014) High throughput screening of dynamic silk-elastin-like protein biomaterials. Adv Funct Mater 24:4303–4310
11. Wright ER, Conticello VP (2002) Self-assembly of block copolymers derived from elastin-mimetic polypeptide sequences. Adv Drug Deliv Rev 54:1057–1073
12. Prince JT, McGrath KP, DiGirolamo CM, Kaplan DL (1995) Construction, cloning, and expression of synthetic genes encoding spider dragline silk. Biochemistry 34:10879–10885
13. Higashiya S, Topilina NI, Ngo SC, Zagorevskii D, Welch JT (2007) Design and preparation of β-sheet forming repetitive and block-copolymerized polypeptides. Biomacromolecules 8:1487–1497
14. Huang W, Krishnaji ST, Tokareva OR, Kaplan D, Cebe P (2014) Influence of water on protein transitions: morphology and secondary structure. Macromolecules 47:8107–8114
15. Krishnaji ST, Huang W, Rabotyagova O, Kharlampieva E, Choi I, Tsukruk VV, Naik R, Cebe P, Kaplan DL (2011) Thin film assembly of spider silk-like block copolymers. Langmuir 27:1000–1008
16. Huang W, Krishnaji S, Rabotyagova Tokareva O, Kaplan D, Cebe P (2014) Influence of water on protein transitions: thermal analysis. Macromolecules 47:8098–8106
17. Xia X-X, Qian Z-G, Ki CS, Park YH, Kaplan DL, Lee SY (2010) Native-sized recombinant spider silk protein produced in metabolically engineered Escherichia coli results in a strong fiber. Proc Natl Acad Sci 107:14059–14063
18. Sambrook J, Russell DW (2001) Molecular cloning. A laboratory manual. Cold Spring Harbor Laboratory Press, Cold Spring Harbor, NY

Chapter 11

Synthesis of Mikto-Arm Star Peptide Conjugates

Jin Mo Koo, Hao Su, Yi-An Lin, and Honggang Cui

Abstract

Mikto-arm star peptide conjugates are an emerging class of self-assembling peptide-based structural units that contain three or more auxiliary segments of different chemical compositions and/or functionalities. This group of molecules exhibit interesting self-assembly behavior in solution due to their chemically asymmetric topology. Here we describe the detailed procedure for synthesis of an ABC Mikto-arm star peptide conjugate in which two immiscible entities (a saturated hydrocarbon and a hydrophobic and lipophobic fluorocarbon) are conjugated onto a short β-sheet forming peptide sequence, GNNQQNY, derived from the Sup35 prion, through a lysine junction. Automated and manual Fmoc-solid phase synthesis techniques are used to synthesize the Mikto-arm star peptide conjugates, followed by HPLC purification. We envision that this set of protocols can afford a versatile platform to synthesize a new class of peptidic building units for diverse applications.

Key words Mikto-arm, Peptide conjugate, Amino acids, Fmoc-solid phase synthesis, HPLC, Amphiphile

1 Introduction

Self-assembly of amphiphilic molecules (e.g., surfactants [1], lipids [2, 3], dendrimers [4], block copolymers [5, 6], or peptide amphiphiles [7]) presents an effective "bottom-up" approach to construct a variety of nanoscale materials. In particular, peptide amphiphiles (PAs) developed by Stupp, Tirrell, and others have been well recognized due to their potential biomedical applications in tissue engineering [8, 9], regenerative medicine [10], immunotherapies [11], and drug delivery [12, 13]. In a typical design of peptide amphiphile, a short hydrophobic alkyl chain is conjugated onto either the *C*- or *N*-terminal end of a short peptide to generate an amphiphilic molecule possessing both the structural features of amphiphilic surfactants and the function of bioactive peptides. This simple combination of two chemically distinct components into a conjugate can be easily achieved using standard Fmoc-solid phase synthesis techniques.

Bradley L. Nilsson and Todd M. Doran (eds.), *Peptide Self-Assembly: Methods and Protocols*, Methods in Molecular Biology, vol. 1777, https://doi.org/10.1007/978-1-4939-7811-3_11,

In contrast to linear diblock conjugates, a Mikto-arm star molecule contains three or more segments of different chemical moieties [14–16]. The added complexity in chemical structure and chain architecture can lead to the formation of structurally more sophisticated aggregates. In one example, Mikto-arm star copolymers with multiple linear polymers originating from a single branching point have been shown to exhibit an interesting phase-separation behavior both in bulk and in solution [16–18]. Lodge, Hillmyer, and coworkers reported a novel class of multi-compartment micelles formed by solution state self-assembly of ABC Mikto-arm star block terpolymers containing three mutually immiscible polymeric components [19–22]. The rich assembly behavior of Mikto-arm star terpolymers has inspired the design and synthesis of peptide-based Mikto-arm star conjugates. The chemically asymmetric star topology, combined with the unique ability of peptides to assume well-defined secondary structures, could open a new avenue for low molecular weight molecular units to aggregate into a variety of classic and novel morphologies.

In this chapter, we describe detailed procedures for synthesizing an ABC Mikto-arm star peptide conjugate comprising three distinct segments linked through a lysine junction. The three chosen segments are (1) a saturated linear hydrocarbon, (2) a hydrophobic and lipophobic fluorocarbon, and (3) a short β-sheet forming peptide sequence (GNNQQNY) derived from the key amyloid-forming region in yeast prion Sup35 (Fig. 1) [23]. To synthesize the fluorocarbon-hydrocarbon-peptide (FHP) conjugate and the other two control molecules (hydrocarbon-peptide (HP) and fluorocarbon-peptide (FP) conjugates) (Fig. 1a), we used a combination of automated and manual Fmoc-solid phase synthesis techniques to obtain the target molecules. The peptide segment was first synthesized using a peptide synthesizer, followed by manual coupling of the remaining hydrocarbon and fluorocarbon. The low molecular weight of these conjugates allows them to be routinely purified as single compounds by HPLC. Therefore, it is our hope that, with this set of synthesis protocols, a great diversity of monodisperse Mikto-arm star peptide conjugates can be easily accessed with high quality.

2 Materials

Prepare all solutions using ultrapure water (sensitivity of 18 MΩ at 25 °C), analytical or HPLC grade solvents and store at room temperature (unless indicated otherwise). Solvent mixtures were prepared in terms of volume ratios (unless specified otherwise).

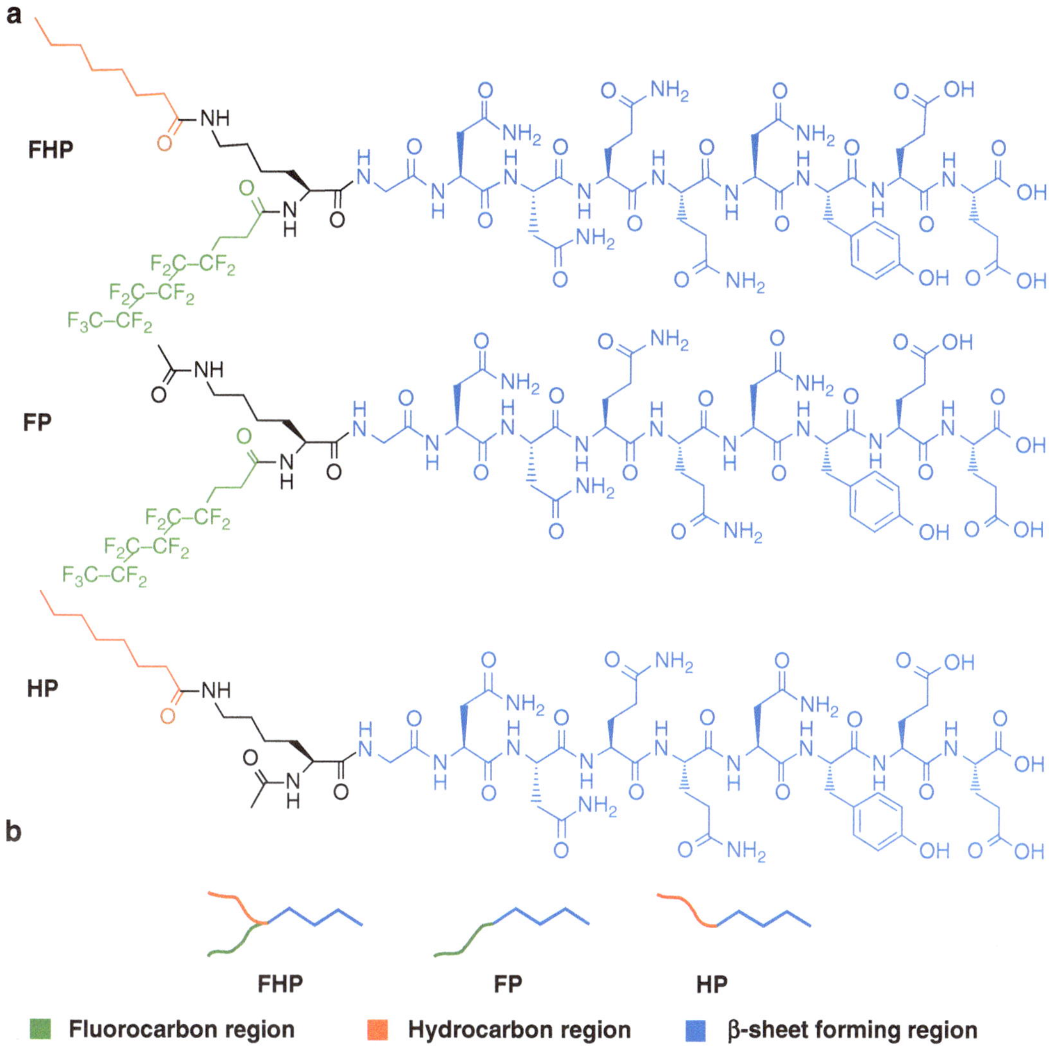

Fig. 1 (**a**) Chemical structures and (**b**) schematic representations of the Mikto-arm star peptide FHP, and the two control peptides, FP and HP, possessing only one of the two hydrophobic blocks. The ABC Mikto-arm star peptide conjugate (FHP) comprising three distinct segments linked through a lysine junction: (1) a saturated hydrocarbon, (2) a hydrophobic and lipophobic fluorocarbon, and (3) a short *β*-sheet forming peptide sequence (GNNQQNY) Sup35 (adapted from ref. [23])

2.1 Reagents and Solutions

1. Fluorenylmethyloxycarbonyl (Fmoc) amino acids (unless otherwise stated) and coupling reagents (*O*-(7-azabenzotriazol-1-yl)-*N*,*N*,*N*′,*N*′-tetramethyluronium hexafluorophosphate (HATU) or *O*-(benzotriazol-1-yl)-*N*,*N*,*N*′,*N*′-tetramethyluronium hexafluorophosphate (HBTU) (Advanced Automated Peptide Protein Technologies (AAPPTec), Louisville, KY). Store at 4 °C.
2. N-Ethyldiisopropylamine (DIEA). Store at room temperature.
3. Resins: Rink Amide MBHA resin or Fmoc-Glu(OtBu)-Wang resin beads (Novabiochem, San Diego, CA). Store at 4 °C.

Other resins can also be employed for different molecular designs and cleavage conditions (*see* **Note 1**).

4. Fmoc deprotection solution: 20% 4-methylpiperidine in *N,N*-dimethyl formamide (DMF). Mix 20 mL 4-methylpiperidine with 80 mL DMF to get 100 mL deprotection solution (*see* **Note 2**).
5. Mtt deprotection solution: 3% of trifluoroacetic acid (TFA) and 5% of triisopropylsilane (TIS) in dichloromethane (DCM). Add 300 μL TFA and 500 μL TIS into 9200 μL DCM to yield a 10 mL deprotection solution (*see* **Note 2**).
6. HBTU coupling solution: 0.4 M HBTU in DMF. Weigh 37.9 g of HBTU and add into 250 mL of DMF.
7. DIEA solution: 2 M DIEA in DMF/DCM (4/1). 87 mL of DIEA/33 mL of DCM/130 mL of DMF.
8. Acetylation solution: 20% acetic anhydride in DMF. Add 4 mL acetic anhydride to 16 mL DMF to get 20 mL solution (*see* **Note 3**).
9. Peptide cleavage solution: Mix 95% of TFA, 2.5% of TIS, and 2.5% of H_2O (*see* **Note 4**).
10. Reverse-phase high performance liquid chromatography (RP-HPLC) basic mobile phases (eluents): 0.1% *v*/*v* ammonium hydroxide in water (phase A) and 0.1% *v*/*v* ammonium hydroxide in acetonitrile (phase B). Add 4 mL ammonium hydroxide to 4 L water or acetonitrile to prepare mobile phase A or B. The acidic mobile phases are 0.1% *v*/*v* TFA in water (phase A) and 0.1% *v*/*v* TFA in acetonitrile (phase B). Store at room temperature (*see* **Note 5**).
11. Kaiser test kit: Reagent A (0.001 M KCN in H_2O/pyridine); Reagent B (ninhydrin, 6% in ethanol); Reagent C (phenol, 80% in ethanol) (*see* **Note 6**). Store at room temperature in brown glass bottle.
12. 2*H*,2*H*,3*H*,3*H*-Perfluorononanoic acid (4,4,5,5,6,6,7,7,8,8,9,9,9-tridecafluorononanoic acid) and octanoic acid. Store at room temperature.
13. Cold anhydrous ethyl ether. Store in an ice bath for 10 min to cool it down (*see* **Note 7**).

2.2 Instruments

1. Automated peptide synthesis instrument: AAPPTec Focus peptide synthesizer (AAPPTec, Louisville, KY).
2. Manual peptide synthesis instruments: Wrist Action Shaker Model 75 (Burrell Scientific Pittsburgh, PA) and Kemtec-Synthware™ Peptide Synthesis Vessel.
3. Rotary vaporization instruments: Rotavac Valve Tec (Heidolph instruments, Cinnaminson, NJ) and Chemglass 100 mL/250 mL round bottom flask.

4. Centrifuge instruments: SORVALL RT1 Centrifuge (Thermo Fisher Scientific, Pittsburgh, PA), BLUE MAX™ 50 mL polypropylene conical tube (Becton Dickinson, Franklin Lakes, NJ), Parafilm "M" laboratory film (Bemis Flexible Packaging, Neenah, WI).
5. Fisher Scientific Ultrasonic Cleaner (Fisher Scientific, Pittsburgh, PA), Fisher STD Vortex Mixer 120 V (Fisher Scientific, Pittsburgh, PA), and Evergreen Scientific 50 mL freestanding centrifuge tubes.
6. Varian ProStar Model 325 HPLC equipped with a fraction collector (Agilent Technologies, Santa Clara, CA).
7. Varian PLRP-S column for preparative separations (100 Å, 10 μm, 150 × 25 mm) and Varian Pursuit XRs C_{18} column (5 μm, 150 × 4.6 mm) for analytical HPLC.
8. HPLC syringes: BD 1 mL TB syringe Luer-Lok™ tip (Becton Dickinson, Franklin Lakes, NJ), BD 10 mL syringe Luer-Lok™ tip (Becton Dickinson, Franklin Lakes, NJ), and BD Precision-Glide needles (Becton Dickinson, Franklin Lakes, NJ).
9. Lyophilization instrument: FreeZone −105 °C 4.5 L freeze dryer (Labconco, Kansas City, MO).
10. Mass spectrometric instruments: Finnigan LCQ ion trap mass spectrometer (Thermo-Finnigan, Waltham, MA) for ESI-MS or Autoflex III Smartbeam (Bruker, Billerica, MA) for MALDI-ToF MS.

3 Methods

The efficient synthesis of oligopeptides was accomplished using a combination of automated and manual Fmoc-solid phase synthesis techniques (*see* **Note 8**). Carry out all procedures at room temperature unless indicated otherwise. All the experiments should strictly follow laboratory safety rules.

3.1 Peptide Synthesis and Purification

3.1.1 Mikto-arm Star Peptide Synthesis

1. Swell the resin of choice, in this case, Fmoc-Glu(OtBu)-Wang resin (*see* **Note 1**). Weigh 0.25 mmol resin (471 mg used, *see* **Note 9**) and suspend in 10–15 mL of DCM in a 50 mL plastic tube and shake the tube for 10–15 min using a shaker (*see* **Note 10**). Transfer the resin to the shaking vessel of an AAPPTec Focus peptide synthesizer (or the corresponding chamber in other commercial peptide synthesizers).
2. Weigh Fmoc-amino acids. Amino acids for each coupling step are fourfold excess of the molar number of the resin. Since we use 0.25 mmol of resin here, the amount of amino acid for each coupling should be 1 mmol (*see* **Note 11**). Prepare each Fmoc-amino acid/DMF solution at a concentration of 0.2 mmol/mL

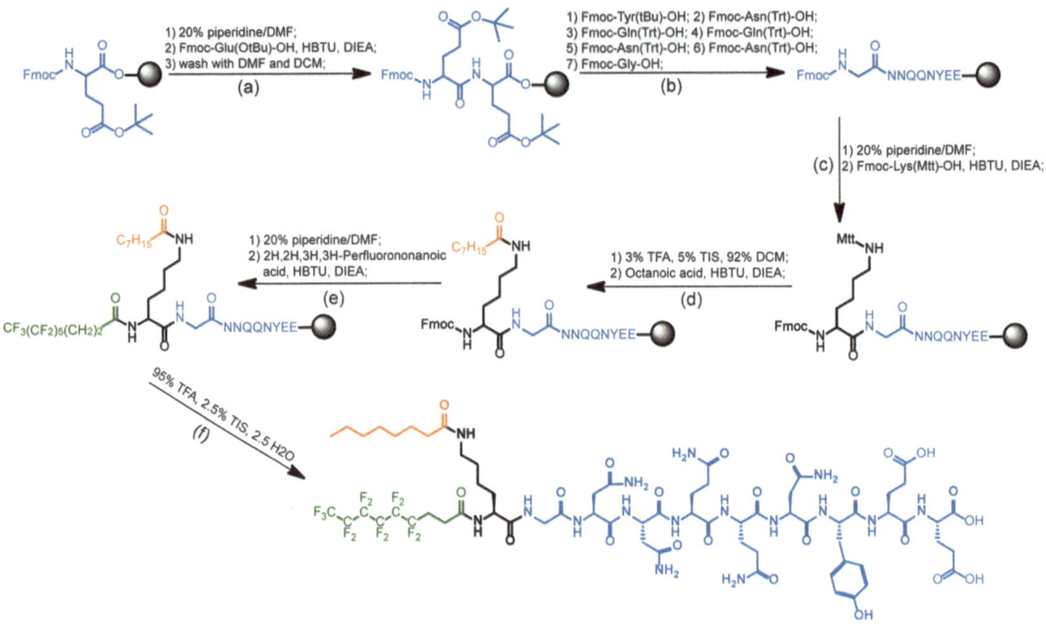

Fig. 2 Schematic illustration of a synthetic route for the Mikto-arm star peptide (FHP) utilizing standard Fmoc-based solid phase peptide synthesis techniques (**a** and **b**). The first part of the synthetic route employs automatic synthesis using a peptide synthesizer, (**c**) the addition of the last amino acid Fmoc-Lys(Mtt)-OH, (**d**) the functionalization of hydrocarbon chain, (**e**) the functionalization of fluorocarbon chain, and (**f**) the cleavage of peptide from the resin are performed manually

(*see* **Note 12**). Fill Fmoc deprotection solution, HBTU coupling solution, DIEA solution (*see* Subheading 2.1), DMF, DCM, and prepared amino acid solutions into the reagent reservoirs of the synthesizer. Set up the program for the automated synthesis of the sequence Fmoc-GNNQQNYEE-Wang.

3. (**Steps 3–5** describe the coupling of the glutamic acid E, sequence of Fmoc-GNNQQNYEE-Wang) (Fig. 2 step a). Deprotect Fmoc by mixing the swelled resin with 10 mL Fmoc deprotection solution (20% of 4-methylpiperidine in DMF) and shaking the vessel for 15 min. Wash the resin three times each with DCM and with DMF after each deprotection. Repeat the process once for deprotecting each Fmoc group.
4. Premix 5 mL Fmoc-Glu(OtBu)-OH/DMF solution, 2.5 mL HBTU coupling solution, and 1.3 mL DIEA solution for 2 min before filling the solution into shaking vessel.
5. Blend the above mixture with the Fmoc-deprotected resin. Start the coupling of a corresponding amino acid by shaking the vessel for 1 h (*see* **Note 13**). After the reaction, wash the resin three times each with DCM and with DMF.

6. Complete the synthesis of the full sequence, Fmoc-GNNQQNYEE-Wang, by repeating **steps 3–5** for each amino acid sequence utilizing AAPPTec Focus peptide synthesizer (Fig. 2 step b) (*see* **Note 14**).
7. Transfer the peptide resin (Fmoc-GNNQQNYEE-Wang) into a peptide synthesis vessel to start manual synthesis (*see* **Note 15**).
8. Add the last amino acid Fmoc-Lys(Mtt)-OH using manual coupling protocols (**steps 8–11**) (Fig. 2 step c). To synthesize a peptide sequence manually, repeat **steps 8–11** for each coupling cycle. Pour 10 mL of 20% 4-methylpiperidine in DMF to the peptide synthesis vessel and shake the vessel for 15 min using a wrist action shaker. Empty the vessel by adding pressure on the top. Repeat the deprotection step once. Wash the resin three times each with DCM and DMF after each deprotection.
9. Take 10–15 beads of resin into a small test tube. Perform the Kaiser test (a very sensitive test for primary amines) by dropping 2 drops of the Kaiser test kit reagent A, 1 drop of reagent B, and 1 drop of reagent C. Heat the tube to 110°C for 5 min (except for Mtt protected peptides since Mtt is deprotected at high temperature) (*see* **Note 16**). Perform the Kaiser test accordingly to determine if the deprotection is completed. If the test yields positive result (the resin beads turning dark blue), the deprotection step is completed. Otherwise, repeat Fmoc deprotection one more time and confirm again with the Kaiser test.
10. Weigh 624.8 mg (1 mmol) Fmoc-Lys(Mtt)-OH (4 equivalents), 379.2 mg HBTU (4 equivalents) and transfer solids to a 50 mL plastic tube. Pour 5–7 mL of DMF, vortex or sonicate to fully dissolve the solids. Add 10 equivalents of DIEA (432 μL) to the tube, vortex to fully mix the solution, and transfer the solution to the peptide synthesis vessel. Shake the vessel for 2 h utilizing the wrist action shaker. Empty the reagents in the vessel by applying pressure to the top. Wash the resin three times each with 10–15 mL of DCM and DMF after the reaction.
11. Perform the Kaiser test at room temperature for 30 min. If it yields a positive result (resin and solution blue), repeat the coupling **step 10** to ensure that Fmoc-deprotected amines are coupled with Fmoc-Lys(Mtt)-OH.
12. (**Steps 12–13** illustrate Fig. 2 step d). Deprotect Mtt groups by treating the resin with Mtt deprotection solution (3% TFA, 5% TIS, 92% DCM) (*see* **Note 17**). Fill the vessel with 10 mL of Mtt deprotection solution and shake the mixture for 10 min. Empty the vessel and wash with 10–15 mL of DCM three times. Repeat 5–7 times (*see* **Note 18**). Perform the Kaiser

test in accordance with **step 11** to ensure the presence of free amine groups. If free amine groups are not present, repeat the deprotection of Mtt groups as delineated in this step.

13. Attach hydrocarbon to the Mtt-deprotected amine group of the lysine side chain by reacting the resin with 4 equivalents of octanoic acid (144 mg). Since octanoic acid is a liquid at room temperature, first weigh the right amount of octanoic acid in a 50 mL plastic tube and follow the coupling process shown in **step 10**. Follow **step 10** to perform manual coupling, but extend the coupling time from 2 to 4 h (*see* **Note 19**). Perform the Kaiser test following **step 9**.
14. Deprotect the N-terminal Fmoc group by following **step 8**. Perform the Kaiser test following **step 9**. Functionalize the other arm after Fmoc deprotection with fluorocarbon by reacting the resin with 4 equivalents of 2*H*,2*H*,3*H*,3*H*-Perfluorononanoic acid (392.1 mg) (Fig. 2 step e) (*see* **Note 20**).
15. Add 10 mL of 95% TFA/2.5% TIS/2.5% H_2O solution to the peptide synthesis vessel and shake the vessel for 3 h to cleave the peptide from the resin (Fig. 2 step f) (*see* **Note 21**). Collect the peptide containing cleavage solution in a 100 mL round bottom flask by adding pressure to the top of the synthesis vessel. Wash the resin with a small amount of pure TFA and collect the remaining peptide in the same round bottom flask (*see* **Note 22**). Dispose of the solid phase resin remaining inside the synthesis vessel as hazardous waste.
16. Concentrate the peptide/TFA solution to less than 5 mL by rotary vaporization (*see* **Note 23**).
17. Precipitate the resulting crude peptides by adding the concentrated solution dropwise to 40 mL of chilled anhydrous ethyl ether in a centrifuge tube. Vortex to fully mix. Use Parafilm to fully wrap the cap of the centrifuge tube and centrifuge the tube at 2400–2600 × *g* for 4–5 min (*see* **Note 24**). Carefully discard the supernatant to obtain the precipitated peptides as semisolids. Wash the precipitates with 35–40 mL of chilled ethyl ether followed by centrifugation. Wash two more times likewise to fully remove TFA. Air-dry the precipitates in a ventilated fume hood.

3.1.2 Peptide Purification and Lyophilization

1. Add 6 mL of 0.1% *v/v* ammonium hydroxide in water and 4 mL of 0.1% *v/v* ammonium hydroxide in acetonitrile to the precipitated peptide (*see* **Note 25**).
2. Vortex or sonicate the solution until the solid precipitates are fully dissolved (*see* **Note 26**).
3. Check for an appropriate eluent to be used in the HPLC system. As a rule of thumb, if the peptide possesses more carboxylic acid groups than amine groups, the base-modified

solvent should be used as the eluent (except for compounds that will be degraded in the basic environment). Alternatively, if the peptide contains more amines, the eluent should be modified with acid (such as 0.1% TFA). *See* **item 10** in Subheading 2.1 for details of these basic or acidic eluents. Here, use basic eluent for this particular peptide sequence containing multiple carboxylic acids.

4. Perform analytical HPLC to analyze the components of resulting crude peptide. Prewash the analytical column with 100% acetonitrile at 1 mL/min for 5 min. Quickly inject 100 μL of a water/acetonitrile (1/1) mixture to wash the analytical injection site. Then, slowly inject 25 μL of the peptide solution for one analytical HPLC run. Run analytical HPLC to approximate the most effective gradient to purify the peptide. The gradient that we use is in Fig. 3a, and the corresponding peak of FHP is labeled in the HPLC trace (*see* **Note 27**). Refer to your HPLC manual for detailed protocols for analysis and purification.

5. Perform preparative HPLC to purify the peptide mixtures. Quickly inject one full 10 mL syringe of water/acetonitrile (1/1) mixture to wash the preparative injection site. Then, slowly inject 10 mL of the peptide solution with a concentration of 10–20 mM using 10 mL syringe for one run. Run preparative HPLC for 35 min using a gradient based on that

(a)

Time (min)	Total flow (mL/min)	A (%)	B (%)
Initials	1	95.0	5.0
2	1	95.0	5.0
20	1	60.0	40.0
30	1	10.0	90.0
35	1	95.0	5.0

(b)

Time (min)	Total flow (mL/min)	A (%)	B (%)
Initials	25	95.0	5.0
2	25	95.0	5.0
20	25	60.0	40.0
30	25	10.0	90.0
35	25	95.0	5.0

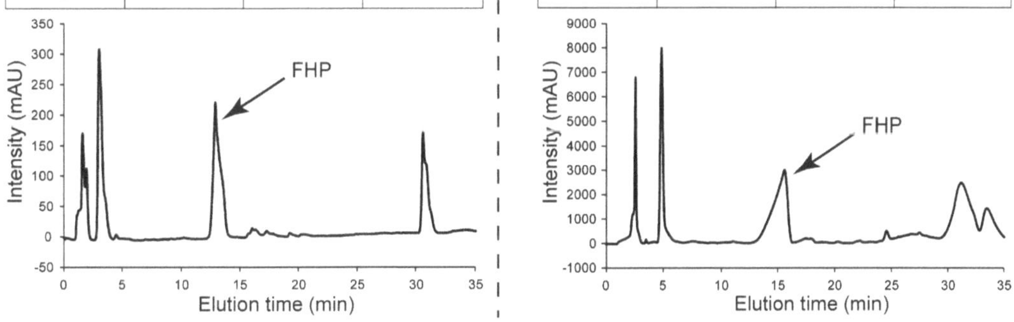

Fig. 3 (**a**) Table of the analytical HPLC gradient (top) and the HPLC trace (bottom) for the peptide mixture cleaved from resin. The peak representing FHP is labeled out. (**b**) Table of the preparative HPLC gradient (top) and the HPLC trace (bottom) for the same mixture. The fractions containing FHP are collected and verified with MALDI-ToF mass spectrometer

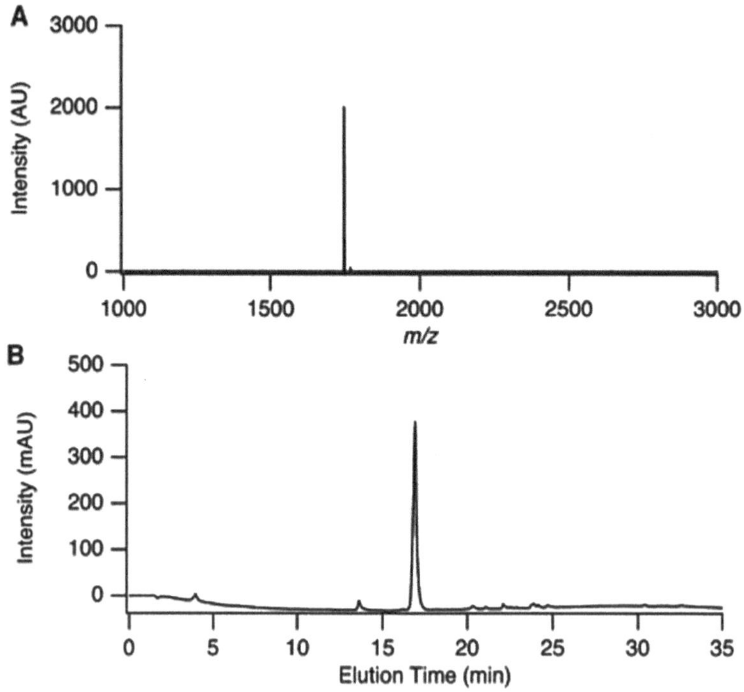

Fig. 4 (**a**) MALDI-ToF mass spectrum and (**b**) the analytical RP-HPLC trace of FHP. The exact mass of FHP reaches 1722.62, and the peak in A estimates the experimental mass to be 1745.4, which is the mass of $[M + Na]^+$. The analytical RP-HPLC trace confirms the purity of the resulting peptide (adapted from ref. [23])

obtained from analytical HPLC in **step 4** (Fig. 3b) (*see* **Note 28**). Collect fractions using a fraction collector.

6. Perform MALDI-ToF to verify the mass of each fraction and find the corresponding fractions of the target peptide (Fig. 4a). The appropriate fractions are collected and concentrated using rotary vaporization to remove the acetonitrile. In our lab, we use α-cyano-4-hydroxycinnamic acid (α-CHCA) as the matrix. Pipet 1 drop of the matrix on an MTP 384 target plate and let it dry for few minutes. Pipet 1 drop of the peptide solution on top of the matrix stain. Wait several minutes for the solution to dry and pipet 1 more drop of α-CHCA on top of the solution stain. Commonly, the peptide solution with a concentration of 1–5 mg/mL is considered to be appropriate for characterizing the peptide. Fix the prepared sample plate on the adapter and insert the adapter into the instrument specified in **item 10** in Subheading 2.2. Follow official protocols for MALDI-ToF to identify molecular weights of all major peaks from the HPLC trace.
7. Transfer the concentrated peptide solution into a 50 mL plastic tube and lyophilize the solution using FreeZone −105°C 4.5 L freeze dryer to attain the dry mass of the pure peptide.

8. Re-dissolve the obtained powder and run analytical HPLC to verify the purity of the peptide (Fig. 4b). Calibrate the concentration of peptides, aliquot it into cryo-vials, re-lyophilize it into powder, and store at −20°C.

4 Notes

1. The type of resin used depends on the molecular design and cleavage conditions. Wang resin is used when the *C*-terminus has a negatively charged carboxylic group, and Rink Amide MBHA resin is used when the *C*-terminus has an amide group.
2. Fmoc deprotection solution can be prepared in large batches and stored in a brown glass bottle at room temperature. However, Mtt deprotection solution should be freshly prepared since long-term storage may lead to the evaporation of TFA or DCM.
3. We find that it is best to prepare the solution fresh each time since long-term storage may induce the hydrolysis of anhydride.
4. Cleavage solutions may vary for different types of resins. In our lab, we typically use a mixture of 95% TFA, 2.5% TIS, and 2.5% H_2O for Rink amide MBHA resin or Fmoc-Glu(OtBu)-Wang resin. Otherwise, cleavage solutions should be prepared according to the protocol. Be cautious when handling TFA, since it is a strongly corrosive acid. All solutions should be prepared in a ventilated fume hood. Wear appropriate personal protective equipment when handling these materials.
5. Acid-modified eluent is usually used in HPLC when peptides carry positive charges (amine groups), while base-modified eluent is used for peptides carrying negative charges (carboxylic acid groups). It is strongly encouraged that eluents be filtered through a 0.45 μm PVDF filter using 200 mL syringe to make sure that they will not cause congestion in HPLC system.
6. The Kaiser test solution is employed to monitor the completion of amino acid coupling in solid phase peptide synthesis. Carefully label the reagents and keep them in a ventilated fume hood due to their toxicity.
7. Cold anhydrous ethyl ether will be utilized for precipitation of peptides after cleavage. Since ethyl ether is odoriferous, avoid inhalation by wearing a standard mask with earloops when using ethyl ether and by placing an ice bath in the fume hood.
8. Automated synthesis was performed on an AAPPTec Focus peptide synthesizer according to technical protocols, while manual synthesis was conducted using peptide synthesis vessels and a wrist action shaker. Although manual synthesis is time-

consuming, the yield is normally much higher than that of the automated synthesis. Therefore, in our lab, we usually employ manual synthesis for peptides containing less than five amino acids to give a higher yield.

9. The substitution of each resin is different. For the resin used here, the substitution reaches 0.53 mmol/g. Experimenters should take caution when weighing the amount of resin.
10. Pre-swelling the resin in a plastic tube prevents small resin beads from blocking the filter of the shaking vessel. Refill and empty errors will occur with a blocked filter and the synthesizer will stop automatic synthesis due to a resulting system error.
11. Since the peptide employs 0.25 mmol of resin, 4 equivalents of amino acids require 1 mmol of Fmoc-amino acid for each coupling cycle. Accordingly, 425.5 mg of Fmoc-Glu(OtBu)-OH for one E, 459.5 mg of Fmoc-Tyr(tBu)-OH for one Y, 1790.1 mg of Fmoc-Asn(Trt)-OH in total for three Ns, 1221.4 mg of Fmoc-Gln(Trt)-OH in total for two Qs, and 297.3 mg of Fmoc-Gly-OH for one G are weighed. Also, it should be noted that commercial Fmoc-Glu(OtBu)-Wang resin already includes the C-terminal glutamic acid.
12. Add 5 mL of DMF to 1 mmol of Fmoc-amino acid powder. Sonication and vortexing will help to dissolve the solids. In our lab, we add 1 mL of additional DMF to each amino acid solution as a dead volume for the synthesizer. For 1 mmol of Fmoc-amino acid, add extra 1 mL of DMF to make a 6 mL solution. For 2 mmol of amino acid, the total solution should be 11 mL ($5 \times 2 + 1$) in volume.
13. Elongating the reaction time increases the yield of the reaction but consumes more time. In our lab, we typically set the reaction time to 1 h for the coupling of one amino acid using a synthesizer, and 2 h for the manual coupling of one amino acid. Elongating reaction time might increase the extent of epimerization under some conditions, but the effect of epimerization is often minimal to affect the purity of the product.
14. Automatic synthesis of a peptide with a synthesizer results in a lower yield of resulting peptide than manual synthesis does. However, for the synthesis of long sequences, automated synthesis saves time and prevents human mistakes.
15. The peptide synthesizer should only be employed for standard synthesis processes. Functionalization of other components should be performed manually.
16. The Kaiser test is a qualitative test for the presence or absence of free primary amino groups, and it effectively tests the completion of a coupling step. The resin beads and the solution turn dark blue (positive result) when the primary amine is

present. Without free primary amine being present, the resin beads remain in their original color and the solution stays light yellow (negative result). Here, after the deprotection of Fmoc, the beads and solution should turn dark blue for the presence of free primary amine group. Otherwise, repeat deprotection one more time and confirm again with the Kaiser test.

17. The coupling of Fmoc-Lys(Mtt)-OH enables the branching of a peptide arm. The orthogonality of Fmoc and Mtt protecting groups allows the sequential addition of two arms of different chemical compositions. Another Fmoc-protected amino acid (Bis-Fmoc-L-lysine) can also be used as a branch point; however, it is only suitable for conjugating the same chemical groups at both branches.
18. The characteristic yellowish color of the Mtt group should be completely removed after this step. If the color does not disappear after 5–7 deprotection cycles, repeat the deprotection process until the color totally fades.
19. The extended reaction time is to achieve a higher yield of the reaction. Extended reaction times can reach up to one full day (also suitable for hydrocarbon coupling in **step 14**).
20. When one or both of the peptide arms is/are not further functionalized with any species (like HP or FP in Fig. 1), acetylation of the arm(s) utilizing 20% acetic anhydride in DMF prevents undesired side reactions of free amine groups. Perform acetylation by treating the resin with 15 mL of 20% acetic anhydride in DMF solution and 150 μL of DIEA. Shake the vessel for 15 min using a wrist action shaker. Empty the vessel and wash with DMF. Repeat the reaction three times. Perform the Kaiser test following **step 9**.
21. Do NOT cap the vessel too tightly. Make sure that the PTFE side of the PTFE-Silicone septa faces the TFA solution. If the silicone side of septa faces the TFA solution, acidic resin cleavage solution will corrode some of the rubber attached to the cap, which significantly affects the purification process of the peptide or potentially ruins the whole reaction.
22. While transferring the TFA solution containing the peptide, utilize a clamp or buttress to prevent solution spillage.
23. Concentrating the TFA solution beforehand facilitates the precipitation process and increases the yield of precipitation.
24. Wrapping the cap of centrifuge tube with parafilm will reduce the escape of ethyl ether odors.
25. Since we have two glutamic acids in our peptide, we use a basic mobile phase (pH = 9) to increase the solubility of crude materials. The total volume of 0.1% *v*/*v* ammonium hydroxide in water and 0.1% *v*/*v* ammonium hydroxide in acetonitrile can

vary according to the amount and the solubility of the peptide precipitates obtained. However, the ratio of acetonitrile is usually less than 50% and the total volume should be a multiple of 10 mL for convenience in preparative HPLC injections.

26. If the precipitates do not dissolve fully, adjust the ratio of basic water/acetonitrile or add more water and acetonitrile to expedite the dissolving process.
27. If the peaks in HPLC trace are not separated enough from each other, you may tune the gradient and re-run the HPLC until the peaks are separated. Analytical HPLC will function as the guidance for running preparative HPLC.
28. Use a preparative HPLC gradient that is similar to the analytical HPLC gradient; change the flow rate to 25 mL/min.

Acknowledgments

This work was supported by the National Science Foundation (DMR/1255281, DMR/1506937, and CHE/1412985) and the National Institutes of Health (NIH/R21CA191740). We acknowledge the JHU Department of Chemistry for the use of mass spectrometer (NSF CHE-0840463).

References

1. Karaborni S, Esselink K, Hilbers PAJ, Smit B, Karthauser J, Vanos NM, Zana R (1994) Simulating the self-assembly of gemini (Dimeric) surfactants. Science 266(5183):254–256
2. Bangham AD, Standish MM, Watkins JC (1965) Diffusion of univalent ions across lamellae of swollen phospholipids. J Mol Biol 13(1):238–252
3. Wong GCL, Tang JX, Lin A, Li YL, Janmey PA, Safinya CR (2000) Hierarchical self-assembly of F-actin and cationic lipid complexes: stacked three-layer tubule networks. Science 288 (5473):2035–2039
4. Percec V, Wilson DA, Leowanawat P, Wilson CJ, Hughes AD, Kaucher MS, Hammer DA, Levine DH, Kim AJ, Bates FS, Davis KP, Lodge TP, Klein ML, DeVane RH, Aqad E, Rosen BM, Argintaru AO, Sienkowska MJ, Rissanen K, Nummelin S, Ropponen J (2010) Self-assembly of janus dendrimers into uniform dendrimersomes and other complex architectures. Science 328(5981):1009–1014
5. Pochan DJ, Chen ZY, Cui HG, Hales K, Qi K, Wooley KL (2004) Toroidal triblock copolymer assemblies. Science 306(5693):94–97
6. Zhang LF, Eisenberg A (1995) Multiple morphologies of crew-cut aggregates of polystyrene-B-poly(Acrylic Acid) block-copolymers. Science 268(5218):1728–1731
7. Hartgerink JD, Beniash E, Stupp SI (2001) Self-assembly and mineralization of peptide-amphiphile nanofibers. Science 294 (5547):1684–1688
8. Harrington DA, Cheng EY, Guler MO, Lee LK, Donovan JL, Claussen RC, Stupp SI (2006) Branched peptide-amphiphiles as self-assembling coatings for tissue engineering scaffolds. J Biomed Mater Res A 78A(1):157–167
9. Webber MJ, Kessler JA, Stupp SI (2010) Emerging peptide nanomedicine to regenerate tissues and organs. J Intern Med 267(1):71–88
10. Silva GA, Czeisler C, Niece KL, Beniash E, Harrington DA, Kessler JA, Stupp SI (2004) Selective differentiation of neural progenitor cells by high-epitope density nanofibers. Science 303(5662):1352–1355
11. Black M, Trent A, Kostenko Y, Lee JS, Olive C, Tirrell M (2012) Self-assembled peptide amphiphile micelles containing a cytotoxic T-cell epitope promote a protective immune

response in vivo. Adv Mater 24 (28):3845–3849

12. Guler MO, Claussen RC, Stupp SI (2005) Encapsulation of pyrene within self-assembled peptide amphiphile nanofibers. J Mater Chem 15(42):4507–4512
13. Soukasene S, Toft DJ, Moyer TJ, Lu HM, Lee HK, Standley SM, Cryns VL, Stupp SI (2011) Antitumor activity of peptide amphiphile nanofiber-encapsulated camptothecin. ACS Nano 5(11):9113–9121
14. Cheetham AG, Zhang P, Lin YA, Lin R, Cui H (2014) Synthesis and self-assembly of a mikto-arm star dual drug amphiphile containing both paclitaxel and camptothecin. J Mater Chem B 2 (42):7316–7326
15. Gao H, Matyjaszewski K (2007) Arm-first method as a simple and general method for synthesis of miktoarm star copolymers. J Am Chem Soc 129(38):11828–11834
16. Khanna K, Varshney S, Kakkar A (2010) Miktoarm star polymers: advances in synthesis, self-assembly, and applications. Polym Chem 1 (8):1171–1185
17. Ludwigs S, Boker A, Voronov A, Rehse N, Magerle R, Krausch G (2003) Self-assembly of functional nanostructures from ABC triblock copolymers. Nat Mater 2(11):744–747
18. Okamoto S, Hasegawa H, Hashimoto T, Fujimoto T, Zhang HM, Kazama T, Takano A, Isono Y (1997) Morphology of model three-component three-arm star-shaped copolymers. Polymer 38(21):5275–5281
19. Li ZB, Kesselman E, Talmon Y, Hillmyer MA, Lodge TP (2004) Multicompartment micelles from ABC miktoarm stars in water. Science 306 (5693):98–101
20. Li ZB, Hillmyer MA, Lodge TP (2006) Morphologies of multicompartment micelles formed by ABC miktoarm star terpolymers. Langmuir 22(22):9409–9417
21. Li ZB, Hillmyer MA, Lodge TP (2006) Control of structure in multicompartment micelles by blending mu-ABC star terpolymers with AB diblock copolymers. Macromolecules 39 (2):765–771
22. Li ZB, Hillmyer MA, Lodge TP (2006) Laterally nanostructured vesicles, polygonal bilayer sheets, and segmented wormlike micelles. Nano Lett 6(6):1245–1249
23. Lin YA, Ou YC, Cheetham AG, Cui HG (2013) Supramolecular polymers formed by ABC miktoarm star peptides. ACS Macro Lett 2(12):1088–1094

Chapter 12

Synthesis and Evaluation of Self-Assembled Nanostructures of Peptide-π Chromophore Conjugates

Tejaswini S. Kale and John D. Tovar

Abstract

Peptides provide a biomolecular scaffold for solubilizing and assembling π-conjugated molecules in aqueous media. The properties of such peptide-π chromophore conjugates can be manipulated by varying the constituent amino acid residues in the peptide backbone as well as the chromophore moieties. Such a precise handle on molecular structure leads to molecules with diverse macromolecular and material properties. We developed a versatile synthetic protocol that leads to a wide range of peptide-π chromophore conjugates in which the chromophores are covalently attached to the peptide backbone such that the chromophore is flanked by two peptide chains. This "trimer" structure lends interesting self-assembly properties to these materials which may be useful for a plethora of biological applications.

Key words Peptides, Peptide-π conjugates, Self-assembly, Stimuli sensitive self-assembly, 1-D nanostructure

1 Introduction

Self-assembly provides a promising approach for building a wide variety of complex nanostructures with varying functionalities [1–4]. Peptides are an interesting class of molecules with a propensity to self-assemble dictated by the H-bonding interactions and the dihedral angles of the constituent amino acid residues. The interactions of peptides in adjacent planes provide avenues for generation of variety of secondary and tertiary structures having geometries including α-helices, β-sheets, and β-turns, to name a few. The formation of these supramolecular structures is dependent on the specific amino acid sequence. This provides a synthetic handle to incorporate specific functionalities into the self-assembled nanostructures. The macroscale properties of such materials can thus be varied by changing the molecular structure of the constituent building blocks. A diverse set of assemblies having distinct secondary and tertiary structures has been designed by varying the peptide composition [5–10]. Peptide-based systems

Bradley L. Nilsson and Todd M. Doran (eds.), *Peptide Self-Assembly: Methods and Protocols*, Methods in Molecular Biology, vol. 1777, https://doi.org/10.1007/978-1-4939-7811-3_12,

have been investigated for various bio-nano-technological applications such as supports for functional and responsive materials and sensors [11, 12].

Ordered arrays of π-conjugated molecules are known to be effective energy and charge transporting materials. For biological applications, these chromophores must be rendered water soluble and hydrolytically stable. In nature, the formation of these arrays may be driven by the self-assembling nature of the surrounding amino acids in peptide or protein segments [4, 5, 13], nucleotides in DNA or RNA fragments [5, 14], or sugars in carbohydrates [15]. To generate such materials in the laboratory, biomolecules that promote hierarchical ordering as well as water solubility may be covalently attached to the π-conjugated unit.

We developed versatile synthetic strategies to incorporate a library of π-conjugated chromophores into peptide backbones easily attainable using "on-resin" dimerization techniques (Scheme 1) [16, 17]. These peptide-π chromophore conjugates self-assemble in response to a change in pH of the environment, assembling into 1-D nanostructures under acidic conditions when acidic amino acid residues are present or under alkaline conditions when basic residues are incorporated into the peptide backbone. The chromophore within the peptide backbone can be utilized to evaluate the nature of the assembly as the photophysical properties of chromophores are sensitive to their local environment. These chromophores can also be utilized as tools, for instance, as energy and electron transport functionalities as they are assembled into ordered arrays within the peptidic nanostructure.

Here, a representative procedure for synthesis of a peptide containing the quaterthiophene core (as a prototypical π-conjugated oligomeric core) is described (Scheme 2). The synthetic methodology can be adopted for a number of different core chromophores [16]. The peptide-π chromophore conjugates were synthesized using the general solid phase peptide synthesis (SPPS) procedure starting from Wang resin preloaded with the first amino acid followed by on-resin *N*-acylation and dimerization under appropriate air-free palladium catalyzed conditions. The peptides were cleaved off the resin and purified to obtain pure peptide-π chromophore conjugates as single molecule components. The photophysical behavior of the chromophores in the peptide backbone in the molecularly dissolved state as well as self-assembled state achieved by varying the pH of the solution were evaluated using spectroscopic techniques. The 1-D nanostructures were imaged using transmission electron microscopy (Fig. 1).

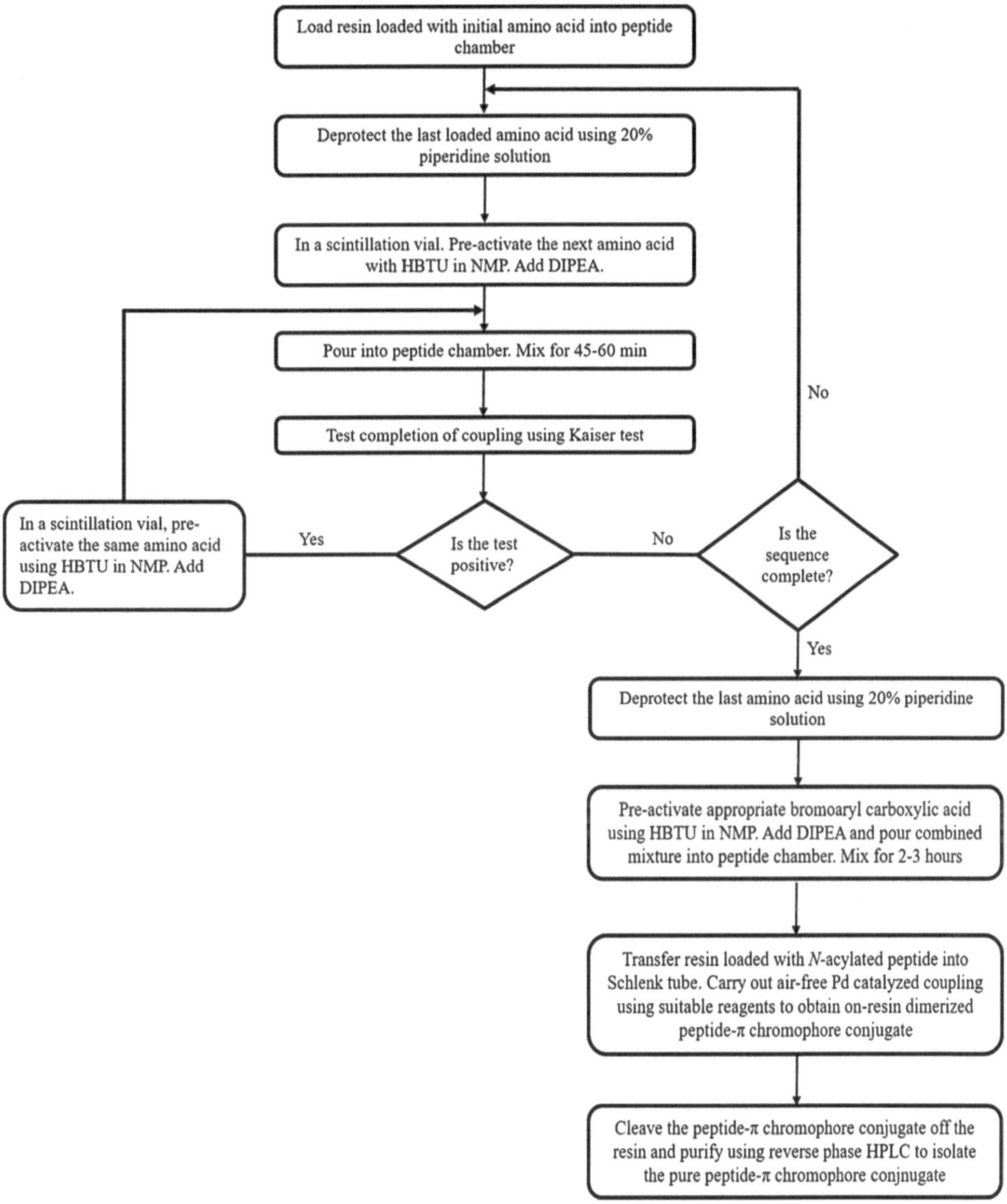

Scheme 1 Flow chart for the general synthesis of peptide-π chromophore-peptide triblock molecules

2 Materials

2.1 Materials for Peptide Synthesis

1. *N*-Methylpyrrolidone (NMP).
2. Wang resin (Wang- Asp(OtBu), amino acid loading of 0.6 mmol/g).
3. Fmoc-protected amino acids.

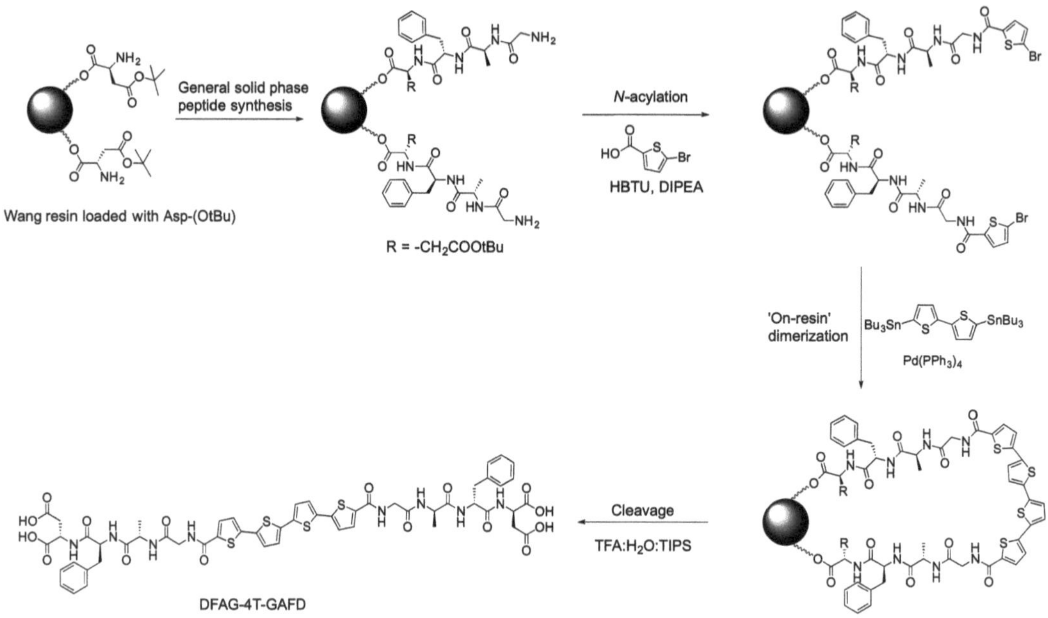

Scheme 2 Synthesis of DFAG-4 T-GAFD peptide-π chromophore-peptide triblock molecule

4. Methanol.
5. O-(Benzotriazol-1-yl) *N*,*N*,*N*′,*N*′-tetramethyluronium hexafluorophosphate (HBTU).
6. Tetrahydrofuran (THF).
7. Dry diethyl ether obtained from an Innovative Technologies Pure Solv solvent purification system and dried over 3 Å molecular sieves.
8. *N*,*N*-dimethylformamide (DMF).
9. Dichloromethane (DCM).
10. Dry *N*,*N*-diisopropylethylamine (DIPEA) dried over 4 Å molecular sieves.
11. Deprotection Solution: 20% piperidine in DMF (v/v).
12. Tetrakis(triphenylphosphine)palladium.
13. 5,5′-bis-tributylstannyl-2,2′-bithiophene, and 5-bromo-2-thiophenecarboxylic acid, which can be prepared using procedures reported previously in the literature [18, 19].
14. Solid state peptide synthesis chamber fitted with a coarse frit (Chemglass Inc., CG-1860-12).
15. A Barnstead Thermolyne Labquake Shaker/Rotator.
16. Cannula.
17. Nitrogen gas.
18. Heating stir plate capable for reaching temperatures of 80 °C and an oil bath.

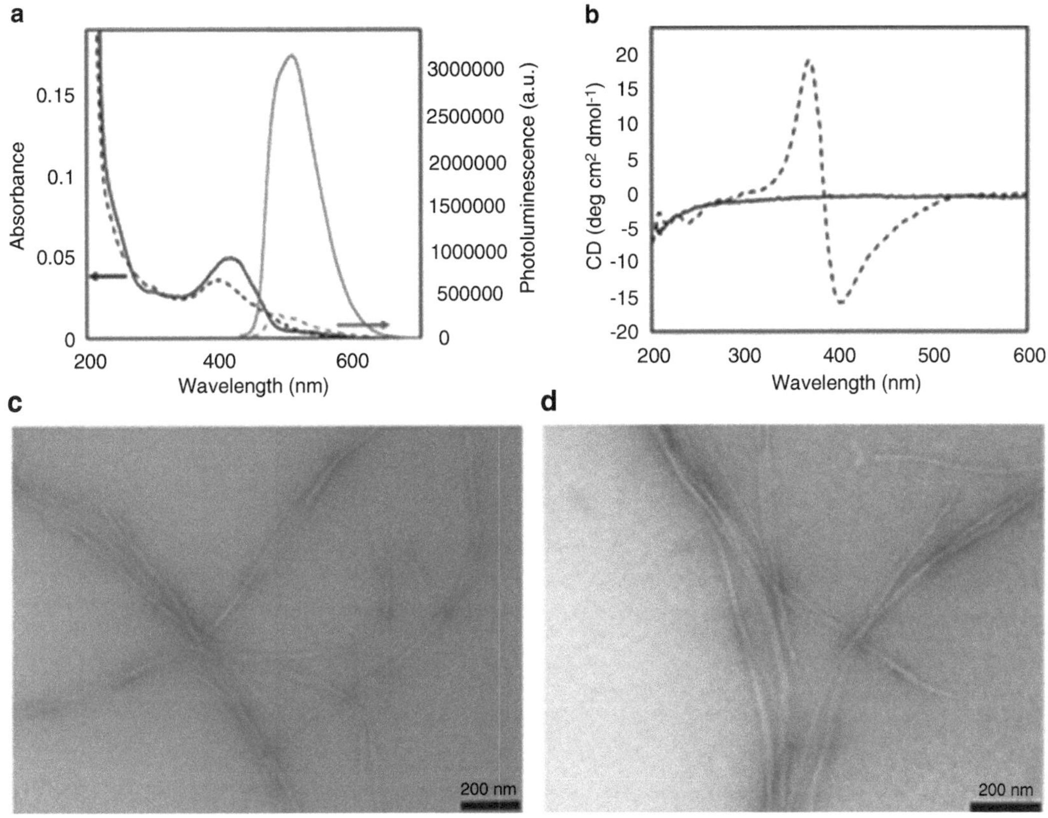

Fig. 1 UV-vis and photoluminescence spectra (**a**), circular dichroism (**b**) and TEM images (**c**, **d**) of DFAG-4T-GFAD peptide-quaterthiophene conjugates. The spectra in (**a**) and (**b**) were recorded at pH 10 (solid lines) and at pH 2 (dashed lines)

19. Isopropanol.
20. Hexanes.
21. Ninhydrin.
22. Scintillation vial.
23. Vortexer.
24. Sonicator.
25. Schlenk Tube.
26. Cleavage solution: 95% TFA, 2.5% triisopropylsilane, 2.5% water (v/v/v).

2.2 Peptide Purification and Characterization

1. Freshly filtered Millipore water (resistivity 18.2 MΩ cm).
2. Reverse phase HPLC and C8 column (Luna 5 μm, 250 × 21.20 mm and 250 × 4.60 mm).
3. HPLC Mobile-phase: Ammonium formate aqueous buffer (1% (v/v), pH 8) prepared using HPLC grade water and acetonitrile.

4. FT NMR spectrometer.
5. Thermo Finnigan LCQ Deca Ion Trap Mass Spectrometer.
6. Mass Spectrometry Solvent: 1:1 MeOH:water solution with 1% ammonium hydroxide.
7. 1 M NaOH in water.
8. 1 M HCl in water.
9. UV-Vis spectrophotometer.
10. Fluorometer with Xenon lamp.
11. CD spectropolarimeter.
12. Transmission Electron Microscope with CCD digital camera.
13. NMR spectrometer. These spectra were acquired using a 1 s presaturation pulse to suppress the water signal.
14. 200 mesh copper grid coated with formvar.
15. TEM stain solution: 2% uranyl acetate solution.
16. Parafilm.
17. Filter Paper.

3 Methods

All solvents were degassed by purging with nitrogen for 30–90 min before use.

3.1 General Solid Phase Peptide Synthesis (SPPS) Procedure

1. Place the appropriate amount of Wang resin preloaded with the first amino acid into a peptide synthesis chamber.
2. Allow the resin to swell in 10 mL dichloromethane (DCM) for 10 min. Remove the DCM by vacuum filtration through the glass frit in the chamber.
3. To the swollen resin beads, add 10 mL of deprotection solution and gently shake the mixture at room temperature for 2 min. Remove the supernatant by vacuum filtration.
4. Again, add 10 mL of deprotection solution to the peptide chamber and agitate the mixture for 10 min at room temperature.
5. Remove the supernatant by vacuum filtration. Wash the resin thrice each with 10 mL NMP, 10 mL methanol, and 10 mL DCM (*see* **Notes 1** and **2**). After the second DCM washing, add 10 mL DCM into the peptide chamber and allow the resin to swell for 10 min.
6. While the resin beads swell, place 3 eq. (based on initial resin loading) of the appropriate Fmoc-protected amino acid relative to resin loading and 2.9 eq. of HBTU in a 20 mL scintillation vial and dissolve these compounds in 10 mL NMP in order to

activate the amino acid prior to addition into the peptide chamber. Sonicate the solution until a clear solution is obtained (*see* **Notes 3** and **4**).

7. To this solution, add 10 eq. of degassed DIPEA (based on initial resin loading) and mix the solution using a vortexer for 30 s (*see* **Notes 5** and **6**).
8. After 10 min swelling, remove the solvent using vacuum filtration and add the externally activated amino acid solution into the peptide chamber. Rinse the vial with ~2 mL NMP and add the washings into the peptide chamber. Agitate the mixture for 45–90 min.
9. After coupling, remove the supernatant using vacuum filtration and wash the resin thrice each with 10 mL NMP, 10 mL methanol, and 10 mL DCM. As previously described, allow the resin to swell in DCM for 10 min before commencing the next step.
10. Monitor the coupling using a Kaiser test on a few dry resin beads [20].
11. Once the coupling is deemed complete, repeat **steps 3–11** until the desired oligopeptide sequence is obtained.
12. Following completion of the oligopeptide sequence, deprotect the last amino acid residue by following **steps 3–5**.

3.2 General N-Acylation Procedure for Peptides

1. In a 20 mL scintillation vial, take 3 eq. of 5-bromothiophene-2-carboxylic acid and 2.9 eq. of HBTU (based on initial resin loading) and dissolve in 10 mL NMP in order to activate the acid prior to addition into peptide chamber. Sonicate the solution until a clear solution is obtained.
2. To this solution, add 10 eq. of degassed DIPEA (based on initial resin loading) and mix the solution using a vortexer for 30 s.
3. Add this solution to the swollen resin beads and agitate the mixture for 2–3 h.
4. After coupling, remove the supernatant using vacuum filtration and wash the resin thrice each with 10 mL NMP, 10 mL methanol, and 10 mL DCM. As previously described, allow the resin to swell in DCM for 10 min before commencing next step.

3.3 General On-Resin Stille Coupling and Cleavage Procedure

1. Transfer the resultant *N*-acylated peptide capped with the aryl bromide (2 eq. of–Br) to a Schlenk tube equipped with a reflux condenser. Dry the resin under vacuum for 30 min.
2. Add $Pd(PPh_3)_4$ (5.0 mol%, per –Br eq., based on initial resin loading) to the reaction vessel and place the flask under vacuum for about 10 min and back fill with nitrogen.
3. Prepare an approximately 15 mM solution of 5,5′-bis(tributylstannyl)-2,2′-bithiophene (0.5 eq based on initial resin

loading) in dry DMF. Degas this solution by bubbling nitrogen for about 30 min (*see* **Notes 7** and **8**).

4. Transfer this solution to the flask via cannula and place the reaction vessel into a preheated oil bath at 80 °C for 16 h. Agitate the reaction mixture by constantly bubbling nitrogen through the solution.
5. After completion of reaction time, cool the reaction mixture to room temperature and transfer the resin beads into a peptide chamber. Remove the supernatant solution using vacuum filtration and wash the beads thrice with 10 mL each of NMP, DMF, water, THF, isopropanol, DCM, and hexanes.
6. Cleave the peptide from the resin by adding 10 mL of cleavage solution and agitating the mixture for 2–3 h (*see* **Notes 9** and **10**).
7. Collect the filtrate using positive nitrogen gas pressure and wash the resin thrice with 5 mL DCM (*see* **Note 11**). Combine the organic layer with the peptide solution and concentrate the combined solution by removing the solvent in vacuo until the volume is reduced by ~90% (*see* **Note 12**).
8. Precipitate the peptide by pouring 90 mL cold (0 °C) diethyl ether into the peptide solution. Isolate the peptide solids by centrifugation and discard the supernatant (*see* **Note 13**).
9. Triturate the peptide solids with 90 mL diethyl ether (room temperature) and isolate the crude peptide solids by centrifugation. Discard the supernatant solution (*see* **Note 14**).
10. Dissolve the peptide solids thus obtained in 5 mL of deionized water and 50 μL of ammonium hydroxide to obtain a clear solution. Lyophilize this solution to obtain the crude peptide.
11. Purify the crude peptide mixture using reverse phase HPLC.

3.4 Determination of Nanostructure Properties

1. UV-vis absorption, circular dichroism and photoluminescence spectroscopies can be used to analyze the photophysical properties of the peptides in the molecularly dissolved state as well as self-assembled state achieved in response to change in pH of solution.
2. Prepare solutions in Millipore water and adjust the concentration to obtain absorbance of ~0.1 absorbance units.
3. The molecularly dissolved state can be achieved by adding 10 μL of 1 M KOH solution to 3 mL peptide solution.
4. The self-assembled state can be obtained by adding 20 μL of 1 M HCl to the same solution (*see* **Note 15**).
5. Samples for CD should be prepared at ~20 times the concentration used for UV-vis experiments.

6. To generate self-assembled peptide nanostructures for TEM imaging, acidify 100 μL of 1 mg/mL solution of peptide in Millipore water using 10 μL of 1 M HCl solution.
7. Place a drop of this solution on clean parafilm.
8. Gently float a 200 mesh copper grid coated with formvar onto the solution and incubated for 5 min at 25 °C.
9. Wick off the excess solution by touching the side of the grid to filter paper.
10. Stain the sample with TEM stain solution and wick off the excess moisture by touching to the side of a filter paper.
11. Allow the grid to dry in air before imaging.

4 Notes

1. When washing the resin beads, observe a 10 s standing period after each solvent addition, to ensure even interaction of beads with solvent, followed by removal of solvent using vacuum filtration. This will ensure that all beads get compressed upon exposure to methanol and similarly, are allowed swelling time in DCM.
2. Every time the resin beads are to be washed after the deprotection step as well as after a coupling step, the insert in the peptide chamber cap should be removed and both the cap and the insert should be rinsed with acetone to remove any solution trapped in between these.
3. Allow all amino acids and coupling reagents to warm up to room temperature before weighing to avoid water condensation.
4. The preactivation of amino acid using HBTU should be carried out such that the solution is ready once the resin swelling time is complete to avoid standing of solution by itself which may lead to epimerization of the amino acid [21].
5. DIPEA may be degassed in an oven dried scintillation vial containing dry 3 Å molecular sieves capped with a rubber septum for 30 min by bubbling nitrogen gas. This solution may be used for up to 24 h after degassing.
6. Sometimes, a color change from colorless to yellow was noted upon addition of DIPEA into preactivated amino acid solution. No impact of whether this change was observed was noted on the final peptide yield.
7. Air-free metal catalyzed couplings were carried out while bubbling nitrogen gas continuously through the reaction mixture. No stir bar was used for agitation purposes as this may damage

the resin beads. A 20 G long stainless steel needle was used for bubbling nitrogen ensuring that the needle reached the base of the Schlenk tube so that all the beads were continuously mixed.

8. The flow rate of nitrogen gas should be maintained at a level so as to ensure that the beads are in continuous motion throughout the reaction to avoid settling of beads as this would lower the extent of reaction and therefore the overall reaction yield. Low flow may also lead to blocking of the needle with beads again leading to settling and low conversions.
9. The color of beads has been observed to change from orange to a dark blue or green color upon addition of TFA mixture when cleaving the peptide off the resin beads. This is attributed to oxidation of the oligothiophene unit by TFA [22]. After completion of cleavage process, the resultant peptide solution as well as the precipitated peptide carries forward this color. However, upon addition of water to the dry solids, the expected orange color from the quaterthiophene core is recovered.
10. For cleaving the peptide from the resin, a separate shaker/ rotator is used when agitating the resin beads in TFA mixture to avoid any TFA vapor from interacting with peptides that are still being synthesized and inadvertently cleaving them albeit in possibly low amounts. Typically, the rotator used for peptide synthesis may be a closed chamber and that used for cleavage may be an open chamber.
11. When collecting the filtrate after cleaving the peptide off the resin, positive pressure of nitrogen gas is preferred over vacuum filtration as this avoids the contact of TFA solution with rubber tubing. Sometimes, pieces of rubber were observed to break off after prolonged exposure to TFA and this may contaminate the final peptide.
12. After washing the resin multiple times to collect all the peptide, the beads remain colored. However, further interaction with cleavage mixture or further washing does not yield any significant amounts of product. Therefore, 2–3 h of treatment with the cleavage mixture is deemed sufficient.
13. If precipitation is not observed upon addition of cold diethyl ether, a larger volume of ether may be required. If this still does not yield precipitate, remove solvent in vacuo to reduce volume by ~95%. Dissolve in alkaline water (ammonium hydroxide) and lyophilize.
14. After the first precipitation of peptide solids using diethyl ether, the supernatant may be slightly colored. Triturating once with diethyl ether usually leads to a colorless supernatant. If this is not observed, a second trituration step may be required

followed by a third step using acetonitrile to ensure all organics not covalently bound to the peptide are removed.

15. Photophysical studies may be carried out in one of two ways. (a) A solution of peptide-chromophore conjugate in Millipore water may be made alkaline using 10 μL of 1 M KOH solution/3 mL peptide solution and this solution may be acidified using 20 μL of 1 M HCl solution, or (b) Two separate solutions may be prepared from a stock solution, one solution taken to basic pH and the other to acidic pH. We find that this does not result in any measureable difference in the optical response of the materials.

Acknowledgments

We thank Johns Hopkins University and the Department of Energy Office of Basic Energy Sciences (DE-SC0004857, J.D.T. peptide nanomaterials synthesis) for generous support.

References

1. Palmer LC, Stupp SI (2008) Molecular self-assembly into one-dimensional nanostructures. Acc Chem Res 41:1674–1684
2. Whitesides G, Grzybowski B (2002) Self-assembly at all scales. Science 295:2418–2421
3. Zhang S (2003) Fabrication of novel biomaterials through molecular self-assembly. Nat Biotechnol 21:1171–1178
4. Aida T, Meijer EW, Stupp S (2012) Functional supramolecular polymers. Science 335:813–817
5. Jatsch A, Schillinger EK, Schmid S, Bauerle P (2010) Biomolecule assisted self-assembly of [small pi]-conjugated oligomers. J Mater Chem 20:3563–3578
6. Hartgerink J, Benaish E, Stupp S (2001) Self-assembly and mineralization of peptide-amphiphile nanofibers. Science 294:1684–1688
7. Rayburn J, Cardoso A, Adams D (2013) The importance of the self-assembly process to control mechanical properties of low molecular weight hydrogels. Chem Soc Rev 42:5143–5156
8. Smith K, Tejeda-Montes E, Poch M, Mata A (2011) Integrating top-down and self-assembly in the fabrication of peptide and protein-based biomedical materials. Chem Soc Rev 40:4563–4577
9. Bowerman C, Nilsson B (2012) Review self-assembly of amphipathic β-sheet peptides: insights and applications. Pept Sci 98:169–184
10. Mandal D, Shirazi A, Parang K (2014) Self-assembly of peptides to nanostructures. Org Biomol Chem 12:3544–3561
11. Matson J, Zhao H, Stupp S (2011) Peptide self-assembly for crafting functional biological materials. Curr Opin Solid State Mater Sci 15:225–235
12. Skilling K, Citossi F, Bradshaw T, Ashford M, Kellam B, Marlow M (2014) Insights into low molecular mass organic gelators: a focus on drug delivery and tissue engineering applications. Soft Matter 10:237–256
13. Tovar JD (2013) Supramolecular construction of optoelectronic biomaterials. Acc Chem Res 46:1527–1537
14. Bauerle P, Emge A (1998) Specific recognition of nucleobase-functionalized polythiophenes. Adv Mater 10:324–330
15. Schmid S, Mena-Osteritz E, Kopyshev A, Bauerle P (2009) Self-assembling carbohydrate-functionalized oligothiophenes. Org Lett 11:5098–5101
16. Sanders AM, Dawidczyk TJ, Katz HE, Tovar JD (2012) Peptide-based supramolecular semiconductor nanomaterials via pd-catalyzed solid-phase "Dimerizations". ACS Macro Lett 1:1326–1329

17. Vadehra GS, Wall BD, Diegelmann SR, Tovar JD (2010) On-resin dimerization incorporates a diverse array of pi-conjugated functionalities within aqueous self-assembling peptide backbones. Chem Commun 46:3947–3949
18. Guo X, Watson MD (2008) Conjugated polymers from naphthalene bisimide. Org Lett 10:5333–5336
19. Kilbinger AFM, Schenning APHJ, Goldoni F, Feast WJ, Meijer EW (2000) Chiral aggregates of α,ω-disubstituted sexithiophenes in protic and aqueous media. J Am Chem Soc 122:1820–1821
20. Kaiser E, Colescott RL, Bossinger CD, Cook PI (1970) Color test for detection of free terminal amino groups in the solid-phase synthesis of peptides. Anal Biochem 34:595–598
21. El-Faham A, Albericio F (2011) Peptide coupling reagents, more than a letter soup. Chem Rev 111:6557–6602
22. Zinger B, Mann KR, Hill MG, Miller LL (1992) Photochemical formation of oligothiophene cation radicals in acidic solution and nafion. Chem Mater 4:1113–1118

Chapter 13

Programmable Fabrication of Multilayer Collagen Nanosheets of Defined Composition

Tao Jiang and Vincent P. Conticello

Abstract

Two-dimensional nanostructures offer significant promise as components for the construction of functional biomaterials. However, the controllable fabrication of these structures remains a challenge. Ideally, one desires to control the composition, structure, and surface functionality of the resultant materials with precision, in order to tailor properties for a particular application and minimize the unintended side effects. We recently reported the synthesis of triple-layer nanosheets from template-driven assembly of a negatively charged collagen-mimetic peptide $\mathbf{CP^-}$ on a preassembled nanosheet of a positively charged collagen-mimetic peptide $\mathbf{CP^+}$ [1]. This process enabled the fabrication of nanosheets of defined composition, internal structure, and surface chemistry using a modified layer-by-layer approach. Herein, we describe the synthesis and purification procedures for these two 45-mer peptides, $\mathbf{CP^+}$ and $\mathbf{CP^-}$, and guidelines for the directed assembly of triple-layer structures, along with routine methods of structural analysis.

Key words Peptides, Self-assembly, Collagen, Nanosheets, Triple helix

1 Introduction

Over the last two decades, the supramolecular self-assembly of de novo designed collagen-mimetic peptides has resulted in the successful fabrication of ordered structures in zero [2–4], one [5–13], and two dimensions [14–16], as well as nanofibrous hydrogels [17–19]. Contrary to initial expectations, synthetic collagens appear to be a versatile scaffold for the incorporation of chemically interactive structural elements that can direct the formation of triple helices of defined register and promote specific modes of hierarchical assembly [20]. Recently, we reported the fabrication of two-dimensional nanosheets from self-assembly of collagen-mimetic peptides that incorporated different epimers of the non-canonical imino acid 4-(*R*/*S*)-aminoproline [21, 22]. Structural analysis of these materials provided insight into the mechanism of self-assembly of the nanosheets and suggested

Bradley L. Nilsson and Todd M. Doran (eds.), *Peptide Self-Assembly: Methods and Protocols*, Methods in Molecular Biology, vol. 1777, https://doi.org/10.1007/978-1-4939-7811-3_13,

methods to control the fabrication of more complex, multicomponent two-dimensional assemblies.

Two structurally related peptide sequences, **CP$^+$** and **CP$^-$** [1], were designed that encompassed an asymmetric block-like architecture containing excess positively charged (PRG) and negatively charged (EOG) triads, respectively. Remarkably, while both peptides formed nanosheets, they displayed significant differences in the kinetics of self-assembly and in thermal stability of the resultant nanostructures [1]. The **CP$^+$** peptide rapidly formed large nanosheets from aqueous solution, while the **CP$^-$** peptide formed small nanosheets only after prolonged incubation at high concentration. These results suggested that the surfaces of a preformed **CP$^+$** nanosheet could template the self-assembly of sheets of the **CP$^-$** peptide, which afforded a triple-layer structure with compositional control. The resultant triple-layer nanosheets displayed negative surface charge from the **CP$^-$** peptide and could be employed to bind oppositely charged species at high occupancy at the surfaces of the nanosheet. These results demonstrated that collagen-mimetic peptides represent useful substrates for the construction of functionalized hybrid materials. However, due to the high content of imino acids and arginine, the peptide synthesis and purification remains challenging to implement. In order to reliably achieve formation of these multilayer collagen structures, it is necessary to perform peptide synthesis with high fidelity and to isolate the target peptides in high purity. Moreover, the fabrication of the triple-layer structure requires careful control of the experimental conditions during self-assembly. In this chapter, we describe the solid-phase peptide synthesis for these two collagen-mimetic peptides as well as guidelines for the templated growth of **CP$^-$** layers on preformed **CP$^+$** nanosheets.

CP$^+$: NH-PRGPRGPRGPRGPRGPRGPRGPOGPOGPOG-POGEOGEOGEOGEOG-COOH.

CP$^-$: NH-PRGPRGPRGPRGPOGPOGPOGPOGEOGEO-GEOGEOGEOGEOGEOG-COOH.

P: proline; O: hydroxyproline; G: glycine; R: arginine; E: glutamic acid.

2 Materials

All solutions were freshly prepared in ultrapure water (18 MΩ cm), and used at ambient temperature (20 °C), unless otherwise specified.

2.1 Solid-Phase Peptide Synthesis

1. Dimethylformamide (DMF).
2. Fluorenylmethyloxycarbonyl (Fmoc)-protected glycine-PEG-PS resin (Applied Biosystems).
3. Amino acid solutions:

0.2 M side-chain-protected Fmoc-amino acids in DMF.
Amino acids used in this study: Fmoc-arginine (Pbf)-OH, Fmoc-glutamic acid (OtBu)-OH, Fmoc-glycine-OH, Fmoc-proline-OH, and Fmoc-hydroxyproline (tBu)-OH (Anaspec).

4. Fmoc-deprotection solution:

   20% piperidine and 0.1 M 1-hydroxybenzotriazole (HOBt) in DMF.

5. Coupling activating solution:

   0.5 M *O*-benzotriazol-*N,N,N′,N′*-tetramethyluronium hexafluorophosphate (HBTU) in DMF.

6. Coupling activating base solution:
   2 M *N,N*-diisopropylethylamine (DIEA) in *N*-methyl pyrrolidone.

2.2 Peptide Cleavage and Deprotection

1. Peptide cleavage and side-chain deprotection cocktail:

   92.5% trifluoroacetic acid (TFA), 2.5% triisopropylsilane, 2.5% 2,2′-ethylenedioxydiethanethiol, and 2.5% water.

2. Acetone.
3. Diethyl ether.
4. Disposable 10 mL polypropylene column with porous polyethylene discs (Thermo).
5. 20 mL glass vial (Scintillation).
6. Wheaton dry-seal vacuum desiccator (Sigma).

2.3 Peptide Purification and Characterization

1. Reverse-phase high-performance liquid chromatography (HPLC) solutions:
   Solvent A: HPLC grade water with 0.1% TFA
   Solvent B: HPLC grade acetonitrile with 0.1% TFA.
2. Liquid nitrogen.
3. Matrix-assisted laser desorption/ionization (MALDI) matrix solution:
   Add 10–20 mg α-cyano-4-hydroxycinnamic acid (CHCA) to 1 mL of 50:50 water/acetonitrile with 0.1% TFA (saturated solution). Vortex 5 min to facilitate dissolution, centrifuge the sample to remove the undissolved matrix. Employ the clear supernatant solution for peptide MS analysis.

2.4 Peptide Self-Assembly and Primary Characterization

1. Slide-A-Lyzer dialysis cassette G2 MWCO = 2000 Da (Thermo).
2. MES buffer:

   20 mM 2-*N*-(morpholino)ethanesulfonic acid (MES) in water, pH 6.0.

3. Diluted MES buffer:

   0.5 mM MES in water, pH 6.0.

4. Hellma quartz cuvettes, demountable cells, 0.1 mm path length.
5. Carbon film 200 mesh copper grid (Electron microscopy sciences).
6. Negative stain solution:

   1% uranyl acetate (Electron microscopy sciences) dissolved in water and filtered through a 0.2 μm syringe filter (Whatman) before use.
7. Mica disc, V-1 grade (2spi).
8. Silicon tip, 3-lever with a force constant of 54–16 N/m (μmasch).
9. Silicon wafer block test grating with step height of 20 ± 1 nM (Ted Pella).

2.5 Instrumentation

1. Solid-phase peptide synthesis: CEM liberty microwave peptide synthesizer.
2. Reverse-phase HPLC: Waters 600 series modular HPLC with an Atlantis C18 semipreparative column (19 × 250 mm OBD 10 μm) for semipreparative HPLC and an Atlantis C18 analytical column (4.6 × 250 mm OBD 5 μm) for analytical HPLC.
3. Peptide lyophilization: FTS Flexi-Dry MP freeze dryers.
4. Peptide mass characterization: Applied biosystem 4700 proteomics analyzer as a matrix-assisted laser desorption/ionization (MALDI) mass spectrometer.
5. Peptide thermal annealing: Precision microprocessor controlled 280 series water bath.
6. Circular dichroism (CD): Jasco J-810 circular dichroism spectropolarimeter.
7. Atomic force microscopy (AFM): JEOL, JSPM-4200 scanning probe microscope.
8. Transmission electron microscope (TEM): Hitachi H7500 transmission electron microscope.

3 Methods

All procedures were performed at room temperature unless indicated otherwise.

3.1 Solid-Phase Peptide Synthesis

The CEM liberty microwave peptide synthesizer adopts the general method of the Fmoc chemistry [23], and uses microwave energy to improve synthesis efficiency [24]. The resin is placed in a reaction vessel, into which reagents and solvents were introduced through application of nitrogen pressure. The general procedure and setup for synthesis are as follows:

1. Input peptide sequences and set synthesis scale to 0.10 mmol (corresponding to a 30 mL reaction-vessel used in this peptide synthesizer). Select the options of *C*-terminal acid and *N*-terminal amine for peptides $\mathbf{CP^+}$ and $\mathbf{CP^-}$. In this study, we used a low-loading (0.19 mmol/g) resin, Fmoc-glycine-PEG-PS for both of the 45-mer peptides.
2. Swell the resin and perform a two-step deprotection to remove Fmoc group from amine group of the attached glycine on resin. First deprotection cycle: transfer Fmoc-deprotection solution (7 mL) to the reaction vessel, and perform an initial 30 s deprotection with 14 Watts of microwave power. Drain the solvent and wash the resin using DMF. A second deprotection cycle: deliver Fmoc-deprotection solution (7 mL) to the reaction vessel, and perform a 3 min deprotection with 14 Watts of microwave power at 75 °C. Drain the solvent and wash the resin.
3. To incorporate amino acids to the amine on resin, transfer fivefold excess amino acid solution (2.5 mL of 0.2 M solutions) and activator (1.0 mL) and activator base (0.5 mL) solutions to the reaction vessel for amino acid coupling. Repeat the coupling process twice for each residue for better efficiency, because the steric effect of imino acid (proline and hydroxyproline) incorporation and the γ-lactam formation from arginine reduce the coupling efficiency. For peptides $\mathbf{CP^+}$ and $\mathbf{CP^-}$, optimized conditions for coupling process is reacting 330 s at 75 °C with 8 Watts of microwave energy. Special coupling condition should be used for arginine (first coupling cycle: 25 min at 25 °C with 0 Watt microwave; second coupling cycle: 330 s at 75 °C with 8 Watts of microwave power) (*see* **Note 1**).
4. Drain the solvent and wash the resin. Repeat steps of Fmoc-deprotection and amino acid coupling using designated amino acids to afford full-length peptides.

3.2 Peptide Cleavage and Deprotection

1. Wash the resin thoroughly using acetone (3 × 15 mL). Dry the resin under vacuum for at least 30 min. Transfer the dried resin into a clean glass vial containing a stir bar.
2. Prepare 10 mL of the peptide cleavage and side-chain deprotection cocktail listed in Subheading 2.2.
3. Slowly add 7 mL of the cleavage reagent to cover the resin in the glass vial, and stir the resin/cocktail mixture slowly in the capped glass vial for 3 h.
4. Transfer the resin/cocktail mixture to the disposable 10 mL polypropylene column with filter. Collect filtrate in a 50 mL falcon tube. Rinse the resin in the column with 3 mL of the cleavage cocktail to recover cleaved peptide left on and between resin beads.

5. Add cold (4 °C) diethyl ether (30 mL) to the falcon tube containing filtrate to precipitate crude peptides from the TFA solution.
6. Centrifuge the falcon tube at 2000 × *g* for 10 min, and remove supernatant. Re-suspend the pellet with additional cold diethyl ether (30 mL) and repeat centrifugation. Dry the peptide pellet in a vacuum desiccator.

3.3 Peptide Purification via Reverse-Phase HPLC

1. Dissolve the crude peptides in 5% solvent B at a concentration of 1–5 mg/mL.
2. Equilibrate the semipreparative C18 column with 5% solvent B for more than 30 min with a flow rate of 20 mL/min. Inject the crude peptide solution (100 μL) into the column, and run a linear gradient of 5–45% solvent B with a steepness at 1%/min and a flow rate at 20 mL/min. Detect the elution of peptides by monitoring UV absorption at 220 nm versus peptide retention time. Collect peptide peaks for subsequent analysis.
3. Use MALDI mass spectrometry to assess components within HPLC fractions. To prepare peptide samples for MALDI, mix 1 μL of CHCA matrix solution with 1 μL of HPLC fractions. Deposit 0.5 μL of the mixture onto the MALDI target plate, and leave it air dry prior to MS assessment (Fig. 1). Purified peptides are analyzed for the presence of impurities via analytical RF HPLC using the same linear gradient as semipreparative RF HPLC, and a flow rate at 1 mL/min (Fig. 1).
4. Combine the HPLC fractions containing pure peptide. Remove most of the solvent via rotary evaporation. Peptide solutions were flash frozen in liquid nitrogen and lyophilized to afford colorless solids.

3.4 Peptide Self-Assembly

1. Dissolve all the peptide powder in up to 10 mL water. Dialyze the peptide solution in a 15 mL dialysis cassette (MWCO = 2000 Da) against 4 L water for 3 × 6 h at room temperature (*see* **Note 2**).
2. Dialyzed peptides are subject to lyophilization. Store the peptide powder at −20 °C for future use. Before use, remove the peptides from cold storage and allow them to equilibrate to room temperature in a desiccator (*see* **Note 3**).
3. Weigh 0.5 mg $\mathbf{CP^+}$ and 1.0 mg $\mathbf{CP^-}$ into separate 1.5 mL Eppendorf tubes. Add 1 mL of MES buffer (20 mM, pH 6.0) to afford samples of $\mathbf{CP^+}$ (0.5 mg/mL) and $\mathbf{CP^-}$ (1.0 mg/mL) at the appropriate concentration. Adjust pH to 6.0 by dropwise addition of a solution of sodium hydroxide (0.1 M).
4. To initiate individual peptide self-assembly, put sample tubes (sealed with para-film) in a water bath at 85 °C for 45 min for peptide thermal annealing. Then turn off heater, and allow the

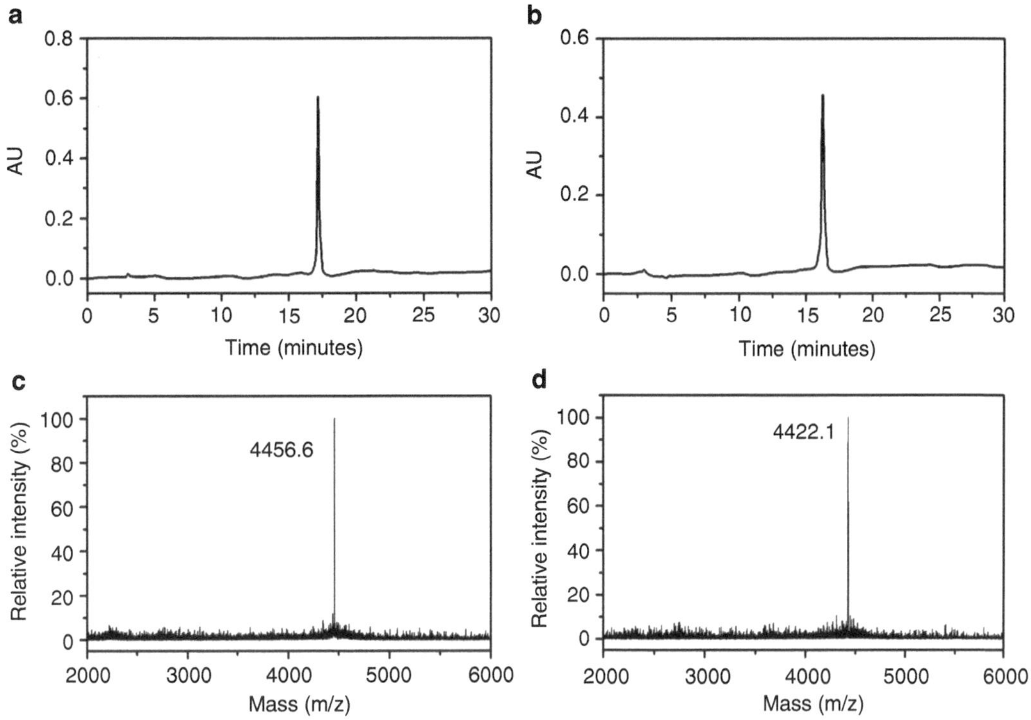

Fig. 1 Analytical RP HPLC trace of purified peptides **CP$^+$** (**a**) and **CP$^-$** (**b**). MALDI mass spectra of purified peptides **CP$^+$** (**c**) and **CP$^-$** (**d**)

water bath to cool down to ambient temperature (with the water bath lid on). Employ TEM and AFM to investigate peptide self-assembly at ambient temperature. The **CP$^+$** solution (0.5 mg/mL) appears cloudy after annealing. Single-layer sheets can be detected from **CP$^+$** solution after one-day incubation (Figs. 2a and 3a). No sheet formation could be observed within weeks from **CP$^-$** solution under the same conditions.

5. The preformed **CP$^+$** sheet can be employed as a substrate to template new layer formation derived from **CP$^-$**. After one-day incubation of the individual peptide solutions, add 50 μL of unassembled **CP$^-$** peptide (0.1 mg/mL) into 50 μL of **CP$^+$** sample containing preassembled nanosheets (peptide concentration, 0.5 mg/mL). Layered structures can be observed that indicate small sheets located on larger sheets (Figs. 2b and 3b).
6. At a stoichiometric ratio of **CP$^-$**:**CP$^+$** more than 2:1, mature triple-layer sheet can be formed. After one-day incubation of the individual peptide solutions, add 50 μL of **CP$^-$** peptide (1.0 mg/mL) into 50 μL of the **CP$^+$** sample with preformed nanosheet (0.5 mg/mL). Incubate the mixture at room temperature for 7 days to promote the slow formation of the mature triple-layer (**CP$^-$/CP$^+$/CP$^-$**) sheets (*see* Fig. 2c and 3c, **Note 4**).

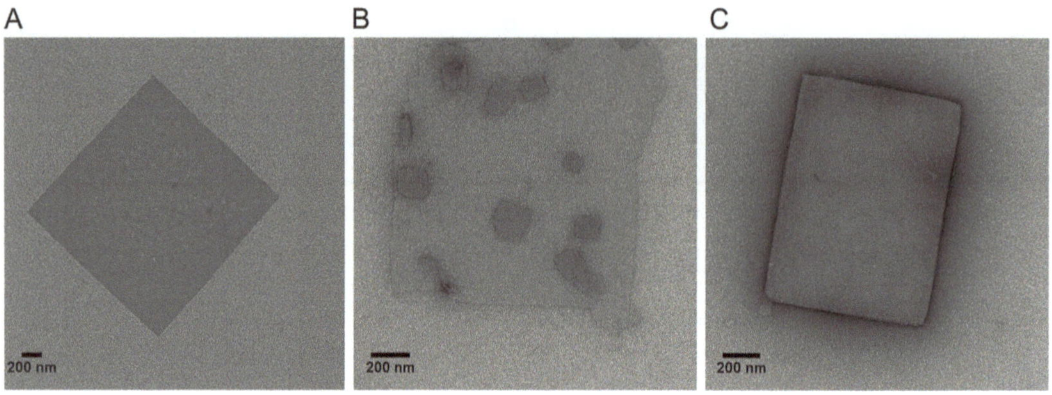

Fig. 2 TEM images of a single-layer nanosheet resulting from self-assembly of peptide **CP⁺** (0.5 mg/mL) (**a**), a multilayer structure resulting from addition of **CP⁻** to preformed nanosheets of **CP⁺** (**CP⁻**/**CP⁺**; 0.1/0.5 mg/mL) (**b**), and mature triple-layer **CP⁻**/**CP⁺**/**CP⁻** nanosheets (**CP⁻**/**CP⁺**; 1.0/0.5 mg/mL) after 7 days incubation (**c**)

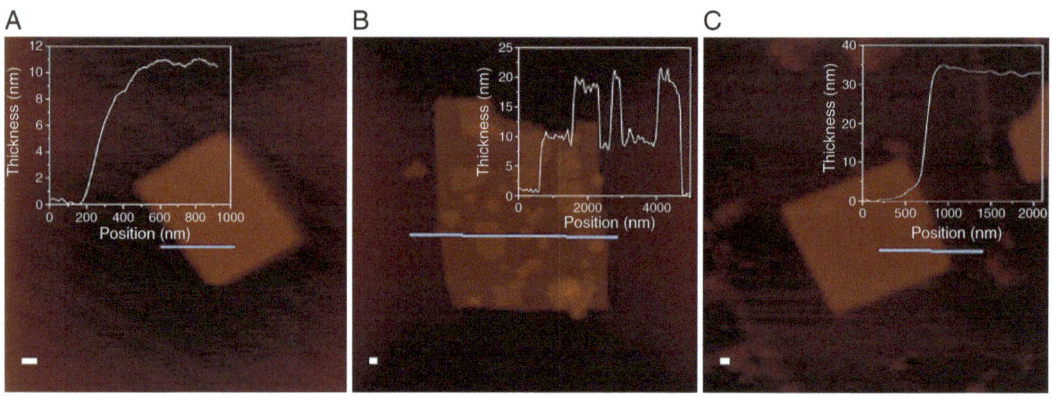

Fig. 3 Representative tapping mode AFM images for single-layer sheet from **CP⁺** (0.5 mg/mL) (**a**), an incomplete multilayer structure resulting from addition of **CP⁻** to preformed nanosheets of **CP⁺** (**CP⁻**/**CP⁺**; 0.1/0.5 mg/mL) (**b**), and mature triple-layer **CP⁻**/**CP⁺**/**CP⁻** nanosheets (**CP⁻**/**CP⁺**; 1.0/0.5 mg/mL) after 7 days incubation (**c**). Scale bars denote 200 nm. Reprinted with permission from [1] Copyright 2015 American Chemical Society

3.5 Primary Characterization

3.5.1 Circular Dichroism (CD) Spectropolarimetry

CD can be employed to identify secondary structure of collagen-mimetic peptides, and assess thermal stability of the triple helices (Fig. 4).

1. Weigh 1 mg **CP⁺** and 1 mg **CP⁻** into separate 1.5 mL Eppendorf tubes. Add 1 mL of MES buffer (20 mM, pH 6.0) to afford CD samples of **CP⁺** and **CP⁻** (1.0 mg/mL). Adjust pH to 6.0 using sodium hydroxide. Perform thermal annealing of the CD samples at 85 °C for 45 min.
2. After samples cool to ambient temperature, store them at 5 °C for 24 h.

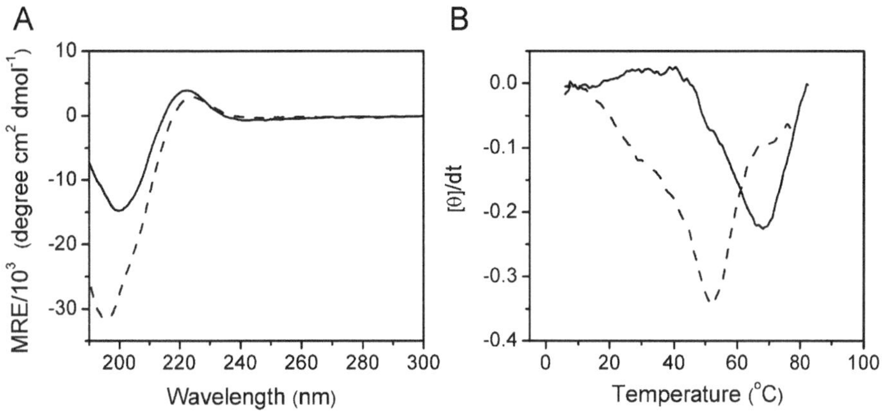

Fig. 4 CD spectra of peptide **CP$^+$** (solid line) and **CP$^-$** (dash line) (**a**). First derivative of the CD melting curves at 223 nm as a function of temperature. The melting temperatures can be estimated at the minimum of the first derivative curves (**b**). Reprinted with permission from [1] Copyright 2015 American Chemical Society

3. Set the temperature of sample chamber in CD spectropolarimetry to 5 °C. Transfer 40 μL of the cold CD samples to demountable CD quartz cell. Seal the edge of cell with parafilm to prevent leakage. Load the cell into CD sample chamber and equilibrate for 10 min.
4. Start CD measurement and record spectra from 300 to 190 nm with a scanning rate at 100 nm/min and a resolution at 0.5 nm. Characteristic CD signatures for collagen triple helix can be detected (Fig. 4a).
5. To assess the thermal stabilities of the triple helices, raise temperature of the sample chamber from 5 to 85 °C with a heating rate at 0.1 °C/min, and record the maximum CD absorption (223 nm in this study) as a function of temperature. Samples of **CP$^+$** and **CP$^-$** will exhibit sigmoidal melting curves.
6. Make the first derivative plot of the melting curves, and estimate melting temperature (T_m) of **CP$^+$** and **CP$^-$** triple helices from the minimum in the plot (Fig. 4b).

3.5.2 Transmission Electron Microscopy (TEM)

The morphology of peptide assemblies is investigated using TEM (Fig. 2).

1. Deposit an aliquot (4 μL) of the peptide assembles (*see* Subheading 3.4) onto a TEM grid, and wait for 1 min for adsorption.
2. Wick away excess liquid using filter paper, and stain the grid with 1% uranyl acetate for 45 s (*see* **Note 5**). Wick away excess stain solution using filter paper.
3. Dry the sample grid in vacuum desiccator before TEM measurement at an accelerating voltage of 75 kV.

3.5.3 Atomic Force Microscopy (AFM)

Height measurements of peptide nanosheets or layered structures are performed using AFM (Fig. 3).

1. Assembled nanosheets and layered structures (*see* Subheading 3.4) will sediment at the bottom of tube. After 1-week incubation at room temperature, carefully remove the clear supernatant using pipette, and re-suspend the precipitates in diluted MES buffer (0.5 mM, pH 6.0).
2. Use thick tape to cleave mica disc several times to achieve an atomically flat surface.
3. Deposit 10 μL of peptide assemblies onto the freshly cleaved mica disc, and wait for 5 min for adsorption.
4. Dry the mica surface under a flow of nitrogen gas.
5. Calibrate the AFM *z*-axis using a block test grating with a known step height, and operate AFM height measurement in the tapping mode.

4 Notes

1. The imino acids proline and hydroxyproline reduce amino acid coupling efficiency due to the steric hindrance associated with the pyrrolidine nitrogen. A trial synthesis using a method based on single rounds of coupling (8 Watts 330 s, 75 °C) for proline, hydroxyproline, and glycine resulted in truncated peptides due to incomplete coupling. In this study, a method based on double coupling is preferred for all residues in $\mathbf{CP^+}$ and $\mathbf{CP^-}$, which produces crude peptides with higher purity. Microwave energy leads to rapid heating at molecular level, and increases reaction rate of coupling [24, 25]. However, higher microwave energy (>13 Watts 330 s, 75 °C) was found to induce reaction vessel overheating that resulted in lower purity of crude peptide. Here we used a microwave power of 8 Watts for peptide synthesis.
2. Residual TFA interferes with peptide self-assembly and inhibits the triple layer formation of the two peptides. Excess TFA was removed from peptides by dialyzing the peptide solutions against water thoroughly.
3. Exposure to moisture will thwart long-term storage of peptide and cause inaccuracy when weighing peptides. Vials of lyophilized peptides were removed from cold storage and were placed in a desiccator first for equilibration to ambient temperature.
4. Fabrication of the triple-layer sheets requires prior synthesis of the $\mathbf{CP^+}$ sheets as a template to promote growth of two $\mathbf{CP^-}$ layers. Thermal annealing of the premixed $\mathbf{CP^+}$ and $\mathbf{CP^-}$

peptides in MES buffer resulted in amorphous aggregates, which probably arises from uncontrolled assembly and precipitation.

5. Other common negative stains have also been used investigated, such as uranyl formate, methylamine tungstate and methylamine vanadate. However, peptide nanosheet structures were not observed in TEM measurements after staining with the negatively charged stains methylamine tungstate or methylamine vanadate. We hypothesize that these negatively charged heavy atom stains react with the peptides and destabilize the nanosheets toward disassembly. In contrast, positively charged heavy atom stains (such as uranyl salts) appear innocuous. Mature sheets of **CP^+** and **$CP^-/CP^+/CP^-$** can also be observed without staining in TEM measurements.

Acknowledgments

We acknowledge the National Science Foundation grant CHE-1012620 and CHE-1412580 for support.

References

1. Jiang T, Vail OA, Jiang Z, Zuo X, Conticello VP (2015) Rational design of multilayer collagen nanosheets with compositional and structural control. J Am Chem Soc 137:7793–7802
2. Pires MM, Chmielewski J (2009) Self-assembly of collagen peptides into microflorettes via metal coordination. J Am Chem Soc 131:2706–2712
3. Przybyla DE, Rubert Perez CM, Gleaton J, Nandwana V, Chmielewski J (2013) Hierarchical assembly of collagen peptide triple helices into curved disks and metal ion-promoted hollow spheres. J Am Chem Soc 135:3418–3422
4. Hsu W, Chen Y-L, Horng J-C (2012) Promoting self-assembly of collagen-related peptides into various higher-order structures by metal-histidine coordination. Langmuir 28:3194–3199
5. Kotch FW, Raines RT (2006) Self-assembly of synthetic collagen triple helices. Proc Natl Acad Sci U S A 103:3028–3033
6. Reimer AE, Feher KM, Hernandez D, Slowinska K (2012) Self-assembly of collagen peptides into hollow microtubules. J Mater Chem 22:7701–7703
7. Luo J, Tong YW (2011) Self-assembly of collagen-mimetic peptide amphiphiles into biofunctional nanofiber. ACS Nano 5:7739–7747
8. Chen C-C, Hsu W, Kao T-C, Horng J-C (2011) Self-assembly of short collagen-related peptides into fibrils via cation-pi interactions. Biochemistry 50:2381–2383
9. Kar K, Ibrar S, Nanda V, Getz TM, Kunapuli SP, Brodsky B (2009) Aromatic interactions promote self-association of collagen triple-helical peptides to higher-order structures. Biochemistry 48:7959–7968
10. Przybyla DE, Chmielewski J (2008) Metal-triggered radial self-assembly of collagen peptide fibers. J Am Chem Soc 130:12610–12611
11. Cejas MA, Kinnney WA, Chen C, Vinter JG, Almond HR, Balss KM, Maryanoff CA, Schmidt U, Breslav M, Mahan A, Lacy E, Maryanoff BE (2008) Thrombogenic collagen-mimetic peptides: self-assembly of triple helix-based fibrils driven by hydrophobic interactions. Proc Natl Acad Sci U S A 105:8513–8518
12. Rele S, Song YH, Apkarian RP, Qu Z, Conticello VP, Chaikof EL (2007) D-periodic collagen-mimetic microfibers. J Am Chem Soc 129:14780–14787
13. Sarkar B, O'Leary LE, Hartgerink JD (2014) Self-assembly of fiber-forming collagen mimetic peptides controlled by triple-helical nucleation. J Am Chem Soc 136:14417–14424

14. Xu F, Khan IJ, McGuinness K, Parmar AS, Silva T, Murthy NS, Nanda V (2013) Self-assembly of left- and right-handed molecular screws. J Am Chem Soc 135:18762–18765
15. Matsui S, Yamazaki CM, Koide T (2012) Surface-modifiable free-floating films formed by multiway connection of collagen-like triple-helical peptides. Macromol Rapid Commun 33:911–915
16. Przybyla DE, Chmielewski J (2010) Metal-triggered collagen peptide disk formation. J Am Chem Soc 132:7866–7867
17. O'Leary LE, Fallas JA, Bakota EL, Kang MK, Hartgerink JD (2011) Multi-hierarchical self-assembly of a collagen mimetic peptide from triple helix to nanofibre and hydrogel. Nat Chem 3:821–828
18. Yamazaki CM, Asada S, Kitagawa K, Koide T (2008) Artificial collagen gels via self-assembly of de novo designed peptides. Biopolymers 90:816–823
19. Rubert Perez CM, Rank LA, Chmielewski J (2014) Tuning the thermosensitive properties of hybrid collagen peptide-polymer hydrogels. Chem Commun 50(60):8174–8176
20. Shoulders MD, Raines RT (2009) Collagen structure and stability. Annu Rev Biochem 78:929–958
21. Jiang T, Xu C, Liu Y, Liu Z, Wall JS, Zuo X, Lian T, Salaita K, Ni C, Pochan D, Conticello VP (2014) Structurally defined nanoscale sheets from self-assembly of collagen-mimetic peptides. J Am Chem Soc 136:4300–4308
22. Jiang T, Xu C, Zuo X, Conticello VP (2014) Structurally homogeneous nanosheets from self-assembly of a collagen-mimetic peptide. Angew Chem Int Ed 53:8367–8371
23. Carpino LA, Han GY (1972) 9-Fluorenylmethoxycarbonyl Amino-Protecting Group. J Org Chem 37:3404–3409
24. Collins JM, Porter KA, Singh SK, Vanier GS (2014) High-efficiency solid phase peptide synthesis (HE-SPPS). Org Lett 16:940–943
25. Marek P, Woys AM, Sutton K, Zanni MT, Raleigh DP (2010) Efficient microwave-assisted synthesis of human islet amyloid polypeptide designed to facilitate the specific incorporation of labeled amino acids. Org Lett 12:4848–4851

Chapter 14

Practical Considerations in the Design and Use of Immunologically Active Fibrillar Peptide Assemblies

Carolina Mora Solano, Yi Wen, Huifang Han, and Joel H. Collier

Abstract

The design, formulation, and immunological evaluation of self-assembling peptide materials is relatively straightforward. Indeed, one of the advantages of synthetic self-assembling peptides is that one can progress from initial concept to in vivo testing in a matter of days. However, because these materials are supramolecular, working with them is not without some practical challenges, and subtle changes in design, synthesis, handling, and formulation can affect the materials' immunogenicity. This chapter is intended to communicate some of these practical aspects of working with these materials that are not always enumerated in conventional research papers. Epitope considerations, peptide synthesis, purification, storage, nanofiber formation, quality control, immunological evaluation, and the overall phenotypic characteristics of the immune responses to be expected from these materials are discussed.

Key words Vaccine, Supramolecular, Immunogenicity, Immunization

1 Introduction

Peptides that self-assemble into nanofibers have been recently developed to elicit therapeutic immune responses, and they are currently being explored for the treatment of a variety of infectious and noninfectious diseases and conditions [1–7]. In this strategy, a peptide sequence with a high propensity for fibrillization is extended at its N-terminus with a peptide epitope or antigen. Multiple different peptides, when mixed in the proper sequence, molar ratios, and solution conditions, can be formulated into multi-epitope nanofibers. These nanofibers raise immune responses without needing supplemental adjuvants [1], and the phenotype of the responses is advantageously noninflammatory [3]. Further, the materials' modular nature greatly facilitates the combination and optimization of epitope mixtures, which in turn allows for the independent adjustment of B cell and T cell responses [4]. Although in general these materials are straightforward to design, synthesize, formulate, and test in vivo, working with them requires attention to

Bradley L. Nilsson and Todd M. Doran (eds.), *Peptide Self-Assembly: Methods and Protocols*, Methods in Molecular Biology, vol. 1777, https://doi.org/10.1007/978-1-4939-7811-3_14,

a number of important practical considerations. These considerations are the focus of this chapter. The self-adjuvanting capacity of these materials is strongly dependent on the organization of the peptides into nanofibers [8], as well as the identity and composition of the various epitope peptides. Failure to produce co-assembled nanofibers with proper epitope mixtures is likely to result in fibers with reduced or absent immunogenicity. Therefore this chapter will provide in-depth protocols for producing and evaluating reliably immunogenic peptide nanofibers. Our group has focused mainly on the fibrillizing peptide Q11 (QQKFQFQFEQQ) [1–4, 7, 8], which will be used as an example here, but other fibrillizing peptides can also function in a similar capacity [8]. Although folded protein antigens can be incorporated into peptide nanofibers and raise specific immune responses [9, 10], here we will focus on all-peptide systems for simplicity. Part 1 and its notes will discuss considerations for the design, synthesis, and preparation of the peptides. Part 2 will focus on quality control of the nanofiber preparations and immunization procedures. Part 3 will focus on assessing immune responses.

2 Materials

2.1 Peptide Design, Synthesis, and Preparation

1. Rink amide AM resin (Loading ≈ 0.7 mmol/g, 200–400 mesh).
2. 9-Fluorenylmethoxycarbonyl (Fmoc) amino acids, appropriately side-chain-protected.
3. 2-(1H-Benzotriazole-1-yl)-1,1,3,3-tetramethyluronium hexafluorophosphate (HBTU).
4. 6-Chloro-1-hydroxybenzotriazole dehydrate (Cl-HOBt).
5. N,N-Dimethylformamide (DMF).
6. Dichloromethane (DCM).
7. Diisopropylethylamine (DIPEA).
8. Deprotection Solution: 20% piperidine in DMF.
9. Acetic acid.
10. Biotin-ONp.
11. 5-(and-6)-Carboxytetramethylrhodamine (5 (6)-TAMRA).
12. 1-Ethyl-3-(3-dimethylaminopropyl)carbodiimide (EDC).
13. N-Hydroxy succinimidyl ester of fluorescein (NHS-fluorescein).
14. Cleavage cocktail: TFA/Water/TIS (0.95/0.025/0.025 vol %).
15. Cleavage cocktail for cysteine-containing peptides TFA/-Water/TIS/EDT (0.94/0.025/0.01/0.025 vol %).
16. Diethyl ether.

2.2 Instrumentation

1. Peptide synthesizer (We use models CS136XT and CS336XT from CS Bio, but self-assembling peptides can also be synthesized by hand).
2. Analytical HPLC with C18 column.
3. Semipreparative HPLC with C18 column.
4. Rotary evaporator.
5. Freeze dryer/lyophilizer.
6. MALDI-TOF mass spectrometer.

2.3 Immunizations: Formulation and Delivery

1. Limulus Amebocyte Lysate (LAL) chromogenic endpoint assay kit.
2. Formvar Carbon film on 400 mesh copper grid.
3. 1% Uranyl acetate in water.
4. Mice of appropriate haplotype for evaluating the epitope of interest.
5. Isoflurane.
6. Peptide immunizing formulations.
7. Sterile ½ cm^3 insulin syringes with 28G 1/2 needles.
8. Lancets and microcentrifuge tubes.
9. Peptides (purified to >90% purity).
10. Sterile, pyrogen-free, clear microcentrifuge tubes.
11. Sterile water.
12. Sterile 10× phosphate buffered saline (Fisher Scientific, Cat#: BP399-500).

2.4 Instruments

1. Balance.
2. Vortexer.
3. Sonicator.
4. Transmission electron microscope.
5. Circular dichroism spectrometer.

3 Methods

3.1 Peptide Design and Preparation

1. Peptide design: Select epitope or epitopes for appending to Q11 (*see* **Note 1** for important considerations in epitope selection and peptide design). We have found that a Ser-Gly-Ser-Gly linker between the epitope domain and assembling domain is appropriate. Therefore the peptides have the primary structure NH_2-*epitope*-SGSG-QQKFQFQFEQQ-Am.

2. Synthesize the peptide using standard Fmoc procedures on Rink amide-AM resin. Couple amino acids in DMF with 4 equivalents each of protected amino acid, HBTU, Cl-HOBt, and DIPEA (*see* **Note 2**) for 45 min. Wash the resin with DMF and DCM between couplings. Deprotect with 20% piperidine in DMF. With regard to scale, we usually synthesize 0.25 mmol for in vivo experiments and 0.125 mmol for preliminary in vitro work. For long or difficult peptides, *see* **Note 3**.
3. Consider N-terminal acetylation, biotinylation, or fluorescent labeling (*see* **Note 4**).
4. After peptide synthesis, cleave the peptide from the resin in 20 mL of cleavage cocktail for 0.25 mmol peptide (*see* **Note 5**). Let the reaction proceed for 1.5–2 h, and swirl gently about every 10–15 min. Collect the cleavage cocktail using a clean polypropylene fritted column and remove about 75% of the TFA by rotary evaporation (*see* **Notes 6** and **7**). Immediately after rotary evaporation add about 75 mL of cold (4 °C) diethyl ether to the concentrated peptide solution. The peptide should now be a fine, snow-like precipitate. Vortex and wash the peptides five times with ether, collecting between washes by centrifugation for 5 min at 3000 RCF (Hazard: *see* **Note 8**). After final wash, decant the ether and dry the peptide under a stream of nitrogen. Break up the pellet during drying to achieve a fine powder. Desiccate under vacuum for at least 2 h to reach complete dryness (*see* **Note 9**).
5. Dissolve the dry peptide in ultrapure water (we use a Millipore Milli-Q system with >18.1 MΩ resistivity). The final concentration should be about 5–10 mg/mL. We find 50 mL conical tubes to be convenient containers. Sonicate in a bath sonicator if precipitate is visible (*see* **Note 10**). Lyophilize the peptide to complete dryness. Seal the cap with parafilm. Store at −80 °C or −20 °C.
6. Assess peptide identity using MALDI mass spectrometry and assess peptide purity using analytical HPLC. For HPLC of Q11-based peptides, we typically inject 20 μL of 10 mg/mL peptide dissolved in neat TFA to ensure monomerization (Hazard: *see* **Note 11**). Linear water/acetonitrile gradients on C18 columns usually resolve the peptide. Analyze fractions with MALDI mass spectrometry.
7. For semipreparative purification of fibrillizing peptides, dissolve peptide in neat TFA at a concentration of about 90 mg/mL. Inject about 1 mL onto a C18 semipreparative column (Hazard: *see* **Note 11**). Collect fractions and analyze by mass spectrometry.
8. Storage: At −20 °C, peptides are stable for at least 2 years as lyophilized powders. When opening stocks, equilibrate at room

temperature for 30 min before opening, and use aseptic procedures to minimize moisture accumulation and bacterial contamination.

3.2 Immunizations: Formulation and Delivery

3.2.1 Formulation

In addition to the epitope considerations described in **Note 1**, we have identified additional design parameters important for inducing robust antigen-specific B cell and T cell immune responses using peptide assemblies. These include:

1. Epitopes must be covalently attached to the nanofibers in order to elicit responses against them. Soluble peptides co-mixed with nanofibers or proteins that cannot assemble into the fibers may not elicit antibody responses [1, 8–10].
2. If a Q11-based peptide contains a B cell epitope but lacks a CD4 T helper epitope, it must be co-assembled with another Q11-based peptide containing a CD4 T helper epitope in order to raise antibody responses [2, 4, 8].
3. If the T cell epitope and a B cell epitope occupy two different peptides (i.e., they are not integrated, as in the $OVA_{323-339}$ epitope), they must be assembled together into the same nanofiber in order to raise antibody responses. If the epitopes are separated into mutually exclusive populations of nanofibers and co-delivered, they may not raise antibody responses [4]. A strategy for generating either co-assembled or separately assembled epitopes has been previously published and is described in Fig. 1 [11].
4. Determine whether nanofibers will need a single type of epitope peptide or multiple co-assembled types of epitope peptides (*see* above and **Note 1**).
5. For single peptide preparations:
   (a) Prepare 8 mM peptide in sterile water.
   (b) Vortex for 1 min and sonicate for 5 min to dissolve fully.
   (c) Incubate the solution at 4 °C overnight for a minimum of 12 h.
   (d) Dilute to 2 mM in 1× PBS. Use 10× PBS and water to achieve precisely 1× PBS. Allow complete fibrillization at room temperature for 3 h before immunization.
6. For co-assembly of two or more peptides:
   (a) The total peptide concentration in the aqueous stock solution should be 8 mM, as for single-peptide formulations.
   (b) To prepare the stock solution, weigh the peptides and combine the dry powders together into a flip-top microcentrifuge tube. Flick the tube with a finger 3–4 times and vortex the dry powder mixture for at least 30 min for

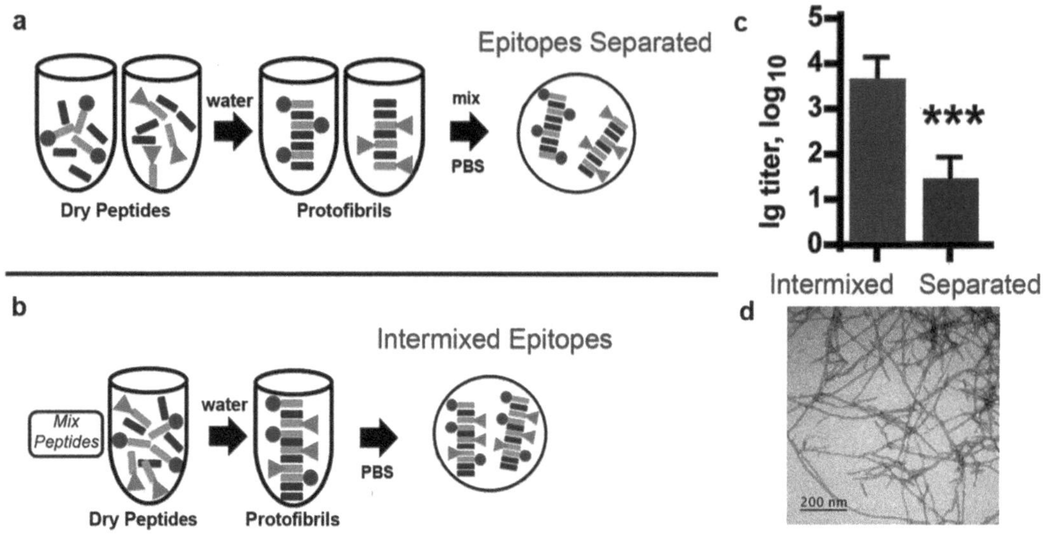

Fig. 1 The intermixing of epitopes within self-assembled peptide nanofibers has a significant influence on the nanofibers' immunogenicity. Epitope peptides that are assembled separately (**a**) do not raise significant antibody responses (**c**), whereas those that are intermixed (**b**) raise antibody responses without adjuvant (**c**) [4, 11]. Nanofibers should have a fibrillar morphology (**d**), with predominant β-sheet character. (**a**–**b**) adapted with permission from Gasiorowski et al., Biomacromolecules, 2011, 12: 3549–3558. (**c**) adapted with permission from Pompano et al., Adv. Healthcare Mater., 2014, 3: 1898–1908

complete mixing. Add water to achieve 8 mM total peptide. For very small amounts of peptide, *see* **Note 12**. Owing to the importance of producing co-assembled fibrils, we have found that the precision that is sacrificed from not making stock solutions with spectrophotometrically verified peptide concentrations is a necessary trade-off to ensure that the peptides are properly mixed from the earliest point possible.

(c) Incubate the solution at 4 °C overnight for a minimum of 12 h.

(d) Dilute to 2 mM total peptide in 1× PBS. Use 10× PBS and water to achieve precisely 1× PBS. Allow complete fibrillization at room temperature for 3 h before immunization.

7. Storage: We have found that some peptide assemblies are extremely stable, with unaltered immunogenicity even when stored at elevated temperatures (e.g., 45 °C for several months) [12]. Indeed this stability is an advantage of peptide assemblies. However, to reduce the chance of bacterial growth we recommend storage as a dry, lyophilized powder at −20 °C. In our experience self-assembled peptides are generally stable in these storage conditions for at least 2 years.

3.2.2 Quality Control: Checks Prior to Immunization

1. **Endotoxin:** In general, peptide self-assemblies contain very low levels of endotoxin. However, given the potential effect of any contaminating endotoxin, all batches of peptide should be tested prior to immunization. Because endotoxin could be inadvertently introduced at any point (synthesis, purification, formulation), it is most useful to measure endotoxin in the final immunization preparation immediately before use in vivo. We use the limulus amebocyte lysate chromogenic endpoint assay (LAL, Lonza). Formulations prepared using the procedures described above typically contain less than 0.5 endotoxin units (EU) per mL of endotoxin, which is within the limit recommended for preclinical studies in rodents (1.5 EU/mL) [13].
2. **Fibrillization:** Whether nanofibers are composed of one peptide or multiple peptides, appropriate fibrillization is critical for raising robust immune responses. Disruption of fibrillization ablates immunogenicity [8]. A simple quality check is to observe nanofiber morphology using Transmission Electron Microscopy (TEM). To prepare TEM samples, dilute immunization formulations (2 mM peptide) to 0.2 mM in PBS and deposit them onto TEM grids. Incubate for 1 min. Wash 3–5 times with 0.2 μm filtered ultrapure water. Apply 1% w/v uranyl acetate in water for 1 min for negative staining. Withdraw liquid quickly with a piece of filter paper (*see* **Note 13**). When the liquid cannot be completely withdrawn owing to thickly deposited nanofibers, further dilute the peptide to 0.1 mM or 0.05 mM, and this may resolve the problem.
3. **Conformation:** Nanofiber formation is based on the organization of the peptides into β-sheet conformations. Circular Dichroism is a simple measure to assess peptide conformation. Dilute peptides to 0.2 mM or 0.1 mM in PBS and scan between 180 and 260 nm. Owing to scattering and absorbance from nanofibers and buffer constituents, respectively, reliable data below 200 nm may be difficult to obtain.

3.2.3 Immunizations (in Mice)

1. Before primary immunization, collect a pre-immune blood draw (<100 μL) to serve as a control for subsequent ELISAs. After immunization we typically perform blood draws weekly or every other week to evaluate antibody responses.
2. For subcutaneous immunizations:
   (a) Using the formulations prepared above and sterile ½ cm^3 insulin syringes with 28G needles, inject 50 μL each into two separate locations on the back of mice under anesthesia (*see* **Notes 14** and **15**).
   (b) To boost, inject 25 μL each of the same formulation into two separate locations on the back. The timing of boosting is typically 3–4 weeks after the primary immunization.

Subsequent boosting can also be administered at 3-week intervals. In our hands, robust antibody responses of titers of 10^4 or greater are usually observed after one or two boosts (*see* **Note 16**).

3. For intraperitoneal immunizations: the prepared formulation is administered using insulin syringes at two intraperitoneal locations (left and right) and with a volume of 50 μL each. Intraperitoneal immunizations are used as a convenient alternative to study antigen uptake and presentation. For peritoneal lavage procedures, *see* Part 3.

3.3 Assessment of Immune Response Phenotypes

In general, peptide assemblies tend to elicit immune responses with the following characteristics:

1. Long-lasting antibody titers. After one or two boosts, if the quality control metrics above have been met and the epitopes have been appropriately selected, generally mice produce anti-peptide antibodies for at least a year (Fig. 2a) [8].
2. Variable T cell responses. The magnitude and quality of T cell responses are dependent on the quality of the T cell epitope and the molar ratio with respect to the B cell epitope (Fig. 2b) [4].
3. Minimal inflammation. Peptide assemblies, while able to raise significant antibody responses, do so with negligible inflammation (Fig. 2c, d) [3].
4. Multiple Ig isotypes [1, 4].

To assess the phenotype of immune response raised by peptide self-assemblies, the following methods are useful. These will be described in brief, as they are widely used immunological techniques, but specific practical considerations regarding peptide assemblies will be noted.

3.3.1 Enzyme-Linked Immunosorbent Assay (ELISA)

After drawing blood, allow the blood to clot for 30 min at RT and then centrifuge blood samples at 2000 × *g* for 15 min to separate the serum. Serum can be stored at −20 °C or −80 °C. For analysis within a few days, serum can be stored at 4 °C. Practical considerations when performing antibody ELISAs after immunization with peptide nanofibers include:

(a) *ELISA coating solutions.* If a protein is the intended target of the antibody response, then make a coating solution of the pure protein in 1× PBS at a concentration of 1 μg/mL protein. If the protein does not coat well, as reflected by low titers of a known positive control, test higher coating densities (up to 20 μg/mL). If peptide is used for coating the ELISA plate, the preferred method is to perform a 2-step coating protocol with streptavidin (SA, 1 μg/mL in 1× PBS), followed by washing off excess SA and coating with biotinylated

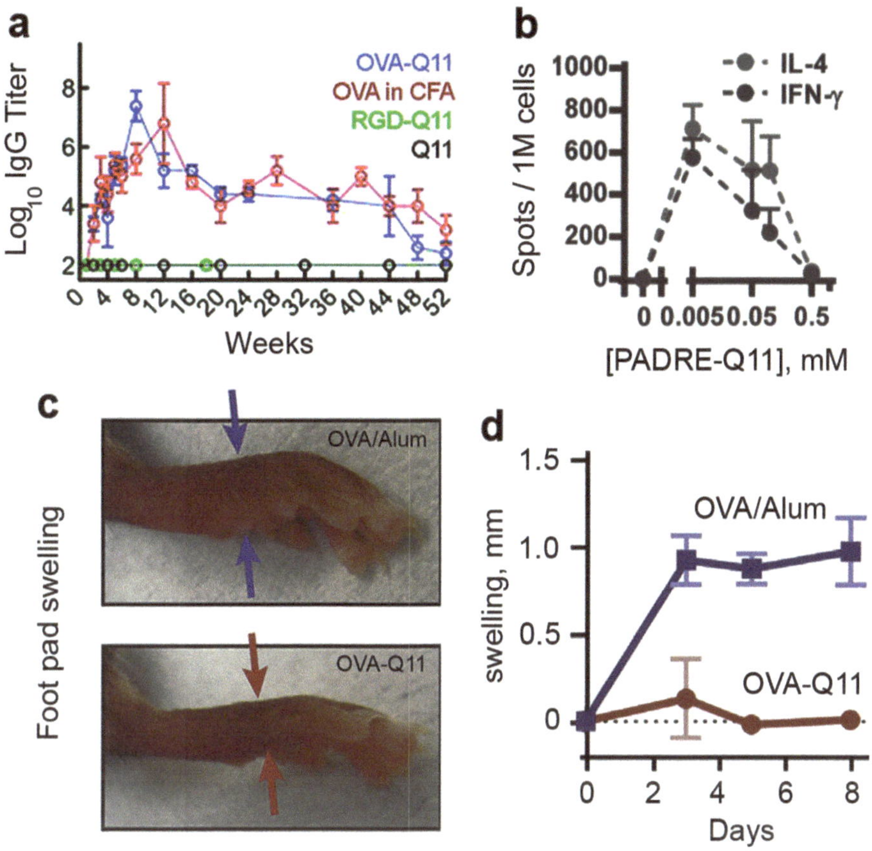

Fig. 2 Overall immune phenotype typically elicited by peptide self-assemblies. Antibody responses against fibrillized peptides containing a competent T/B cell epitope such as $OVA_{323-339}$ are long-lived, lasting for at least a year (**a**) [8]. T cell responses are variable, depending on the presence and amount of appropriate T cell epitopes (**b**, in this case PADRE is the T cell epitope; shown are ELISPOT measurements of IFNγ and IL4 secreted from T cells from mice immunized with varying amounts of PADRE in the peptide nanofibers [4]). Although significant antibody titers are raised by peptide assemblies, there is little to no associated inflammation compared to conventional adjuvants such as alum, shown here for foot pad injections (**c–d**) [3]. (**a**) reproduced with permission from Rudra et al., ACS Nano, 2012, 6: 1557–1563. (**b**) reproduced with permission from Pompano et al., Adv. Healthcare Mater., 2014, 3: 1898–1908. (**c–d**) reproduced with permission from Chen et al., Biomaterials, 2013, 34: 8776–8785

epitope peptide (without Q11) at 20 μg/mL in 1× PBS. If biotinylated peptide is not available, then use unlabeled epitope peptide or fibrillized peptide (containing Q11) by first making a stock solution of 1 mg/mL peptide in water and then diluting to 20 μg/mL in 1× PBS. The main disadvantage of using unlabeled peptide solutions is that the positioning of the peptide epitope on the ELISA plate cannot be as controlled as for SA + biotinylated peptides. Also, in some cases antibodies may be generated against the SGSG-Q11 fragment of a fibrillizing peptide, especially at early time points. These antibodies may contribute a small amount to the titers measured against epitope-Q11 peptides when such peptides are used as the coating material.

(b) *Control sera.* It is important to collect pre-immune serum to detect background antibody binding to ELISA plates. Pooled naive serum serves as a negative control for the assay and must be included in every ELISA plate that is run. For positive controls, sera from mice with previously determined titers is useful. Control serum samples can be aliquoted and stored at −80 °C. These controls should maintain consistent titers across different ELISA experiments if they are stored and analyzed properly.

(c) *Determination of endpoint titers.* Subtract the absorbance of background wells (those coated only with PBS) from the absorbance of sample wells (controls and experimental samples). Use pooled naive mouse serum as a negative control to determine the signal strength arising from nonspecific binding. Create a series of tenfold dilutions for both the control and experimental sera. Run the ELISA in at least triplicate for each sample, and determine the titer cutoff for each serum dilution by calculating the average absorbance from all naive (negative control) serum samples +3 standard deviations. The titer for each experimental serum sample is defined as the highest tenfold dilution of serum that gives absorbances larger than the corresponding cutoff values calculated from the negative control sera.

(d) *Ig Isotype ELISA.* To identify different antigen-specific Ig isotypes (for example: IgM, IgG1, IgG2b, IgG2c, and IgG3 in C57BL/6 mice), perform ELISA as above but using isotype-specific detection antibodies [4].

3.3.2 Enzyme-Linked ImmunoSpot (ELISPOT)

ELISPOTassays are capable of detecting single cytokine-secreting cells (B or T cells) that are antigen- or peptide epitope-specific. Cell suspensions from organs of interest, for example, the spleen, bone marrow, or lymph nodes, are stimulated with peptide ex vivo on an ELISPOT plate that contains a high-protein binding PVDF or nylon membrane. Stimulated cells produce cytokine that binds to the membrane and are subsequently detected using anti-cytokine antibodies. After detection, plates are dried and cytokine spots are counted using ImmunoSpot Analyzer (Cellular Technology, Ltd.). Here, we outline some of the most important practical considerations specific to nanofiber vaccines:

(a) *Positive and negative ELISPOT controls.* Negative control wells should be stimulated with complete media alone, while positive control wells should be stimulated with 10 μg/mL phytohemagglutinin (PHA, from Sigma), anti-CD3 antibody, or another appropriate T cell activator. Negative control wells serve as a baseline of actively secreting T cells prior to peptide stimulation, while positive control wells measure the viability of the cells and the successful implementation of the ELISPOT technique. ELISPOTs can be affected by the skill level of the researcher and so these controls are critical. In our experience, wells stimulated with media alone elicit <10 cytokine spots per million cells for both IL-4 and IFNγ. In our hands, for cells isolated from C57BL/6 mice, PHA elicits a higher proportion of IFNγ spots compared to IL-4.

(b) *Cell plating density and conditions.* Cell numbers can be adjusted depending on the expected frequency of antigen-specific T cells or on the strength of the immune response. The more cells plated, the higher the number of activated cells that can be detected. Low cell numbers may prevent detection due to low frequency of antigen-specific T cells, while high cell numbers may affect cell culture viability. We find that about 5×10^5 cells per well (in 200 μL volume) is appropriate for detecting responses to peptide nanofibers in previously immunized mice that are boosted 1 week before the ELISPOT. Also, we recommend re-filtering the cell suspensions just prior to plating. This will reduce smears on the ELISPOT membrane caused by cell clumping.

(c) *Stimulation with self-assembling peptides.* Stimulate cells with 1–10 μM peptide in complete media. For non-fibrillized peptides of medium-to-high affinity (for example, $OVA_{323-339}$), 5 μM peptide is appropriate. For higher affinity peptides (for example, PADRE), 1 μM peptide may provide better results, and for peptide nanofibers (for example, OVA-Q11), 10 μM peptide is optimal in our experience. For all peptides, prepare stock solutions of 1 mM peptide in PBS by first dissolving peptide in water at slightly greater than 1 mM peptide concentration, then adding 10× PBS and any additional water to achieve 1 mM final peptide concentration in 1× PBS. Dilute this peptide stock to 20–200 μM in media and add it to culture wells to achieve the desired concentration of stimulating peptide. The concentrated stock in PBS can be stored at −20 °C for up to 2 years. The working solution in medium should be used immediately. We typically stimulate cells for 48 h at 37 °C (*see* **Notes 17** and **18**).

3.3.3 Analysis of Antigen Presenting Cells Using Flow Cytometry

Peptide internalization by antigen presenting cells (APCs) and the subsequent activation of the APCs can be studied using flow cytometry, using cells isolated from the peritoneal space. Typically, mixed nanofibers containing 0.02 mM of fluorescence-conjugated peptide are prepared by co-assembling labeled and unlabeled peptide (*see* **Notes 4** and **12**). At predetermined time points after IP immunization (6 h, 18 h, 24 h, or 48 h), inject 2 mL of Hank's balanced salt solution (HBSS) intraperitoneally. Massage the abdomen 40 times and withdraw about 1.2–1.5 mL of fluid from the IP space. Collect the cells by centrifugation and wash them once with flow buffer (PBS containing 2% fetal bovine serum). After blocking the Fc receptor with 2.4G2 antibody (BD Bioscience, Cat#: 553142), stain the cells with fluorescence-conjugated antibodies against F4/80 (Biolegend, Cat#: 123128), MHCII (Biolegend, Cat#: 107606), CD11c (Biolegend, Cat#: 117318), CD11b (BD Bioscience, Cat#: 557657), CD40 (Biolegend, Cat#: 124611), CD80 (Biolegend, Cat#: 104729), and CD86 (BD Bioscience, Cat#:553692). Resuspend cells in flow buffer with 1 μg/mL DAPI, and analyze them using an appropriate flow cytometer. We use an LSRII analyzer (BD Bioscience). Macrophages are gated as $F4/80^+$ and dendritic cells are gated as $F4/80^-CD11c^+CD11b^+$. In each cell population, positive fluorescence signals from nanofibers indicate internalization of the nanofibers. Elevated geometric mean fluorescence intensities of MHCII, CD40, CD80, and CD86 indicate activation.

4 Notes

1. *Epitope selection*: Generally, self-assembled peptide nanofibers will not raise antibody responses unless B cell epitopes and CD4 T helper epitopes are present together in the same nanofiber. A B cell epitope is the component of an antigen that is specifically recognized by antibodies and by B cells (via B cell receptors). A T cell epitope is the portion of an antigen that is internalized, processed, and presented by antigen-presenting cells in the Class-I or Class-II major histocompatibility complex (MHC-I, MHC-II) for subsequent recognition via T cell receptors. In our hands we have found that it is necessary to incorporate CD4 T cell epitopes along with B cell epitopes in the same peptide nanofibers in order to raise detectable antibody responses [2, 4, 8]. The presence of CD8 T cell epitopes allows for cytotoxic T cell responses [14]. Epitopes can be incorporated into the nanofiber scaffold by appending to the Q11 domain in one continuous peptide [1, 8], by co-assembling two or more separate peptides containing either a B cell epitope or a T cell epitope so that the final nanofiber contains both [11, 4], via covalent binding of proteins to the fibers [9], or by assembling expressed proteins into the nanofiber via the β-tail system [10]. B cells can recognize linear or discontinuous amino acid sequences as epitopes. In contrast, T cells bind linear peptides that are appropriately presented within MHC molecules [5]. To predict whether a candidate

sequence contains B cell or T cell epitopes, online resources such as the Immune Epitope Database are extremely useful (www.iedb.org) [15–19].

2. A fourfold excess of the amino acid generally results in couplings with >99% yield in our hands. This is crucial since an accumulated decrease in the coupling efficacy can negatively affect the final yield and purity of even small peptides [20].
3. For long or difficult peptides, Fmoc amino acids may need to be double coupled to the growing resin-bound sequence and require longer coupling time. DMSO can be added to improve resin swelling [21].
4. Additional N-terminal modifications can be considered, including:
   (a) Acetylation. Removes the positive charge on the N-terminus of peptides, thus mimicking natural proteins. In some cases, it increases peptide stability by preventing N-terminal degradation [22, 23]. If an epitope is selected that has been previously described in the literature, in general it is best to keep the N-terminal character originally reported, whether acetylated or not.
   (b) Biotinylation. Biotin-labeled peptides are commonly used in immunoassays [3, 24], histocytochemistry [25], and fluorescence-based flow cytometry [26]. To couple biotin to the N-terminal of completed peptides on-resin, we use 4 equivalents of pre-activated biotin 4-nitrophenyl ester (Biotin-ONp) in DMF. It is added to the N-terminally deprotected peptide resin and reacted overnight while rocking.
   (c) Fluorescent labels (e.g., NHS-Fluorescein, TAMRA). N-terminal on-resin conjugation of fluorophores is useful for producing labeled peptides for flow cytometry and localization studies [3, 27–29]. Couple NHS-Fluorescein as for biotinylation above, and use EDC coupling to conjugate TAMRA.
5. The cocktail would turn yellow to pale yellow, but it should not turn red. If it does, add more cocktail or more TIS while stirring until the redness disappears. This should be resolved in the first 10 min. After about 5–10 min commonly there is crystalline precipitation of cleaved protecting groups.
6. Possible Instrument Damage: TFA can damage the vacuum pump. Be sure that the primary and secondary traps are fully cooled before proceeding with evaporation. Prepare rotary evaporator using a dry ice/acetone mixture in the cold trap condenser. Allow the trap to cool for 20–30 min before evaporation.

7. For rotary evaporation, use an unheated water bath and do not allow bumping. Not executing the rotary evaporation correctly will negatively impact the ultimate solubility and self-assembly behavior of the peptide. The peptide should not be evaporated to dryness during this step. Some TFA must remain to keep the peptide in a monomeric state. Rotovapping to dryness can lead to aggregation during workup and purification that may be difficult to reverse. The peptide precipitate should form slowly and progressively, and it should be white with the consistency of a fine suspension (not large chunks). A rule of thumb is to remove TFA until about 25% of its volume remains. This amount allows subsequent precipitation in diethyl ether but retains the peptide in a low aggregation state.
8. Explosion Hazard: Centrifuging ether is dangerous. This step should be performed in a fume hood using an appropriate centrifuge. Use the smallest quantity of ether possible and ensure that there is no ether outside the tubes. Do not centrifuge several tubes at once. Do not centrifuge for longer than the allotted time, and stop the centrifuge immediately if it is unbalanced at all. Discard tube if leaking. Ensure there is no open flame, spark source, or hotplate nearby.
9. It may make the peptide difficult to dissolve in water if it is not completely dry.
10. If the peptide is hydrophobic and does not dissolve in water easily, add about ¼ to ½ of total volume of acetonitrile. After dissolution of the peptide, remove the acetonitrile using a centrifugal concentrator under vacuum. It is critical to completely dissolve the peptide at this step, as insoluble aggregates can compromise subsequent self-assembly if not addressed at this point.
11. HAZARD: TFA is corrosive. When injecting TFA into the HPLC, wear appropriate personal protective gear.
12. If desired peptide quantities are too small to be weighed accurately, dissolve in water an excess amount of peptide that can be accurately weighed, and use this solution to dissolve the other peptides. For example, to prepare a 0.05 mM PADRE-Q11:1 mM OVA-Q11:0.95 mM Q11 mixture, prepare 0.2 mM PADRE-Q11 in sterile water and add the appropriate volume of this solution to dry OVA-Q11 and Q11 peptides to achieve 8 mM total peptide. Add water if final concentration adjustment is needed. Then incubate this solution overnight and dilute to 2 mM for immunizations as described. Work quickly because even in pure water some amount of fibrillization can occur, though the kinetics of assembly are much slower than in buffered salt solutions.

13. Incomplete removal of liquid from the grid may lead to the precipitation of salts, which can compromise the quality of TEM images.
14. For routine work we use C57BL/6 mice purchased from Harlan Laboratories (Envigo). Mice are maintained in a specific pathogen-free (SPF) facility with controlled temperature and light/dark-cycle. The mice are fed water and food ad libitum. All animal work is done under a protocol approved by the Institutional Animal Care and Use Committee. We typically immunize mice that are between 6 and 12 weeks old.
15. Isoflurane should be administered inside a chemical fume hood, and an isoflurane scavenging device should be employed.
16. If antibody responses are not as high as desired, alterations in the prime/boost regimen should be attempted. These include additional boosting, larger immunizing doses, different routes of immunization, or adjusting the T:B epitope ratios [4].
17. Do not move the ELISPOT plates during stimulation as this can cause artifacts such as blurred spots and streaks.
18. Low-affinity epitopes require higher concentrations of stimulating peptide to generate spots. We typically limit the concentration of stimulating nanofiber peptides to below 200 μM, as we have not seen any adverse effects on cell viability below this concentration [3]. However, this represents a rather high concentration compared to typical peptide stimulation in ELISPOT. Varying peptide concentrations between 0.1 and 100 μM peptide should usually reveal the optimal stimulating dose.

Acknowledgments

Work in our group on immunologically active self-assembled materials has been funded by the National Institutes of Health: The National Institute of Biomedical Imaging and Bioengineering (NIBIB) under grant number 1R01EB009701, the National Institute of Allergy and Infectious Diseases (NIAID) under grant number 1R01AI118182, and the National Institute for Arthritis and Musculoskeletal and Skin Diseases (NIAMS) under grant number R21AR066244. The contents of this article are solely the responsibility of the authors and do not necessarily represent the official views of these agencies. None of the authors have financial conflicts of interest to disclose.

References

1. Rudra JS, Tian YF, Jung JP et al (2010) A self-assembling peptide acting as an immune adjuvant. Proc Natl Acad Sci U S A 107:622–627
2. Rudra JS, Mishra S, Chong AS et al (2012) Self-assembled peptide nanofibers raising durable antibody responses against a malaria epitope. Biomaterials 33:6476–6484
3. Chen J, Pompano RR, Santiago FW et al (2013) The use of self-adjuvanting nanofiber vaccines to elicit high-affinity B cell responses to peptide antigens without inflammation. Biomaterials 34:8776–8785
4. Pompano RR, Chen J, Verbus EA et al (2014) Titrating T-cell epitopes within self-assembled vaccines optimizes CD4+ helper T cell and antibody outputs. Adv Healthc Mater 3:1898–1908
5. Mora Solano C, Collier JH (2014) Engaging adaptive immunity with biomaterials. J Mater Chem B 2:2409–2421
6. Huang Z-H, Shi L, Ma J-W et al (2012) A totally synthetic, self-assembling, adjuvant-free MUC1 glycopeptide vaccine for cancer therapy. J Am Chem Soc 134:8730–8733
7. Wen Y, Collier JH (2015) Supramolecular peptide vaccines: tuning adaptive immunity. Curr Opin Immunol 35:73–79
8. Rudra JS, Sun T, Bird KC et al (2012) Modulating adaptive immune responses to peptide self-assemblies. ACS Nano 6:1557–1564
9. Hudalla GA, Modica JA, Tian YF et al (2013) A self-adjuvanting supramolecular vaccine carrying a folded protein antigen. Adv Healthc Mater 2:1114–1119
10. Hudalla GA, Sun T, Gasiorowski JZ et al (2014) Gradated assembly of multiple proteins into supramolecular nanomaterials. Nat Mater 13:829–836
11. Gasiorowski JZ, Collier JH (2011) Directed intermixing in multicomponent self-assembling biomaterials. Biomacromolecules 12:3549–3558
12. Sun T, Han H, Hudalla GA et al (2015) Thermal stability of self-assembled peptide vaccine materials. Acta Biomater 30:62–71
13. Malyala P, Singh M (2008) Endotoxin limits in formulations for preclinical research. J Pharm Sci 97:2041–2044
14. Chesson CB, Huelsmann EJ, Lacek AT et al (2013) Antigenic peptide nanofibers elicit adjuvant-free CD8(+) T cell responses. Vaccine 32:1174–1180
15. Vita R, Zarebski L, Greenbaum JA et al (2010) The immune epitope database 2.0. Nucleic Acids Res 38:D854–D862
16. Kim Y, Ponomarenko J, Zhu Z et al (2012) Immune epitope database analysis resource. Nucleic Acids Res 40:W525–W530
17. Nielsen M, Marcatili P (2015) Prediction of antibody epitopes. Methods Mol Biol 1348:23–32
18. Liu J, Zhang W (2014) Databases for B-cell epitopes. Methods Mol Biol 1184:135–148
19. Tung C-W (2014) Databases for T-cell epitopes. Methods Mol Biol 1184:123–134
20. Nagarkar RP, Schneider JP (2008) Synthesis and primary characterization of self-assembled peptide-based hydrogels. Methods Mol Biol 474:61–77
21. Hyde C, Johnson T, Sheppard RC (1992) Internal aggregation during solid phase peptide synthesis. Dimethyl sulfoxide as a powerful dissociating solvent. J Chem Soc Chem Commun:1573–1575. https://doi.org/10.1039/C39920001573
22. Arnesen T (2011) Towards a functional understanding of protein N-terminal acetylation. PLoS Biol 9:e1001074. https://doi.org/10.1371/journal.pbio.1001074
23. Wallace RJ (1992) Acetylation of peptides inhibits their degradation by rumen micro-organisms. Br J Nutr 68:365–372
24. Sélo I, Négroni L, Créminon C et al (1996) Preferential labeling of α-amino N-terminal groups in peptides by biotin: application to the detection of specific anti-peptide antibodies by enzyme immunoassays. J Immunol Methods 199:127–138
25. Howl J, Wang X, Kirk CJ et al (1993) Fluorescent and biotinylated linear peptides as selective bifunctional ligands for the V1a vasopressin receptor. Eur J Biochem 213:711–719
26. Buranda T, Lopez GP, Keij J et al (1999) Peptides, antibodies, and FRET on beads in flow cytometry: a model system using fluoresceinated and biotinylated beta-endorphin. Cytometry 37:21–31
27. Farley RA, Tran CM, Carilli CT et al (1984) The amino acid sequence of a fluorescein-labeled peptide from the active site of (Na,K)-ATPase. J Biol Chem 259:9532–9535
28. Richard JP, Melikov K, Vives E et al (2003) Cell-penetrating peptides. A reevaluation of the mechanism of cellular uptake. J Biol Chem 278:585–590
29. Futaki S, Suzuki T, Ohashi W et al (2001) Arginine-rich peptides. An abundant source of membrane-permeable peptides having potential as carriers for intracellular protein delivery. J Biol Chem 276:5836–5840

Chapter 15

Microwave-Assisted Synthesis and Immunological Evaluation of Self-Assembling Peptide Vaccines

C. Brent Chesson, Rojelio Elias Alvarado, and Jai S. Rudra

Abstract

Self-assembling peptides spontaneously associate into functional supramolecular scaffolds, which have found numerous biomedical applications. These molecular assemblies have applications in nerve regeneration, wound healing, and both prophylactic and therapeutic vaccination. They can also be useful tools for proliferation assays, sustained culture of difficult cell lines, or activation of cell lines for immunoassays. This protocol will describe the basic peptide synthesis and purification of model self-assembling peptide immunogen and methods for vaccinating mice, collecting lymph nodes, and stimulating cells ex vivo.

Key words Self-assembling peptides, Vaccine, Nanofiber

1 Introduction

In the last decade, a rapidly growing field of research is vaccine design based on natural and synthetic biomaterials [1, 2]. Peptide-based biomaterials are attractive due to their biocompatibility, ability to fold into specific structures, and ease of synthesis [3, 4]. The primary sequence can be manipulated to impart structure, function, and/or smartness (e.g., stimulus responsiveness) into the scaffold making them highly versatile [5]. Self-assembling peptide vaccines require the synthesis of a single domain, which assembles into higher order structures eliminating the need for complex chemical reactions [6]. Multivalent presentation of antigens on nanofiber scaffolds engages immune cell receptors providing a significant boost in binding affinity leading to robust immune responses without the need for added adjuvants [7, 8].

Rudra et al. first reported that when an antigen is linked to a self-assembling peptide, the resulting nanofibers elicit adjuvant-free antibody (Ab) and responses in mice [7]. The Ab response was comparable in magnitude to complete Freund's adjuvant (CFA), one of the strongest adjuvants known. Using orthogonal chemistries available for functionalizing self-assembling peptides, Hudalla

Bradley L. Nilsson and Todd M. Doran (eds.), *Peptide Self-Assembly: Methods and Protocols*, Methods in Molecular Biology, vol. 1777, https://doi.org/10.1007/978-1-4939-7811-3_15,

et al. demonstrated that whole proteins could be conjugated to pre-formed peptide nanofibers without loss of adjuvanting potential [9]. Furthermore, controlled mixing of peptides linked to B or T cell antigens has been shown to generate tailored vaccine constructs that elicit Ab and cellular responses of different magnitudes and phenotypes [10]. As opposed to physical conjugation, simple mixing of antigens and PRR agonists or live-virus vaccines into shear-thinning peptide hydrogels has also been demonstrated to induce robust Ab and cellular immunity in mice [11–13]. Self-assembling peptides can also be encapsulated within polymeric or inorganic particles for targeted mucosal delivery without loss of immunogenicity [14]. Our lab and others have shown that peptide nanofiber vaccines afford Ab- or cell-mediated protection in mouse models of malaria, influenza, cancer, cocaine addiction, and tuberculosis [15–18]. The synthetic nature of self-assembling peptide vaccines allows for high purity, minimal contamination, and validation of the product through biophysical and biochemical analysis, a major advantage for their evaluation and approval by regulatory agencies.

In this protocol we describe the synthesis and purification of a self-assembling model vaccine OVA-KFE8 by microwave-assisted synthesis, an alternative to manual SPPS, where OVA is a peptide antigen derived from chicken egg ovalbumin (aa 257–264) and KFE8 is an amphipathic peptide that assembles into β-sheet rich nanofibers. The antigenic and self-assembling domains are separated by the spacer sequence GGAAY known to be a cleavage site for proteases. The protocol also describes preparation of peptide nanofiber vaccines for vaccination, immunizing mice in the footpad, collection of lymph nodes, and ex vivo stimulation of lymphocytes for assessment of cytokine production.

2 Materials

Peptide synthesis is commonly carried out on automated synthesizers that carry out steps for washing, deprotection, and chain elongation with precision. These steps can be carried out by hand although the process can be tedious. For optimal results, all reagents should be prepared fresh with high quality solvents (ACS grade and above) unless otherwise indicated. Protective gear should be worn as outlined in the user's safety program for hazardous materials and the reagent manufacturer's material safety data sheet (MSDS). Refer to these guidelines for proper decontamination, handling, and disposal of reagents or any hazardous materials.

2.1 Microwave-Assisted Peptide Synthesis and Purification

1. Single mode microwave synthesizer configured for manual peptide cleavage equipped with a fiber optic temperature probe (Accent microwave-assisted peptide cleavage system, CEM Corporation).

Table 1
Suggested Fmoc-AA-OH solutions

Amino acid	Molecular weight	Volume (mL)	Mass (g)
Fmoc-Ala-OH	311.3	21	1.31
Fmoc-Asn(Trt)-OH	596.7	11	1.32
Fmoc-Glu(OtBu)-OH	425.5	21	1.79
Fmoc-Gly-OH	297.3	21	1.25
Fmoc-Ile-OH	353.4	21	1.49
Fmoc-Leu-OH	353.4	11	0.78
Fmoc-Lys(Boc)-OH	468.5	32	3
Fmoc-Phe-OH	387.4	53	4.11
Fmoc-Ser(tBu)-OH	383.4	11	0.85
Fmoc-Tyr(tBu)-OH	459.6	11	1.02

2. Reaction vessel: 30 mL Teflon Body reaction vessel with a tapered bottom and 45 μm porous glass frit pressed into its bottom. Equipped with capped Luer fitting (CEM Corporation).
3. Wang resin preloaded with leucine, scale = 0.25 mmol (0.833 g, 0.30 mmol/g substitution, 100–200 mesh).
4. Wash solvents: 4 L of DMF and 100 mL of DCM.
5. Solution for Fmoc removal: 500 mL of 20% piperidine in DMF in a 0.5 L bottle. This solution can be stored for up to a month.
6. Fmoc-AA-OH solutions: 0.2 mM amino acid in DMF solutions of each of the Fmoc-amino acids comprising the sequence (FKFEFKFEGGAAYSIINFEKL) in separate 50 mL polypropylene conical tubes (Table 1). For best results these solutions should be fresh, but most amino acids can be stored in solution for up to 2 weeks.
7. Activator solution: 100 mL of 0.5 M 1-Hydroxybenzotriazole hydrate [HOBt] in DMF in a 250 mL amber bottle. Cap the bottle and shake to ensure complete dissolution of the activator. It may take up 10 min to fully go into solution.
8. Activator base solution: 35 mL of 2.0 M Diisopropylcarbamide [DIC] in DMF in a 500 mL amber bottle.
9. Desiccator.

2.2 Peptide Cleavage and Purification

1. Equipment: Centrifuge and Lyophilizer.
2. Cleavage Cocktail: To 9.5 mL of trifluoroacetic acid (TFA) add 0.25 mL each of water and triisopropylsilane (TIS), to give

10 mL total volume of cleavage cocktail. For best results, this mixture should be prepared just prior to performing the peptide cleavage (*see* **Note 1**).

3. 100 mL cold diethyl ether.
4. Extract Clean™ SPE filter columns (8 mL) with frit.

2.3 HPLC Purification

1. Purification of peptide products requires a high-pressure liquid chromatography system.
2. Solvent A: HPLC grade water (H_2O) containing 0.1% TFA.
3. Solvent B: HPLC grade acetonitrile (ACN) containing 0.1% TFA.
4. Reverse-phase C18 column (analytical and semipreparative scale, *see* **Note 2**).
5. Collection tubes.
6. Centrifugal evaporator and lyophilizer (Labconco).
7. MALDI-TOF Mass spectrometer.
8. MALDI-TOF sample matrix solution: saturated α-cyano-4-hydroxycinnamic acid suspended in a 1:1 mixture of HPLC-grade H2O:HPLC-grade ACN (*see* **Note 3**).

2.4 Preparation of Nanofiber Vaccines and Immunization of Mice

1. Sterile H_2O.
2. Sterile phosphate-buffered saline.
3. Control Adjuvant (Alum or Incomplete Freund's adjuvant).
4. Mice (C57BL/6 or BalbC strain).
5. 25-G needles.
6. Inhalation anesthesia, such as isoflurane.
7. Jar for anesthesia (Wheaton clear glass 16 oz straight).
8. Cotton pads.

2.5 Lymph Node Extraction and Ex Vivo Stimulation

1. Surgical scissors, forceps, pins.
2. Sterile syringes.
3. Cell strainers (40 μm mesh size).
4. Sterile Petri dishes.
5. 15 mL falcon tubes.
6. Ethanol (70%).
7. Complete RPMI (cRPMI): RPMI 1640 media, containing: 10% Fetal Bovine Serum, 2 mM L-glutamine, 1 mM sodium pyruvate, 10 mM HEPES, 100 U/mL penicillin, 100 μg/mL streptomycin, 50 μM 2-mercaptoethanol.

3 Methods

3.1 Automated Microwave-Assisted Peptide Synthesis of OVA-KFE8 Nanofiber Peptide

3.1.1 Instrument Setup

1. Prior to beginning the synthesis verify that all the regularly scheduled maintenance procedures have been performed on the system.
2. Place a dip tube filter on each of the positions necessary for the synthesis and load the following reagents on to the system: wash solvents (DMF), solution for Fmoc removal, activator, and activator base solution.
3. Verify the following: waste container has sufficient capacity, ample nitrogen supply, all necessary reagents are loaded on the system, and all bottles are properly tightened.
4. Weigh 0.833 g of Fmoc-Leu preloaded Wang resin into 30 mL reaction vessel, and screw the vessel and drainage lines into the holder and the fiber optic probe. Place the vessel assembly in the microwave cavity.
5. Open the software and program the method to run the peptide from the N-terminus to the C-terminus: FKFEFKFEGGAAYSIINFEKL.
6. In Method Editor select the following options: Synthesis Scale = 0.25, Resin Type = standard (check box for pre-loaded), Resin Cycle = 0.25 Resin-Swelling. Press Run to start the method.
7. The system will go through an initialization to prepare for the synthesis, and then the system will perform all the steps necessary for the synthesis without user intervention including the resin transfer to the reaction vessel, resin swelling, Fmoc removal, amino acid coupling, peptide chain elongation, and finally return of the resin to the tube with the completed peptide as per the method programmed.
8. When the synthesis is complete transfer the peptide resin to a 10 mL extraction vessel using DCM to wash the tube. Remove the solvent by vacuum filtration.
9. Add 1–2 mL of DCM to the resin to wash. Remove the solvent by gravity filtration.
10. Dry the peptide by desiccation for 2 h or overnight to ensure all DCM has been removed.

3.2 Peptide Release and Side Chain Deprotection

1. Add 10 mL of freshly prepared cleavage cocktail along with dried peptide resin to a flask. Leave at room temperature and occasionally stir during cleavage process. Cleavage time will depend on the sequence of the peptide but at least 1 h is required.

2. Remove the resin by transferring solution into an 8 mL Extract Clean ™ column with a frit filter and collect filtrate into a 50 mL centrifuge tube. Wash any residual peptide from the resin with small amounts of TFA (1 mL max).
3. Add diethyl ether to the cocktail to precipitate the peptide. Ice-cold ether can be used to help precipitate difficult peptides out of solution.
4. Centrifuge to pellet the crude peptide (3 min at 200 rcf) and carefully decant the supernatant. Rinse the peptide with additional diethyl ether and repeat the centrifugation to isolate the precipitate for a total of three washes.
5. Allow peptide to dry and evaporate any residual diethyl ether by leaving centrifuge tube uncapped in chemical fume hood. Crude dried peptide can then be analyzed by MALDI-TOF MS, HPLC analysis, or stored long term at −80 °C.

3.3 HPLC Purification

Peptides are purified by reverse phase HPLC (RP-HPLC) over an increasing gradient of acetonitrile (ACN). Peptides that are more hydrophobic will have longer retention times requiring a higher gradient of ACN. Software gradient methods are tailored to each specific peptide, which will require the use of smaller analytical columns to devise an appropriate gradient and run time before scaling the purification method up to larger columns.

A detailed description of HPLC is beyond the scope of this protocol and many articles are available to provide more information about purifying synthetic peptides. We suggest the following parameters for the gradient separation of the OVA-KFE8 peptide:

1. Connect a C-18 preparative reversed phase column to the solvent delivery system and equilibrate the column using the following initial conditions (*see* **Note 2**).

   Solvent A: 0.1% TFA in HPLC grade water.
   Solvent B: 0.1% TFA in HPLC grade acetonitrile.
2. Run an isocratic hold of 15% B for 5 min set at a flow rate of 10 mL/min.
3. Set gradient separation of sample to 15–70% B in 25 min at a flow rate of 10 mL/min.
4. Set the UV absorption to 215 nm to detect for peptide bonds and 240 nm to detect for aromatic rings in the polypeptide.
5. Peptide will elute approximately at 48–52% solvent B. Make sure that the fraction collected is retrieved near the peak of elution as indicated by the HPLC chromatogram.
6. Confirm purity of collected fraction through MALDI-TOF MS analysis (*see* **Note 3**—indicate matrix compound used).

3.4 Preparation of Nanofiber Vaccines and Footpad Immunization

The final preparation should be allowed to incubate at RT for 4 h prior to injection into mice. C57BL6 or BalbC strain of mice can be used.

1. Dissolve OVA-KFE8 peptide in DI water to prepare a stock solution (4 mM). It might take a bit of vortexing to solubilize the peptide as it might be viscous.
2. Very gently, spin the tube so that liquid sticking to the sides and in the cap is pulled down and incubate at 4 °C overnight.
3. Add an equal volume of sterile PBS the next day to prepare a working concentration solution (2 mM) and vortex (1 min) to mix. Addition of PBS leads to spontaneous gelation and increased viscosity.
4. Gently spin the tube again to make sure no peptide material is sticking to the sides. Do this as gently as you can and for very short time as it doesn't take much to pull the solution down.
5. Unwrap syringes just prior to use from their sterile packing and gently remove the needle along with the needle cap. DO NOT remove the needle cap at this point.
6. Draw 0.1 mL of the nanofiber vaccine preparation by placing the syringe tip into the inoculum and pulling the plunger (*see* **Note 4**).
7. Invert the syringe, remove the needle cap, and prime the plunger until all air is removed and a small drop of liquid appears at the end of the needle (*see* **Note 4**).
8. Anesthetize the mouse by isoflurane inhalation. Grasp the foot to be inoculated (Fig. 1a) just above the toes with the forceps (*see* **Note 5**).
9. Lay the needle on top of the foot with the bevel side up and insert the needle just under the skin between the walking pads and the heel (Fig. 1a).
10. Deliver the inoculum and wait for 5 s before slowly removing the needle. This will minimize fluid leakage from the injection site.
11. Release the foot with the forceps and return the mouse to the cage.
12. Mice can be boosted at day 5 if one is assessing short-term effector responses or day 30 for memory recall responses.
13. Sacrifice the mice and harvest organs 3–4 days after the last boost.

3.5 Lymph Node Extraction

1. Pipette ~4 mL cRPMI into each 15 mL falcon tube (*see* **Note 6**).
2. Wet fur entirely on sacrificed mouse using 70% ethanol.

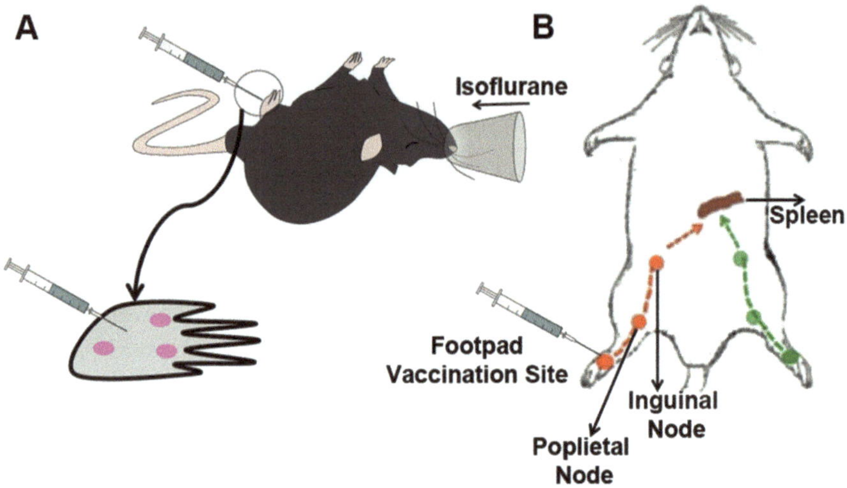

Fig. 1 Schematic showing the mouse footpad and injection. The maximum amount that can be injected into a footpad is 50 μL (0.05 mL) and only one hind foot is used. Restrain the animal with one hind foot isolated and inject subcutaneously into the center of the hind foot forming a small bleb at the injection site. To avoid the blood vessel that runs the length of the foot, inject just off the midline parallel to the vessel. Post injection, the vaccine drains to the draining lymph nodes (popliteal and inguinal) and to the spleen. The positions of the nodes shown are reference only and not anatomically accurate

3. Flay the mouse so a visual indention is present along the base of the neck and cut open the body cavity leaving the peritoneal fascia intact.
4. Pull skin laterally to expose the superficial inguinal lymph nodes and pin skin in place (Fig. 1b).
5. Superficial inguinal lymph nodes are visible just below a small layer of fat at the intersection of vasculature present on the skin (Fig. 1b).
6. Popliteal lymph nodes can be excised by pinning the leg taunt with the dorsal side of the leg exposed. Pull the skin away from the muscle and pin in place.
7. Expose the popliteal lymph node by gently squeezing the biceps femoris and semitendinosus proximal to the pelvis. Lymph node should be about halfway up the hind leg (Fig. 1b).
8. Axillary nodes can be found under the pocket of tissue formed by the pectoralis major. Pin removed skin taunt and gently lift up on the pectoralis to remove the axillary node.
9. Lymph nodes of each mouse are pooled together and each mouse is considered a separate sample regardless of groupE.
10. Using the thumb rest of a plunger from 5 mL syringe (discard barrel), grind tissue thoroughly through the cell strainer (one plunger/well) (Fig. 2) (*see* **Note 7**).

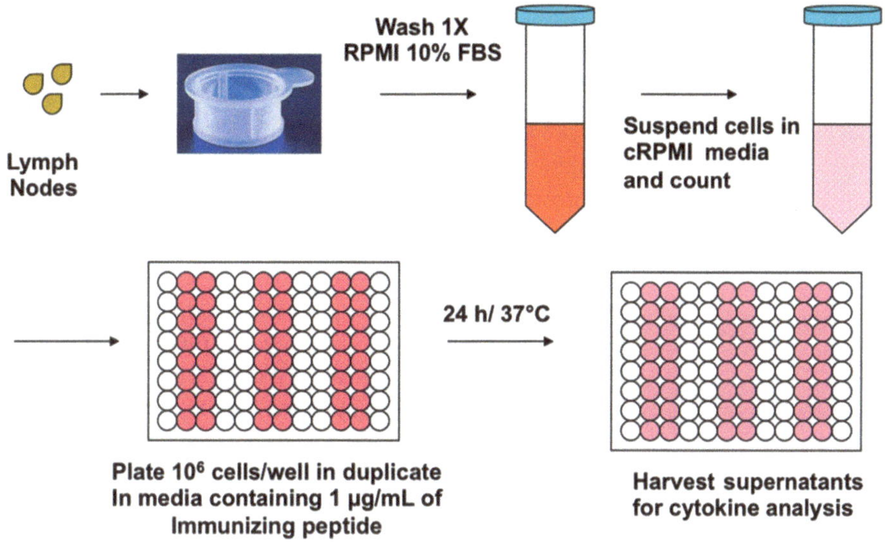

Fig. 2 Schematic showing the preparation of lymphocytes for ex vivo stimulation. The lymph nodes are processed through a cell strainer (40–70 μm) and washed with RPMI media and counted. The cells are then plated into 96 well plates (10^6 cells/well) for stimulation using the immunizing antigen for 24–72 h and the supernatant is collected for evaluation of cytokine production by ELISA

11. Using the syringe, transfer the cell suspension into 15 mL conical tube (Fig. 2).
12. Centrifuge tube(s) at 423 × *g* (4 °C for 5 min) and decant the supernatant. Wash the cell pellet with 10 mL of PBS. Centrifuge again, decant, and re-suspend in cRPMI media (1×10^6 cells per 100 μL).

3.6 Ex Vivo Stimulation

1. Prepare stimulation media by dissolving OVA peptide in cRPMI media (1 μg/mL). Stimulation media can be prepared in advance (1–3 days). Control wells receive cRPMI media without OVA.
2. Add 100 μL of cell suspension from 6 into a 96-well plate in duplicate per mouse and centrifuge to pellet the cells (4 °C 423 × *g* for 5 min). "Flick" the media into the sink DO NOT pipette the media out as this could lead to loss of cells.
3. Add 115 μL of stimulation media per well and incubate at 37 °C for 72 h for cytokine analysis by ELISA (*see* **Note 8**).
4. Following 72 h of incubation, remove the plates from the incubator and centrifuge to pellet the cells (RT 300 × *g* for 5 min). Carefully tilt the plate and gently collect 100 μL of the supernatant so as to not disturb the cell pellet (*see* **Note 9**).

4 Notes

1. Optimum cleavage conditions depend on the number of amino acids present, sequence, side chain protecting groups, and type of linker attached to the resin. Appropriate scavengers should be determined to provide optimum cleavage conditions. The TFA/TIS/Water (95:2.5:2.5) is a good general cocktail that will work for most sequences.
2. C18 columns are the best general columns for peptide purification although we also recommend purchasing a C8 column for more hydrophobic peptides. We typically use columns with 21.2 mm × 250 mm; 7 μm particle size. All solutions going into your HPLC must be the highest analytical grade possible so that no contaminants interfere with your system or sample. Water should be ultrapure and acquired by purifying deionized water from a system that achieves a sensitivity of 18 MΩ.

   α-Cyano-4-hydroxycinnamic acid is a suitable matrix compound for use with nearly all peptides with MALDI-TOF MS. The resin should be mixed with 1:1 HPLC grade H_2O: ACN but will remain insoluble. Resin suspensions should be prepared as needed in small amounts.
3. Multiple doses or a single dose can be injected using one syringe.
4. We describe the protocols for vaccination in the footpads, although other routes of vaccination can be performed, this allows us to collect the antigen draining lymph nodes and enables the assessment of immune responses using immunological techniques.
5. Media for washing cells, stimulation media, and required consumables must be prepared and accounted the day before the organ harvest to ensure smooth operations. All media and PBS must be sterile filtered and surgical equipment should be sterilized and decontaminated between animals.
6. Once the lymph nodes are crushed into single cell suspensions, it is important to maintain the tubes on ice to prevent nonspecific activation and apoptosis.
7. Cytokine analysis by flow cytometry utilizes a cytotoxic reagent, *Brefeldin* A also known as GolgiPlug™, which should limit your stimulation to 12 h maximum. We have found 6 h is optimal for cytokine expression and cell viability.
8. Presence of cells in the supernatant will lead to erroneous results when using ELISA for cytokine analysis.

References

1. Mora-Solano C, Collier JH (2014) Engaging adaptive immunity with biomaterials. J Mater Chem B Mater Biol Med 2(17):2409–2421
2. Khan TA, Reddy ST (2014) Immunological principles regulating immunomodulation with biomaterials. Acta Biomater 10(4):1720–1727
3. Petkau-Milroy K, Brunsveld L (2013) Supramolecular chemical biology; bioactive synthetic self-assemblies. Org Biomol Chem 11 (2):219–232
4. Bowerman CJ, Nilsson BL (2012) Self-assembly of amphipathic beta-sheet peptides: insights and applications. Biopolymers 98 (3):169–184
5. Mart RJ, Osborne RD, Stevens MM, Ulijn RV (2006) Peptide-based stimuli-responsive biomaterials. Soft Matter 2(10):822–835
6. Jung JP, Gasiorowski JZ, Collier JH (2010) Fibrillar peptide gels in biotechnology and biomedicine. Biopolymers 94(1):49–59
7. Rudra JS, Tian YF, Jung JP, Collier JH (2010) A self-assembling peptide acting as an immune adjuvant. Proc Natl Acad Sci U S A 107 (2):622–627
8. Rudra JS, Tripathi PK, Hildeman DA, Jung JP, Collier JH (2010) Immune responses to coiled coil supramolecular biomaterials. Biomaterials 31(32):8475–8483
9. Hudalla GA, Modica JA, Tian YF, Rudra JS, Chong AS, Sun T, Mrksich M, Collier JH (2013) A self-adjuvanting supramolecular vaccine carrying a folded protein antigen. Adv Healthc Mater 2(8):1114–1119
10. Pompano RR, Chen JJ, Verbus EA, Han HF, Fridman A, McNeely T, Collier JH, Chong AS (2014) Titrating T-cell epitopes within self-assembled vaccines optimizes CD4+helper T cell and antibody outputs. Adv Healthc Mater 3(11):1898–1908
11. Li XD, Galliher-Beckley A, Huang HZ, Sun XZ, Shi JS (2013) Peptide nanofiber hydrogel adjuvanted live virus vaccine enhances cross-protective immunity to porcine reproductive and respiratory syndrome virus. Vaccine 31 (41):4508–4515
12. Wang HM, Luo Z, Wang YCZ, He T, Yang CB, Ren CH, Ma LS, Gong CY, Li XY, Yang ZM (2016) Enzyme-catalyzed formation of supramolecular hydrogels as promising vaccine adjuvants. Adv Funct Mater 26 (11):1822–1829
13. Grenfell RFQ, Shollenberger LM, Samli EF, Harn DA (2015) Vaccine self-assembling immune matrix is a new delivery platform that enhances immune responses to recombinant HBsAg in mice. Clin Vaccine Immunol 22 (3):336–343
14. Snook JD, Chesson CB, Peniche AG, Dann SM, Paulucci A, Pinchuk IV, Rudra JS (2016) Peptide nanofiber-CaCO3 composite microparticles as adjuvant-free oral vaccine delivery vehicles. J Mater Chem B 4(9):1640–1649
15. Rudra JS, Mishra S, Chong AS, Mitchell RA, Nardin EH, Nussenzweig V, Collier JH (2012) Self-assembled peptide nanofibers raising durable antibody responses against a malaria epitope. Biomaterials 33(27):6476–6484
16. Huang ZH, Shi L, Ma JW, Sun ZY, Cai H, Chen YX, Zhao YF, Li YM (2012) A totally synthetic, self-assembling, adjuvant-free MUC1 glycopeptide vaccine for cancer therapy. J Am Chem Soc 134(21):8730–8733
17. Chesson CB, Huelsmann EJ, Lacek AT, Kohlhapp FJ, Webb MF, Nabatiyan A, Zloza A, Rudra JS (2014) Antigenic peptide nanofibers elicit adjuvant-free CD8(+) T cell responses. Vaccine 32(10):1174–1180
18. Rudra JS, Ding Y, Neelakantan H, Ding C, Appavu R, Stutz S, Snook JD, Chen H, Cunningham KA, Zhou J (2016) Suppression of cocaine-evoked hyperactivity by self-adjuvanting and multivalent peptide nanofiber vaccines. ACS Chem Neurosci 7(5):546–552

Chapter 16

Preparation and Screening of Catalytic Amyloid Assemblies

Zsofia Lengyel, Caroline M. Rufo, and Ivan V. Korendovych

Abstract

Aggregation of proteins into amyloids has long been recognized as one of the major contributors to disease and aging. Amyloids are known to catalyze their own formation but they have been considered the rock-bottom thermodynamic minimum of the protein fold without much functionality. We have recently demonstrated that aggregation of short peptides in the presence of metal ions gives rise to efficient catalytic activity. Here we present a detailed protocol for the synthesis and purification of these peptides and the preparation of amyloid-like fibrils. Then we describe an easy-to-perform, high-throughput assay to measure their hydrolytic activity.

Key words Self-assembly, Peptides, Amyloids, Catalysis, Hydrolysis

1 Introduction

Amyloid assemblies represent a highly stable, β-sheet rich arrangement of polypeptides. In many cases, the ability of proteins to misfold and aggregate into such assemblies has been implicated in aging and disease [1]. The polymeric (and potentially heterogeneous) nature of amyloids greatly hinders their studies, but it is becoming increasingly evident that even the simplest of amyloid arrangements can result in multiple structural forms, akin to the folding of soluble proteins [2]. Moreover, amyloids have been hypothesized to serve as primordial ancestors of modern proteins [3, 4]. If the polypeptide assemblies present during early stages of evolution had "protein-like" character, then it is possible that they were able to facilitate chemical transformations. Thus, we hypothesized that self-assembly of short, functionalized peptides into amyloid structures, held together through non-covalent interactions, would allow for creation of catalysts.

In order to test this hypothesis, we designed 7-residue peptides that can self-assemble into amyloid-like fibrils and are capable of

Zsofia Lengyel and Caroline M. Rufo contributed equally to this work.

Bradley L. Nilsson and Todd M. Doran (eds.), *Peptide Self-Assembly: Methods and Protocols*, Methods in Molecular Biology, vol. 1777, https://doi.org/10.1007/978-1-4939-7811-3_16, © Springer Science+Business Media, LLC, part of Springer Nature 2018

Table 1
Esterase activity of designed peptides at pH 8 in the presence of 1 mM Zn^{2+}

Peptide	Sequence	k_{cat}/K_M (M^{-1} s^{-1})
1	Ac-LHLHLQL-$CONH_2$	30 ± 3
2	Ac-LHLHLYL-$CONH_2$	13 ± 5
3	Ac-LHLHLKL-$CONH_2$	12 ± 2
4	Ac-LHLHLRL-$CONH_2$	18 ± 4
5	Ac-IHIHIRI-$CONH_2$	22 ± 8
6	Ac-VHVHVRV-$CONH_2$	26 ± 4
7	Ac-IHIHIQI-$CONH_2$	62 ± 2
8	Ac-VHVHVQV-$CONH_2$	32 ± 2

enzyme-like catalysis in the presence of zinc cation (Table 1) [5]. Zn^{2+} stabilizes fibril formation and acts as a cofactor to catalyze ester hydrolysis. Reaction rate enhancements greater than a factor of 10,000 were observed after evaluating fewer than 10 designed peptides [5]. Moreover, the catalytic efficiencies of the fibrils are on par with those of some natural enzymes by weight. We also found that mixing some of these peptides with different sequences resulted in a synergistic effect on their catalysis, providing an opportunity for high-throughput screening of multiple arrangements of functional groups in a single fibril. The catalytic activities of these self-assembling catalysts are impressive given the very small number of peptides screened, and this discovery opens new ways for designing bioinspired catalytic materials. We and others have already shown that this approach can be easily expanded to other reactions and peptide sequences [6–13]. The ease with which we were able to discover catalysts supported by amyloid-like assemblies suggests that one should consider catalytic production of toxic species as one possible source of amyloid toxicity.

In this chapter, we present the synthesis and purification of peptides and their assembly into amyloid-like fibrils. We provide a detailed protocol for kinetic characterization of the hydrolytic activity of self-assembling peptides and peptide mixtures using *p*-nitrophenyl acetate (*p*NPA) as a substrate (Scheme 1). *p*NPA provides an established benchmark for hydrolytic reactions (e.g., ester, amide, lactone hydrolysis, and carbon dioxide hydration) and many enzymes are assayed using this reaction. Thus it allows for comparison with natural enzymes and designed protein catalysts. The kinetic assay is easy to perform, robust, and reproducible. It can be carried out in a high-throughput manner using inexpensive instrumentation that measures changes in UV-Vis absorbance.

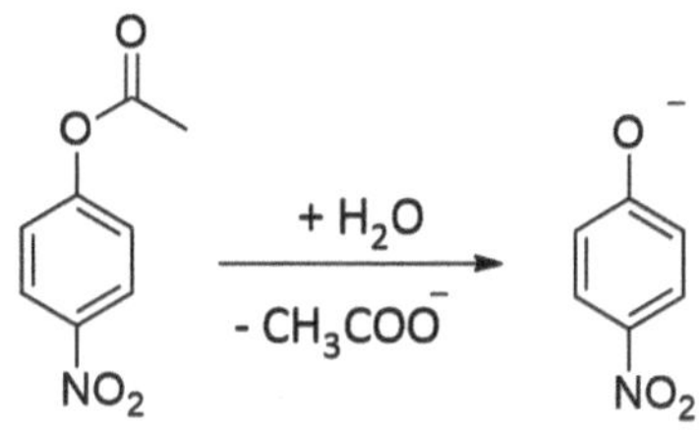

Scheme 1 Hydrolysis of *p*-nitrophenylacetate

2 Materials

Prepare all solutions using ultrapure water (Milli-Q water), analytical grade reagents, and HPLC grade organic solvents. Prepare and store all reagents at room temperature unless indicated otherwise. This protocol is optimized for pH 8, however it can be adopted for other conditions (*see* **Notes 1**, **3**, and **13**).

2.1 Peptide Synthesis and Purification

1. Dimethylformamide (DMF).
2. 5% Piperazine solution in DMF: Weigh out 12.85 g piperazine and dissolve it in 250 mL of DMF at 50 °C. Add 3.35 g hydroxybenzotriazole (HOBt) to the solution if the peptide sequence has glutamate and/or aspartate.
3. N,N-Diisopropylethylamine (DIEA).
4. [2-(6-Chloro-1H-benzotriazol-1-yl)-1,1,3,3-tetramethyluronium-hexafluorophosphate] (HCTU).
5. Methanol.
6. Rink amide MBHA resin.
7. Fmoc-protected amino acids.
8. Cleavage solution: Trifluoroacetic acid (TFA)/triisopropylsilane (TIS)/H_2O (95:2.5:2.5, v/v/v).
9. Nitrogen gas.
10. Ice-cold methyl *tert*-butyl ether.
11. Centrifuge.
12. Solvent A: 0.1% TFA in Milli-Q water.
13. Solvent B: 90% CH_3CN, 9.9% Milli-Q water, 0.1% TFA.
14. Reaction vessel.
15. Reverse-phase high performance liquid chromatography (HPLC) system.
16. C4 preparative column and C18 analytical column.
17. Matrix assisted laser desorption ionization instrument with a time of flight detector (MALDI-TOF).
18. Lyophilizer.

2.2 Peptide Stock Preparation and Kinetic Assay

1. Tris–HCl buffer: 1 M Tris–HCl, pH 8. Weigh out 4.85 g of Tris base and transfer to a 50 mL conical tube. Add 30 mL of water and adjust pH using HCl. Add water up to 40 mL. Readjust the pH if necessary (*see* **Note 1**).
2. $ZnCl_2$ solution: 50 mM $ZnCl_2$. Weigh out 1.70 g of anhydrous $ZnCl_2$, add 1 mL of 1 M HCl and 249 mL of water; mix the solution until $ZnCl_2$ dissolves (*see* **Note 2**).
3. Working buffer: 25 mM Tris, 1 mM $ZnCl_2$, pH 8. Measure about 30 mL of water into a 50 mL falcon tube, add 1 mL of 1 M Tris solution, and mix. Then add 0.80 mL of 50 mM $ZnCl_2$ and add water up to 40 mL (*see* **Note 3**).
4. 10 mM HCl (*see* **Note 4**).
5. Urea: 8 M urea. Weigh out 2.40 g of urea and add water up to 5 mL.
6. *p*-Nitrophenyl acetate (*p*NPA) stock solution: 100 mM *p*NPA. Weigh out 0.036 g of *p*NPA and dissolve it in 2 mL of acetonitrile. Store at 4 °C (*see* **Note 5**).
7. Isopropanol.
8. Acetonitrile.
9. 96-well plate (clear, flat bottom).
10. UV-Vis spectrophotometer.
11. Plate reader.
12. Multichannel pipette.

3 Methods

3.1 Peptide Synthesis and Purification

1. Weigh out 233 mg (0.1 mmol, 0.43 meq/g) of Rink-amide resin and place it into a reaction vessel. Swell the resin for 30 min in DMF at room temperature.
2. Remove the Fmoc group from the resin by adding 2 mL of the 5% piperazine solution for 5 min at 65 °C.
3. Wash the resin four times for 30 s each using DMF.
4. Activate the amino acid (0.3 mmol, 3 eq.) in a separate vial using 1 mL of HCTU (0.28 mmol, 2.8 eq.) in DMF and 104 μL of DIEA (0.6 mmol, 6 eq.).
5. Add the activated amino acid to the resin and let it couple for 7 min at 65 °C.
6. Repeat **steps 2–5** until the desired sequence is synthesized.
7. Acylate the N-terminus of the peptide by incubating it with a mixture of acetic anhydride (56.6 μL, 0.6 mmol, 6 eq.), DIEA (116 μL, 0.63 mmol, 6.3 eq.), and DMF (1 mL) for 5 min at room temperature.

8. Wash the resin twice for 30 s each using DMF.
9. Wash the resin with methanol twice for 1 min each.
10. Filter the resin and dry it under vacuum for at least 30 min.
11. Transfer the resin into a vial and cleave/deprotect the peptide by stirring the resin with 5 mL of TFA/TIS/H_2O (95:2.5:2.5, v/v) mixture for 2 h at room temperature.
12. Filter the solution through a glass wool and remove TFA by passing a gentle stream of nitrogen.
13. In order to separate protecting groups from the peptide, precipitate the peptide by adding 20 mL of dry ice cold methyl *tert*-butyl ether to the solution acquired in **Step 12**. Centrifuge the precipitate for 5 min at 1750 × *g* and decant the supernatant.
14. Wash the precipitate twice with 20 mL of dry ice cold methyl *tert*-butyl ether.
15. Dry the precipitate by passing a gentle stream of nitrogen.
16. Purify the peptide to at least 95% purity on an HPLC system with C4 preparative column by applying a linear gradient of solvent A and solvent B. Confirm peptide purity using HPLC with a C18 analytical column and check the identity of peptides using MALDI-TOF.
17. Lyophilize pure peptide and store it at −20 °C until needed.

3.2 Preparation of Peptide Stock Solutions

1. Dissolve a small amount of solid peptide in 1 mL of 10 mM HCl (pH 2) (*see* **Note 6**). This solution is referred to as a pH 2 peptide stock.
2. Adjust the concentration of the pH 2 peptide stock to 1 mM. Use UV-Vis spectrophotometer to determine peptide concentration (*see* **Note 7**).
3. Prepare pH 8 stock solution by mixing 180 μL of the pH 2 peptide stock with 20 μL of isopropanol and 1.8 mL of the working buffer (*see* **Note 8**). The final concentration of peptide is 90 μM.

3.3 Kinetic Assay

1. Turn on the plate reader and set up a kinetic reading protocol with the wavelength of 405 nm, 15 min scan time, and 30 s kinetic interval time at room temperature.
2. Pipette 50 μL of the working buffer (blank) or 50 μL of the pH 8 peptide stock solution (90 μM peptide) into different wells on a 96-well plate.
3. Prepare *p*NPA substrate solutions with different concentrations of the substrate. Measure out ingredients shown in Table 2, immediately mix the solutions, and pour them into separate reagent reservoirs (*see* **Notes 9** and **10**).

Table 2
Preparation of *p*NPA substrate solutions

	V (Tris working buffer), mL	*V* (acetonitrile), μL	*V* (100 mM *p*NPA stock solution), μL	Final [*p*NPA] in the reaction mixture, μM
Solution 1	4.9	87	13	195
Solution 2	4.9	75	25	375
Solution 3	4.9	62	38	570
Solution 4	4.9	50	50	750

4. Immediately add 150 μL of the substrate solutions into the wells using a multichannel pipette and place the plate into the plate reader.
5. Start the kinetics reading.
6. Plot the acquired absorbance vs. time data and obtain change of absorbance per unit of time. To calculate initial rate (v_0), divide this number by the extinction coefficient of *p*-nitrophenol (16600 M^{-1} cm^{-1} at pH 8) (*see* **Notes 11–13**).

$$v_0 = \frac{\Delta A}{t}/\varepsilon$$

7. Account for the hydrolysis of substrate in buffer by subtracting initial rate in blank sample from the initial rate of peptide sample.
8. To obtain kinetic parameters (k_{cat} and K_M), plot the initial rate vs. substrate concentration and fit the acquired data to the Michaelis–Menten equation, where $[E]_0$ is the concentration of catalyst (*see* **Note 14**) and $[S]_0$ is the initial concentration of substrate at the onset of the reaction:

$$v_0 = \frac{k_{cat}\,[E]_0[S]_0}{K_M + [S]_0}$$

To obtain k_{cat}/K_M values, fit the linear portion of the Michaelis–Menten plot to the following equation:

$$v_0 = \frac{k_{cat}}{K_M}[E]_0[S]_0$$

3.4 Preparation of Peptide Mixtures

1. Prepare 4 mM peptide stock solutions in 10 mM HCl (pH 2 stock) following the instructions described in Subheading 3.2.
2. Mix two peptide stock solutions in different percent ratios (0, 8, 16, 24, 32, 40, 48, 55, 70, 85, and 100%) to produce 100 μL samples.

3. Add 300 μL of 8 M urea to each of the 100 μL peptide samples and let the mixture sit for 15 min (*see* **Note 15**).
4. Add 3.6 mL of working buffer into this solution and mix well by inverting the tube several times. The resulting is a pH 8 peptide stock with a final peptide concentration of 100 μM. Mixing of the already formed fibrils by pipetting should be avoided to ensure reproducibility as the fibrils tend to stick to plasticware.
5. Pipette 50 μL of working buffer (blank) or 50 μL of the pH 8 peptide solution (100 μM) into different wells on a 96-well plate (*see* **Note 16**).
6. Follow **steps 3–8** described in Subheading 3.3 to perform the kinetics assay on these peptide samples.

4 Notes

1. To screen the peptide activity at different pH, use the following buffers: MES for pH 6–6.5, HEPES for pH 7–7.5, TAPS for pH 9–9.5, and CAPS for pH 10–10.5.
2. The shelf life of this solution is at least 1 year.
3. Zinc precipitates as $Zn(OH)_2$ at higher pH. Lower the concentration of the zinc stock solution to 0.5 mM at pH 9 or higher. Always make this solution fresh, as precipitation may occur within hours.
4. First make a 1 M HCl solution using concentrated HCl (12 M), then it can be further diluted to 10 mM.
5. The shelf life of this solution is 1–2 months.
6. Peptide stocks at pH 2 are stable for at least several weeks.
7. Calculate the expected absorbance using Beer's law. In order to measure the absorbance of sample, dilute 10 μL of the peptide stock solution using 10 mM HCl (the extent of dilution depends on the extinction coefficient of the peptide). Calculate the extinction coefficient of the peptides, using the following equation and the extinction coefficient of amino acids at 214 nm (Table 3) [14].

$$\varepsilon_{\text{peptide}}\left(\text{M}^{-1}\ \text{cm}^{-1}\right) = \varepsilon_{\text{peptidebond}} \times n_{\text{peptidebond}} + \sum\nolimits_{i=1}^{20} \varepsilon_{\text{aminoacid}(i)} \times n_{\text{aminoacid}(i)}$$

8. Always make the pH 8 solution immediately before the experiment.

Table 3
Extinction coefficients of amino acids at 214 nm

Amino acid	Extinction coefficient at 214 nm (M^{-1} cm^{-1})
A	32
R	102
N	136
D	58
C	225
Q	142
E	78
G	21
H	5,125
I	45
L	45
K	41
M	980
F	5,200
P	2,675 (or 30 if at N-terminus)
S	34
T	41
W	29050
Y	5,375
V	43
Peptide bond	923

9. Zn^{2+} in the buffer solution also hydrolyzes *p*NPA, so *p*NPA stock solution should be the last component added to the substrate solutions.
10. Other substrate concentrations can be used as long as the substrate is soluble in water (solubility is $2.9*10^{-3}$ M at 25 °C).
11. The path length of the absorbing solution in a well as measured by plate reader is not fixed and it depends on the volume of the solution, plate reader model, well dimensions, etc. Before taking kinetic measurements, calculate an empirical conversion coefficient: measure the absorbance of *p*-nitrophenol solution in your plates using plate reader and then in cuvette using

UV-Vis spectrophotometer (1 cm path length). Use the following equation to calculate the conversion coefficient to normalize the observed absorbance to the 1 cm path length:

$$\text{Conversion coefficient} = \frac{A_{\text{UV-Vis}}}{A_{\text{plate reader}}}$$

12. When acquiring change of absorbance per unit of time, fit only the linear portion of the data set to a straight line.
13. When working at different pH, use the following extinction coefficients of the product: 1700, 4300, 9100, 12700, and 16600 M^{-1} cm^{-1} at pH 6, 6.5, 7, 7.5, and 8, respectively. Use the extinction coefficient of 18700 M^{-1} cm^{-1} at pH 8.5 and above.
14. The final concentration of peptide is 22.5 μM, however concentration of catalyst is 11.25 μM because two peptides coordinate one zinc ion to form one catalytic unit.
15. It is important to mix the peptide stocks at pH 2 in the presence of urea. Urea precludes peptide association and ensures effective mixing at the molecular level.
16. For the blank, mix 100 μL of 10 mM HCl, 300 μL of 8 M urea, and 3.6 mL of working buffer.

Acknowledgments

This work was supported in part by a grant number 1332349 from NSF-EFRI, GM119634 from the NIH, ORAU Ralph E. Powe Junior Faculty Enhancement award and a Humboldt Fellowship to I.V.K. We thank Prof. Olga Makhlynets for constructive suggestions.

References

1. Tipping KW, van Oosten-Hawle P, Hewitt EW, Radford SE (2015) Amyloid fibres: inert end-stage aggregates or key players in disease. Trends Biochem Sci 40(12):719–727
2. Sawaya MR, Sambashivan S, Nelson R, Ivanova MI, Sievers SA, Apostol MI, Thompson MJ, Balbirnie M, Wiltzius JJW, McFarlane HT, Madsen AØ, Riekel C, Eisenberg D (2007) Atomic structures of amyloid cross-β spines reveal varied steric zippers. Nature 447:453–457
3. Greenwald J, Riek R (2012) On the possible amyloid origin of protein folds. J Mol Biol 421 (4–5):417–426
4. Carny O, Gazit E (2005) A model for the role of short self-assembled peptides in the very early stages of the origin of life. FASEB J 19:1051–1055
5. Rufo CM, Moroz YS, Moroz OV, Stöhr J, Smith TA, Hu X, DeGrado WF, Korendovych IV (2014) Short peptides self-assemble to produce catalytic amyloids. Nat Chem 6:303–309
6. Friedmann MP, Torbeev V, Zelenay V, Sobol A, Greenwald J, Riek R (2015) Towards prebiotic catalytic amyloids using high throughput screening. PLoS One 10(12): e0143948
7. Pappas CG, Sasselli IR, Ulijn RV (2015) Biocatalytic pathway selection in transient tripeptide nanostructures. Angew Chem Int Ed 54 (28):8119–8123

8. Wang C, Sun Y, Wang J, Xu H, Lu JR (2015) Copper(II)-mediated self-assembly of hairpin peptides and templated synthesis of CuS nanowires. Chem Asian J 10(9):1953–1958
9. Li D, Jones EM, Sawaya MR, Furukawa H, Luo F, Ivanova M, Sievers SA, Wang W, Yaghi OM, Liu C, Eisenberg DS (2014) Structure-based design of functional amyloid materials. J Am Chem Soc 136(52):18044–18051
10. Peralta MDR, Karsai A, Ngo A, Sierra C, Fong KT, Hayre NR, Mirzaee N, Ravikumar KM, Kluber AJ, Chen X, G-y L, Toney MD, Singh RR, Cox DL (2015) Engineering amyloid fibrils from beta-solenoid proteins for biomaterials applications. ACS Nano 9(1):449–463
11. Zhang C, Xue X, Luo Q, Li Y, Yang K, Zhuang X, Jiang Y, Zhang J, Liu J, Zou G, Liang X-J (2014) Self-assembled peptide nanofibers designed as biological enzymes for catalyzing ester hydrolysis. ACS Nano 8 (11):11715–11723
12. Fry HC, Liu Y, Dimitrijevic NM, Rajh T (2014) Photoinitated charge separation in a hybrid titanium dioxide metalloporphyrin peptide material. Nat Commun 5:8
13. Makhlynets OV, Gosavi PM, Korendovych IV (2016) Short self-assembling peptides are able to bind to copper and activate oxygen. Angew Chem Int Ed 55:9017–9020
14. Kuipers BJH, Gruppen H (2007) Prediction of molar extinction coefficients of proteins and peptides using UV absorption of the constituent amino acids at 214 nm to enable quantitative reverse phase high-performance liquid chromatography–mass spectrometry analysis. J Agric Food Chem 55:5445–5451

Chapter 17

Self-assembly of Filamentous Cell Penetrating Peptides for Gene Delivery

Dawei Xu, Louis DeRidder, Brendan Elmore, and He Dong

Abstract

Cell penetrating peptides (CPPs) have been proven to be an effective vector to deliver a variety of membrane-impermeable macromolecules, such as DNAs, siRNAs, and proteins. Conventional single-chain CPPs typically suffer from severe protease degradation and fast clearance rate for in vivo therapeutic delivery application. In this chapter, we show that supramolecular assembly of de novo designed cationic multidomain peptides (MDPs) leads to nanostructured filaments with increased proteolytic stability and potent membrane activity necessary for improved transfection efficiency.

Key words Self-assembly, Cell penetrating peptide, Gene delivery, Membrane activity

1 Introduction

The development of gene delivery vehicles offers great opportunity to treat a variety of diseases by correcting malfunctioned genetic information. As one of the most promising gene delivery vehicles, CPPs attracted great attention due to their unique ability to cross cell membranes to deliver a variety of therapeutic cargos [1–3]. However, their susceptibility to proteases and fast clearance rate in vivo [4, 5] has been a significant concern for their further application as systemic gene delivery vehicles. Here we demonstrate a new strategy to assemble single-chain cationic peptides into finite nanofibers that are stable over months under physiological conditions [6]. These assembled peptides, referred as Filamentous Cell Penetrating Peptides (FCPPs) have improved stability and potent membrane activity. Compare to traditional synthetic polymer [7, 8] and small peptides [9, 10], FCPPs strike a good balance between toxicity and stability [11] and have been shown to be highly efficient for the delivery of both DNA plasmids [12] and anticancer

Dawei Xu and Louis DeRidder contributed equally to this work.

Bradley L. Nilsson and Todd M. Doran (eds.), *Peptide Self-Assembly: Methods and Protocols*, Methods in Molecular Biology, vol. 1777, https://doi.org/10.1007/978-1-4939-7811-3_17, © Springer Science+Business Media, LLC, part of Springer Nature 2018

drugs [13]. In this chapter, although we are primarily focusing on the activity measurement based on gene delivery, the fundamental structure–activity relationship established here is expected to provide important guidelines for the design and further optimization of membrane active supramolecular peptide-based delivery systems for various other biomedical applications, e.g., drug and vaccine delivery and antimicrobial therapy development.

2 Materials

All samples should be prepared with ultrapure water (18 MΩ) and analytical grade reagents at room temperature unless indicated otherwise. All chemical and biological waste disposal regulations should be strictly followed.

2.1 Peptide Synthesis and Purification

1. Amino acids and resin: Fmoc-Lys(Boc)-OH, Fmoc-Gln(Trt)-OH, Fmoc-Trp(Boc)-OH, and Rink Amide MBHA resin LL (100–200 mesh) (Novabiochem).
2. Synthesis reagents: 2-(6-Chloro-1-H-benzotriazole-1-yl)-1,1,3,3-tetramethylaminium hexafluorophosphate (HCTU), 2-(1H-Benzotriazole-1-yl)-1,1,3,3-tetramethyluronium hexafluorophosphate (HBTU) (Novabiochem); 5(6)-carboxyfluorescein, piperidine, N, N-diisopropylethylamine (DIPEA), and triisopropylsilane (TIS); Kaiser Test Kit (AAPPTec).
3. Solvents: Diethyl ether, trifluoroacetic acid (TFA), dimethylformamide (DMF), acetonitrile (ACN), Tris (1 M, pH 7.4).
4. Peptides were synthesized on a PS3 peptide synthesizer (Protein Technologies, Inc.).
5. Reversed-phase HPLC was carried out using a HITACHI L-7100 pump, Waters 717 plus auto-sampler, and a Higgins analytical column (proto 300 C4 10 μm, 250 × 10 mm).
6. HPLC organic mobile phase: 0.05% (v/v) TFA in acetonitrile.
7. HPLC aqueous mobile phase: 0.05% (v/v) TFA in DI water.
8. Desalting column: Agilent varipure IPE.
9. Lyophilizer: FreeZone Plus 2.5 (Labconco).
10. 20 mM Tris buffer diluted from Tris buffer (1 M, pH 7.4) in DI water.
11. Agilent 8453 UV-vis spectrophotometer.

2.2 Circular Dichroism (CD) Spectroscopy

1. Jasco J-715 spectrometer.
2. Quartz cell with a path length at 1 mm.

2.3 Transmission Electron Microscopy (TEM)

1. 2 wt% uranyl acetate negative staining solution: Dissolve 20 mg of uranyl acetate (Ted Pella Inc.) in 1 mL of DI water. After sonication for 1 min, centrifuge the solution at 42 × *g*. Collect the supernatant and store it at room temperature (*see* **Note 1**).
2. Holey carbon grid (Ted Pella Inc. #01824).
3. JEOL 2010 high-resolution transmission electron microscope.

2.4 Agarose Gel Electrophoresis

1. Electrophoresis running buffer: dilute 20 mL of 50× TAE electrophoresis buffer (Fisher Scientific) into 980 mL of ultra-pure water.
2. Agarose.
3. Ethidium bromide.
4. Mini-Sub® Cell GT Cell (Bio-Rad).
5. ChemiDoc™ MP System (Bio-Rad).

2.5 Cell Uptake Determination

1. Hoechst 33258 (Fisher Scientific): Dissolve 1.8 mg powder in 0.1 mL DMSO and dilute it in Tris buffer (20 mM, pH 7.4) to reach a final concentration at 30 μM as the stock solution.
2. Nikon TE200 epifluorescence microscope.

2.6 Transfection

1. Plasmid DNA encoding a 5.2 kb firefly luciferase (pCMV-luc) (Elim Biopharmaceuticals).
2. Fetal bovine serum (VWR).
3. Phosphate-buffered saline (KH_2PO_4 144.0 mg/L, NaCl 9000 mg/L, Na_2HPO_4-$7H_2O$ 795.0 mg/L, pH 7.4).
4. Dulbecco's modified Eagle culture medium (ATCC).
5. BCA Protein Assay Kit (Thermo Fisher Scientific Inc.).
6. Glo lysis buffer (Promega).
7. Luciferase assay kit (Promega).
8. Micro-plate reader (Vitor^2 1420 Multilabel Counter, PerkinElmer).
9. 96 well plates (CytoOne CC7682-7596).

3 Methods

3.1 Peptide Design

The peptide sequence used in the study was based on the previously designed Multidomain Peptides (MDPs) which have been established as a new class of finite self-assembling peptide nanofibers that are stable over months under physiological condition [6]. MDPs were designed to have a general formula of $K_x(QL)_yK_x$ (amino acid single letter code K: Lysine, Q: Glutamine, L: Leucine). The end products reflect an energetic balance between the intermolecular hydrogen bonding and hydrophobic interactions among the (QL) repeating units and the electrostatic repulsion among the lysine residues. All MDP nanofibers display a very high positive charge density due to the accumulation of the lysine residues at the fiber-solvent interface. The supercharged nanofiber offers great opportunity to perturb the cell membrane to facilitate the delivery of a variety of therapeutic cargoes [12, 13]. The peptide used in the current study, $K_{10}(QW)_6$, consists of six repeating units of alternating hydrophilic (Q)-hydrophobic (W) residues and ten lysine residues at the N-terminus (KKKKKKKKKKQ WQWQWQWQWQW). Tryptophan (W) was selected as the hydrophobic residue because of the high prevalence of the aromatic residues found in transmembrane proteins. The total number of lysine residues was fixed to ten to provide sufficient charge flexibility required for efficient DNA complexation and membrane perturbation.

3.2 Synthesis and Purification of Peptides

The synthesis of MDPs followed the standard Fmoc-solid phase peptide synthesis method [13].

3.2.1 Peptide Synthesis

1. Perform solid phase peptide synthesis on a 50 μmol scale on a PS3 automatic peptide synthesizer (Protein Technologies). Swell MBHA-rink amide resin in 5 mL of DMF for 30 min and transfer the resin to the reaction vessel on the synthesizer. Set up an automatic synthesis program to perform the following operations.
2. Remove the Fmoc group on the resin by adding 3 mL of 20% (*v/v*) piperidine/DMF and react for 5 min. Repeat the deprotection step once.
3. Wash the resin 6 times with 3 mL of DMF to remove residual piperidine.
4. Premix 200 μmol of HCTU with 200 μmol of Fmoc-protected amino acid for dissolution in 5 mL of DMF containing 0.4 M DIPEA. Transfer the solution to the reaction vessel and allow coupling reaction to proceed for 45 min.
5. Wash the resin with 3 mL of DMF and repeat once.
6. Repeat **steps 2** through **5** for each amino acid coupling.
7. Upon completion of the synthesis, acetylate the N-terminus of the peptide by adding 250 μL of acetic anhydride and 50 μL of DIPEA in 3 mL of DMF to the resin and react for 1 h.
8. Remove small amounts of peptide resins (~ tip of a spatula) and transfer to a test tube. Add 0.5 mL of Kaiser test kit solution to the test tube and heat it in boiled water bath for 2 min. A pale yellow color indicates the absence of free primary amino groups and completion of the coupling reaction.

9. Cleave acetylated peptide from the resin and remove the side chain protecting groups by adding 5 mL of a mixture of TFA/TIS/H_2O (95/2.5/2.5 by volume) to the resin and let reaction proceed for 3 h.
10. Collect the cleavage solution and rinse the resin twice with 1 mL of TFA. Combine the washes with the cleavage solution.
11. Evaporate TFA and triturate the residual peptide with cold diethyl ether (*see* **Note 2**).
12. Separate peptide precipitates by centrifugation at 3300 × *g* for 5 min, wash the peptides with 30 mL of cold diethyl ether, and repeat it for three times.
13. Dry peptides under vacuum for 12 h for further HPLC purification.

3.2.2 Peptide Purification

1. Dissolve crude peptides in DI water to reach a concentration at 10 mg/mL.
2. Purify peptides using a preparative reverse phase C18 column with a linear gradient of binary water/acetonitrile mixed solvent containing 0.05% TFA. Monitor elution peaks at 280 nm.
3. Equilibrate desalting columns with 4 mL of methanol and 4 mL of acetonitrile (*see* **Note 3**). Load HPLC fraction onto the desalting column and start elution with acetonitrile/DI water mixture (1:1 by volume).
4. Collect elution fraction in a centrifuge tube, freeze it in liquid nitrogen and lyophilize for 3 days.

3.3 Self-assembly and Structural Characterization of Filamentous Peptides

3.3.1 Formulation of Self-assembled Filamentous Peptides

1. Weigh out lyophilized peptide powder on a microbalance and dissolve it in Tris buffer (pH = 7.4, 20 mM) as the stock solution.
2. Quantify the concentration of peptide stock solution by UV spectroscopy at 280 nm using the extinction coefficient of tryptophan residue at 5560 L mol^{-1} cm^{-1}.
3. Store the stock solution at 4°C for future use (*see* **Note 4**).

3.3.2 Secondary Structure Characterization of Self-assembly Filamentous Peptides

1. Dilute peptide stock solution in Tris buffer (pH = 7.4, 20 mM) to reach a final concentration at 100 μM.
2. Add 200 μL of peptides into a CD cuvette. Acquire spectra from 250 nm to 190 nm at room temperature with a scan rate of 100 nm/min, a response time of 2 s, and a bandwidth of 1 nm.
3. Convert the molar residual ellipticity using the formula $\theta = (\mathrm{mDeg}^*1000)/(c^*n^*\mathrm{l})$, where c is the concentration of the peptide solution expressed in mM, n is the number of amino acids in the peptide sequence, and l is the path length of the cell used in mm.

3.3.3 Nanostructure Characterization of Self-assembled Filamentous Peptides

1. Dilute peptides in Tris buffer (pH = 7.4, 20 mM) to reach a final concentration of 100 μM.
2. Deposit 10 μL of the sample solution onto a holey carbon grid (TED PELLA 01824). After 1 min, carefully remove the excess solution with filter paper and stain the sample by adding 10 μL of 2 wt% uranyl acetate solution for 1 min.
3. Remove excessive staining solution with filter paper and allow the TEM sample to dry for 18 h before imaging.
4. To prepare the peptide–DNA complex, dilute the peptide into Tris buffer (pH = 7.4, 20 mM) to reach a final concentration of 640 μM.
5. Mix 10 μL of peptide solution with an equal volume of DNA (0.108 mg/mL) to reach a *N/P* ratio of 20:1. Incubate the solution for 1 h at room temperature.
6. Deposit 10 μL of the solution onto a holey carbon grid (TED PELLA 01824). After 1 min, carefully remove the excess solution with filter paper and stain the sample with 10 μL of 2 wt % uranyl acetate solution for 1 min (*see* **Note 5**).
7. Remove excessive staining solution with filter paper and allow the TEM sample to dry for 18 h before imaging. Figure 1 shows the nanostructure of peptides and peptide–DNA complex.

3.4 Determination of Membrane Activity

The membrane activity of MDPs is evaluated by monitoring the fluorescence of DNA-bound membrane-impermeable dye, Hoechst 33258 [14].

1. Dissolve 1.8 mg Hoechst 33258 in 100 μL of DMSO and dilute into 10 mL of Tris buffer (20 mM, pH 7.4).
2. Seed HeLa cells in a confocal dish at a density of 5×10^4 cells/well and incubate for 24 h.
3. Replace the medium with 100 μL of fresh DMEM containing 10% deactivated FBS.
4. Add Hoechst 33258 stock solution to the cell culture media to reach a final concentration at 3 μM.
5. After 5 min, add 10 μL of peptide solution (160 μM) to the cell culture dish and record fluorescence under fluorescence microscope for 20 min.
6. Process fluorescence images using Image J and plot the change of fluorescence using Igor Pro 6.04 (Fig. 2).

3.5 Determination of Peptide–DNA Complex Stability

Peptide–DNA complex stability is evaluated through agarose gel electrophoresis using a 15 well Bio-Rad Mini-Sub® Cell GT system (Fig. 3). Agarose gel electrophoresis was used to test the binding affinity between DNA plasmid (*see* **Note 6**) and $K_{10}(QW)_6$ [15].

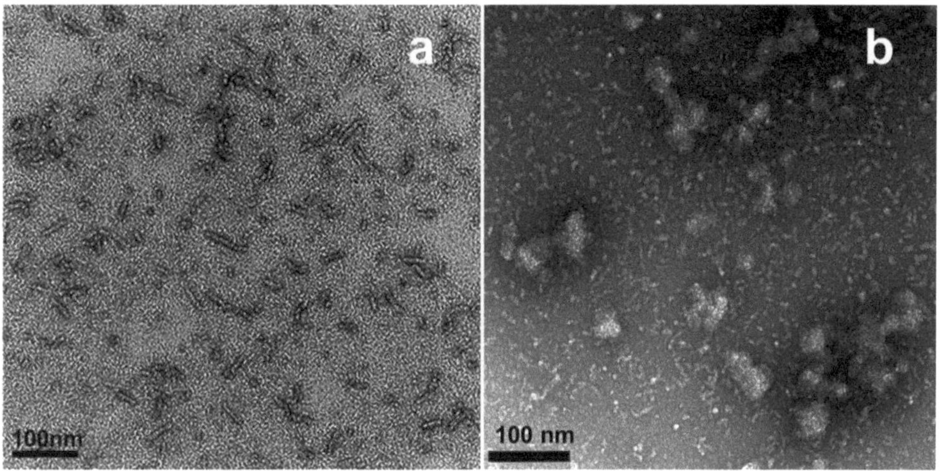

Fig. 1 (**a**) TEM images of short fibrils formed by $K_{10}(QW)_6$. (**b**) TEM image of $K_{10}(QW)_6$–luciferase complex in the form of near spherical aggregates. $N/P = 20/1$

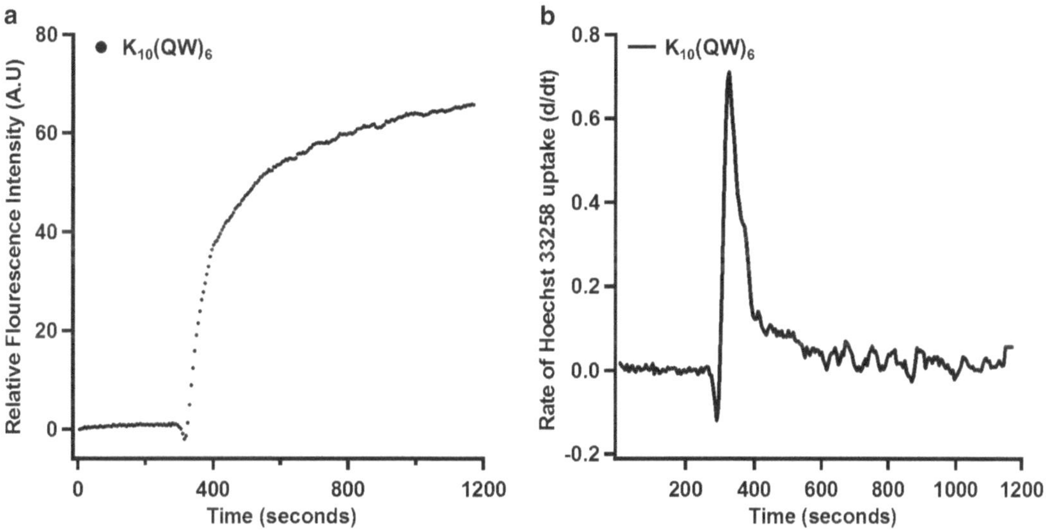

Fig. 2 (**a**) Kinetics of Hoechst 33258 uptake. The arrow indicates the starting point of the $K_{10}(QW)_6$ addition to the cultured HeLa cells which were co-incubated with Hoechst 33258. $K_{10}(QW)_6$: 16 μM; (**b**) rate of cell uptake by taking the derivatives of the fluorescence intensity change as a function of time

1. Prepare gels using 0.8% (w/v) agar in 1× TAE buffer containing ethidium bromide at a concentration of 0.5 μg/mL.
2. Prepare peptide/DNA complexes by sequentially adding 20 mM Tris buffer, DNA plasmid (*see* **Note 7**), and peptide solution to achieve *N/P* ratios ranging from 0.5 to 16.
3. Equilibrate samples at room temperature for 30 min (*see* **Note 8**) and load onto wells using agarose gel loading buffer.
4. Perform electrophoresis in 1× TAE buffer at 100 mV for 30 min. Image the gels using a UV transilluminator.

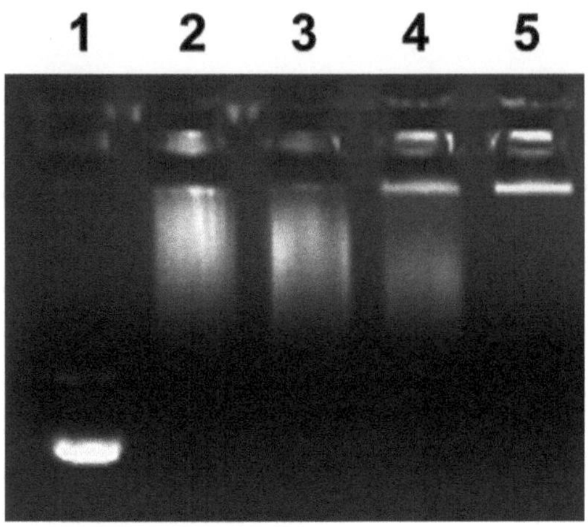

Fig. 3 Agarose gel showing the stability of peptide and plasmid complexes incubating with trypsin. Lane 1: DNA plasmid. Lane 2: DNA plasmid treated with trypsin. Lane 3–5: $K_{10}(QW)_6$/DNA with *N/P* ratio of 2, 5, 10, respectively, in the presence of trypsin

3.6 Gene Transfection

1. Culture HEK 293 and HeLa cells in DMEM with 10% deactivated FBS with 5% of CO_2 at 37°C. Change culture media every 2 days.
2. For transfection measurements, seed HEK293 cells and HeLa cells in a 96-well plate at a density of 10,000 cells/well and incubate for 24 h (*see* **Note 9**).
3. Incubate the DNA–peptide complex at an *N/P* ratio of 10 at room temperature for 1 h to form stable complexes.
4. Add the DNA–peptide complexes (0.5 μg DNA per well) to cell culture dish and incubate for 4 h.
5. Remove the media and replace with new DMEM containing 10% deactivated FBS and incubate for 20 h (*see* **Note 10**).
6. Remove culture media and add 50 μL of 1× Glo lysis buffer. After two cycles of freezing (−80 °C for 30 min) and thawing at room temperature, centrifuge cell lysates at 178 × *g* for 15 min to remove cell debris.
7. Isolate the supernatant and determine luciferase expression according to the manufacturer's protocol. Briefly, mix 20 μL of cell lysate supernatant with 200 μL of Luciferase Assay Reagent and monitor chemiluminescence intensity by a luminometer (Vitor^2 1420 Multilabel Counter, PerkinElmer) instantaneously upon mixing [16].
8. Normalize the relative light unit (RLU) by the protein concentration in the supernatant, which is determined by BCA protein

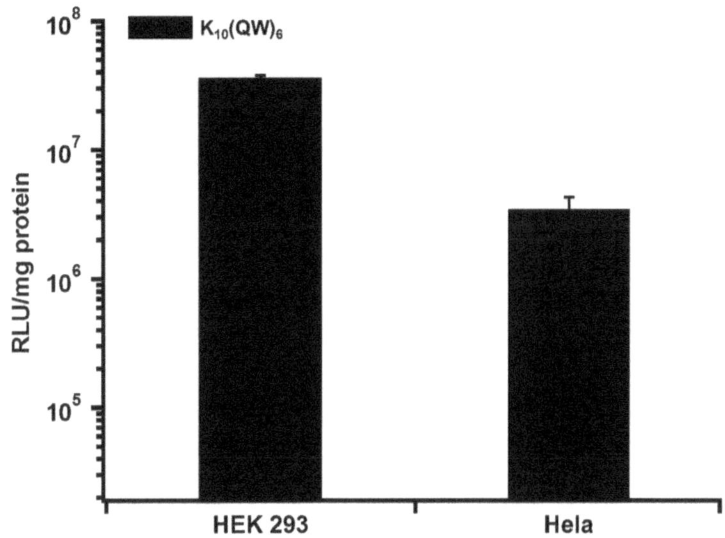

Fig. 4 Luciferase gene transfection efficiency of $K_{10}(QW)_6$ in both HEK 293 and Hela cell lines. *N/P* ratio = 10. Luciferase gene/per well = 0.54 μg

assay. Briefly, mix 10 μL of cell lysates supernatant with 200 μL of working reagent included in the BCA protein assay kit and incubate at 37°C for 30 min. Measure the absorbance at 562 nm on a micro-plate reader (Vitor^2 1420 Multilabel Counter, PerkinElmer). Figure 4 shows the expression level in RLU/mg protein [16].

4 Notes

1. Uranyl acetate solution was freshly prepared and centrifuged before use. This will minimize salt precipitation, which may affect TEM image quality.
2. Precipitated peptides from the post cleavage reaction should be thoroughly washed with 20 mL of cold diethyl ether for at least 4 times.
3. The desalting column was equilibrated using methanol and acetonitrile sequentially. The column should be rinsed by water and acetonitrile mixture to achieve maximum recovery rate of the peptides attached onto the column.
4. Peptide stock was freshly prepared and may not be used after 2 weeks of storage. Typically, the peptide stock solution is prepared at 640 μM and subject to incubation at 4 °C for 24 h before mixing with DNA plasmids for transfection experiment.

5. For TEM sample preparation, excessive drying of peptide solutions should be avoided before applying negative staining solution. The uranyl acetate stain should be added within 1 min after liquid is removed from the deposited peptide solution.
6. Accurate quantification of DNA plasmids is very important to achieve reproducible results. Plasmid concentration should be measured by UV spectroscopy immediately before transfection experiments.
7. When preparing DNA working solutions, the frozen stock solution was fully defrosted and vortexed before use.
8. The mixture of peptides and DNA plasmids was gently vortexed. The incubation time of the mixture is no less than 1 h to ensure a stable complex formation.
9. Initial seeding density is typically around 10^4 cells/mL. High cell density may affect the transfection efficiency.
10. Cell culture media were freshly mixed with deactivated FBS to achieve high and reproducible transfection efficiency.

Acknowledgments

Clarkson University is acknowledged for the support of this work. We thank the Clarkson-Trudeau Partnership for providing seed fund to support this project. This study was supported by the National Science Foundation (DMR 1654426).

References

1. Frankel AD, Pabo CO (1988) Cellular uptake of the tat protein from human immunodeficiency virus. Cell 55:1189–1193
2. Green M, Ishino M, Loewenstein PM (1989) Mutational analysis of HIV-1 Tat minimal domain peptides: identification of trans-dominant mutants that suppress HIV-LTR-driven gene expression. Cell 58:215–223
3. Vives E, Brodin P, Lebleu B (1997) A truncated HIV-1 Tat protein basic domain rapidly translocates through the plasma membrane and accumulates in the cell nucleus. J Biol Chem 272:16010–16017
4. Jiang T, Olson ES, Nguyen QT, Roy M, Jennings PA, Tsien RY (2004) Tumor imaging by means of proteolytic activation of cell-penetrating peptides. Proc Natl Acad Sci U S A 101:17867–17872
5. Fosgerau K, Hoffmann T (2015) Peptide therapeutics: current status and future directions. Drug Discov Today 20:122–128
6. Dong H, Paramonov SE, Aulisa L, Bakota EL, Hartgerink JD (2007) Self-assembly of multidomain peptides: balancing molecular frustration controls conformation and nanostructure. J Am Chem Soc 129:12468–12472
7. Som A, Tezgel AO, Gabriel GJ, Tew GN (2011) Self-activation in de novo designed mimics of cell penetrating peptides. Angew Chem Int Ed 50:6147–6150
8. Som A, Reuter A, Tew GN (2012) Protein transduction domain mimics: the role of aromatic functionality. Angew Chem Int Ed 51:980–983
9. Lu H, Wang J, Bai Y, Lang JW, Liu S, Lin Y, Cheng J (2011) Ionic polypeptides with unusual helical stability. Nat Commun 2:206
10. Yin L, Song Z, Kim KH, Zheng N, Tang H, Lu H, Gabrielson N, Cheng J (2013) Reconfiguring the architectures of cationic helical polypeptides to control non-viral gene delivery. Biomaterials 34:2340–2349

11. Xu D, Jiang L, Singh A, Dustin D, Yang M, Liu L, Lund R, Sellati TJ, Dong H (2015) Designed supramolecular filamentous peptides: balance of nanostructure, cytotoxicity and antimicrobial activity. Chem Commun 51:1289–1292
12. Xu D, Dustin D, Jiang L, Samways DSK, Dong H (2015) Designed filamentous cell penetrating peptides: probing supramolecular structure-dependent membrane activity and transfection efficiency. Chem Commun 51:11757–11760
13. Yang M, Xu D, Jiang L, Zhang L, Dustin D, Lund R, Liu L, Dong H (2014) Filamentous supramolecular peptide-drug conjugates as highly efficient drug delivery vehicles. Chem Commun 50:4827–4830
14. Bukhari M, Deng H, Jones N, Towne Z, Woodworth CD, Samways DSK (2015) Selective permeabilization of cervical cancer cells to an ionic DNA-binding cytotoxin by activation of P2Y receptors. FEBS Lett 589:1498–1504
15. Zheng N, Yin L, Song Z, Ma L, Tang H, Gabrielson NP, Lu H, Cheng J (2014) Maximizing gene delivery efficiencies of cationic helical polypeptides via balanced membrane penetration and cellular targeting. Biomaterials 35:1302–1314
16. Manthorpe M, Cornefert-Jensen F, Hartikka J, Felgner J, Rundell A, Margalith M, Dwarki V (1993) Gene therapy by intramuscular injection of plasmid DNA: studies on firefly luciferase gene expression in mice. Hum Gene Ther 4:419–431

Chapter 18

Biogelx: Cell Culture on Self-Assembling Peptide Gels

Mhairi M. Harper, Michael L. Connolly, Laura Goldie, Eleanore J. Irvine, Joshua E. Shaw, Vineetha Jayawarna, Stephen M. Richardson, Matthew J. Dalby, David Lightbody, and Rein V. Ulijn

Abstract

Aromatic peptide amphiphiles can form self-supporting nanostructured hydrogels with tunable mechanical properties and chemical compositions. These hydrogels are increasingly applied in two-dimensional (2D) and three-dimensional (3D) cell culture, where there is a rapidly growing need to store, grow, proliferate, and manipulate naturally derived cells within a hydrated, 3D matrix. Biogelx Limited is a biomaterials company, created to commercialize these bio-inspired hydrogels to cell biologists for a range of cell culture applications. This chapter describes methods of various characterization and cell culture techniques specifically optimized for compatibility with Biogelx products.

Key words Hydrogels, Self-assembly, Cell culture, Biomaterials

1 Introduction

Molecular gels have enormous potential as supramolecular biomaterials with precisely tunable properties [1–6]. There have been major developments in particular in the application of self-assembled peptide-based materials as instructive matrices for cell growth. These peptide-based biomaterials are of particular interest as a result of their programmability, bioactivity, and biodegradability. A number of strategies exist to design supramolecular biomaterials based on self-assembled peptides and their derivatives [7]. These are typically inspired by structural motifs found in biological systems, including coiled-coils, β-sheets, and β-hairpins. Other designs combine peptides with non-peptidic components such as polymers or aliphatic/aromatic tails to induce self-assembly [8]. Significant progress in applying these systems in three-dimensional cell culture and tissue engineering has been made by the groups of Zhang [9] (resulting in the commercial product Puramatrix™), Stupp [10] (focused on peptide amphiphiles

Bradley L. Nilsson and Todd M. Doran (eds.), *Peptide Self-Assembly: Methods and Protocols*, Methods in Molecular Biology, vol. 1777, https://doi.org/10.1007/978-1-4939-7811-3_18,

which display high densities of bioactive epitopes), and others [11–13].

The above examples typically use peptide sequences of at least 10 residues. While they are much simpler compared to naturally occurring protein-based materials, they are still relatively complex, expensive, and not easily scalable as they rely on solid-phase synthesis techniques for their production. Much simpler peptide derivatives provide a potentially useful alternative as cell culture scaffolds. In particular, low molecular weight hydrogelators composed of short (di- or tri-) peptides modified at the N-terminus with aromatic structures (such as fluorenylmethoxycarbonyl, Fmoc) can form self-supporting nanostructured hydrogels which are much simpler in composition. The first example of this type of self-assembling systems was reported in 1995 [14]. Since then a number of research groups have contributed to the further development of fundamental understanding and application of these materials [15–17]. Aromatic peptide amphiphiles are now a relatively mature separate class of building blocks for low molecular weight hydrogels [18]. Some of these hydrogels are stable at physiological conditions and have been shown to support 2D and 3D cell culture by mimicking features of the extracellular matrix. In particular, Fmoc-di-phenylalanine (Fmoc-FF) forms nanostructured gels at physiological conditions by a combination of H-bonding and π-stacking interactions. These gels were shown to hold promise as matrices for cell growth, especially when co-assembled with functional surfactant-like components that carry hydrophilic chemical groups, such as Fmoc-serine, or cell adhesive peptides, such as Fmoc-RGD [19–22].

Since the potential of these gels as matrices for cell growth was demonstrated, a range of cell culture applications have been studied [11, 12]. Biogelx Limited, a UK-based biomaterials company that designs tunable peptide hydrogels, was established to commercialize these hydrogels. Biogelx currently develops and supplies highly tunable, cell-matched biomaterials, capable of revolutionizing the way cell biologists control and manipulate cell behavior in the laboratory. It offers a range of hydrogel platforms that are three-dimensional, hold 90–99% water, and are composed of a nanoscale matrix structure that morphologically resembles that of the extracellular matrix (Fig. 1). Biogelx gels are composed of short, yet simple, di- or tri-peptides modified at the N-terminus with the aromatic structure, Fmoc. Through a combination of H-bonding and π-stacking interactions these peptide building blocks co-assemble to form nanoscale fibers. The gels' fiber density controls the gel stiffness which allows the cell biologist a level of control of cell behavior, as the gels can be tuned to meet the needs of a given cell type.

Biogelx products are supplied as lyophilized powder (Biogelx Powder). These powders are rehydrated with water to form a

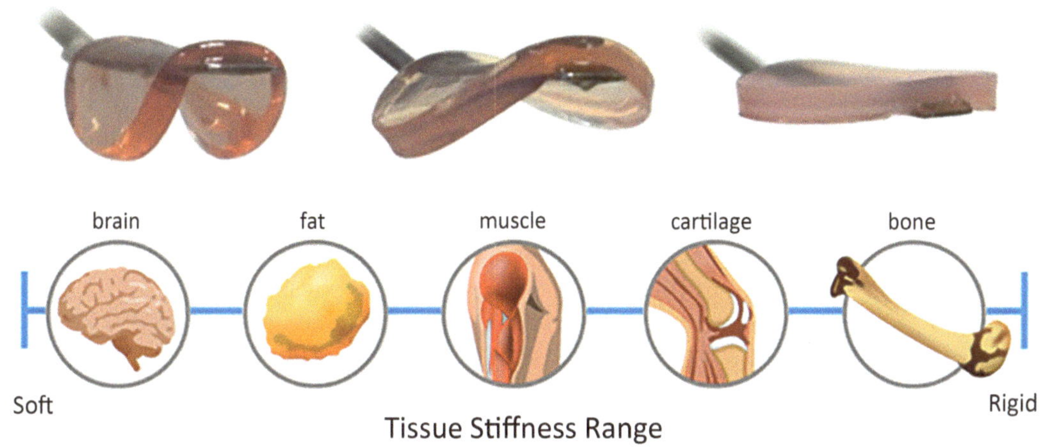

Fig. 1 Images of a range of Biogelx's 3D in vitro hydrogels, tuned to match different tissue type environments

Pre-gel solution. The gelation of these products is triggered by divalent cations, such as calcium, which cross-link the gel fibers to set the gel. The gelation process is triggered instantly when the Biogelx Pre-gel solution comes into contact with cell culture medium.

Cell types successfully tested with these hydrogels include human/bovine chondrocytes, human dermal fibroblasts, 3T3 cells, MG64 cells, human bone marrow stem cells, human adipose derived stem cells, embryonic stem cells, primary human hepatocytes, cortical neurons and human pericytes [23].

This chapter details the methods used for cell culture preparation and applications of Biogelx Powder. This includes techniques such as microscopy imaging and cell viability techniques as well as cell recovery, immunostaining, and RNA extraction methods.

2 Materials

2.1 Biogelx Powder: Preparation and Guidelines for Use Materials

1. Biogelx Powder (Biogelx, UK).
2. Cell incubator (conditions 37 °C with a humidified atmosphere of 5% CO_2).
3. Cell culture multi-well inserts with 1.0 μm pore size (Greiner Bio-One, UK).
4. Cell culture plates (96, 48, 24, 12, or 6 well).
5. Tissue culture medium for cells of interest.
6. Sterile water.
7. Parafilm.

2.2 AFM Materials

1. Mica Sheet.
2. Distilled water.

3. AFM (in our case a Veeco MultiMode with NanoScope IIID Controller Scanning Probe Microscope was used for analysis.) (Digital Instruments, Santa Barbara, CA, USA; Veeco software Version 6.14r1).
4. The AFM measurements should be obtained using a sharp silicon probe (TESP; nominal length (lnom) = 125 μm, width (wnom) = 40 μm, tip radius (Rnom) = 8 nm, resonant frequency (υnom) = 320 kHz, spring constant (knom) = 42 N m^{-1}; Veeco Instruments SAS, Dourdan, France).

2.3 TEM Materials

1. Carbon-coated copper grids (No. 400).
2. Filter Paper.
3. Negative Stain—Nanovan: 1% aqueous methylamine vanadate, obtained from Nanoprobes.
4. TEM (in our case a LEO 912 energy filtering transmission electron microscope operating at 120 kV fitted with 14 bit/2 K Proscan CCD camera).

2.4 LIVE/DEAD® Staining Materials

1. 1× Phosphate Buffer Saline (PBS).
2. LIVE/DEAD® Viability/Cytotoxicity assay kit for mammalian cells (Life Technologies Cat. # L3224).
3. Staining solution contains 4 μM ethidium homodimer-1 (EthD-1) and 2 μM calcein AM in PBS (10 mL PBS to 20 μL EthD-1 and 5 μL calcein AM). This solution must be protected from light.
4. Foil.
5. Glass Slide.
6. Coverslip.
7. Mounting Agent—Prolong® Gold Antifade Reagent (Molecular Probes, Cat. # P36930).
8. Fluorescence microscope.

2.5 Phalloidin Staining

1. 1× Phosphate Buffered Saline (PBS).
2. 4% Paraformaldehyde (PFA).
3. Permeabilization (Perm) buffer: 0.25% Triton-X in PBS.
4. BSA solution: 1 g Bovine Serum Albumin (BSA) in 100 mL of 1× PBS with Triton-X solution.
5. Foil.
6. Distilled water.
7. Phalloidin solution: Phalloidin Dylight 594 (Thermo) or any equivalent Phalloidin staining solution. DyLight working solution is prepared by diluting 1 μL stock solution in 300 μL PBS. Each sample requires approximately 600 μL.

8. Mounting agent: Prolong® Gold Antifade reagent with DAPI (Thermo) or use any glycerol-based mounting medium for preserving fluorescence containing DAPI for nucleic acid staining. DAPI stained mounting agent is recommended as opposed to separate DAPI staining as these give optimized staining results.
9. Glass Slide.
10. Coverslip.

2.6 MTS Assay Materials

1. CellTiter 96® Aqueous One Solution Cell Proliferation Assay (MTS) (Promega Cat. # G3582).
2. Eppendorf tube (1.5 mL).
3. Tissue culture medium for cells of interest.
4. Cell culture medium (user's desired medium).
5. Cell incubator (conditions 37 °C with a humidified atmosphere of 5% CO_2).
6. Centrifuge.
7. 96-Well cell culture plates.
8. Plate reader.

2.7 Cell Recovery Materials

1. Centrifuge tube (15 mL).
2. Phosphate Buffered Saline (PBS).
3. 1% BSA solution: 1 g Bovine Serum Albumin (BSA) in 100 mL of 1× PBS solution Centrifuge Trypsin.

2.8 RNA Extraction Materials

This method is based on using a combination of the Invitrogen TRIzol and Purelink protocols.

1. Eppendorf tube (1.5 mL).
2. TRIzol (Invitrogen).
3. Cell incubator (conditions 37 °C with a humidified atmosphere of 5% CO_2).
4. Centrifuge.
5. Chloroform.
6. Glycoblue.
7. Isopropanol.
8. Ethanol.
9. Distilled RNAse-free water.

2.9 RNA Extraction Method Using Columns Materials

This method is based on using a combination of the Qiagen QIAshredder and RNeasy plus mini kit.

1. RNA stabilizer—RNAlater® solution (Thermo Fisher Scientific).

2. Qiagen QIAshredder (Qiagen).
3. RNeasy plus mini kit (Qiagen).

3 Methods

Biogelx gels are supplied as a lyophilized powder (Biogelx Powder), which is rehydrated with water to prepare a Pre-gel solution of the user's desired concentration. This Pre-gel is added to a well plate, along with cell culture medium which promotes gelation. Cells can be cultured inside the gel (3D culture) or on top (2D culture). Gelation is initiated by cell culture medium and/or salt-containing buffers. The material will remain in the "Pre-gel" state until medium and/or salt-containing buffers are added.

All procedures should be carried out at room temperature unless otherwise specified. Additional notes are provided where further information about procedures is required.

3.1 Biogelx Powder: Preparation and Guidelines for Use

3.1.1 Preparation of Pre-gels from Biogelx Powders

1. Remove glass vial containing Biogelx Powder from freezer, ensure the outside of the vial is dry and allow the powder to reach room temperature before opening the vial.
2. To open, remove the flip-tear-up seal and rubber stopper.
3. In a new sterile vial, weigh the required quantity of Biogelx Powder (Table 1). Should a gel be required that is of lower or higher stiffness than those detailed in the table, the quantity of Biogelx Powder may be scaled appropriately at the user's discretion.
4. Ensure that all of the Biogelx Powder is located at the bottom of the vial and then carefully pipette 5 mL of sterile water to prepare the Pre-gel Solution (Fig. 2). Volume of Pre-gel Solution prepared can be scaled up or down as required.
5. Fully dissolve the Biogelx Powder by applying vortex mixing and sonication for approximately 30 s. If any air bubbles are present in the Pre-gel Solution, remove these by applying sonication to the solution for 10 s.
6. Store the Pre-gel Solution at 4 °C until required for cell culture.
7. Replace the unused Biogelx Powder in the −20 °C freezer by first replacing the rubber stopper and then wrapping in Parafilm or similar material to ensure the container is protected against moisture and air contamination.

3.1.2 Plating Volume Protocol Structure

Cell culture using Biogelx Powder can be carried out using both 2D and 3D cell culture methods (Fig. 3). When performing 2D culture, two formats can be used: the insert and well method. Alternatively, when culturing cell within a 3D format three methods are available: insert, well, and sphere methods.

Table 1
Weight of Biogelx Powder to prepare gels of a certain stiffness

Stiffness range of gel required (kPa)	Weight of Biogelx Powder (milligrams) for 5 mL volumes (mg)
0.8–1.1	22
3.0–4.0	43
8.0–9.2	67

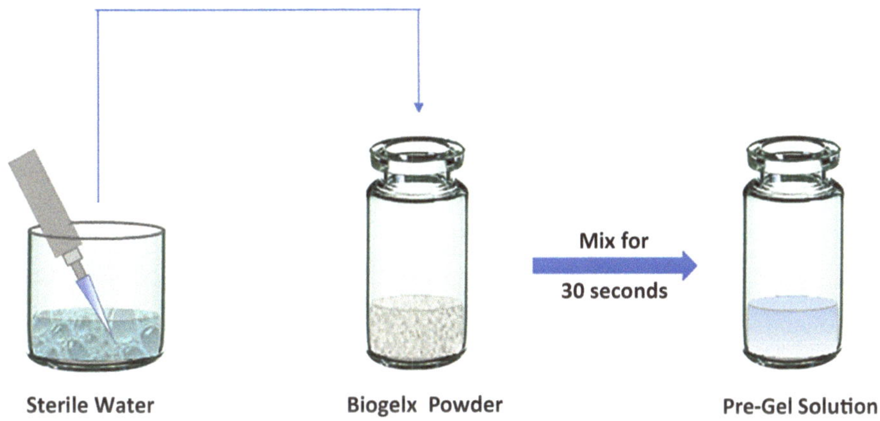

Fig. 2 Schematic of the preparation of Pre-gel solution from Biogelx Powder

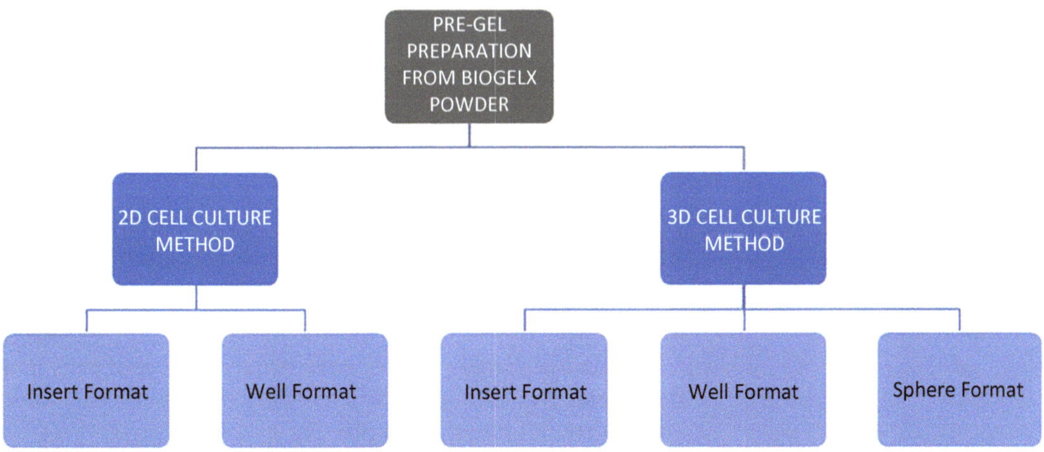

Fig. 3 Organization chart demonstrating the formats available for both 2D and 3D cell culture using Biogelx Powder

Table 2
Volumes of Pre-gel solution and cell culture medium required for well plate sizes performing 2D cell culture with inserts

Culture plate size	Volume of Pre-gel per insert (μL)	Volume of cell culture medium (μL) added outside the insert	Volume of cell culture medium (μL) added to surface of gel
24 Well	100	1000	100
12 Well	300	1400	300
6 Well	1000	2000	1000

3.1.3 2D Cell Culture Method

Pre-Gel Preparation Method

1. Prior to use, place the cell medium and Pre-gel solution in an incubator or water bath for 30 min until the solution(s) reaches 37 °C.
2. Gently mix the Pre-gel solution using a pipette to produce a homogeneous solution. During mixing the pipette tip should not be removed from the solution. If air bubbles are present in the solution, remove these by placing the solution in a bath sonicator for 10 s.

2D Cell Culture Method (Insert Format)

1. Gently pipette the required volume of Pre-gel solution into each well plate insert. Volumes of Pre-gel solution and cell culture medium required are shown in Table 2 for each plate size. Add cell medium into each well outside of the insert, so that the Pre-gel solution in each insert is in contact (through the membrane) with medium from underneath. Refer to Fig. 4 for guidance.
2. Place the culture plate containing the inserts, medium outside of each insert and Pre-gel solution in the cell incubator at 37 °C with a humidified atmosphere of 5% CO_2 for 15 min. At this stage, complete gelation has not occurred.
3. To promote gelation, gently pipette cell medium dropwise onto the center of the gel (Refer to Table 2 for cell culture medium volumes).
4. Place the culture plate in the incubator for a minimum of 2 h without adding cells. During this time, the medium will diffuse through the Pre-gel promoting gelation.
5. Centrifuge the trypsinized cells and resuspend the cell pellet with cell medium, to an appropriate concentration of cells in a final volume, typically 4–16 × 10^4 cells/cm^2 final concentration for most cell types (*see* **Note 2**).
6. Remove the old medium from the gel surface within the insert making sure the pipette tip does not touch the surface of the gel (*see* **Note 1**).

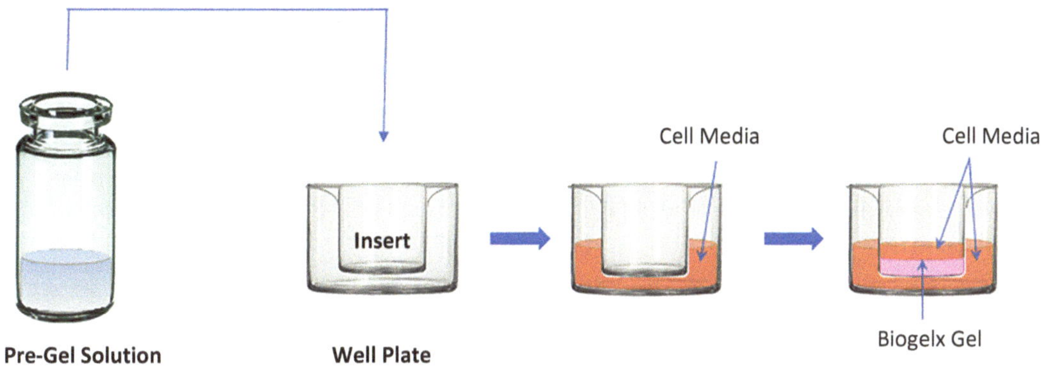

Fig. 4 Schematic of preparation for 2D cell culture in Biogelx gels using insert format

Table 3
Volumes of cell suspension required for well plate sizes using 2D cell culture with inserts

Culture plate size	Volume of prepared cell suspension (μL) on top of the gel	Volume of cell culture medium (μL) added outside the insert
24 Well	100	1000
12 Well	300	1400
6 Well	1000	2000

7. Gently pipette the prepared cell suspension onto the top of the gel in each insert. Replace the old cell medium outside the insert with fresh medium. Volumes are shown in Table 3 for each well plate size (*see* **Note 1**).
8. Medium should be replaced every 24 h during the first 2 days of incubation, then every second day after that (*see* **Note 1**).
9. The gels in these well plates can be kept in the incubator for further use.

2D Cell Culture Method (Well Format)

1. Pipette the Pre-gel solution into the bottom of the well plate. Volumes of Pre-gel solution required are shown in Table 4 for each plate size used. Refer to Fig. 5 for guidance.
2. Place the well plate containing the Pre-gel solution in the cell incubator at 37 °C with a humidified atmosphere of 5% CO_2 for the time specified in Table 4. At this stage, gelation has not occurred.
3. To promote gelation, gently pipette cell medium dropwise onto the center of the gel (Refer to Table 4 for cell culture medium volumes). The Pre-gel and medium should form two distinct layers.

Table 4
Volumes of Pre-gel solution and cell culture medium required for well plate sizes performing 2D cell culture in well format

Culture plate size	Volume of Pre-gel solution per well (μL)	Volume of cell culture medium (μL) onto surface of gel	Incubation time before media addition (minutes)
96 Well	100	150	15
48 Well	200	250	15
24 Well	500	600	30
12 Well	800	1000	30
6 Well	3000	4000	60

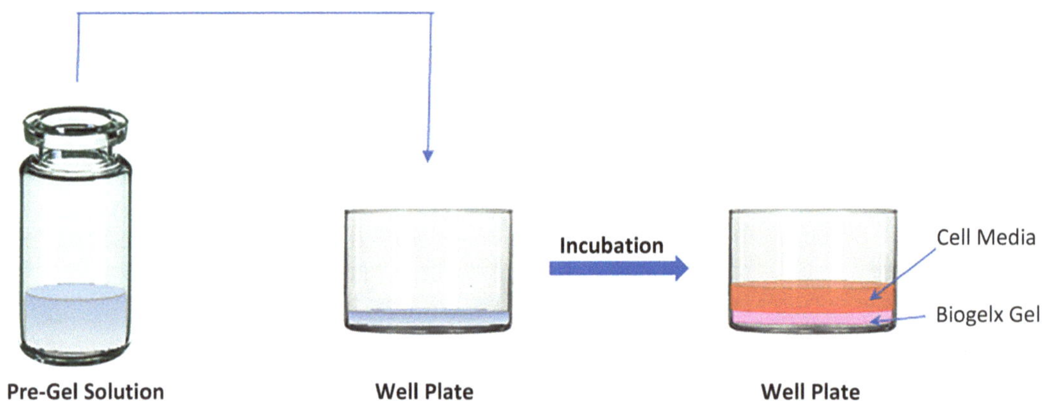

Fig. 5 Schematic of preparation for 2D cell culture in Biogelx gels using well format

4. Place the culture plate in the incubator for a minimum of 2 h without adding cells. During this time, the medium will diffuse through the Pre-gel promoting gelation.
5. Centrifuge trypsinized cells and resuspend the cell pellet with cell culture medium at an appropriate concentration of cells in a final volume, typically 4–16 × 10^4 cells/cm^2 final concentration for most cell types (*see* **Note 2**).
6. After the appropriate incubation period, gently remove the cell medium from top of the gel and replace with the prepared cell suspension. Cell suspension volumes required are shown in Table 5 for each plate size used.
7. Medium should be replaced with fresh medium every 24 h during the first 2 days of incubation, then every second day after that.

Table 5
Volumes of cell suspension required for well plate sizes using 2D cell culture in well format.

Culture plate size	Volume of prepared cell suspension (μL) on top of the gel
96 Well	150
48 Well	250
24 Well	600
12 Well	1000
6 Well	4000

3.1.4 3D Cell Culture Method

Pre-gel and Cell Preparation

1. Prior to use, place the cell medium and Pre-gel solution in an incubator or water bath for 30 min until the solution(s) reaches 37 °C.
2. Gently mix the Pre-gel solution using a pipette to produce a homogeneous solution. During mixing the pipette tip should not be removed from the solution. If air bubbles are present in the solution, remove these by placing the solution in a bath sonicator for 10 s.
3. Meanwhile, trypsinize cells and centrifuge so as to obtain a cell pellet. Determine the cell density.
4. Remove the required volume of cells to a sterile tube to give a final cell density of $0.5–1 \times 10^6$ cells/mL in Pre-gel solution (*see* **Note 2**).
5. Centrifuge the cell-medium mix for 5 min at $12{,}000 \times g$ to again obtain a cell pellet and remove medium, leaving a maximum of 10% medium in which to resuspend the cell pellet.
6. Disturb the pellet through heavy shaking to allow breaking up of the cell mass. To this, add the required volume of Pre-gel solution ($0.5–1 \times 10^6$ cells/mL of Pre-gel solution).
7. Mix by carefully pipetting up and down to allow even distribution of cells in the Pre-gel solution (*see* **Note 3**).

3D Cell Culture Method (Insert Format)

1. Gently pipette the prepared Pre-gel/cell solution to each well plate insert required. Add cell culture medium into the well (outside the insert) and incubate at 37 °C with a humidified atmosphere of 5% CO_2 for 15 min. At this stage, gelation has not occurred.
2. Volumes for Pre-gel/cell solution and medium for each culture plate size are shown in Table 6. Refer to Fig. 4 for guidance.
3. After this period, add medium gently to the surface of the gel (Refer to Table 7 for cell culture medium volumes) and continue incubation.

Table 6
Volumes of Pre-gel/cell solution and Cell Culture Medium required for performing 3D cell culture with inserts

Culture plate size	Volume of Pre-gel/cell solution per insert (μL)	Volume of cell culture medium (μL) outside the insert
24 Well	100	1000
12 Well	300	1400
6 Well	1000	2000

Table 7
Volumes of cell culture medium required for 3D cell culture with inserts

Culture plate size	Volume of cell culture medium added to the gel surface (μL)
24 Well	100
12 Well	300
6 Well	1000

4. Replace the medium surrounding the insert (Table 6) and on the surface of the gel (Table 7) after 2 h (*see* **Note 1**).
5. Incubate the prepared well plates at 37 °C with a humidified atmosphere of 5% CO_2. After 24 h of incubation, replace medium with fresh medium (surface of the gel and outside the inserts).
6. Replace the medium every 24 h during the first 2 days of incubation, then every second day after that.

3D Cell Culture Method (Well Format)

1. Gently pipette the prepared Pre-gel/cell solution into the bottom of the well plate required. Volumes required for each plate size used are shown in Table 8. Refer to Fig. 5 for guidance.
2. Place the culture plate containing the Pre-gel/cell solution in the cell incubator at 37 °C with a humidified atmosphere of 5% CO_2 for 15 min. At this stage, gelation has not occurred.
3. To promote complete gelation, gently add medium dropwise onto the center of the gel (Refer to Table 9 for cell culture medium volumes). The Pre-gel and medium should form two distinct layers. Incubate the prepared well plates at 37 °C with a humidified atmosphere of 5% CO_2.
4. Replace the medium on the surface of the gel after 2 h (*see* **Note 1**).

Table 8
Volumes of Pre-gel/cell solution required for well plate sizes performing 3D cell culture in well format

Culture plate size	Volume of Pre-gel/cell solution per well (μL)
96 Well	100
48 Well	200
24 Well	500
12 Well	800
6 Well	2500

Table 9
Volumes of cell culture medium required for well plate sizes using 3D cell culture in well format

Culture plate size	Volume of cell culture medium added to the gel surface (μL)
96 Well	150
48 Well	250
24 Well	600
12 Well	1000
6 Well	3000

5. Incubate the prepared well plates at 37 °C with a humidified atmosphere of 5% CO_2. After 24 h of incubation, replace the medium.
6. Replace medium every 24 h during the first 2 days of incubation, then every second day after that.

3D Cell Culture Method (Sphere Format)

1. Add cell medium to each of the desired wells in the required well plate (Table 10 shows volume required for each well plate size).
2. Then gently add the Pre-gel/cell solution to each well, using a pipette in which the tip has had 1 cm of the end cut from it. This allows for the Pre-gel/cell solution to be delivered to the well in a ball-like structure (*see* **Note 4**). Refer to Fig. 6 for guidance.
3. Incubate the prepared well plates at 37 °C with a humidified atmosphere of 5% CO_2 for 1 h until gel is fully formed.
4. After 24 h of incubation, replace the medium in the well (*see* **Note 1**). Replace the medium every 48 h thereafter.

Table 10
Volumes of cell culture medium and Pre-gel/cell solution required for performing 3D cell culture in sphere format

Culture plate size	Volume of cell culture medium (μL) in the well	Volume of Pre-gel cell solution per well (μL)
96 Well	150	50
48 Well	500	100
24 Well	1500	300
12 Well	2000	300

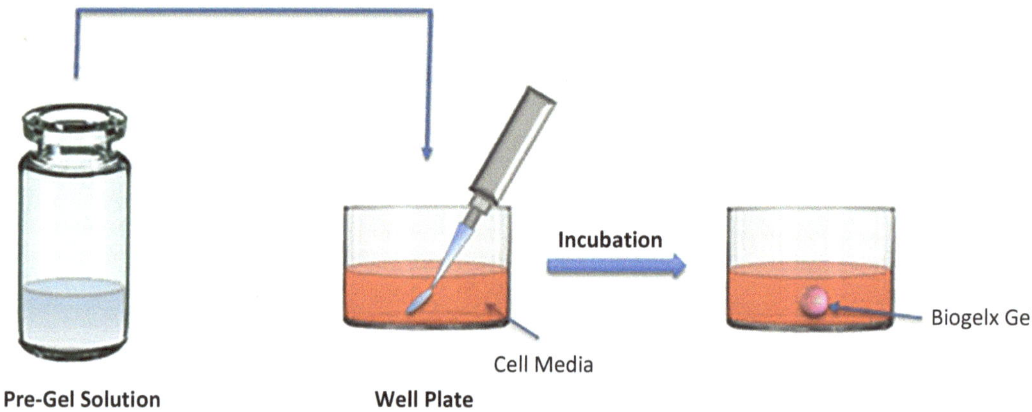

Fig. 6 Schematic of preparation for 3D cell culture in Biogelx gels using sphere format

3.2 Microscopy Imaging Techniques for Biogelx Gel

3.2.1 Atomic Force Microscopy (AFM)

1. To image the fibers in the Biogelx Pre-gel, rehydrate the Biogelx Powder according to Subheading 3.1.1.
2. Place 100 μL of solution on a trimmed, freshly cleaved mica sheet attached to an AFM support stub.
3. After 1 min remove any excess sample by capillary action and rinse the surface with 200 μL of distilled water.
4. To image the Biogelx gel structure (prepare according to Subheading 3.1), place the gel on mica sheet for 1 min and rinse twice with 200 μL of distilled water.
5. Leave the samples to air-dry overnight in a dust-free environment, prior to imaging.
6. Scan the mica surface in air under ambient conditions using an AFM, operated in tapping mode. Typical scanning parameters are as follows: tapping frequency 308 kHz, integral and proportional gains 0.3 and 0.5, respectively, set point 0.5–0.8 V,

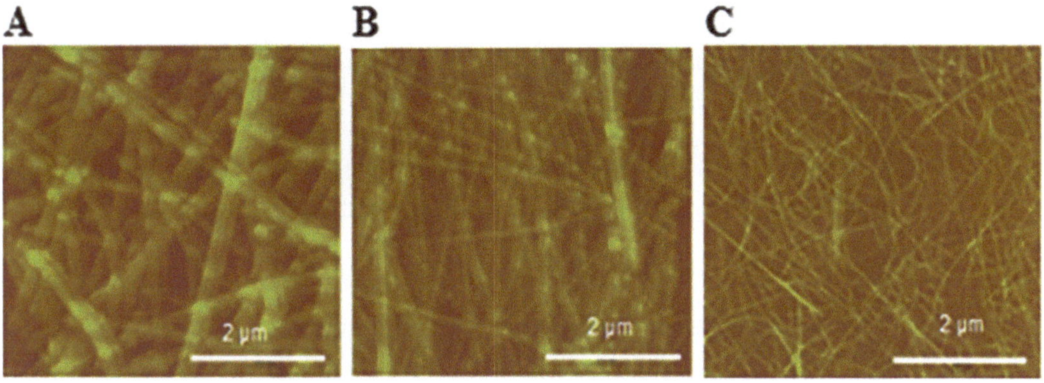

Fig. 7 AFM images of (**a**) soft, (**b**) stiff, and (**c**) rigid gel demonstrate differential fibrous morphologies

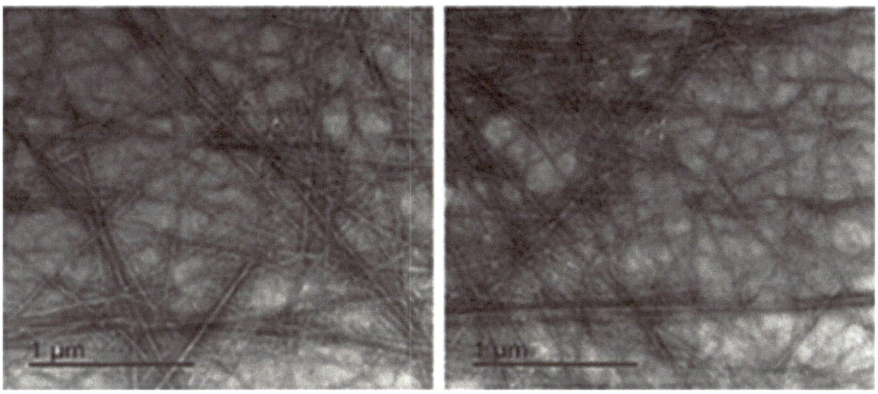

Fig. 8 TEM images of 4 kPa Biogelx Pre-gel solution

and scanning speed 1.0 Hz. Figure 7 shows AFM images of the fibrous morphologies of Biogelx gels comprising three different strengths.

3.2.2 Transmission Electron Microscopy (TEM)

1. Glow discharge carbon-coated copper grids (No. 400) for 30 s before making contact with the hydrogels.
2. Place the shiny side down on the surface of the Pre-gel solution or gel (gels should be prepared according to Subheading 3.1 for less than 5 s.
3. Remove the excess sample by blotting with a filter paper and then place 10 mL of negative stain on the top of the sample on the grid and allow to dry for 10 min.
4. Image the dried sample using a TEM.

Figure 8 shows images obtained using TEM of Biogelx Pre-gel solutions demonstrating the fibrous morphologies of the hydrogel.

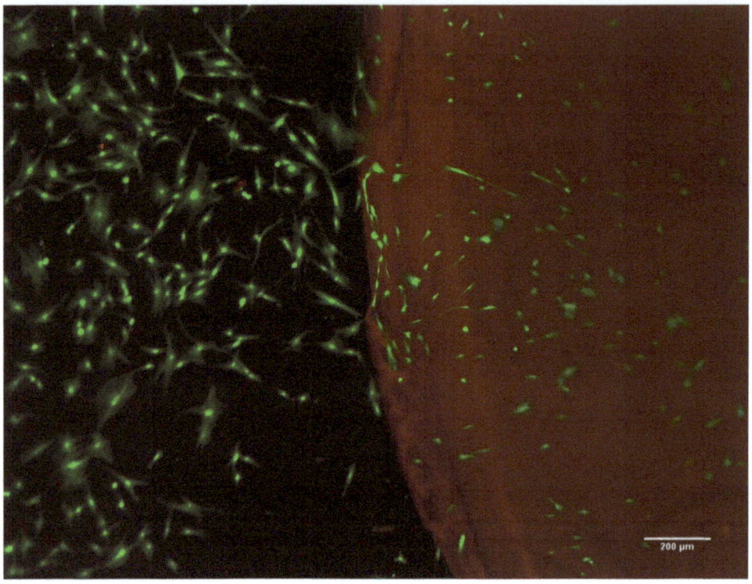

Fig. 9 LIVE/DEAD® staining images of human mesenchymal stem cells (hMSCs) at plastic–hydrogel interface (hydrogel shown in red for visualization). Scale bar represents 200 μm

3.3 Cell Viability Assay Methods

LIVE/DEAD® Staining should be carried out after cell culture has been performed (Subheading 3.1).

3.3.1 LIVE/DEAD® Staining for Biogelx Gel

1. Remove the medium from the gel surface and then wash the gel with 1× PBS, warm (37 °C).
2. Add LIVE/DEAD® staining to the top of the gel, ensuring surface is fully covered, and wrap the plate with foil.
3. Incubate the gels with staining solution for 20 min at 37 °C.
4. Remove the stain and wash with 1× PBS to remove unbound reagents.
5. Remove the gel and place on a glass slide.
6. Put a drop of mounting agent on top of the gel and place another coverslip on top ensuring all air bubbles are removed.
7. Immediately observe hydrogels with cells using a fluorescence microscope. Live cells will show green fluorescent calcein, while dead cells will be detected by red EthD-1 binding to DNA.

Figure 9 shows the cell viability of human mesenchymal stem cells (hMSCs) grown in 2D culture using Biogelx gels and visualized by LIVE/DEAD® staining. A comparable number of live human mesenchymal stem cells hMSCs were observed on hydrogel as were on the plastic surface.

3.3.2 Phalloidin Staining for Biogelx Gel

Phalloidin Staining should be carried out after cell culture has been performed (Subheading 3.1).

1. Remove the cell culture medium from the gels and wash twice in 1× phosphate buffered saline (PBS).

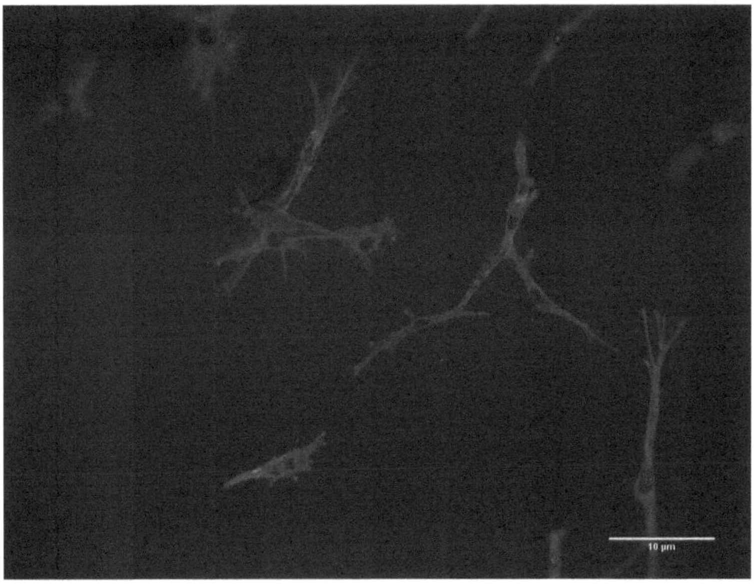

Fig. 10 Phalloidin staining of bone marrow stem cells following culture in Biogelx gel using Phalloidin 488 staining solution and DAPI mounting agent. Scale bar represents 10 μm

2. To fix the cells, cover the gel with 4% paraformaldehyde (PFA) in the well plate for 15 min at 4 °C.
3. Wash with 2 mL PBS for 3 × 5 min.
4. Permeabilize the cells in 0.25% Triton-X in PBS for 5 min at room temperature.
5. Wash with 2 mL PBS for 3 × 5 min.
6. To block the cells, add 1% BSA in PBST to the gel surface for between 30–60 min and incubate at 37 °C. Wash once with PBS.
7. Remove BSA solution and add the phalloidin solution (F-actin), ensuring the solution covers the gel surface. Cover with foil and incubate at room temperature for 20 min.
8. Remove staining solution and wash with 2 mL PBS for 3 × 5 min. Then, wash once with distilled water and remove any excess by blotting.
9. Mount samples using a drop of mounting agent on a coverslip for imaging (Fig. 10).

3.3.3 MTS (Cell Metabolic Assay) for Biogelx Gel

MTS assay should be carried out after cell culture has been performed (Subheading 3.1).

1. Prior to use thaw the CellTiter 96® Aqueous One Solution Reagent.
2. Remove the medium surrounding the hydrogel and transfer the gel into an eppendorf tube.

3. Add 300 μL of cell culture medium to each eppendorf tube and mix thoroughly using a pipette to break up the gel (*see* **Note 5**).
4. Add 60 μL of CellTiter 96® Aqueous One Solution Reagent into each eppendorf tube and mix thoroughly using a pipette. One sample without cells (only with culture medium and CellTiter 96® Aqueous One Solution Reagent) should be used as control to correct for any background absorbance.
5. Incubate the eppendorf tube at 37 °C with a humidified atmosphere of 5% CO_2 atmosphere for 4 h (*see* **Note 6**).
6. After 4 h, centrifuge the eppendorf tube at 12,000 × *g* for 4 min.
7. Transfer 120 μL of supernatant from each sample into a 96-well plate.
8. Record the absorbance at 490 nm using a 96-well plate reader.
9. Use a calibration curve (with different cell densities) to convert the optical density values (absorbance) into actual cell number.

3.4 Cell Recovery from Biogelx Gel

Cell recovery should be carried out after cell culture has been performed (Subheading 3.1).

1. After culture period transfer the gels into 15 mL centrifuge tube.
2. Add 10 mL of PBS or 1% BSA in PBS and mechanically disrupt the hydrogel by gently pipetting the mixture (*see* **Note 5**).
3. Centrifuge the tube (12,000 × *g* for 5 min) and discard the supernatant (*see* **Note 7**).
4. Resuspend the pellet in 3 mL trypsin and incubate at 37 °C for 5 min. This will encourage cells that remain attached to hydrogel to separate.
5. After 5 min, add 10 mL of fresh PBS and gently agitate the trypsin and PBS mixture using a pipette.
6. Centrifuge the tube again at 12,000 × *g* for 5 min.
7. Discard the supernatant, the cells will remain in the bottom of the centrifuge tube as a pellet. Figure 11 shows bone marrow stem cells recovered from Biogelx gels after 3 days in culture.

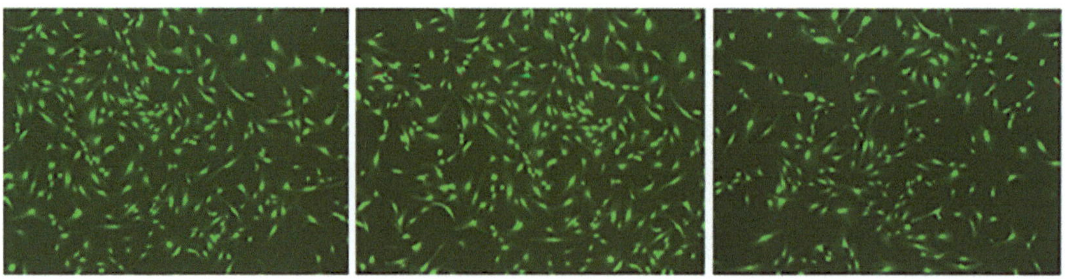

Fig. 11 LIVE/DEAD® staining results of recovered bone marrow stem cells after 1 day in monolayer culture following culture in Biogelx gel for 3 days. Scale bar represents 50 μm

3.5 RNA Extraction for Biogelx Gel

This method is based on using a combination of the Invitrogen TRIzol and Purelink protocols.

3.5.1 RNA Extraction Method Without Using Columns

1. Transfer the cell-seeded hydrogel samples into an eppendorf tube and add TRIzol. TRIzol volumes should be adapted depending on cell number. Vortex the tube in order to disrupt the gel (*see* **Note 8**).
2. Incubate the prepared samples at room temperature for a minimum of 10 min to allow for cell lysis.
3. Centrifuge samples at 12,000 × *g* for 15 min at 4 °C to pellet the excess gel, while retaining RNA in solution.
4. Transfer the supernatant to a clean tube and add 100 μL of chloroform. Subject the tube to vigorous mixing for 20 s and incubate the sample at room temperature for 3 min.
5. Centrifuge the sample at 12,000 × *g* for 15 min at 4 °C to enable phase separation. Transfer the upper aqueous phase of the sample to a clean tube and mix 5 μL of glycoblue (3 mg/mL) and 250 μL of isopropanol to the sample.
6. Incubate the sample at −20 °C for at least 1 h and further centrifuge at 12,000 × *g* for 15 min at 4 °C.
7. Remove the supernatant using a pipette without disturbing the resultant blue pellet in the bottom of the tube.
8. Wash the precipitated RNA pellet twice with ethanol (70%) and centrifuged at 7500 × *g* for 5 min at 4 °C.
9. Without disturbing the pellet, remove the ethanol by pipette. Ensure that all ethanol in the tube is removed by air-drying the pellet in the tube.
10. Resuspend the pellet in distilled RNAse-free water (20 μL) and incubated at 60 °C for 10 min to enable the pellet to fully resuspend.
11. Store samples at −80 °C overnight and then heat at 60 °C for 10 min for a better solubilization before the determination of RNA concentration (*see* **Note 9**).

3.5.2 RNA Extraction Method Using Columns

This method is based on using a combination of the Qiagen QIAshredder and RNeasy plus mini kit.

1. Aspirate the medium from the gel surface and add an RNA stabilizer.
2. Freeze the culture plate at −20 °C until ready to proceed with extraction.
3. Follow instructions as per column-based kit (for example, Qiagen QIAshredder kit) to collect the homogenized lysate, through loading the gel onto the QIAshredder homogenizer and centrifuge to collect the RNA.

4. Follow instructions as per RNA purification kit (for example, RNAeasy plus mini kit) to purify collected RNA. The quality and quantity of this extracted RNA should be assessed before it is subjected to cDNA synthesis.

4 Notes

1. Care should be taken when removing the medium on top of the gel; to avoid direct contact with the gel, carefully place the pipette tip at the top of the medium and remove the medium slowly. Do not use an aspirator at any point.
2. The cell number stated is a recommend guideline and can be adjusted depending on the user's cell type or application.
3. Ensure the pipette tip does not leave the Pre-gel solution when mixing as this can introduce air bubbles into the gel structure.
4. Pipette tip should be fully immersed before Pre-gel/cell solution is released into the medium, and the plunger of the pipette depressed in one smooth motion to form a single gel "sphere"—do not add dropwise.
5. Volumes mentioned in here are for the gels made in 12-well plate inserts; therefore, volumes need to be adjusted for the other well plates or insert gels accordingly.
6. Recommend mixing of the solution by pipette is performed every hour to ensure uniform distribution of CellTiter 96® Aqueous One Solution Reagent.
7. The pellet at the bottom of the tube contains cells and at this stage some hydrogel may remain attached to the cell pellet.
8. RNA extraction requires the use of a phenol-based solvent, chloroform, isopropyl alcohol, and ethanol. Appropriate care should be taken. Gloves and lab coat should be worn at all times and phenol and chloroform should be handled in a fume cupboard/laminar flow hood.
9. The quality and quantity of extracted RNA should be assessed before it is subjected to cDNA synthesis.

References

1. Silva GA, Czeisler C, Niece KL, Beniash E, Harrington DA, Kessler JA, Stupp SI (2004) Selective differentiation of neural progenitor cells by high-epitope density nanofibers. Science 303:1352–1355
2. Dankers PYW, Harmsen MC, Brouwer LA, Van Luyn MJA, Meijer EW (2005) A modular and supramolecular approach to bioactive scaffolds for tissue engineering. Nat Mater 4:568–574
3. Banwell EF, Abelardo ES, Adams DJ, Birchall MA, Corrigan A, Donald AM, Kirkland M, Serpell LC, Butler MF, Woolfson DN (2009) Rational design and application of responsive alpha-helical peptide hydrogels. Nat Mater 8:596–600

4. Gelain F, Bottai D, Vescovi A, Zhang SG (2006) Designer self-assembling peptide nanofiber scaffolds for adult mouse neural stem cell 3-dimensional cultures. PLoS One 1:e119
5. Collier JH (2008) Modular self-assembling biomaterials for directing cellular responses. Soft Matter 4:2310–2315
6. Webber MJ, Appel EA, Meijer EW, Langer R (2016) Supramolecular biomaterials. Nat Mater 15:13–26
7. Whitesides GM, Mathias JP, Seto CT (1991) Molecular self-assembly and nanochemistry: a chemical strategy for the synthesis of nanostructures. Science 254:1312–1319
8. Mart RJ, Osborne RD, Stevens MM, Ulijn RV (2006) Peptide-based stimuli-responsive biomaterials. Soft Matter 2:822–835
9. Zhang S, Holmes T, Lockshin C, Rich A (1993) Spontaneous assembly of a self-complementary oligopeptide to form a stable macroscopic membrane. Proc Natl Acad Sci U S A 90:3334–3338
10. Hartgerink JD, Beniash E, Stupp SI (2001) Self-assembly and mineralization of peptide-amphiphile nanofibers. Science 294:1684–1688
11. Dasgupta A, Mondal JH, Das D (2013) Peptide hydrogels. RSC Adv 3:9117–9149
12. Caliari SR, Burdich JA (2016) A practical guide to hydrogels for cell culture. Nat Methods 13 (5):405–414
13. Seow WY, Hauser CAE (2014) Short to ultrashort peptide hydrogels for biomedical uses. Mater Today 17(8):381–388
14. Vegners R, Shestakova I, Kalvinsh I, Ezzell RM, Janmey PA (1995) Use of a gel-forming dipeptide derivative as a carrier for antigen presentation. J Pept Sci 1:371–378
15. Orbach R, Alder-Abramovich L, Zigerson S, Mironi-Harpaz I, Seliktar D, Gazit E (2009) Self-assembled Fmoc-peptides as a platform for the formation of nanostructures and hydrogels. Biomacromolecules 10(9):2646–2651
16. Babu SS, Praveen VK, Ajayaghosh A (2014) Functional π-gelators and their applications. Chem Rev 144(4):1923–2129
17. Alakpa EV, Jayawarna V, Lampel A, Burgess KV, West CC, Bakker SCJ, Roy S, Javid N, Fleming S, Lamprou DA, Yang J, Miller A, Urquhart AJ, Frederix PWJM, Hunt NT, Péault B, Ulijn RV, Dalby MJ (2016) Tunable supramolecular hydrogels for selection of lineage-guiding metabolites in stem cell cultures. Chem 1:1–22
18. Fleming S, Ulijn RV (2014) Design of nanostructures based on aromatic peptide amphiphiles. Chem Soc Rev 43:8150–8177
19. Jayawarna V, Ali M, Jowitt TA, Miller AF, Saiani A, Gough JE, Ulijn RV (2006) Nanostructured hydrogels for 3D cell culture through self-assembly of Fmoc-dipeptides. Adv Mater 18:611–614
20. Mahler A, Reches M, Rechter M, Cohen S, Gazit E (2006) Rigid, Self-assembled hydrogel composed of a modified aromatic dipeptide. Adv Mater 18:1365–1370
21. Cheng G, Castelletto V, Jones RR, Connona CJ, Hamley IW (2010) Hydrogelation of self-assembling RGD-based peptides. Soft Matter 7:1326–1333
22. Yang ZM, Liang GL, Wang L, Xu B (2006) Using a kinase/phosphatase switch to regulate a supramolecular hydrogel and forming the supramolecular hydrogel in vivo. J Am Chem Soc 128:3038–3043
23. Jayawarna V, Richardson SM, Hirst A, Hodson NW, Saiani A, Gough JE, Ulijn RV (2009) Introducing chemical functionality in fmoc-peptide gels for cell culture. Acta Biomater 5:934–943

Part III

Methods Related to Preparing, Characterizing and Inhibiting Pathological Amyloid Peptide Self-Assembly

Chapter 19

Production and Use of Recombinant Aβ for Aggregation Studies

Tiernan T. O'Malley, Sara Linse, and Dominic M. Walsh

Abstract

The amyloid β-protein (Aβ) is believed to play a central role in Alzheimer's disease (AD) pathogenesis and there is great interest in understanding the process of Aβ aggregation, its underlying mechanism and the species generated during aggregation and their biological activity. Although Aβ has been studied for more than 30 years, analysis of its aggregation has been hampered by structural and chemical impurities. Here we provide a detailed protocol for the expression and purification of chemically and structurally homogeneous Aβ monomer. We also describe a method to produce covalent Aβ dimers linked by phenolic coupling of tyrosine residues.

Key words Alzheimer's disease, Amyloid β-protein, Dityrosine, Dimer, Fibrillogenesis, Monomer, Aggregation

1 Introduction

Aggregation of the amyloid β-protein (Aβ) is believed to play an initiating role in the pathogenic process that causes Alzheimer's disease (AD) [1]. In vivo, Aβ occurs as a family of related peptides with a common amino acid core of ~30 residues (reviewed in [2]). All members of the Aβ family are amphipathic and have a proclivity to self-associate and aggregate, often forming thermodynamically stable amyloid fibrils [3]. Due to the high costs of synthetic Aβ, systematic studies of aggregation and the use of peptides of varying primary sequences were limited. Moreover, sequence inhomogeneity in synthetic Aβ affects the underlying mechanism and overall rate of aggregation [4]. Here we detail an inexpensive and simple system which can be used to generate naturally occurring Aβ primary structures [5–7] and Aβ containing designed mutations [8, 9]. This method allows for the production of large quantities of chemically and structurally pure Aβ [5, 9]—prerequisites for kinetic and structural experiments. In addition, we have optimized a system to monitor Aβ fibrillogenesis which yields highly

Bradley L. Nilsson and Todd M. Doran (eds.), *Peptide Self-Assembly: Methods and Protocols*, Methods in Molecular Biology, vol. 1777, https://doi.org/10.1007/978-1-4939-7811-3_19,

reproducible results [7, 10]. We and others have used these advances to connect the macroscopic features of Aβ fibril formation to their microscopic determinants through the analysis of experimental kinetic data [11–14] and to study the structure of Aβ monomer and dimer [9].

Besides variations in Aβ primary structure, evidence suggests that post-translational modifications such as the formation of Aβ dimers may also be important in AD [2, 15–18]. In nature, there are few ways by which covalent Aβ dimers can be generated. The most likely mechanism of dimer formation involves coupling of two Aβ monomers by a dityrosine bond. Here we describe a method for the production of dityrosine Aβ that can be used to study kinetics of aggregation [9] and as a standard for use in immunoassays [19]. In this chapter we describe protocols for the preparation of Aβ42 monomer and Aβ42 dityrosine dimer; however, it is important to note that these methods are applicable to many other Aβ sequences. The establishment of a facile method to produce a multitude of highly pure Aβ peptides coupled with a robust protocol to monitor fibrillogenesis has already revealed important information [11–13] and offers a vehicle for further discovery when applied to the study of complex mixtures of Aβ species with different primary sequences and posttranslational modifications.

2 Materials

2.1 Expression and Purification of Aβ (M1-42)

1. Glycerol stock of chemically competent BL21* DE3 pLysS *E. coli* cells (*see* **Notes 1–3**).
2. pET-Sac vector containing a gene encoding for wild type, human Aβ(M1-42) (available from Addgene plasmid repository—www.addgene.org) (*see* **Note 3**).
3. Agar plates containing ampicillin and chloramphenicol at a final concentration of 50 mg/L and 38 mg/L, respectively.
4. Autoclaved glass beads.
5. Sterile super optimal broth with catabolite repression (S.O.C.) (Thermo Fischer, Cat. #15544-034).
6. Lysogeny broth (LB) medium containing ampicillin and chloramphenicol at 50 μg/mL and 38 μg/mL, respectively. LB (5 g NaCl, 5 g Bacto Tryptone and 2.5 g Bacto Yeast Extract) is made up to 500 mL using H_2O and autoclaved at 121 °C for 21 min in a 2.5 L baffled flask. Smaller, 250 mL flasks containing 50 mL of LB are prepared for overnight starter cultures. Once autoclaved, the broth is allowed to cool to room temperature and 0.5 mL of a 50 g/L ampicillin solution prepared in milliQ water (MQ H_2O) and sterile filtered, and 0.5 mL of a 30 g/L chloramphenicol solution, prepared in 100% ethanol

and sterile filtered, added to 500 mL of LB. Fifty microliters of chloramphenicol and 50 μL of ampicillin solution are added to 50 mL of LB.

7. Induction solution: aqueous 1 M isopropyl β-D-1-thiogalactopyranoside (IPTG) stored as 1 mL aliquots at −20 °C.
8. Preparation of DE23 anion exchange resin. Resin is resuspended in 0.1 M HCl (15 mL per gram of dried resin), washed ten times with MQ H_2O, then once with 0.5 M NaOH, and finally ten times with MQ H_2O. The resin is allowed to settle between all steps and the supernatant decanted. The remaining base is neutralized by resuspending the resin in 250 mM tris containing 50 mM EDTA (unbuffered). Finally the resin is resuspended in 25 mM tris–HCl, pH 8.5 containing 5 mM EDTA at a 50:50 mix of buffer to resin. For long-term storage, 0.01% w/v benzalkonium chloride is added and the resin kept at 4 °C. Before use, resin is homogenized by gentle inversion and washed with 25 mM tris–HCl, pH 8.5 containing 5 mM EDTA until the pH of the flow through is 8.5.
9. Mixed bed resin (Sigma Aldrich, Cat. #M8032).
10. Buchner funnel and G3 filter paper (Whatman, Cat. #1003-917).
11. Anion exchange buffer (Buffer A): 25 mM tris–HCl, pH 8.5 containing 5 mM EDTA.
12. Low salt wash solution—Buffer A containing 25 mM NaCl.
13. Elution solution—Buffer A containing 50 mM NaCl.
14. High salt wash solution—Buffer A containing 500 mM NaCl.
15. High salt/urea regeneration solution—Buffer A 500 mM NaCl and 8 M urea.
16. Inclusion body denaturant: Buffer A containing 8 M urea. All urea-containing solutions were deionized prior to using (*see* **Note 4**).
17. Dialysis: Snake skin tubing, 3500 MWCO (Thermo Fischer, Cat. #88242).
18. Sonicator: Branson digital sonifier model 250.
19. Centrifuge: Sorvall SS-34 rotor.
20. UV-Vis spectrophotometer.
21. Heating plate set at 42 °C for heat shock.
22. Water bath set at 37 °C.
23. Bunsen burner.

2.2 Size Exclusion Chromatography (SEC)

1. A Superdex 75 16/60 is used for purification of bacterial extract and two Superdex 75 10/300 columns linked in tandem for isolation of samples for aggregation assays. All columns

are attached to a Bio-Rad Duo Flow FPLC eluted at 0.8 or 0.5 mL/min in a cold room at 4 °C. Columns must be thoroughly cleaned before and/or after SEC using Aβ peptides (*see* **Note 6**).

2. Aβ disaggregation solution: 50 mM tris–HCl, pH 8.5 containing 7 M guanidine HCl and 5 mM EDTA.
3. SEC elution buffer (for initial size purification): 50 mM ammonium bicarbonate, pH 8.5—the pH of which is attained by dropwise addition of 28–30% ammonium hydroxide. Buffer is then passed through a 0.22 μm filter and degassed under vacuum overnight.
4. SEC elution buffer (for aggregation studies): 20 mM sodium phosphate pH 8.0 containing 200 μM EDTA and 0.01% sodium azide (depending on the experiment other buffers may be used, *see* **Note 7**). Phosphate buffer is prepared by combining 94.7 mL of a 200 mM Na_2HPO_4 solution with 5.3 mL of a 200 mM NaH_2PO_4 solution. To this, 0.4 mL of 500 mM EDTA and 1 mL of 10% sodium azide solution is added and the solution made up to 1 L with MQ H_2O. As with the preparative buffer, the experimental SEC buffer is passed through a 0.22 μm filter and degassed as above.
5. UV-Vis spectrophotometer.
6. Quartz cuvette: Starna Cells 160 μL cuvette with 1 cm path length (Cat #: 16.160-Q-10/Z15).

2.3 Dityrosine Cross-Linking of Aβ (M1-42)

1. 10 mM hydrogen peroxide (H_2O_2) solution: 56.5 μL of 30% w/w H_2O_2 diluted to 50 mL using MQ H_2O.
2. Horse radish peroxidase (HRP) solution: HRP (Thermo Fischer Cat. #31490) dissolved in MQ H_2O to yield a 2 mg/mL solution.
3. Tissue culture incubator at 37 °C.

2.4 Continuous Thioflavin T Dye Binding

1. 100× Thioflavin T (ThT) stock solution: 1 mM ThT in SEC elution buffer (for aggregation studies).
2. Fluorescence plate reader, such as a CLARIOstar plate reader.
3. Black, clear bottom, 96-well half-area low-binding polystyrene plate.
4. Low-fluorescence adhesive cover.

2.5 Negative Contrast Electron Microscopy

1. Dumont N4AC Tweezers (Dumoxel Biology, Cat. #72870-D).
2. Acrodisc Filters 33 mm/0.22 μM (Millipore, Cat. #SLGP033RS). MQ water should be placed in a 5 mL syringe fitted with a 0.22 μM filter.

3. EM grids: 200 mesh Copper Grids, Formvar/Carbon film (Electron Microscopy Sciences Cat. #EMS200-Cu).
4. 0.5% glutaraldehyde solution: 6.25 μL of EM Grade 8% aqueous glutaraldehyde (Electron Microscopy Science, Cat. #16020) diluted to 10 mL with MQ immediately before use. Two milliliters of this solution is passed through a 0.22 μm filter and collected into a clean eppendorf tube.
5. 2% (w/v) uranyl acetate stain (Electron Microscopy Sciences Cat# 22400). Dissolve 1 g in 25 mL of 100% ethanol and stir for 30 min. Shield solution from light with aluminum foil. Dilute to a final volume of 50 mL with MQ H_2O and store in two 50 mL tubes covered with aluminum foil.

3 Methods

3.1 Expression and Purification of Aβ (M1-42) Peptide

3.1.1 *E. coli* Transformation

1. Thaw competent *E. coli* Bl21* DE3 pLysS cells (40 μL cell suspension) on ice.
2. Thaw DNA plasmid stock from freezer.
3. Beside a Bunsen flame and using a sterile pipette tip, add 1 μg of DNA to the *E. coli* suspension, gently flick tube to mix.
4. Incubate on ice for ~30 min.
5. Heat shock bacteria at 42°C for 45 s and cool immediately on ice for 10 min.
6. Add 400 μL of S.O.C. to bacterial suspension and incubate at 37 °C for 30 min.
7. Transfer 50–100 μL of bacterial suspension to an agar plate.
8. Add 5–10 sterile glass beads and shake several times to distribute cells.
9. Incubate overnight at 37 °C (*see* **Note 8**) in a dry incubator or warm room.
10. Plates with colonies can be used immediately or sealed with parafilm and stored at 4 °C.

3.1.2 Colony Selection and Propagation

1. Sterilize and seal the tip of a glass Pasteur pipette using a Bunsen burner flame.
2. Use a sealed pipette tip to collect a small well-isolated single colony (*see* **Note 9**).
3. Transfer the pipette to a 250 mL flask containing 50 mL of LB and agitate for 30 s.
4. Remove and discard pipette.
5. Incubate the inoculated culture with agitation at 37 °C overnight (max. 16 h).

3.1.3 Growth of Transformed Bacteria and Induction of Protein Expression

1. After overnight bacterial culture, determine the extent of bacterial growth in the starter culture by measuring the OD_{600} (>2.0 desired).
2. Add 5 mL of starter culture to 500 mL LB media and agitate at 37 °C.
3. Measure OD_{600} at regular intervals until a value of ~0.6–1.0 (~2–2.5 h) is attained.
4. When OD_{600} is ~0.6–1.0, add 125 μL 1 M IPTG solution (final concentration 400 μM).
5. Monitor OD_{600} readings every 30 min until growth plateaus (~3–4 h).

3.1.4 Collection of Bacteria

1. Harvest the cells by centrifugation (10 min at 10,000 rpm (16,900 × *g*); SORVAL-SLA-3000; 350 mL Sorvall centrifuge bottle).
2. Decant the supernatant and scrape bacterial pellets into a 50 mL conical tube and suspend in 25 mL buffer A (per pellet from 1 L culture) and freeze overnight at −20 °C.

3.1.5 Sonication and Isolation of Inclusion Bodies

1. Thaw the pellet. Sonicate for 30 s at 4 W and 30% amplitude and transfer to centrifuge tube.
2. Centrifuge the suspension for 10 min at 18,000 rpm (38,724.2 × *g*) in a Sorvall SS-34 rotor.
3. Decant the supernatant and resuspend the pellet in 25 mL buffer A. Repeat **steps 1–3** two times, i.e., a total of 3 sonication steps.
4. Resuspend the pellet in 12.5 mL inclusion body denaturant and sonicate for 30 s at 4 W and 30% amplitude.
5. Centrifuge denatured inclusion bodies for 10 min at 18, 000 rpm (38,724.2 × *g*) and 4 °C in a Sorvall SS-34 rotor to remove debris.
6. Collect supernatant and store on ice.

3.1.6 Purification of Aβ (M1-42) from Solubilized Inclusion Bodies

1. Dilute the 12.5 mL of denatured inclusion bodies with 37.5 mL of buffer A (1:4 dilution).
2. Add DE23 resin (*see* **Note 10**) and incubate with gentle rocking for 30 min at room temperature.
3. Filter under vacuum using a Buchner funnel.
4. Wash the resin with 25 mL of buffer A, stir well.
5. Apply vacuum and collect the flow through.
6. Wash the resin with 25 mL of low salt wash solution and then elute with 4 × 50 mM NaCl in buffer A (*see* **Note 11**).
7. Analyze a sample of each fraction using SDS-PAGE and silver staining (*see* **Note 12**).

8. Pool fractions containing Aβ and transfer into 2 × 20 cm long pieces of snake skin tubing.
9. Dialyze 50 mL of pooled fractions into 5 L of 50 mM ammonium bicarbonate, pH 8.5, with three changes (to achieve a dilution of NaCl, tris and EDTA of 1×10^6) (*see* **Note 13**).
10. Lyophilize samples and store at −80 °C.

3.2 SEC Purification of Recombinant Aβ Peptide

1. Dissolve the dried peptide at a concentration of ~10 mg/mL in Aβ disaggregation solution (*see* **Note 13**) and incubate at room temperature overnight.
2. Inject 5 mL of Aβ solution onto a Superdex 75 16/60 column and elute at 0.8 mL/min in 50 mM ammonium bicarbonate, pH 8.5 (*see* **Note 13**).
3. Monitor UV absorbance at 280 nm and collect the Aβ monomer peak in 1.2 mL fractions. Pool the peak fractions and measure the absorbance at 275 nm (Abs_{275}). The concentration of the Aβ can be determined using the extinction coefficient 1361 M^{-1} cm^{-1} [9] (*see* **Note 13**).
4. Freeze and lyophilize volumes containing 1 mg of peptide. Peptide mass and purity (>99%) are assessed by MALDI-TOF mass spectrometry, reverse phase HPLC, and SDS-PAGE with silver staining (Fig. 1).

3.3 Preparation of Dityrosine (DiY) Cross-Linked Aβ (M1-42)

1. Dityrosine cross-linking of peptides is performed essentially as described previously [9]. Dissolve 5 mg of Aβ(M1-42) in 7 M GuHCl, incubate overnight at room temperature and purify by gel filtration using two Superdex 75 10/300 columns linked in tandem and eluted in 50 mM ammonium bicarbonate, pH 8.5 (*see* **Note 13**).
2. Collect 0.5 mL fractions, pool the Aβ monomer peak fractions, measure the Abs_{275}, and determine the concentration.
3. Make a 1:20 dilution of a 2 mg/mL HRP solution (2.2 μM final concentration) and a 1:40 dilution of a 10 mM H_2O_2 solution (250 μM final concentration) into the Aβ monomer solution. Adjust with 50 mM ammonium bicarbonate (pH 8.5) to give a final Aβ concentration of 40 μM and incubate at 37 °C for ~14 h [20].
4. Quench the reaction on ice, freeze the solution, and lyophilize. Dissolve the lyophilizate in 7 M GuHCl (3 mL) and incubate overnight at room temperature.
5. Isolate dityrosine cross-linked Aβ dimer ($[A\beta(M1\text{-}42)]_{DiY}$) from unreacted monomer, HRP and aggregated Aβ using a Superdex 75 16/60 column eluted with 50 mM ammonium bicarbonate, pH 8.5 (*see* **Note 14**).

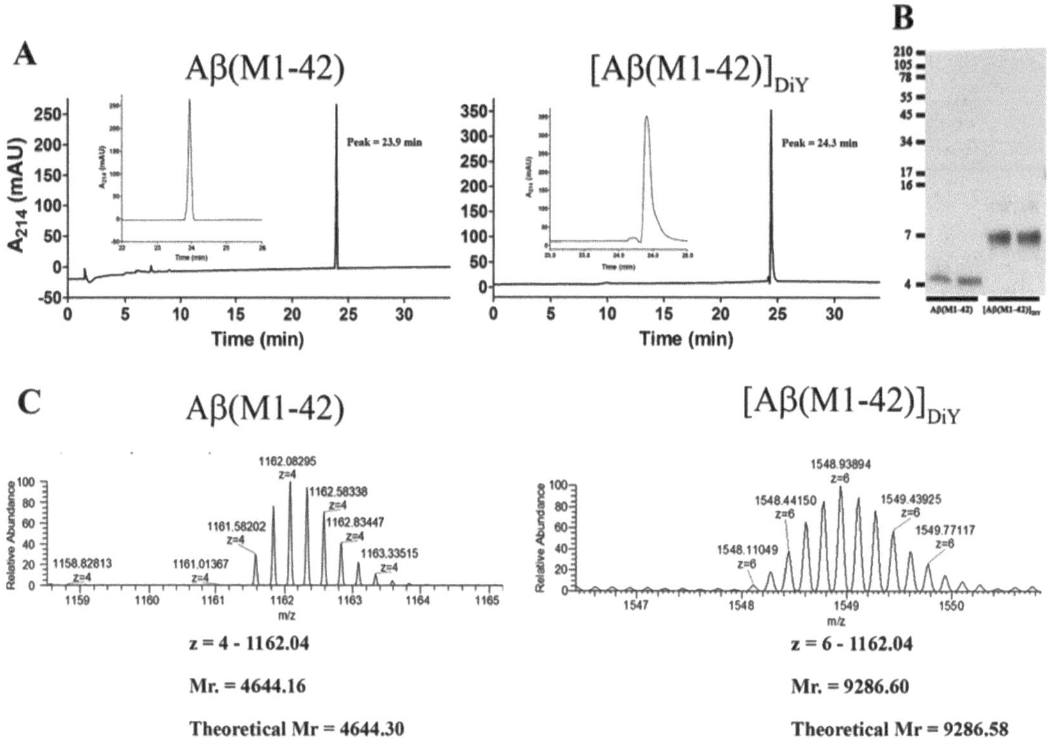

Fig. 1 Analysis of peptide purity and mass. The purity of recombinantly produced Aβ is assessed by (**a**) reverse phase HPLC and (**b**) SDS-PAGE with silver staining. For RP-HPLC analysis of each peptide, 25 μL of a ~0.2 μg/μL solution (in 0.1% trifluoroacetic acid; TFA) is injected onto a Capcell CN column and eluted with a linear gradient of 5–65% acetonitrile in 0.1% TFA over 35 min (**a**). The same sample preparation (12 μL) was treated with 12 μL of 2× sample buffer and duplicate 10 μL lots (~1 μg) electrophoresed on a 16% polyacrylamide tris-tricine gel. Gels are visualized with silver staining. (**c**) Peptide mass is determined by ion-trap mass spectrometry using a Q Exactive hybrid quadrupole-orbitrap connected to a Dionex 3000 LC eluting from 100% buffer A (0.075% NH_4OH in 5% acetonitrile/H_2O) to 50% buffer B (0.025% NH_4OH in 95% acetonitrile/H_2O) over 20 min. Monoisotopic mass spectra are given for the most abundant isotope and charge state present ($m/z_{A\beta(M1\text{-}42)} = 4+$, Expected $Mr_{A\beta(M1\text{-}42)} = 4644.30$, Observed $Mr_{A\beta(M1\text{-}42)} = 4644.16$; $m/z_{[A\beta(M1\text{-}42)]DiY} = 6+$, Expected $Mr_{[A\beta(M1\text{-}42)]DiY} = 9286.58$, Observed $Mr_{[A\beta(M1\text{-}42)]DiY} = 9286.60$)

6. Collect 0.5 mL fractions, pool dimer containing peak fractions, and determine concentration by measuring the absorbance at 283 nm and using the extinction coefficient for dityrosine cross-linked Aβ, $\varepsilon = 6226\ M^{-1}\ cm^{-1}$ [9].
7. The above protocol typically yields 1 mg of $[A\beta(M1\text{-}42)]_{DiY}$ for every 4 mg of starting Aβ(M1-42) monomer.

3.4 Continuous Thioflavin T Dye Binding Assay

1. ThT assay is performed essentially as described previously [8]. Dissolve 1 mg of Aβ monomer or dimer peptide in 0.55 mL 20 mM sodium phosphate, pH 8.5 containing 200 μM EDTA, 0.02% NaN_3 and 7 M GuHCl and incubate overnight at room temperature.

2. Centrifuge at 16,000 × *g* for 30 min prior to injection onto SEC.
3. Inject 0.5 mL of peptide solution onto two Superdex 75 10/300 columns linked in tandem and elute in 20 mM sodium phosphate, pH 8.0 containing 200 μM EDTA and 0.02% NaN_3. Collect 0.5 mL fractions.
4. Prepare a working stock of SEC isolated peptide at 10.1 μM with 20 mM sodium phosphate, pH 8.0 containing 200 μM EDTA and 0.02% NaN_3.
5. A working stock containing ThT is prepared by combining 1980 μL of 10.1 μM Aβ solution with 20 μL of 100 × ThT stock (1 mM), to yield 10 μM ThT and 10 μM Aβ.
6. Prepare serial dilutions of the Aβ ThT solution by combining 1000 μL of solution containing 20 μM ThT with 1000 μL of dilution buffer to yield a 1:2 dilution, and make further dilutions as required (*see* **Note 15**).
7. Using reverse pipetting technique, transfer 6× 100 μL of each Aβ concentration to the wells of a black, clear bottom, 96-well half-area low-binding polystyrene plate.
8. Include a blank without any Aβ (i.e., dilution buffer with 10 μM ThT) alongside Aβ-containing wells. The outer 36 edge wells of the plate are not used for samples, but are filled with buffer and plates sealed using a low fluorescence adhesive cover.
9. Incubate samples at 37 °C in a CLARIOstar plate reader and record the fluorescence at 5 min intervals (Ex_{440nm}; Em_{480nm}) (*see* **Note 16**).

3.5 Negative Contrast Electron Microscopy

1. Transfer 10 μL of sample to duplicate 200 mesh, formvar coated, copper electron microscopy grid (Electron Microscopy Sciences, Cat. #EMS-200-Cu). Incubate for 60 s.
2. Add 10 μL of 0.5% gluteraldehyde solution and incubate for exactly 60 s.
3. Carefully wick the solution onto blotting paper making sure not to touch the face of the grid.
4. Wash the grid with 3 drops of MQ H_2O passed through a syringe fitted with a 0.22 μm filter. If necessary, carefully wick solution onto blotting paper (*see* **Note 17**).
5. Add 10 μL of uranyl acetate solution to the grid, and incubate for exactly 120 s. Thereafter, carefully wick solution onto blotting paper.
6. Allow the grid to dry for at least 5 min.
7. Grids are visualized using a Techni G^2 BioTWIN, or similar electron microscope (FEI). Scan the grid at ~12,000×

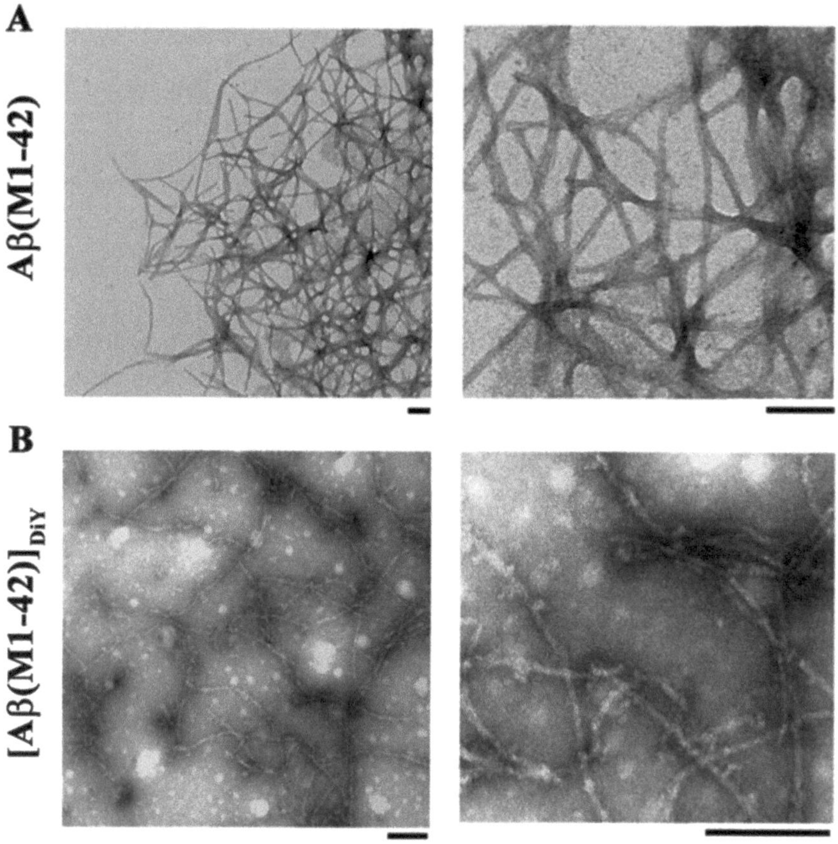

Fig. 2 Both Aβ(M1-42) monomer and dityrosine dimer aggregate to form Aβ fibrils. SEC-isolated Aβ(M1-42) and $[A\beta(M1\text{-}42)]_{DiY}$ were diluted to 10.1 μM with 20 mM sodium phosphate pH 8.0 containing 200 μM EDTA and 0.02% NaN_3. Six 100 μL replicates of each peptide were transferred to a 96-well, half-area plate incubated at 37 °C for 18 h. Peptide samples were analyzed by negative-stain electron microscopy. Size bars = 100 nm

magnification, in a serpentine fashion making note of the relative abundance and morphology of species seen.

8. Representative images (Fig. 2) are taken of several fields using an AMT 2 k charged coupled-device camera, or similar imager.

4 Notes

1. All steps involving bacteria are performed using standard aseptic techniques.
2. Preparation of chemically competent BL21* DE3 pLysS *E. coli* cells: add 1 μL of stockBL21* DE3 pLysS cells (Promega Cat. #L1195) to 50 mL of autoclaved LB media containing 38 μg/mL of chloramphenicol. Incubate with shaking at 37 °C until

OD_{600} reaches ~0.6. Thereafter, transfer cells to a sterile 50 mL tube and centrifuge at 2500 rpm (1258 × *g*) (Eppendorf 5810R centrifuge with A-4-81 rotor) for 10 min at 4 °C. Decant supernatant and gently resuspend the pellet in 20 mL of ice-cold Tfb1 buffer*. Let the cells stand on ice for 10 min and repeat centrifugation. Decant supernatant and gently resuspend cells pellet in 2 mL of ice-cold Tfb2**. Prepare 40 μL aliquots and store at −80 °C.

*Tfb1 contains 30 mM potassium acetate, 50 mM manganese chloride, 100 mM potassium chloride, 10 mM calcium chloride, and 15% glycerol, and the pH is adjusted to 5.8 with 0.1 M acetic acid. ** Tfb2 contains 10 mM sodium-MOPS, 75 mM calcium chloride, 10 mM potassium chloride, and 15% glycerol, and the pH is adjusted to 7.0 with 0.1 M NaOH.

3. A number of *E. coli* strains are suitable in which to express Aβ. For instance, for Aβ(M1-42) we have used *E. coli* strain BL21* DE3 PLysS, and for Aβ(M1-40) we have used *E. coli* strain T7 Express (C2566, New England BioLabs Inc., Ipswich, MA, USA, previously called ER2566). To ensure optimal and reproducible yields of other Aβ variants it may be necessary to test expression in a selection of *E. coli* strains.
4. The gene encoding Aβ42 contains an AUG "start" codon which gives rise to a methionine that precedes the Asp1 start site of Aβ; therefore, we refer to this form of Aβ as Aβ (M1-42)—the sequence of which is: M D A E F R H D S G Y E V H H Q K L V F F A E D V G S N K G A I I G L M V G G V V I A.
5. Solutions of urea may contain iso-cyanic acid citation is as follows: Sun et al. [21] a thermal breakdown product of urea that can cause carbamylation of primary amines in proteins including Aβ. Therefore, all solutions containing urea are prepared just prior to use and deionized using mixed bed resin.
6. SEC columns are washed as described previously [22]. Columns are transferred from elution buffer into MQ H_2O, washed with 1 column volume (CV) of 50% formic acid; 2 CV of MQ H_2O; 1 CV 0.5 M NaOH; 2 CV MQ H_2O; 1 CV 1 M unbuffered tris; 2 CV MQ H_2O, and stored in 20% ethanol.
7. The last gel filtration step is performed using the desired experimental buffer as running buffer. However, for studies under conditions where Aβ is more aggregation prone, it is recommended to perform SEC in a "low aggregation buffer" and adjust the solution composition after SEC isolation. For example, when a pH lower than 8 is desired, gel filtration is performed at pH 8, and the pH is adjusted afterward while the monomer is kept in an ice box. When higher salt concentration is desired, gel filtration is at moderate ionic strength (e.g., in

20 mM sodium phosphate or other buffer of choice), and salt is added from a concentrated stock to the isolated monomer in an ice bath.

8. Typically, we obtain 25–500 small, round colonies distributed evenly over the agar surface.
9. It is important to pick the smallest well-isolated colonies. These are the ones that most likely express high levels of Aβ. Do not add bacteria from more than one colony in each flask; a low-level expressing clone will tend to grow fast and take over the culture at the expense of higher expressing clones that tended to grow more slowly. If the intention is to grow multiple 500 mL day cultures, use several colonies in individual overnight cultures. Even though BL21 DE3* pLyS tends to give very uniform expression levels, this limits losses that would arise should a single colony be used that may be a low-level expresser.
10. DE23 anion exchange resin is prepared as a 50% slurry in buffer A. Apply 25 mL of slurry to a Buchner funnel and vacuum dry. To equilibrate the resin and remove residual benzalkonium chloride, rinse 5× with 25 mL of anion exchange buffer A. Using pH paper, validate that the pH of the last rinse solution is 8.5.
11. Ion exchange resin may be regenerated by washing with high salt buffer and then high salt/urea regeneration buffer.
12. Aβ monomer migrates at approximately 4.5 kDa on 16% polyacrylamide tris tricine gels. Using the silver staining protocol of Shevchenko et al., it is possible to readily detect 100 ng of Aβ [5, 23].
13. Ammonium bicarbonate is used during purification because it is a volatile buffer and the amount will be reduced after lyophilization. Since the two components may be reduced to different amounts, it is important to check the pH of samples that have been dissolved after being lyophilized from ammonium bicarbonate. We used 50 mM ammonium bicarbonate, however lower molarities such as 25 mM also work and save on reagent.
14. Aβ monomer elutes from a Superdex 75 16/60 column at ~83 mL, whereas dimer elutes at ~74 mL (peak volumes). Using two Superdex 75 30/300 columns linked in tandem, Aβ monomer elutes at ~24.5 mL and dimer at ~21 mL (peak volumes). Depending on batch size, larger columns may be used, but when larger columns are employed Aβ monomer and dimer will elute in larger volumes.
15. A logarithmic dilution series is also possible, decreasing concentration by 20% each step. At least six (and preferably at least

nine) concentrations spanning at least one order or magnitude are required for mechanistic studies.

16. Other BMG plate readers can be used, for example, the FLUOstar Galaxy or Omega versions.
17. Rotate grid 90° and slowly push a drop of MQ H_2O from a syringe with filter attached over the grid and allow it to fall onto blotting paper below. Repeat this three times.

Acknowledgments

This work was supported by grants to DMW from the National Institute of Health (AG046275) and the BrightFocus Foundation and to SL from the European Research Council and the Swedish Research Council. We thank Dr. Gunnar Brinkmalm for mass spec. analysis.

References

1. Walsh DM, Hartley DM, Selkoe DJ (2003) The many faces of Aβ: structures and activity. Curr Med Chem Immunol Endocrinol Metab Agents 3:277–291
2. Welzel AT, Maggio JE, Shankar GM, Walker DE, Ostaszewski BL, Li S, Klyubin I, Rowan MJ, Seubert P, Walsh DM, Selkoe DJ (2014) Secreted amyloid β-proteins in a cell culture model include N-terminally-extended prptides that impair synaptic plasticity. Biochemistry 53:3908–3921
3. Walsh DM, Teplow DB (2012) Alzheimer's disease and the amyloid β-protein. Prog Mol Biol Transl Sci 107:101–124
4. Finder VH, Vodopivec I, Nitsch RM, Glockshuber R (2010) The recombinant amyloid-beta peptide Aβ1-42 aggregates faster and is more neurotoxic than synthetic Aβ1-42. J Mol Biol 396:9–18
5. Walsh DM, Thulin E, Minogue AM, Gustavsson N, Pang E, Teplow DB, Linse S (2009) A facile method for expression and purification of the Alzheimer's disease-associated amyloid β-peptide. FEBS J 276:1266–1281
6. Szczepankiewicz O, Linse B, Meisl G, Thulin E, Frohm B, Sala Frigerio C, Colvin MT, Jacavone AC, Griffin RG, Knowles T, Walsh DM, Linse S (2015) N-terminal extensions retard Aβ42 fibril formation but allow cross-seeding and coaggregation with Aβ42. J Am Chem Soc 137:14673–14685
7. Meisl G, Kirkegaard JB, Arosio P, Michaels TC, Vendruscolo M, Dobson CM, Linse S, Knowles TP (2016) Molecular mechanisms of protein aggregation from global fitting of kinetic models. Nat Protoc 11:252–272
8. Cukalevski R, Boland B, Frohm B, Thulin E, Walsh DM, Linse S (2012) Role of aromatic side chains in amyloid beta-protein aggregation. ACS Chem Neurosci 3:1008–1016
9. O'Malley TT, Oktaviani NA, Zhang D, Lomakin A, O'Nuallain B, Linse S, Benedek GB, Rowan MJ, Mulder FA, Walsh DM (2014) Abeta dimers differ from monomers in structural propensity, aggregation paths and population of synaptotoxic assemblies. Biochem J 461:413–426
10. Hellstrand E, Boland B, Walsh DM, Linse S (2010) Amyloid β-protein aggregation produces highly reproducible kinetic data and occurs by a two-phase process. ACS Chem Neurosci 1:13–18
11. Cohen SI, Linse S, Luheshi LM, Hellstrand E, White DA, Rajah L, Otzen DE, Vendruscolo M, Dobson CM, Knowles TP (2013) Proliferation of amyloid-β42 aggregates occurs through a secondary nucleation mechanism. Proc Natl Acad Sci U S A 110:9758–9763
12. Arosio P, Knowles TP, Linse S (2015) On the lag phase in amyloid fibril formation. Phys Chem Chem Phys 17:7606–7618
13. Arosio P, Cukalevski R, Frohm B, Knowles TP, Linse S (2014) Quantification of the concentration of Aβ42 propagons during the lag phase by an amyloid chain reaction assay. J Am Chem Soc 136:219–225

14. Meisl G, Yang X, Hellstrand E, Frohm B, Kirkegaard JB, Cohen SI, Dobson CM, Linse S, Knowles TP (2014) Differences in nucleation behavior underlie the contrasting aggregation kinetics of the Aβ40 and Aβ42 peptides. Proc Natl Acad Sci U S A 111:9384–9389
15. Shankar GM, Li S, Mehta TH, Garcia-Munoz-A, Shepardson NE, Smith I, Brett FM, Farrell MA, Rowan MJ, Lemere CA, Regan CM, Walsh DM, Sabatini BL, Selkoe DJ (2008) Amyloid-β protein dimers isolated directly from Alzheimer's brains impair synaptic plasticity and memory. Nat Med 14:837–842
16. Roher AE, Chaney MO, Kuo YM, Webster SD, Stine WB, Haverkamp LJ, Woods AS, Cotter RJ, Tuohy JM, Krafft GA, Bonnell BS, Emmerling MR (1996) Morphology and toxicity of Aβ(1-42) dimer derived from neuritic and vascular amyloid deposits of Alzheimer's disease. J Biol Chem 271:20631–20635
17. Mc Donald JM, O'Malley TT, Liu W, Mably AJ, Brinkmalm G, Portelius E, Wittbold WM 3rd, Frosch MP, Walsh DM (2015) The aqueous phase of Alzheimer's disease brain contains assemblies built from approximately 4 and approximately 7 kDa Aβ species. Alzheimers Dement 11:1286–1305
18. Vigo-Pelfrey C, Lee D, Keim P, Lieberburg I, Schenk DB (1993) Characterization of β-amyloid peptide from human cerebrospinal fluid. J Neurochem 61:1965–1968
19. Yang T, O'Malley TT, Kanmert D, Jerecic J, Zieske LR, Zetterberg H, Hyman BT, Walsh DM, Selkoe DJ (2015) A highly sensitive novel immunoassay specifically detects low levels of soluble Aβ oligomers in human cerebrospinal fluid. Alzheimers Res Ther 7:14
20. Moir RD, Tseitlin KA, Soscia S, Hyman BT, Irizarry MC, Tanzi RE (2005) Autoantibodies to redox-modified oligomeric Aβ are attenuated in the plasma of Alzheimer's disease patients. J Biol Chem 280:17458–17463
21. Sun S, Zhou J-Y, Yang W, Zhang H. Inhibition of Protein Carbamylation in Urea Solution Using Ammonium Containing Buffers. Analytical biochemistry. 2014;446:76–81. https://doi.org/10.1016/j.ab.2013.10.024
22. Shankar GM, Welzel AT, McDonald JM, Selkoe DJ, Walsh DM (2011) Isolation of low-n amyloid β-protein oligomers from cultured cells, CSF, and brain. Methods Mol Biol 670:33–44
23. Shevchenko A, Wilm M, Vorm O, Mann M (1996) Mass spectrometric sequencing of proteins silver-stained polyacrylamide gels. Anal Chem 68:850–858

Chapter 20

Disaggregation of Aβ42 for Structural and Biochemical Studies

Hyewon Chung, Elliot J. Crooks, Martine Ziliox, and Steven O. Smith

Abstract

The amyloid-β (Aβ) peptides that form the amyloid fibrils associated with Alzheimer's disease are generated by sequential proteolysis of the amyloid precursor protein by β- and γ-secretase. The two predominant Aβ peptides, Aβ40 and Aβ42, differ by two amino acids, are soluble as monomers at low concentration (and/or low temperature) and are normally cleared from the brain parenchyma. In order to study the structure and assembly of these peptides, they are often synthesized using solid-phase peptide synthesis and purified. Here, we outline the method we use to prepare monomeric Aβ for structural and biochemical studies.

Key words Aβ peptides, Aβ40, Aβ42, Alzheimer's disease, FTIR, NMR

1 Introduction

Amyloid β (Aβ) peptides are the major constituent of senile plaques, a hallmark of Alzheimer's disease (AD) [1]. The peptides are derived from the amyloid precursor protein (APP), a ubiquitously expressed type 1 transmembrane protein [2, 3]. With age or in the presence of familial mutations, the peptides aggregate with the longer Aβ42 peptide having the highest occurrence in the insoluble plaques found in the brains of AD patients [4].

Depending on the concentration, the Aβ peptides associate to form nuclei that give rise to fibrils through monomer addition or form soluble oligomers that aggregate to form protofibrils, which convert into mature fibrils [5]. Although Aβ fibrils were first thought to be responsible for AD-induced neurotoxic events, the soluble Aβ oligomers have since been shown to more strongly correlate with clinical symptoms associated with the disease [6, 7]. Different Aβ oligomers have been described, ranging from dimers to dodecamers and high molecular weight assemblies. These have been characterized in vitro, in transgenic mouse brains [6, 8–10] and to some extent in brains of AD patients [11].

Bradley L. Nilsson and Todd M. Doran (eds.), *Peptide Self-Assembly: Methods and Protocols*, Methods in Molecular Biology, vol. 1777, https://doi.org/10.1007/978-1-4939-7811-3_20,

Structural studies on Aβ assemblies have largely come from in vitro measurements of synthetic peptides corresponding to Aβ40 and Aβ42. The starting point for these studies is the generation of monomeric Aβ. There are several methods for preparing Aβ peptides for structural and biochemical studies [12]. Aβ42 peptide has a stronger propensity to aggregate than Aβ40 and is the predominant form found in amyloid plaques. Here we focus on the disaggregation of Aβ42 to generate the monomeric form; the same approach is applicable to Aβ40.

1.1 Stabilization of Aβ42 Monomer and Conversion to Fibrils

Monomeric Aβ can be stabilized at low temperature and low concentration, and has predominantly a random coil structure. Increasing the temperature leads to aggregation. There are several methods for determining the amount of monomeric Aβ and the conversion to larger aggregates. Fluorescence and NMR spectroscopy are two of these methods that can be used over a wide range of concentrations and temperatures in solution, and are discussed below.

For fluorescence studies, thioflavin T has been widely used to characterize the time dependence of fibril formation. Thioflavin T exhibits a large increase in fluorescence intensity at 490 nm when bound to Aβ42 fibrils [13]. Thioflavin T does not bind to monomeric Aβ, but does associate with Aβ oligomers. Fig. 1 presents the changes in thioflavin T fluorescence at a concentration of 200 μM and 37 °C using a plate reader, which facilitates fluorescence measurements at high Aβ42 concentrations. Dilution of Aβ42 in the fluorescence measurements may alter the monomer-oligomer equilibrium. At concentrations above ~150–200 μM, there is a rapid increase in thioflavin T fluorescence followed by a plateau before a second rise in fluorescence. The initial rise in thioflavin T fluorescence is attributed to the weak association of Aβ oligomers at early times in the aggregation process [5]. The initial rise is not seen if the solution is diluted, which shifts the monomer-oligomer equilibrium toward monomer.

NMR spectroscopy provides a high-resolution approach for following the monomer concentration over a wide range of concentrations and temperature. At 4 °C, Aβ42 is predominantly monomer at concentrations up to at least 200 μM [14]. The intensity of the NMR resonances in the ^{1}H-^{15}N heteronuclear single quantum correlation (HSQC) spectrum of Aβ42 at 4 °C is a sensitive measure of monomeric peptide (Fig. 2a). As the temperature is increased or as the Aβ monomers are allowed to sit, they aggregate and the intensity in the HSQC NMR spectrum decreases due to line broadening [15]. Fig. 2b illustrates the extent of monomer loss as a function of time at 4 °C as measured by the NMR signal intensity. At higher temperatures, the intensity very rapidly drops as oligomers and protofibrils form as shown by fluorescence studies (*see* Fig. 1).

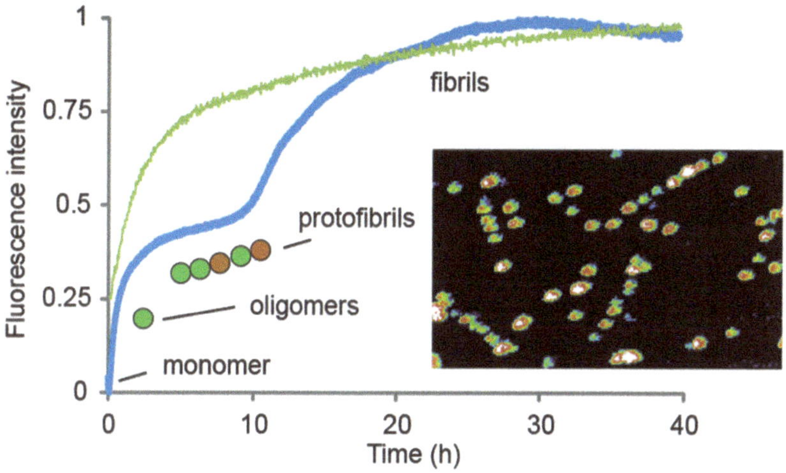

Fig. 1 Fluorescence spectroscopy of Aβ42. Monomeric Aβ42 above a concentration of ~50 μM exhibits an initial rapid increase in fluorescence as oligomers of Aβ form (blue curve). There is then a lag phase in fluorescence as the oligomers laterally associate into protofibrils. During the lag phase, β-sheet secondary structure develops in the oligomers, which serves as the template for rapid fibril growth. If the solution of Aβ42 monomers contain aggregated peptide with β-structure, then a rapid increase of fluorescence is observed from time zero (green curve). These aggregates seed fibril growth. The inset shows an AFM image of capturing individual oligomers and oligomers laterally associating to form protofibrils during the lag phase. The data shown here were obtained using the NaOH pretreatment

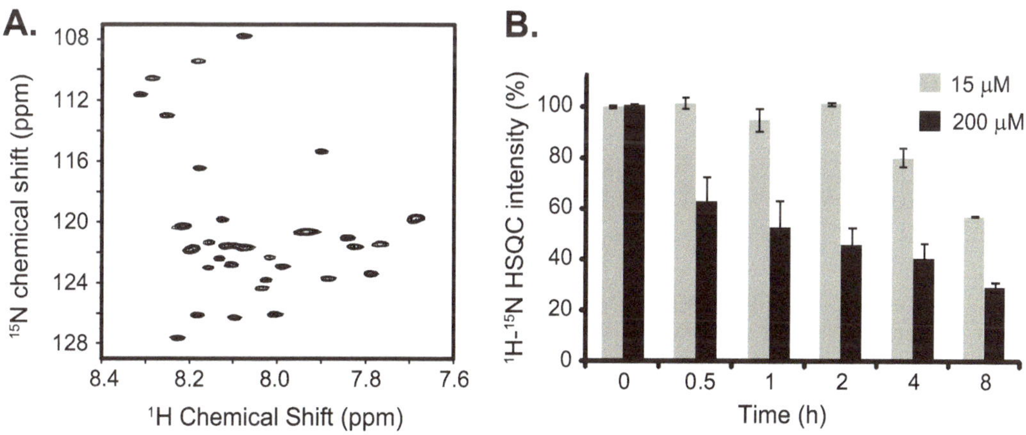

Fig. 2 Solution NMR spectroscopy of Aβ42 monomers. (**a**) ^{1}H-^{15}N HSQC spectrum of Aβ42 obtained at 4 °C. Each contoured spot in the spectrum corresponds to an NH resonance in the Aβ peptide. As a function of time and temperature, the Aβ peptide aggregrates. Aggregation leads to broadening of the resonances and loss of NMR intensity. (**b**) Histograms of the NMR intensity in the ^{1}H-^{15}N HSQC spectrum for Aβ as a function of time at two different Aβ concentrations (15 μM, grey; 200 μM, black). The data shown here were obtained using the NaOH pretreatment

Finally, it is difficult to distinguish monomeric Aβ from a mixture of monomers and dimers. Using fluorescence spectroscopy, Garai and Frieden [16] have observed a rapid change in fluorescence of Aβ42 labeled with tetramethylrhodamine that they associated with the rapid association of monomers to dimers and trimers. Using NMR spectroscopy, diffusion measurements show that the monomeric form of Aβ is likely an equilibrium of monomer and dimer [14]. The rapid initial loss of fluorescence in Fig. 1 and the loss of NMR intensity in Fig. 2b are associated with the formation of low MW oligomers (tetramers to hexamers) starting from monomer or a monomer-dimer mix [5].

1.2 Alkaline Pretreatment

Raising pH above pH 10.5 will disaggregate the Aβ42 peptide [12]. The peptide solution can then be lowered to neutral pH and the peptide will remain monomeric if the temperature of the solution is kept low (e.g., 4 °C). This method can be combined with filtration (e.g., size exclusion chromatography) or ultracentrifugation to remove remaining insoluble aggregates [12].

1.3 HFIP Pretreatment

The Aβ peptides have a very hydrophobic C-terminus that is derived from the N-terminal transmembrane sequence of the amyloid precursor protein. In aqueous solutions, the hydrophobic C-terminus aggregates, which is the reason the monomeric form is difficult to obtain for structural and biochemical studies. As a result, organic solvents are widely used to solubilize the Aβ peptides. Two typical solvents for Aβ are hexafluoroisopropanol or hexafluoro-2-propanol (HFIP) and dimethylsulfoxide (DMSO) [17]. HFIP is the stronger solvent and is able to monomerize Aβ at relatively high peptide concentrations. Removal of the solvent by evaporation induces the Aβ peptide to adopt specific secondary structure, α-helical structure may form when the Aβ has been solubilized in HFIP and β-sheets form when the Aβ has been solubilized in DMSO (Fig. 3a). Consequently, the approach of evaporating solvent and re-dissolving in buffer does not work. It is possible to freeze-dry solutions of Aβ peptide in HFIP and remove the HFIP under vacuum to obtain a fluffy powder. In Fig. 3b, we show that residual HFIP remains bound to the Aβ peptide even under these conditions. Re-dissolving Aβ42 that has been pretreated with HFIP in 50 mM NaOH (which also disaggregates the peptide) and re-lyophilizing the sample does not remove this bound HFIP. Residual solvent (HFIP or DMSO) influences the Aβ aggregation process [14, 18, 19]. The influence is minimized as the amount of solvent is reduced.

The above considerations led Stine and coworkers [17] to develop a two-step protocol. The first step involves monomerization of the Aβ peptide with HFIP, followed by removal by evaporation and placing under vacuum. The second step then is to dissolve the sample in DMSO to make a stock solution with a concentration of 5 mM. The stock solution is diluted to 100 μM or lower (2% residual solvent or less) before use. In this protocol, there is both residual HFIP and DMSO in the sample. Broersen et al. (2011)

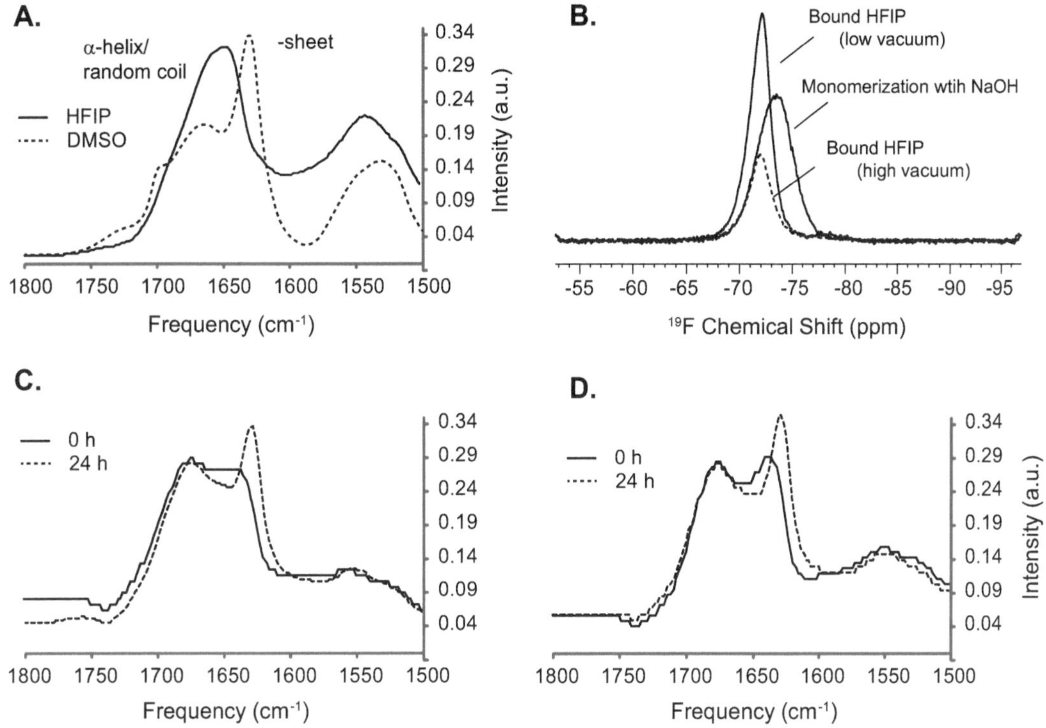

Fig. 3 (**a**) FTIR spectra of Aβ42 as a probe of secondary structure. Aβ42 was dried onto a germanium FTIR plate from solutions of HFIP or DMSO. B. Solid-state NMR spectroscopy of Aβ42 as a probe of tightly bound HFIP. Aβ42 (2.5 mgs) was dissolved in HFIP (500 μL), frozen in liquid nitrogen, and placed under vacuum (24 mTorr) for 48 h. The resulting fluffy powder was loaded into a 2.5 mm NMR rotor for magic angle spinning (MAS) NMR experiments. ^{13}C and ^{19}F MAS NMR spectra were obtained of the dry Aβ42 peptide. The amount of Aβ peptide was determined on the basis of the integrated intensity of ^{13}C resonances of specifically labeled amino acids incorporated into the Aβ sequence compared to the intensity of resonances from a known amount of internal and external ^{13}C-labeled alanine. The amount of bound HFIP was determined by obtaining ^{19}F MAS NMR spectra of the Aβ42 peptide with a MAS frequency of 25 kHz (black trace) and comparing the ^{19}F signal to that from a known quantity of HFIP subsequently added to the NMR rotor (not shown). The ^{19}F signal corresponds to 5 ± 3 HFIP molecules bound to each molecule of Aβ42. The sample was then dissolved in 50 mM NaOH and re-lyophilized. The ^{19}F spectrum (red trace) has broadened—likely the result of the transition from helical to random coil structure of the Aβ42 peptide. However, the amount of bound HFIP does not change appreciably as indicated by the integrated signal area. A parallel experiment using Aβ42 (2.8 mgs) dissolved in HFIP (500 μL), frozen in liquid nitrogen and placed under high vacuum (5×10^{-6} Torr) using a turbomolecular pump connected to a roughing pump for 4 days (dashed line) indicates that some of the tightly bound HFIP can be removed with very high vacuum. (**c**, **d**) Comparison of Aβ42 fibril formation using FTIR spectra of Aβ42 monomerized with HFIP, lyophilized, monomerized with alkaline pretreatment and diluted into buffer (**c**) or of Aβ42 monomerized with just alkaline pretreatment and dilution into buffer (**d**) show that both the preparations yield monomeric Aβ at $t = 0$ h (solid lines) and fibrils after 24 h (dashed lines)

have described a method that makes use of a desalting column to remove residual DMSO [19].

Our preferred method is to avoid both HFIP and DMSO, and simply use an alkaline pretreatment when possible. In cases where it is advantageous to aliquot samples for replicate experiments, we prefer HFIP since most of the HFIP can be removed under high vacuum. In Fig. 3c, d, we show using FTIR spectroscopy that both alkaline pretreatment and HFIP disaggregation followed by alkaline pretreatment yield comparable results.

2 Materials

The experimental protocols described below will require access to equipment typically found in a laboratory equipped for molecular biology and protein purification. Avoid adding Aβ solutions to glass containers, as glass increases Aβ aggregation. For characterization of the Aβ42 of the peptide aggregation state, instrumentation for NMR and fluorescence spectroscopy is needed (*see* **Note 1**).

2.1 Synthetic Aβ42 Peptide

1. Purified Aβ42 peptide (MW 4514 g/mol) (*see* **Note 2**).
2. 50 mM NaOH.
3. Low salt phosphate buffer: 10 mM NaCl (58.44 MW, 0.584 g/L), 10 mM phosphate buffer pH 7.4 (6.1 mM Na_2HPO_4—anhydrous (141.96 MW, 0.866 g/L), 3.9 mM NaH_2PO_4-$2H_2O$ (156.01 MW, 0.608 g/L)) (*see* **Note 3**).
4. Cellulose acetate filters 0.2 μm and Luer-lock sterile syringes.
5. DMSO, anhydrous.
6. HFIP.

2.2 Fluorescence Spectroscopy

1. Thioflavin T solution: 1 mg of thioflavin T (MW 318.9 g/mol) in 1.5 mL of low salt buffer. Keep the thioflavin T solution in the dark.
2. Fluorometer with plate reading capabilities.
3. 96-well microtiter plates.

2.3 NMR Spectroscopy

1. Standard NMR tubes with Teflon inserts (*see* **Note 4**).
2. NMR spectrometer.

3 Methods

3.1 Preparation of Aβ42 Monomer Using NaOH

1. Assess the purity of Aβ42 peptides by analytical reverse phase HPLC, and matrix-assisted laser desorption or electrospray ionization mass spectrometry (*see* **Note 2**).
2. Dissolve purified, lyophilized Aβ42 (1.5–2 mgs) in 50 mM NaOH (60–80 μL) at a concentration of 2.2 mM (let sit at room temperature for 1 h).
3. Dilute the solution in low salt buffer at low temperature (4 °C) and titrate to pH 7.4. The dilution will determine the final concentration.

4. Filter the Aβ solutions with a 0.2-micron cellulose acetate filter to remove insoluble aggregates that can nucleate and influence aggregation.
5. Determine the final concentration of Aβ monomer and adjust as appropriate (*see* **Note 5**).

3.2 Preparation of Aβ42 Monomer Aliquots Using HFIP

1. Dissolve the Aβ peptide overnight in HFIP in 1.5 mL polypropylene microcentrifuge tubes, freeze in liquid nitrogen, and place under vacuum (*see* **Note 6**).
2. Dissolve the freeze-dried Aβ peptide in 50 mM NaOH (starting from Subheading 3.1, **step 2** and continuing through **step 5**) (*see* **Note** 7). Use the monomer within 1–2 h depending on the concentration of Aβ (*see* Fig. 2b).

3.3 Fluorescence of Monomer and Conversion to Fibrils

1. Prepare 150–1000 μL of sample from **step 5** in Subheading 3.1 or **step 2** in Subheading 3.2 for the fluorescence assay. For each well in the plate reader, you need between 50 and 200 μL of sample. We typically use 75 μL. Run replicates (typically 3–5) of each sample. As a result, you need to start with 150–1000 μL of sample.
2. Make sure the thioflavin T stock is dissolved well at room temperature. Cool to 4 °C before adding to the Aβ samples.
3. Add the appropriate amount of concentrated thioflavin T solution to the Aβ monomer solution to obtain a peptide: thioflavin T molar ratio of 1:20.
4. Load ~75 μL of Aβ solution with thioflavin T in each well of the plate reader and run at least in triplicates. Use the thioflavin T solution without added Aβ as a control.
5. Select the temperature and measurement conditions on the fluorometer. For monitoring thioflavin T fluorescence, we use excitation and emission wavelengths of 446 and 490 nm, respectively.

3.4 NMR of Monomer and Conversion to Fibrils

1. Precool the NMR probe and NMR tubes (with Teflon inserts) to 4 °C prior to use. It is important to shim and calibrate the NMR instrument prior to adding your sample. This could be carried out on your actual sample. However, the monomeric peptide may start to aggregate depending on the concentration and temperature to be used.
2. Add the solution of Aβ monomer from **step 5** in Subheading 3.1 or **step 2** in Subheading 3.2 to the NMR tube. Add 10% deuterium oxide (D_2O) that has been cooled to 4 °C for the NMR lock signal.
3. Insert the NMR tube with the Aβ sample into the NMR instrument and acquire spectra as a function of time.

4 Notes

1. We have used both a Horiba Jobin Yvon Fluorolog FL3–22 spectrofluorometer and a SpectraMax M2 microplate reader for fluorescence measurements. The microplate reader has several advantages over Fluorolog spectrofluorometer. Replicate measurements can easily be set up and time points can be made continuously without opening the sample compartment. Measurements can readily be made over a range of concentrations. The HSQC NMR measurements can be made on a wide range of solution NMR instruments. The instruments require two channels (^1H and ^{15}N). A cryoprobe significantly improves sensitivity, which allows concentrations as low as 10 μM to be measured.
2. Aβ42 peptides are most often synthesized by solid-phase peptide synthesis and purified by high performance liquid chromatography (HPLC) using linear water-acetonitrile gradients containing 0.1% (v/v) trifluoroacetic acid. Other chapters in this volume discuss the methods for synthesis in detail. Nevertheless, there are two potential problems with the purity of synthetic Aβ peptides. The first is when the amino acids do not properly couple during synthesis and there is loss of an amino acid. This loss can be minimized during synthesis by double couplings or increasing the coupling time at difficult positions (typically in the hydrophobic C-terminus). The loss of an amino acid typically leads to a peak at ~110 mass units less than the molecular ion in the mass spectrum. It also manifests itself in a long leading edge in the HPLC chromatogram during purification. One can eliminate the peptides eluting at the leading edge of the HPLC chromatogram during purification. Another potential problem is oxidation of the Met35 side chain. This oxidation can easily be checked by mass spectrometry or NMR spectroscopy using ^{13}C5-Met35-labeled Aβ.
3. The specific buffer used depends on the application. The low salt (or no salt) buffer is useful for TEM and NMR measurements. The buffers are typically filtered after they are made up to remove any particulates, which is again important for TEM and NMR measurements.
4. Teflon inserts are essential to eliminate glass-induced aggregation of the peptide.
5. The peptide concentration can be determined in several ways. Absorption spectroscopy can be used to estimate the peptide concentration based on the absorption at 280 nm. There is one tyrosine in Aβ42 with an extinction coefficient of 1490 M^{-1} cm^{-1} at 280 nm. Access to high-field NMR instrumentation allows one to determine the concentration of Aβ

monomers on the basis of the signal intensity and calibrating relative to the signal intensity observed from known concentrations of amino acids.

6. HFIP is difficult to remove completely by vacuum. Typical roughing pumps will achieve 10–20 mTorr vacuums. Under these conditions, a substantial amount of HFIP (roughly 5 ± 3 molecules of HFIP per molecule of Aβ) remains bound to the Aβ peptide. Removing the HFIP by evaporation prior to placing under vacuum makes it more difficult to remove, possibly as a result of forming a film that has trapped HFIP.
7. The Aβ peptide that was dissolved in HFIP and freeze-dried to remove solvent can be redissolved in 50 mM NaOH or in DMSO. With DMSO, one can make a concentrated solution (5 mM) and dilute to 50 μM Aβ (or lower concentration) with buffer for use. A dilution to 50 μM Aβ results in 1% residual DMSO. Above this level, the influence of the DMSO on the Aβ aggregation kinetics is measureable. With DMSO, the solutions must be kept at room temperature or higher due to the high melting point of DMSO. This protocol results in both DMSO and HFIP in the sample. ^{19}F NMR studies comparable to those shown in Fig. 3b reveal that the DMSO effectively strips most of the bound HFIP from the peptide (data not shown). Broersen et al. (2011) have described a method that makes use of a desalting column to remove residual DMSO [19].

Acknowledgments

This work was supported by NIH-NSF instrumentation grants (S10 RR13889 and DBI-9977553), a grant from the NIH to S. O.S (AG 27317). We gratefully acknowledge the W.M. Keck Foundation for support of the NMR facilities in the Center of Structural Biology at Stony Brook.

References

1. Glenner GG, Wong CW (1984) Alzheimer's disease: initial report of the purification and characterization of a novel cerebrovascular amyloid protein. Biochem Biophys Res Commun 120:885–890
2. Kang J, Lemaire HG, Unterbeck A, Salbaum JM, Masters CL, Grzeschik KH, Multhaup G, Beyreuther K, Mullerhill B (1987) The precursor of Alzheimer's disease amyloid A4 protein resembles a cell-surface receptor. Nature 325:733–736
3. Selkoe DJ (2004) Alzheimer disease: mechanistic understanding predicts novel therapies. Ann Intern Med 140:627–638
4. Portelius E, Bogdanovic N, Gustavsson M, Volkmann I, Brinkmalm G, Zetterberg H, Winblad B, Blennow K (2010) Mass spectrometric characterization of brain amyloid beta isoform signatures in familial and sporadic Alzheimer's disease. Acta Neuropathol 120:185–193
5. Fu Z, Aucoin D, Davis J, Van Nostrand WE, Smith SO (2015) Mechanism of nucleated conformational conversion of Aβ42. Biochemistry 54(27):4197–4207
6. Lesne S, Koh MT, Kotilinek L, Kayed R, Glabe CG, Yang A, Gallagher M, Ashe KH (2006) A

specific amyloid-β protein assembly in the brain impairs memory. Nature 440:352–357

7. McDonald RJ, Craig LA, Hong NS (2010) The etiology of age-related dementia is more complicated than we think. Behav Brain Res 214:3–11
8. Pham E, Crews L, Ubhi K, Hansen L, Adame A, Cartier A, Salmon D, Galasko D, Michael S, Savas JN et al (2010) Progressive accumulation of amyloid-beta oligomers in Alzheimer's disease and in amyloid precursor protein transgenic mice is accompanied by selective alterations in synaptic scaffold proteins. FEBS J 277:3051–3067
9. Shankar GM, Leissring MA, Adame A, Sun XY, Spooner E, Masliah E, Selkoe DJ, Lemere CA, Walsh DM (2009) Biochemical and immunohistochemical analysis of an Alzheimer's disease mouse model reveals the presence of multiple cerebral Aβ assembly forms throughout life. Neurobiol Dis 36:293–302
10. Kawarabayashi T, Younkin LH, Saido TC, Shoji M, Ashe KH, Younkin SG (2001) Age-dependent changes in brain, CSF, and plasma amyloid beta protein in the Tg2576 transgenic mouse model of Alzheimer's disease. J Neurosci 21:372–381
11. Hayden EY, Teplow DB (2013) Amyloid beta-protein oligomers and Alzheimer's disease. Alzheimers Res Ther 5:60
12. Teplow DB (2006) Preparation of amyloid beta-protein for structural and functional studies. Methods Enzymol 413:20–33
13. LeVine H (1999) Quantification of β-sheet amyloid fibril structures with thioflavin T. Methods Enzymol 309:274–284
14. Fu Z, Aucoin D, Ahmed M, Ziliox M, Van Nostrand WE, Smith SO (2014) Capping of Aβ42 oligomers by small molecule inhibitors. Biochemistry 53:7893–7903
15. Yamaguchi T, Matsuzaki K, Hoshino M (2011) Transient formation of intermediate conformational states of amyloid-beta peptide revealed by heteronuclear magnetic resonance spectroscopy. FEBS Lett 585:1097–1102
16. Garai K, Frieden C (2013) Quantitative analysis of the time course of a beta oligomerization and subsequent growth steps using tetramethylrhodamine-labeled a beta. Proc Natl Acad Sci U S A 110:3321–3326
17. Stine WB, Dahlgren KN, Krafft GA, LaDu MJ (2003) In vitro characterization of conditions for amyloid-β peptide oligomerization and fibrillogenesis. J Biol Chem 278:11612–11622
18. Pachahara SK, Adicherla H, Nagaraj R (2015) Self-Assembly of Aβ40, Aβ42 and Aβ43 peptides in aqueous mixtures of fluorinated alcohols. PLoS One 10:1–18
19. Broersen K, Jonckheere W, Rozenski J, Vandersteen A, Pauwels K, Pastore A, Rousseau F, Schymkowitz J (2011) A standardized and biocompatible preparation of aggregate-free amyloid beta peptide for biophysical and biological studies of Alzheimer's disease. Protein Eng Des Sel 24:743–750

Chapter 21

Preparation of Stable Amyloid-β Oligomers Without Perturbative Methods

Samuel A. Kotler and Ayyalusamy Ramamoorthy

Abstract

Soluble amyloid-β (Aβ) oligomers have become a focal point in the study of Alzheimer's disease due to their ability to elicit cytotoxicity. A number of recent studies have concentrated on the structural characterization of soluble Aβ oligomers to gain insight into their mechanism of toxicity. Consequently, providing reproducible protocols for the preparation of such oligomers is of utmost importance. The method presented in this chapter details a protocol for preparing an Aβ oligomer, with a primarily disordered secondary structure, without the need for chemical modification or amino acid substitution. Due to the stability of these disordered Aβ oligomers and the reproducibility with which they form, they are amenable for biophysical and high-resolution structural characterization.

Key words Amyloid-β, Alzheimer's disease, Oligomer, Protein aggregation, Purification, NMR

1 Introduction

Alzheimer's disease (AD) is characterized by neurodegeneration and atrophy of affected cerebral areas. This deterioration progresses as the disease progresses, and is generally accompanied with amyloid plaques and neurofibrillary tangles, made up of insoluble collections of amyloid-β (Aβ) peptide and tau protein, respectively [1, 2]. The protein "misfolding" occurring therein may disrupt basic cellular processes, including calcium ion homeostasis, enzyme function, and glucose usage, all of which result in cell death [2–6]. This is normally seen in conjunction with inflammatory responses, which help induce neural degeneration [3]. The resultant neuronal cell death due to such processes is the root of AD. While the mechanism of cytotoxicity remains a topic of intense debate, mounting evidence points to aggregated structures of the Aβ peptide [7–11].

Numerous studies now implicate soluble Aβ oligomers as the pathogenic agent in AD, resulting in intensified efforts to identify such species in biological fluids (such as human cerebrospinal fluid)

Bradley L. Nilsson and Todd M. Doran (eds.), *Peptide Self-Assembly: Methods and Protocols*, Methods in Molecular Biology, vol. 1777, https://doi.org/10.1007/978-1-4939-7811-3_21,

[3]. Furthermore, strengthened efforts to biophysically characterize Aβ oligomers have become increasingly important, with the goal of understanding how structural morphology of oligomeric species relates to their toxic effect. Methods to prepare and characterize Aβ oligomers have been previously reported [11–15]; however, detailed high-resolution structural information remains difficult to obtain. Most of the structural models of Aβ oligomers that do exist bear a structural resemblance closer to the fibrils of Aβ [7, 11, 16, 17], with few exceptions [18]. Despite this, there is significant evidence that some of the most toxic oligomers have a considerably different structure that may be closer to Aβ monomer (rather than the fibrils) [9, 10, 19, 20].

The main difficulty in structurally characterizing Aβ oligomers results from their transient nature; oligomers ultimately aggregate to large amyloid fibrils (which have been extensively studied) [21–23]. Thus, it is of utmost importance to prepare stable oligomeric Aβ species synonymous with a species that elicits cytotoxicity [9, 10, 20]. The method of preparation presented here yields oligomers that share many of the structural qualities of the primarily disordered and toxic Aβ oligomers previously studied. Importantly, this method is: i) highly reproducible; the disordered oligomers form at almost 10% of the total peptide concentration, and ii) results in stable oligomers that persist in solution for at least 3 weeks, allowing for extensive high-resolution structural characterization by techniques such as NMR [24]. Furthermore, while other protocols exist for preparing Aβ oligomers (*see* References 11–15), the current method does not require perturbative methods, such as chemical- or photo-crosslinking, freezing, amino acid substitution, or any other type of protein engineering to stabilize the oligomer.

2 Materials

Unless otherwise stated, all solutions are prepared using ultrapure water (sensitivity of 18.2 MΩ cm at 25 °C) and analytical grade reagents. All reagents are prepared and stored at room temperature (if not otherwise stated) and waste materials are disposed of using appropriate regulations.

2.1 Equipment

1. MS 3 Digital Shaker (IKA).
2. NanoDrop 2000 UV-Vis Spectrophotometer.
3. Tabletop microcentrifuge.
4. Incubator.
5. Benchtop lyophilizer (i.e., a freeze-dry system).
6. NMR spectrometer (field suitable for 2D experiments).

7. High-resolution magic angle spinning (HR-MAS) probe (e.g., Agilent 3.2 mm Nanoprobe).
8. 3.2 mm zirconium HR-MAS rotor.

2.2 Sample Components

1. 10 mM sodium phosphate buffer: Prepared at pH 7.4 as described in Ref. 25.
2. 2.5% sodium azide (NaN_3) solution: 2.5 g in 100 mL water/buffer.
3. Aggregation buffer: 10 mM sodium phosphate and 0.01% NaN_3.
4. 0.22 μm Costar Spin-X microcentrifuge tube filters (*see* **Note 1**).
5. Pierce BCA (bicinchoninic acid) protein assay kit.
6. Deuterium oxide (D_2O): used in buffers for NMR experiments.

2.3 Preparing Disaggregated $A\beta_{1-40}$

1. Amyloid-β(1–40) ($A\beta_{1-40}$): prepared by solid-phase peptide synthesis using Fmoc chemistry (*see* **Note 2**).
2. 1% ammonium hydroxide solution: Dilute ammonium hydroxide in water to 1% (v/v).
3. Liquid nitrogen (LN_2).

3 Methods

3.1 Disaggregation of $A\beta_{1-40}$

This disaggregation protocol was developed from previously published protocols [26, 27].

1. Suspend approximately 1 mg of dry $A\beta_{1-40}$ peptide in 1 mL of 1% ammonium hydroxide solution for 1 h at 4 °C. Gently vortex and sonicate to aid with peptide dissolution.
2. Aliquot 100–300 μL of 1 mg/mL peptide solution into microcentrifuge.
3. Flash freeze the solution aliquots using LN_2 and freeze-dry samples for 24 h. Store the aliquots at −80 °C.

3.2 Preparing Aggregated Samples of $A\beta_{1-40}$

1. Dissolve 0.1–0.3 mg of disaggregated $A\beta_{1-40}$ (*see* **step 3** of Subheading 3.1) in aggregation buffer at a peptide concentration of 1 mg/mL (231 μM) (Fig. 1b). Verify $A\beta_{1-40}$ concentration by UV-Vis ($\varepsilon_{280} = 1490\ M^{-1}\ cm^{-1}$) and adjust as needed.
2. Place the freshly dissolved $A\beta_{1-40}$ sample in an incubator adjusted to 37 °C and allow the sample to aggregate under shaking conditions for 48 h (with the digital shaker set to 1000 rpm). This sample is saved to use as preformed seeds (Fig. 1b).

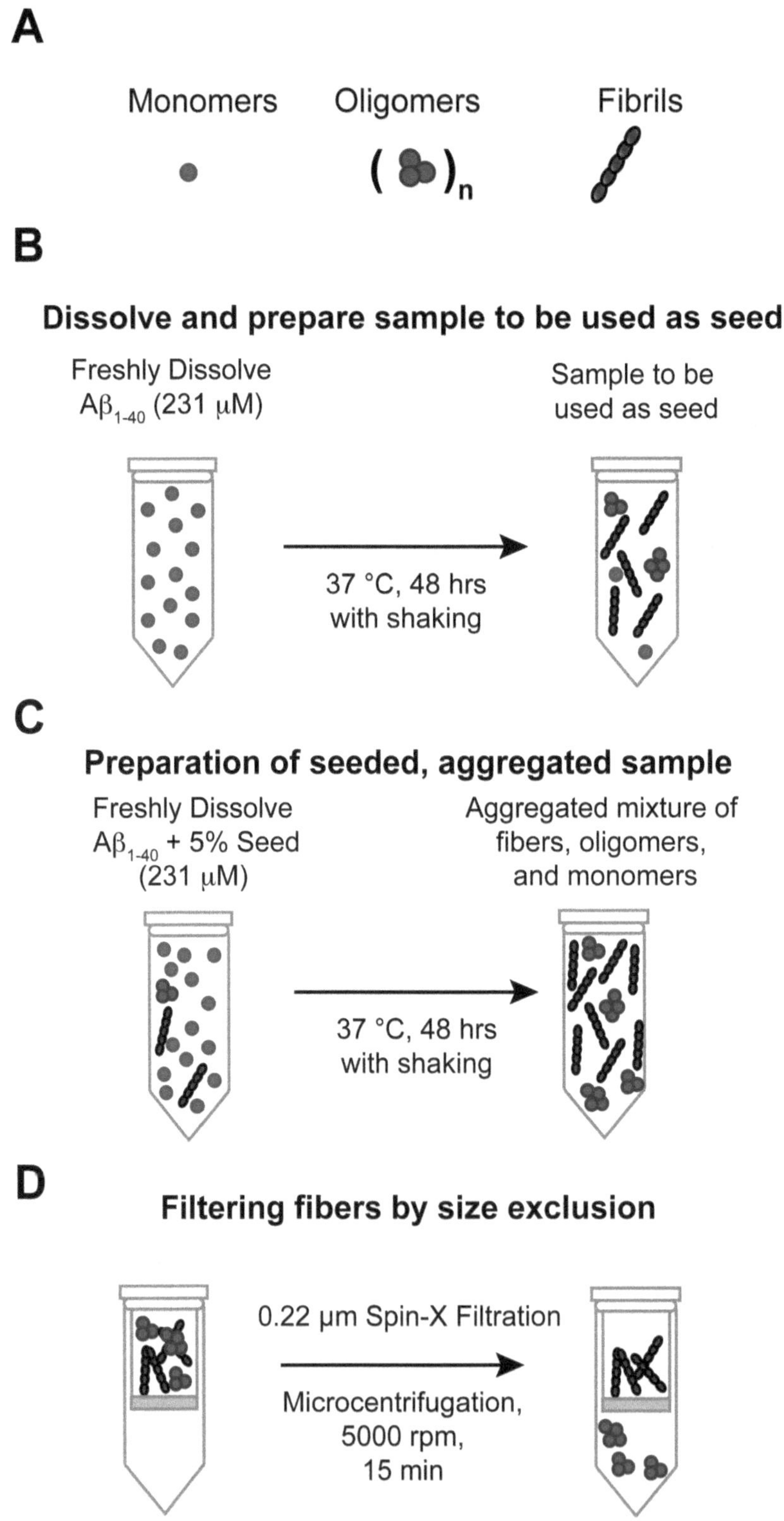
A
Monomers
Oligomers
Fibrils
B
Dissolve and prepare sample to be used as seed
Freshly Dissolve
Aβ1-40 (231 μM)
Sample to be used as seed
37 °C, 48 hrs with shaking
C
Preparation of seeded, aggregated sample
Freshly Dissolve
Aβ1-40 + 5% Seed (231 μM)
Aggregated mixture of fibers, oligomers, and monomers
37 °C, 48 hrs with shaking
D
Filtering fibers by size exclusion
0.22 μm Spin-X Filtration
Microcentrifugation, 5000 rpm, 15 min

3. Prepare a freshly dissolved Aβ_{1-40} sample in 10 mM sodium phosphate, pH 7.4 at 1 mg/mL (verify Aβ_{1-40} concentration by UV-Vis and adjust as needed). Add the preformed aggregated seed prepared in **step 2** at 5% of the total Aβ_{1-40} concentration (Fig. 1c).
4. Incubate the seeded sample for 48 h at 37 °C under shaking conditions (1000 rpm) (Fig. 1c).
5. Following the round of seeded aggregation (*see* **step 3**), the Aβ_{1-40} sample will contain a mixture of amyloid fibrils and non-β-sheet oligomers (typically around 7–10% of the total Aβ_{1-40} concentration as measured by the BCA assay (*see* **Note 3**)). The solution should appear mildly opaque due to the formation of amyloid fibrils.

3.3 Isolating the Non-β-sheet Aβ_{1-40} Oligomers

1. Load the aggregated Aβ_{1-40} sample (prepared as described in Subheading 3.2) onto the 0.22 μm tube filter and microcentrifuge the sample for 15 min at 2300 × *g* (Fig. 1d).
2. The filtrate will appear clear and contains the disordered Aβ_{1-40} oligomers. These oligomers are stable for at least 3 weeks (*see* Ref. 24).
3. The retentate contains the fibrillar Aβ_{1-40} species that are too large to pass through the filter and can be re-suspended in 100 μL of the aggregation buffer for further biophysical characterization (*see* **Note 4**) (Fig. 1d).

3.4 Preparation of Isolated Aβ_{1-40} Oligomers for NMR Experiments

The yield of the disordered Aβ_{1-40} oligomer (ca. 20 μM) is relatively low for NMR measurements; therefore, it is necessary to concentrate the oligomer species.

1. Flash freeze either the aggregated mixture or isolated, disordered oligomer of Aβ_{1-40} with LN_2 and freeze-dry for 24 h.
2. Redissolve the freeze-dried sample at half of its original volume to effectively double the concentration (verify oligomer concentration by the BCA assay) (*see* **Notes 5** and **6**).
3. Affirmation of oligomer presence can be verified by biophysical characterization (CD, fluorescence, DLS, etc.); however, to characterize distinct structural differences between a random coil monomer and a globular, disordered oligomer, magic angle spinning NMR experiments are necessary (*see* Ref. 24) (Fig. 2).

Fig. 1 Schematic representation for the preparation of stable, disordered Aβ_{1-40} oligomers. (**a**) Cartoon representation of different types of Aβ aggregates. The disordered Aβ_{1-40} oligomers are prepared by making Aβ_{1-40} seeds (**b**), using seeded aggregation to create a mixture of oligomeric and fibrillar Aβ_{1-40} species (**c**), and separation of the disordered Aβ_{1-40} oligomers from the amyloid fibrils (**d**)

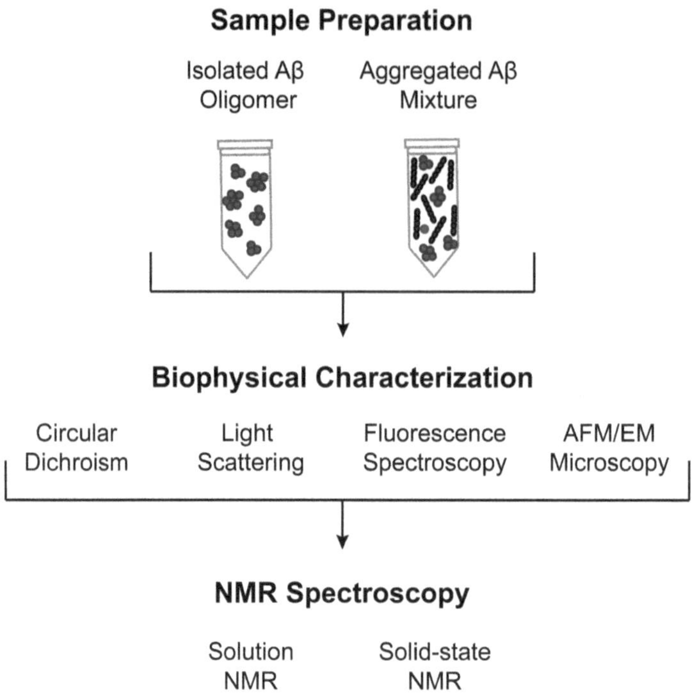

Fig. 2 Workflow to structurally characterize Aβ (and other amyloids) oligomers at high resolution. Elucidating size, morphology, and secondary structure through biophysical characterization prior to NMR measurements aids greatly in choosing the type of NMR experiments to complete in addition to understanding the nature of the oligomers before analyzing NMR results

4 Notes

1. A 100 kDa molecular weight cutoff microcentrifuge tube filter can be used as well.
2. $A\beta_{1-40}$ and $A\beta_{1-42}$ can be prepared by expression in *E. coli* as well (*see* Ref. 27, 28); however, the method presented in this chapter utilized Aβ samples from either synthesis in-house or by purchasing it from commercial vendors (e.g., Genscript). There are excellent protocols available describing synthesis of amyloidogenic peptides [29, 30].
3. It is important to use the BCA assay for measuring peptide concentration at this stage as A_{280} measurements do not accurately report on the actual concentration of aggregated Aβ present (*see* Ref. 13).
4. Resuspending the $A\beta_{1-40}$ fibrils is achieved by gently pipetting the buffer solution up and down on the membrane of the tube filter.
5. Use D_2O at a ratio of 9:1 (v:v) H_2O:D_2O for NMR measurements.

6. Freeze-drying of amyloid oligomers is commonly used to structurally characterize powder samples of intermediate species of Aβ and other amyloid-forming peptides. Previous studies have demonstrated that the freeze-drying process does not measurably perturb oligomer morphology (*see* References 11, 16, 24, 31).

Acknowledgments

Research on amyloid-beta is supported by funds from the National Institutes of Health (AG048934 to A.R.).

References

1. Chiti F, Dobson CM (2006) Protein misfolding, functional amyloid, and human disease. Annu Rev Biochem 75:333–366
2. Mucke L, Selkoe DJ (2012) Neurotoxicity of amyloid β-protein: synaptic and network dysfunction. Cold Spring Harb Perspect Med 2 (7):a006338
3. Benilova I, Karran E, De Strooper B (2012) The toxic Aβ oligomer and Alzheimer's disease: an emperor in need of clothes. Nat Neurosci 15 (3):349–357
4. Kotler SA, Walsh P, Brender JR, Ramamoorthy A (2014) Differences between amyloid-β aggregation in solution and on the membrane: insights into elucidation of the mechanistic details of Alzheimer's disease. Chem Soc Rev 43(19):6692–6700
5. Butterfield SM, Lashuel H (2010) Amyloidogenic protein-membrane interactions: mechanistic insight from model systems. Angew Chem Int Ed Engl 49(33):5628–5654
6. Jang H et al (2013) Alzheimer's disease: which type of amyloid-preventing drug agents to employ? Phys Chem Chem Phys 15(23):8868
7. Chimon S et al (2007) Evidence of fibril-like β-sheet structures in a neurotoxic amyloid intermediate of Alzheimer's β-amyloid. Nat Struct Mol Biol 14(12):1157–1164
8. Cohen SIA et al (2015) A molecular chaperone breaks the catalytic cycle that generates toxic Aβ oligomers. Nat Struct Mol Biol 22 (3):207–213
9. Serra-Vidal B et al (2014) Hydrogen/Deuterium Exchange-Protected Oligomers Populated during Aβ Fibril Formation Correlate with Neuronal Cell Death. ACS Chem Biol 9 (11):2678–2685
10. Ladiwala AR et al (2012) Conformational differences between two amyloid β oligomers of similar size and dissimilar toxicity. J Biol Chem 287(29):24765–24773
11. Ahmed M et al (2010) Structural conversion of neurotoxic amyloid-beta(1-42) oligomers to fibrils. Nat Struct Mol Biol 17(5):561–567
12. Chimon S, Ishii Y (2005) Capturing intermediate structures of Alzheimer's beta-amyloid, Abeta(1-40), by solid-state NMR spectroscopy. J Am Chem Soc 127(39):13472–13473
13. Jan A, Hartley DM, Lashuel H (2010) Preparation and characterization of toxic Abeta aggregates for structural and functional studies in Alzheimer's disease research. Nat Protoc 5 (6):1186–1209
14. Lopes DHJ, Sinha S, Rosensweig C, Bitan G (2012) Application of photochemical cross-linking to the study of oligomerization of amyloidogenic proteins. Methods Mol Biol 849:11–21
15. Rosensweig C et al (2012) Preparation of stable amyloid β-protein oligomers of defined assembly order. Methods Mol Biol 849:23–31
16. Sarkar B et al (2014) Significant structural differences between transient amyloid-β oligomers and less-toxic fibrils in regions known to harbor familial Alzheimer's mutations. Angew Chemie Int Ed 53(27):6888–6892
17. Lendel C et al (2014) A Hexameric peptide barrel as building block of amyloid-β protofibrils. Angew Chem Int Ed Engl 53 (47):12756–12760
18. Yu L et al (2009) Structural characterization of a soluble amyloid beta-peptide oligomer. Biochemistry 48(9):1870–1877
19. Ono K, Condron MM, Teplow DB (2009) Structure-neurotoxicity relationships of amyloid beta-protein oligomers. Proc Natl Acad Sci U S A 106(35):14745–14750

20. Luo J, Wärmländer SKTS, Gräslund A, Abrahams JP (2014) Alzheimer peptides aggregate into transient nanoglobules that nucleate fibrils. Biochemistry 53(40):6302–6308
21. Tycko R (2011) Solid-state NMR studies of amyloid fibril structure. Annu Rev Phys Chem 62:279–299
22. Schütz AK et al (2015) Atomic-resolution three-dimensional structure of amyloid β fibrils bearing the osaka mutation. Angew Chemie Int Ed 54(1):331–335
23. Vilar M, Wang L, Riek R (2012) Structural studies of amyloids by quenched hydrogen-deuterium exchange by NMR. Methods Mol Biol 849:185–198
24. Kotler SA et al (2015) High-resolution NMR characterization of low abundance oligomers of amyloid-β without purification. Sci Rep 5:11811
25. Sodium Phosphate (2006) Cold Spring Harb Protoc 2006(1):pdb.rec8303
26. Bitan G, Teplow DB (2005) Preparation of aggregate-free, low molecular weight amyloid-beta for assembly and toxicity assays. Methods Mol Biol 299:3–9
27. Ryan TM et al (2013) Ammonium hydroxide treatment of Aβ produces an aggregate free solution suitable for biophysical and cell culture characterization. PeerJ 1:e73
28. Weber DK, Sani M-A, Gehman JD (2014) A routine method for cloning, expressing and purifying Aβ(1-42) for structural NMR studies. Amino Acids 46(10):2415–2426
29. Zhang TO, Grechko M, Moran SD, Zanni MT (2016) Isotope-Labeled Amyloids via Synthesis, Expression, and Chemical Ligation for Use in FTIR, 2D IR, and NMR Studies. Methods Mol Biol 1345:21–41
30. Nilsson MR, Nguyen LL, Raleigh DP (2001) Synthesis and purification of amyloidogenic peptides. Anal Biochem 288(1):76–82
31. Walsh P, Neudecker P, Sharpe S (2010) Structural properties and dynamic behavior of nonfibrillar oligomers formed by PrP(106-126). J Am Chem Soc 132(22):7684–7695

Chapter 22

Discriminating Strains of Self-Propagating Protein Aggregates Using a Conformational Stability Assay

Heather H. C. Lau, Angus Lau, and Joel C. Watts

Abstract

Prions and other self-propagating protein aggregates can exist as distinct strains, which are thought to represent different conformations of aggregates. There is growing evidence that protein aggregate strains may be important for understanding the biology of common neurodegenerative disorders, such as Alzheimer's disease and Parkinson's disease. While methodology for discriminating prion strains is in widespread use, there is a paucity of tools for comparing the conformational properties of aggregates composed of β-amyloid (Aβ) peptide or α-synuclein protein, particularly when present in complex samples such as brain extracts. The conformational stability assay (CSA) is a simple technique that measures the relative resistance of protein aggregates to chemical denaturation. While originally developed to differentiate prion strains, the CSA has since been adapted for use with other protein aggregates. Here, we describe the CSA in detail and outline its utility for distinguishing prion strains as well as unique conformational states of Aβ and α-synuclein aggregates.

Key words Protein aggregates, Protein misfolding, Strains, Prions, Amyloid, α-Synuclein, Alzheimer's disease, Parkinson's disease, Creutzfeldt-Jakob disease, Self-propagation

1 Introduction

The presence of misfolded, aggregated proteins in the brain is a hallmark of many human neurodegenerative disorders, including Alzheimer's disease (AD), Parkinson's disease (PD), and Creutzfeldt-Jakob disease (CJD). In each disease, a different protein adopts a conformation that is enriched in β-sheet content and is prone to aggregation. In prion diseases such as CJD, the cellular prion protein (PrP^C) undergoes a conformational change into a disease-associated, infectious conformer termed PrP^{Sc}. In PD, the α-synuclein protein polymerizes into aggregates called Lewy bodies. In AD, the initiating event is thought to be the misfolding and aggregation of the β-amyloid (Aβ) peptide.

Heather H. C. Lau and Angus Lau contributed equally to this work.

Bradley L. Nilsson and Todd M. Doran (eds.), *Peptide Self-Assembly: Methods and Protocols*, Methods in Molecular Biology, vol. 1777, https://doi.org/10.1007/978-1-4939-7811-3_22,

Emerging evidence suggests that aggregates composed of Aβ or α-synuclein may become self-propagating during disease [1–4]. This occurs when protein aggregates attain the ability to catalyze their own formation by templating the conversion of their properly-folded counterparts into additional copies of the misfolded conformer. This is reminiscent of the prion diseases in which the production of PrP^{Sc} from PrP^{C} underlies the infectious nature of prions [5]. The formation of self-propagating prion-like protein aggregates composed of α-synuclein or Aβ may in part explain why PD and AD are progressive disorders as well as the stereotypical spread of protein aggregates observed within the brains of patients [6]. There has been considerable debate over whether self-propagating proteins aggregates composed of α-synuclein or Aβ should be referred to as "prions" or by other terms, such as "prion-like," "prionoids," or "proteinaceous nucleating particles" [7–9], to discriminate them from the authentic, infectious prions that cause prion diseases such as CJD. Regardless of the nomenclature, there is a growing consensus that self-propagating, prion-like protein aggregates may play a key pathological role in several neurodegenerative diseases.

A widely observed phenomenon in the prion diseases is the existence of distinct prion "strains," which are encoded by different conformations of PrP^{Sc} aggregates [10]. Recent studies have revealed that other self-propagating protein aggregates, including those composed of Aβ or α-synuclein, may also exist as different strains [11–22]. The existence of distinct strains of protein aggregates may potentially help to explain the clinical and pathological diversity observed among AD and PD patients. Many methods have been developed that are capable of distinguishing unique strains of PrP^{Sc}. One of these, the conformational stability assay (CSA), has been useful for demonstrating that a continuum of PrP^{Sc} conformations is possible and that the stability of PrP^{Sc} aggregates may be related to the rate of prion disease progression [23–25].

In the CSA, samples containing PrP^{Sc} aggregates are exposed to increasing concentrations of a chemical denaturant such as guanidine hydrochloride (GdnHCl) (Fig. 1). In the absence of GdnHCl, PrP^{Sc} is insoluble in nonionic detergents and will appear in the pellet fraction following ultracentrifugation. Once a critical GdnHCl concentration is reached, the aggregates become solubilized and will instead be present in the supernatant fraction. Following removal of the supernatant, the pellets are resuspended in loading buffer and then analyzed by SDS-PAGE and immunoblotting. The disappearance of PrP^{Sc} in the pellet fraction is quantified by densitometry and then the $[GdnHCl]_{1/2}$ value, the concentration of GdnHCl required to denature 50% of the PrP^{Sc}, is calculated using computer software. The $[GdnHCl]_{1/2}$ value is strain-specific. For example, the RML strain of PrP^{Sc} exhibits a

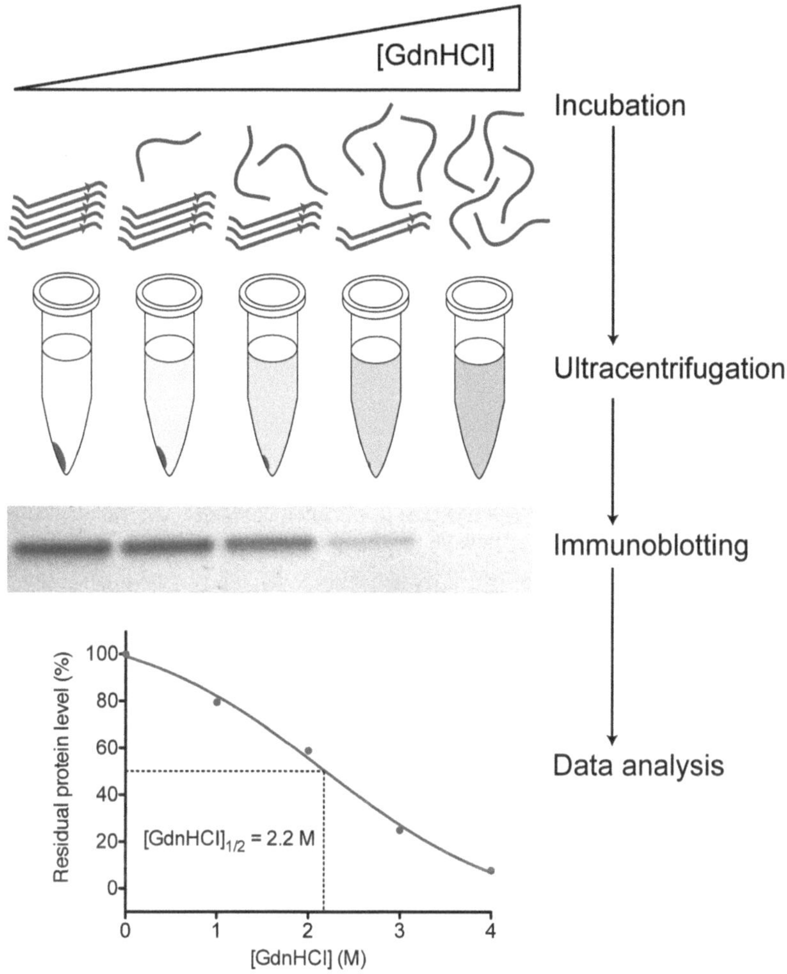

Fig. 1 Schematic representation of the conformational stability assay (CSA). Samples containing protein aggregates are incubated with various concentrations of GdnHCl and then ultracentrifuged to separate the insoluble and soluble fractions. Residual levels of protein aggregates in the pellet fraction are determined by immunoblotting and then the concentration of GdnHCl required to solubilize 50% of the aggregates (the $[GdnHCl]_{1/2}$ value) is calculated. Different $[GdnHCl]_{1/2}$ values are indicative of distinct protein aggregate conformations or strains

$[GdnHCl]_{1/2}$ value of ~1.6 M when propagated in transgenic mice expressing bank vole PrP [26] whereas chronic wasting disease (CWD) prions and prions from a sporadic CJD patient with the VV2 disease subtype possess much higher stabilities ($[GdnHCl]_{1/2}$ values of ~2.4 and ~3.0 M, respectively) when propagated in the same mice (Fig. 2).

We have found that the CSA can also be used to distinguish strains of protein aggregates composed of either Aβ or α-synuclein. In patients with two different genetic forms of AD caused by the "Arctic" or "Swedish" mutations, distinct conformations of Aβ aggregates can be discerned in the brain using the CSA (Fig. 3).

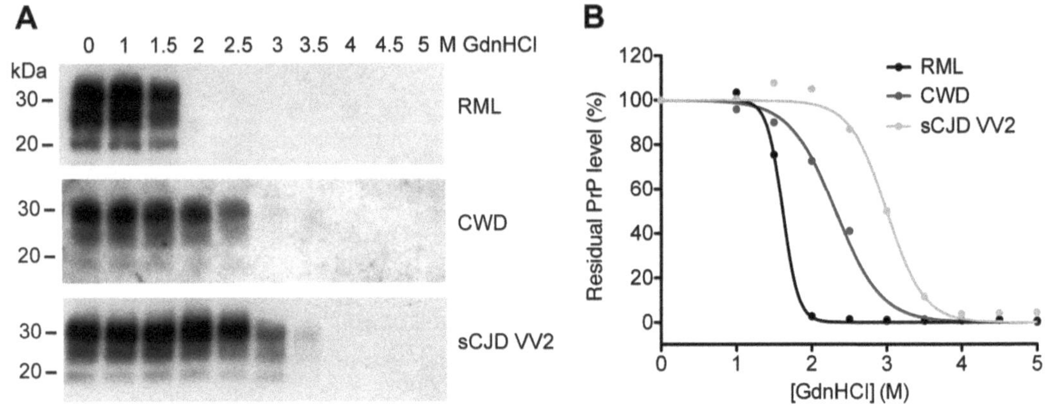

Fig. 2 Distinguishing strains of PrP^{Sc} in bank vole PrP transgenic mice using the CSA. (**a**) Transgenic mice expressing bank vole PrP with methionine at codon 109 were inoculated with RML, CWD, or sporadic CJD (sCJD) VV2 prions and then CSAs were performed using brain homogenates from clinically ill animals [26]. Prior to ultracentrifugation, the samples were digested with PK. PK-resistant PrP^{Sc} was detected using the antibody HuM-P. (**b**) Denaturation curves were constructed for the three samples and the $[GdnHCl]_{1/2}$ values were calculated to be 1.6, 2.3, and 3.0 M for RML, CWD, and sCJD VV2 prions, respectively

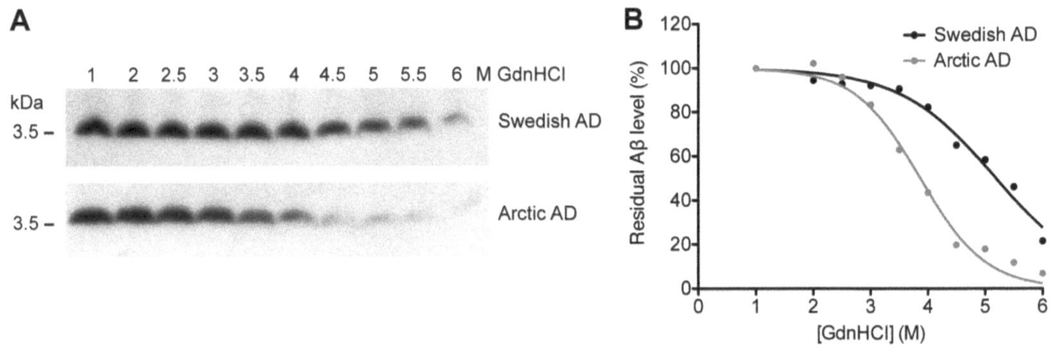

Fig. 3 Discrimination of distinct strains of Aβ aggregates in Alzheimer's disease patients using the CSA. (**a**) Brain samples from AD patients with the Swedish or Arctic mutations were subjected to the CSA [15]. Prior to ultracentrifugation, the samples were digested with PK. Residual Aβ levels were detected by immunoblotting using the antibody 6E10. (**b**) The denaturation curves reveal that Aβ aggregates in the Arctic AD patient are less stable ($[GdnHCl]_{1/2} = 3.9$) than the Aβ aggregates in the Swedish AD patient ($[GdnHCl]_{1/2} = 5.2$), indicating the existence of distinct strains of Aβ aggregates

When the Arctic and Swedish AD brain homogenates were intracerebrally injected into susceptible transgenic mice, distinct Aβ pathologies were observed, supporting the notion that the two samples contain unique strains of self-propagating Aβ aggregates [15]. The CSA is also useful for delineating the conformational properties of α-synuclein aggregates generated by polymerization of recombinant α-synuclein (Fig. 4). Thus, the CSA is a simple technique capable of discriminating different conformations of protein aggregates in a variety of biological samples, including

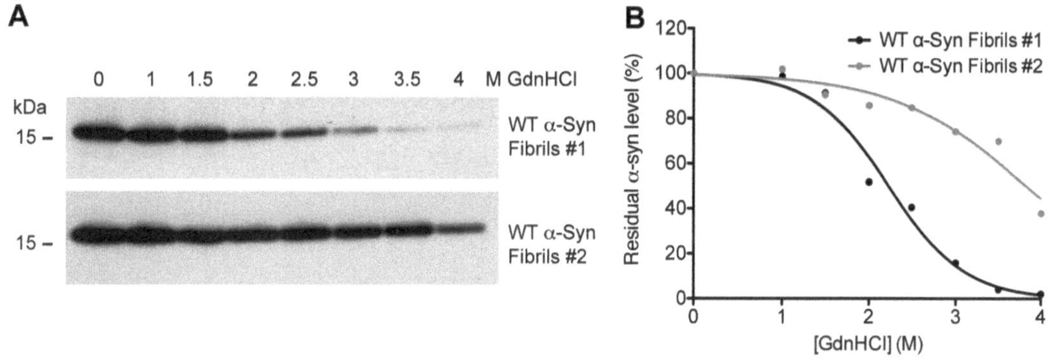

Fig. 4 Differentiating strains of recombinant α-synuclein aggregates using the CSA. (**a**) Wild-type full-length human recombinant α-synuclein was fibrillized using two different buffer conditions and then the fibrils were subjected to the CSA. Samples were not digested with PK. Residual α-synuclein levels were detected by immunoblotting using the antibody Syn1. (**b**) Denaturation curves for the fibrils formed under the two conditions were distinct, with calculated $[GdnHCl]_{1/2}$ values of 2.2 and 3.8 M

complex samples such as brain homogenates. It should be noted that while different $[GdnHCl]_{1/2}$ values obtained using the CSA imply distinct protein aggregate conformations, similar or identical $[GdnHCl]_{1/2}$ values between two samples do not guarantee the presence of a single unique strain. For example, two clearly distinct strains of CWD PrP^{Sc} exhibit identical conformational stabilities [27]. Therefore, the CSA is best applied in combination with orthogonal techniques when characterizing the strain-specified properties of protein aggregates.

In this chapter, we outline step-by-step instructions for performing the CSA on protein aggregates contained in brain homogenates or composed of recombinant protein. A thorough discussion of the most important steps in the assay including sample preparation, GdnHCl treatment and isolation of protein aggregates, visualization of results, and data analysis is provided, as well as troubleshooting tips.

2 Materials

Prepare all solutions with ultrapure water (dH_2O; prepared by purifying deionized water to attain a resistivity of 18 MΩ cm at 25 °C) and analytical-grade reagents. Prepare and store all reagents at room temperature (unless otherwise indicated). Follow all waste disposal regulations when disposing waste materials. No sodium azide was added to the reagents.

2.1 Brain Homogenization (Not Required If Using Recombinant Proteins)

1. 1× Phosphate Buffered Saline (PBS) without calcium or magnesium: 2.67 mM potassium chloride, 1.47 mM potassium phosphate monobasic, 8.06 mM sodium phosphate dibasic, 138 mM sodium chloride.

2. Homogenization tubes containing homogenization beads (2 mL tubes and 7 mL tubes). We use the CK14 soft tissue homogenization tubes containing 1.4 mm zirconium oxide beads (Bertin).
3. Homogenizer. We use the Minilys apparatus (Bertin).
4. Microbalance or scale.
5. Frozen brain samples.

2.2 Detergent Extraction

1. Detergent extraction buffer (10× DB): 5% (wt/vol) sodium deoxycholate, 5% (vol/vol) NP-40 in 1× PBS.
2. Halt Phosphatase Inhibitor (100×; ThermoFisher Scientific).
3. Pierce Universal Nuclease (ThermoFisher Scientific).

2.3 Conformational Stability Assay (CSA)

1. 8 M GdnHCl: Dilute 8 M stock solution with dH_2O to make 2, 3, 4, 5, 6, and 7 M solutions.
2. Recombinant Proteinase K (PK; 20 mg/mL stock).
3. 200 mM phenylmethylsulfonyl fluoride (PMSF) solution prepared in methanol. Store in aliquots at −20 °C.
4. 30% (wt/vol) N-Lauroylsarcosine (Sarkosyl) sodium salt solution (Sigma-Aldrich).
5. 4× Bolt lithium dodecyl sulfate (LDS) sample buffer (ThermoFisher Scientific). Dilute to 1× in dH_2O and add β-mercaptoethanol to a final concentration of 2.5% (vol/vol).
6. A shaking microtube incubator such as the Eppendorf ThermoMixer F1.5 (*see* **Note 1**).
7. A tabletop ultracentrifuge such as the Beckman Coulter Optima MAX-XP and a rotor for snap-cap microtubes such as the TLA-55.
8. Polypropylene ultracentrifuge microtubes (Beckman Coulter).
9. A refrigerated microtube centrifuge.
10. A digital dry bath for boiling samples.

2.4 SDS-Page

1. 10- and 12-well 10% or 4–12% Bolt Bis-Tris Plus gradient gels (ThermoFisher Scientific, *see* **Note 2**).
2. 20× Bolt MES SDS Running Buffer (ThermoFisher Scientific, *see* **Note 3**).
3. Bolt Mini Gel Tank (ThermoFisher Scientific, *see* **Note 3**).
4. Power supply for electrophoresis.
5. Novex Sharp Pre-stained Protein Standard (ThermoScientific).

2.5 Immunoblotting Components

1. Immobilon-P Polyvinylidene fluoride (PVDF) membranes, 0.45 μm (Millipore).
2. Western blot transfer buffer (10×): 250 mM Tris, 1.92 M glycine in dH_2O. Adjust pH to 8.0. For working solution,

dilute to 1× in dH_2O and add methanol to a final concentration of 20% (vol/vol).

3. Grade 703 Western blotting filter paper (VWR).
4. Bolt Mini Blot Transfer Modules with blotting sponges (ThermoFisher).
5. 10× Tris-buffered saline (TBS): 250 mM Tris, 1.37 M NaCl, 27 mM KCl in dH_2O. Adjust pH to 7.4.
6. 1× TBS with 0.05% (vol/vol) Tween-20 (TBST).
7. Methanol.
8. Blocking buffer: 5% (wt/vol) skim milk powder in TBST. Store at 4 °C.
9. 16% (wt/vol) paraformaldehyde. For working solution, dilute to 0.4% (vol/vol) in 1× PBS.
10. Plastic containers for soaking immunoblotting components.
11. Square Petri dishes or similar containers for performing antibody incubations, membrane washes, etc.
12. Humanized anti-PrP recombinant Fab HuM-P [28].
13. Purified anti-Aβ mouse monoclonal antibody 6E10 (BioLegend).
14. Anti-α-synuclein mouse monoclonal antibody Syn-1 (also called Clone 42) (BD Biosciences).
15. Horseradish peroxidase (HRP)-conjugated goat anti-mouse secondary antibody (Bio-Rad).
16. HRP-conjugated goat anti-human IgG F(ab')2 secondary antibody.
17. Western Lightning ECL Pro (PerkinElmer).
18. HyBlot CL autoradiography film (Denville Scientific Inc.).

2.6 Computer Software

1. Photoshop (Adobe Systems Inc.).
2. Curve-fitting software, such as GraphPad Prism (GraphPad Software).
3. Excel (Microsoft Corporation).
4. Software for performing densitometry, such as ImageJ [29].

3 Methods

3.1 Preparation of Brain Homogenates

If you are using purified or recombinant protein aggregates, skip to Subheading 3.2.

1. Weigh the homogenizing tube and tare the scale.
2. Add the frozen brain containing the protein aggregates of interest to the homogenizing tube and then weigh it to calculate the weight of the brain (*see* **Note 4**).

3. Add 9 volumes of chilled PBS to give you a 10% (wt/vol) brain homogenate. For example, if the brain section weighs 0.4 g, add 3.6 mL of PBS (*see* **Note 5**).
4. Place the homogenizing tube containing the brain and the beads into the homogenizer apparatus. Set the speed on the instrument to the highest value (5000 rpm for the Minilys) and then homogenize the sample for 30 s.
5. Place the homogenizing tube on ice for 5 min and then homogenize for an additional 30 s.
6. Aliquot the homogenate into labeled microtubes and then store at −80 °C until needed.

3.2 The Conformational Stability Assay

1. Perform a detergent extraction by mixing nine volumes of sample (brain homogenate or purified/recombinant protein aggregates) with 1 volume of 10× DB. *See* **Note 6** for suggested volumes. If you are analyzing brain homogenates containing α-synuclein aggregates, add Halt Phosphatase Inhibitor (if you will be probing for phosphorylated α-synuclein) and Pierce Universal Nuclease (*see* **Notes 7** and **8**).
2. Incubate on ice for 20 min with periodic vortexing.
3. Centrifuge samples at 1500 × *g* for 5 min at 4 °C. Collect the supernatant into a new tube and discard the pellet. Keep the supernatant on ice. This step is not necessary if using purified or recombinant protein aggregates.
4. Dilute the 8 M GdnHCl stock solution in dH_2O to final concentrations of 2, 3, 4, 5, 6, and 7 M. These 2× stocks will be used to create final GdnHCl concentrations of 1–4 M.
5. Add 20 μL of detergent-extracted sample from **steps 1–3** to each of 8 ultracentrifuge microtubes.
6. Add 20 μL of 2× GdnHCl stocks to seven of the tubes to give final GdnHCl concentrations of 1, 1.5, 2, 2.5, 3, 3.5, or 4 M. To the eighth tube add 20 μL of dH_2O to give a final concentration of 0 M GdnHCl. If you need to test higher concentrations of GdnHCl (i.e., 4.5, 5, 5.5, and 6 M, *see* **Note 9**), use 10 μL of sample and add the following components: for 4.5 M, add 30 μL of 6 M GdnHCl; for 5 M, add 25 μL of 8 M GdnHCl and 5 μL of dH_2O; for 5.5 M, add 27.5 μL of 8 M GdnHCl and 2.5 μL of dH_2O; for 6 M, add 30 μL of 8 M GdnHCl.
7. Incubate the samples at room temperature (22 °C) for 2 h with shaking (800 rpm) using a microtube thermal mixer. If you do not have a thermal mixer, *see* **Note 10**.
8. Normalize the GdnHCl concentration in all samples to 0.4 M by adding the following to each tube (in this order) (*see* **Note 11**):

(a) Samples containing PrP or Aβ aggregates: add (500 − x) μL of PBS, then 60 μL of 10× DB, and then x μL of 8 M GdnHCl, where x = 0, 2.5, 5, . . ., 22.5, 25, or 30 μL for the 6, 5.5, 5, . . ., 1.5, 1, and 0 M GdnHCl tubes, respectively. The final volume in each tube is 600 μL.

(b) Samples containing α-synuclein aggregates: add (410 − x) μL of PBS, then 50 μL of 10× DB, and then x μL of 8 M GdnHCl, where x = 5, 7.5, 10, . . ., 17.5, 20, or 25 μL for the 4, 4.5, 3, . . ., 1.5, 1, and 0 M GdnHCl tubes, respectively. The final volume in each tube is 500 μL.

9. Samples containing PrP or Aβ aggregates are next digested with PK (*see* **Note 12**). For samples containing α-synuclein aggregates, skip to **step 10**.

   (a) Dilute the 20 mg/mL PK stock solution to 5 mg/mL using PBS.

   (b) Add 2.4 μL of diluted PK to each tube to give a final PK concentration of 20 μg/mL.

   (c) Incubate at 37 °C for 1 h with shaking (600–650 rpm) using a thermal mixer (*see* **Notes 1** and **13**).

   (d) Add 6 μL of 200 mM PMSF to give a final concentration of 2 mM and then vortex.

   (e) Add 43 μL of 30% (wt/vol) sarkosyl solution to give a final concentration of 2% and then mix by vortexing.

   (f) Continually rotate or periodically vortex the sample at 4 °C for ~10 min until the white precipitate disappears (*see* **Note 14**).

10. Ultracentrifuge the samples at 100,000 × g for 1 h at 4 °C (*see* **Note 15**).

11. Carefully remove and discard the supernatant (*see* **Note 16**).

12. Add 40 μL of 1× LDS sample buffer (*see* **Note 2**) to the 0–4 M GdnHCl tubes (i.e., the tubes that originally contained 20 μL of detergent-extracted sample). Add 20 μL of 1× LDS sample buffer to the 4.5–6 M GdnHCl tubes (i.e., the tubes that originally contained 10 μL of detergent-extracted sample).

13. Resuspend the pellets by vortexing and then boil using a digital dry bath set at 95 °C for 5–10 min. Briefly centrifuge the samples to bring down any condensate that forms on the tube lid.

14. Store the samples at −20 °C if not proceeding immediately to SDS-PAGE.

3.3 SDS-Page

1. Prepare 1× Bolt MES SDS running buffer and place a Bolt 4–12% or 10% gel (*see* **Note 3**) into the Bolt Mini Gel tank after removing the comb and the gel tape at the bottom of the cassette. Fill up the inner chamber with running buffer so

that the buffer covers the wells of the gel. Pour sufficient buffer in the outer chamber so that the buffer level is higher than the gel foot. Thoroughly rinse the wells with running buffer.

2. Load 5–10 μL of Novex Sharp Standard into the first lane.
3. Load 10–15 μL of samples per well.
4. Place the lid on the gel apparatus and electrophorese the samples at 165 V for 30–35 min, or until the dye front has migrated to approximately 1 cm from the bottom of the gel.

3.4 Immunoblotting (see Note 17)

1. Prepare 1× transfer buffer containing 20% (vol/vol) methanol (*see* **Note 18**).
2. Cut a PVDF membrane to the size of the gel (~7.5 × 8.5 cm). Activate the membrane by placing it in methanol for 1–2 min and then transfer to a plastic container (*see* **Note 19**) containing 1× transfer buffer.
3. Soak the blotting sponges and blotting paper (2 pieces per gel, cut to a size of ~7.5 × 9 cm) in 1× transfer buffer (*see* **Note 20**).
4. Remove the gel cassette from the electrophoresis apparatus and thoroughly rinse with dH_2O. Gently pry open the cassette using a gel knife so that the gel adheres to one side of the cassette. Cut off the wells and the gel foot using a razor blade, and then carefully transfer the gel to a plastic container with 1× transfer buffer in it (*see* **Note 21**).
5. Lay one presoaked blotting sponge on the cathode (negative) core of the Bolt Mini Blot Module and then one piece of blotting paper. Smooth out any bubbles by gently using a gel roller or similar apparatus (*see* **Note 22**).
6. Gently place the gel on the blotting paper and carefully remove any bubbles.
7. Place the activated PVDF membrane on the gel and carefully remove any bubbles.
8. Place a second piece of pre-wet blotting paper on top of the membrane, smooth out any bubbles, and then place a second sponge on top.
9. Place the anode (positive) core of the Mini Blot Module over the top of the transfer sandwich and gently apply pressure to seal the module.
10. Place the module into the Bolt Mini Gel tank and then fill up the interior of the blotting module with 1× transfer buffer.
11. Fill up the outer chamber of the apparatus with dH_2O (it is not necessary to use transfer buffer).
12. Place the lid on the gel apparatus and transfer the samples at 20 V for 1 h.

13. Following transfer, remove the Mini Blot Module from the gel apparatus and carefully disassemble the transfer sandwich. An efficient transfer has occurred if the pre-stained protein standard has entirely migrated to the PVDF membrane (i.e., it is no longer visible on the gel).
14. For PrP- or Aβ-containing samples, skip to **step 15**. For samples containing α-synuclein aggregates, it is advisable to cross-link the proteins to the membrane as follows (*see* **Note 23**):
    (a) Transfer the membrane to a square Petri dish containing 0.4% (vol/vol) paraformaldehyde in PBS.
    (b) Incubate the membrane at room temperature on a rocker for 30 min.
    (c) Discard the formaldehyde solution and quickly rinse the membrane two times with 1× TBST.
15. Transfer the membrane to a square Petri dish containing blocking buffer and then incubate with orbital agitation for a minimum of 1 h.
16. Discard the blocking buffer and incubate the membranes with primary antibody diluted in blocking buffer overnight with rocking at 4 °C. For PrP, use the antibody HuM-P at a dilution of 1:10,000 (*see* **Note 24**); for Aβ, use the antibody 6E10 at a dilution of 1:4000; for α-synuclein, use the antibody Syn-1 at a dilution of 1:10,000 (*see* **Note 25**).
17. Wash the membrane 3 times with 1× TBST for 10 min.
18. Incubate with HRP-conjugated secondary antibodies, diluted in blocking buffer, for 1–2 h with rocking. For 6E10 and Syn1 antibodies, use a goat anti-mouse secondary antibody at a dilution of 1:10,000; for the HuM-P antibody, use a goat anti-human F(ab')2 secondary antibody at a dilution of 1:10,000.
19. Wash the membrane 3 times with 1X TBST for 10 min.
20. Remove excess TBST from the membrane by blotting with a Kimwipe. Add equal volumes of the two ECL components to cover the surface of the membrane. Using a pipette, gently spray the membrane with ECL for ~1 min (*see* **Note 26**).
21. Remove excess ECL from the membrane by blotting with a Kimwipe and then wrap the membrane in plastic wrap (*see* **Note 27**).
22. Place the wrapped membrane into an X-ray cassette and expose to X-ray film (*see* **Note 28**).

3.5 Data Analysis

1. Scan the X-ray film and save it as a digital image.
2. If necessary, convert the image to grayscale using Photoshop and, if desirable, adjust the image contrast using the Auto Contrast function in Photoshop.
3. Import the image into software capable of performing densitometry, such as ImageJ. Invert the image so that the background is black and the protein bands are white.
4. Draw a rectangular region of interest around the brightest band and then perform densitometry on each band using the same region of interest. Also using the same region of interest, determine the background signal by placing the rectangle on an area of the blot with no protein signal.
5. Copy and paste the densitometry values into Excel.
6. Subtract the background signal from each of the measured protein signals.
7. Normalize the background-corrected signals by dividing each value by the signal obtained with the lowest GdnHCl concentration used. Multiply each value by 100 so that the starting value for the denaturation is 100%.
8. Create a new "XY" graph in GraphPad Prism and enter the GdnHCl concentrations used in the "X" column.
9. Copy and paste the normalized values into the "Y" column.
10. For data analysis, choose "Non-linear regression (curve fit)" under "XY analyses" and then "log(inhibitor) vs. response—Variable slope (4 parameters)" under "Dose-response—Inhibition." The $[\text{GdnHCl}]_{1/2}$ value will be the "LogIC50" value (*see* **Note 29**).

4 Notes

1. If a thermal mixer is not available, just incubate the samples without shaking in a dry bath or a water bath set at 37 °C.
2. Other gel-loading buffers, such as SDS loading buffer, should also work if standard SDS-PAGE gels are being used.
3. Other gel systems and running buffers should work equally as well.
4. Half mouse brains fit fine into the 2 mL homogenizing tubes. If homogenizing whole brains, it is advisable to using the 7 mL homogenizing tubes.
5. If the volume necessary to give a 10% (wt/vol) homogenate is too large for the homogenizing tube, add less, homogenize, and then add the remaining volume after homogenization. For example, when using the 2 mL homogenizing tubes, we

typically add 1 mL of PBS, homogenize, and then add the homogenate to a 15 mL Falcon tube containing the remaining volume of PBS.

6. For each concentration of GdnHCl to be tested, 20 μL of detergent-extracted brain homogenate is required. If GdnHCl concentrations of 1–4 M in 0.5 M steps as well as a 0 M sample are being tested, you will need 160 μL of the detergent-extracted brain homogenate. To account for pipetting error, it is best to prepare ~200 μL of the extract. To ensure that this volume is obtained, multiply by 1.5 to calculate the final volume of 10% brain homogenate plus 10× DB needed (since some of the volume will be taken up by the pellet fraction following centrifugation). For example, to get ~200 μL of detergent-extracted brain homogenate, combine 270 μL of 10% brain homogenate and 30 μL 10× DB.
7. Since the majority of disease-associated α-synuclein aggregates are phosphorylated [30], it is advisable to add phosphatase inhibitor to prevent dephosphorylation of the aggregates during the CSA when analyzing brain extracts. This step is not strictly necessary when using an antibody for immunoblotting that detects total α-synuclein, such as Syn-1.
8. Because the α-synuclein aggregates are not subjected to PK digestion during the CSA, the solubilized pellets following ultracentrifugation can produce "streakiness" during SDS-PAGE. Including a nonspecific nuclease during the detergent extraction step greatly improves this problem.
9. It is usually necessary to use the extended range of GdnHCl concentrations when analyzing Aβ aggregates. We have found that the vast majority of PrP^{Sc} and α-synuclein aggregates (especially those present in brain extracts) are completely solubilized by 4 M GdnHCl.
10. Instead of using a thermal mixer set to 22 °C, the tubes can be placed in an Eppendorf tube rack and attached to any orbital shaker (such as a microplate shaker) and then incubated at room temperature.
11. The 8 M GdnHCl solution used for normalization must be added last. If it is added first, the concentration of GdnHCl will be transiently much higher in the tube and we have found that this can result in the undesirable solubilization of protein aggregates in the tubes containing lower GdnHCl concentrations.
12. Although it is not strictly necessary, adding PK to samples containing aggregates that are PK-resistant (such as the majority of PrP^{Sc} and Aβ aggregates) greatly improves resolution during immunoblotting.

13. A white precipitate will likely form during the PK digestion. It does not affect the performance of the digestion or the CSA.
14. The precipitate that forms likely results from the coprecipitation of the detergent buffer and GdnHCl when incubated at 37 °C. Re-solubilization occurs readily when the samples are incubated on ice with frequent vortexing. It is essential that complete re-solubilization occur before proceeding to the ultracentrifugation step. Otherwise, the precipitate will appear in the ultracentrifuge pellet.
15. If an ultracentrifuge is unavailable, it may be possible to use a standard microcentrifuge set at maximum speed (>20,000 × *g*). Longer centrifugations may be necessary to achieve complete pelleting of the aggregates. Thus, optimization of the protocol at these lower speeds is required.
16. This is the trickiest step in the CSA. The ultracentrifuge pellet is typically very small or not visible, and comes off the side of the tube very easily. Removal of the supernatant can be achieved by either (1) slow, gentle aspiration using a gel loading tip with the vacuum set on low power; or (2) careful pipetting. In our experience, the performance of the CSA is not substantially affected if it is necessary to leave a few microliters of supernatant to ensure that the pellet is not lost.
17. For recombinant protein aggregates, we have also had good success with using Coomassie Blue-stained gels to analyze the CSA.
18. Other transfer buffers may also work. We always include 20% (vol/vol) methanol in the transfer buffer, but this may not be necessary for all proteins.
19. We use GladWare plastic food storage containers for soaking the sponges, blotting papers, and membranes prior to transfer.
20. The 1× transfer buffer used for soaking the transfer components can be reused for subsequent transfers.
21. To prevent the gel from ripping, pour some 1× transfer buffer on it, use the gel knife to carefully lift one of the bottom corners of the gel, and then gently "peel" the gel off the cassette from the bottom to the top using your gloved fingers.
22. We frequently use the cap from a Sharpie marker to smooth out the blotting paper and membrane.
23. It has been shown that cross-linking α-synuclein species to the membrane following transfer prevents them from detaching during subsequent incubation and washing steps and greatly increases the sensitivity of immunoblotting [31].
24. The HuM-P antibody is not commercially available. Other PrP antibodies such as 3F4 (BioLegend; for detecting hamster or

human PrP^{Sc}), 6D11 (BioLegend; for detecting PrP^{Sc} from many different species), or 8H4 (Abcam; for detecting PrP^{Sc} from many different species) should work equally as well.

25. It may also be possible to use phospho-specific α-synuclein antibodies such as EP1536Y (Abcam) to detect disease-specific α-synuclein aggregates using the CSA.
26. We find that excellent results can be obtained using low volumes of ECL (800 μL of each of the two components) and constant pipetting of the solution over the membrane.
27. The plastic wrap prevents excess ECL from coming into contact with the X-ray film (which can cause black splotches on the film).
28. Although we most often use X-ray film, Western blot imaging systems should also work perfectly well.
29. For optimum results, it is often useful to constrain the bottom value of the curve to "0" and the top value to "100."

References

1. Prusiner SB (2012) A unifying role for prions in neurodegenerative diseases. Science 336:1511–1513
2. Jucker M, Walker LC (2013) Self-propagation of pathogenic protein aggregates in neurodegenerative diseases. Nature 501:45–51
3. Guo JL, Lee VM (2014) Cell-to-cell transmission of pathogenic proteins in neurodegenerative diseases. Nat Med 20(2):130–138
4. Goedert M (2015) NEURODEGENERATION. Alzheimer's and Parkinson's diseases: the prion concept in relation to assembled Abeta, tau, and alpha-synuclein. Science 349 (6248):1255555
5. Colby DW, Prusiner SB (2011) Prions. Cold Spring Harb Perspect Biol 3:a006833
6. Polymenidou M, Cleveland DW (2012) Prion-like spread of protein aggregates in neurodegeneration. J Exp Med 209:889–893
7. Aguzzi A, Rajendran L (2009) The transcellular spread of cytosolic amyloids, prions, and prionoids. Neuron 64:783–790
8. Hardy J, Revesz T (2012) The spread of neurodegenerative disease. N Engl J Med 366:2126–2128
9. Walker LC, Jucker M (2015) Neurodegenerative diseases: expanding the prion concept. Annu Rev Neurosci 38:87–103
10. Collinge J, Clarke AR (2007) A general model of prion strains and their pathogenicity. Science 318:930–936
11. Petkova AT, Leapman RD, Guo Z, Yau WM, Mattson MP, Tycko R (2005) Self-propagating, molecular-level polymorphism in Alzheimer's beta-amyloid fibrils. Science 307:262–265
12. Paravastu AK, Qahwash I, Leapman RD, Meredith SC, Tycko R (2009) Seeded growth of beta-amyloid fibrils from Alzheimer's brain-derived fibrils produces a distinct fibril structure. Proc Natl Acad Sci U S A 106:7443–7448
13. Lu JX, Qiang W, Yau WM, Schwieters CD, Meredith SC, Tycko R (2013) Molecular structure of beta-amyloid fibrils in Alzheimer's disease brain tissue. Cell 154(6):1257–1268
14. Heilbronner G, Eisele YS, Langer F, Kaeser SA, Novotny R, Nagarathinam A, Aslund A, Hammarstrom P, Nilsson KP, Jucker M (2013) Seeded strain-like transmission of beta-amyloid morphotypes in APP transgenic mice. EMBO Rep 14(11):1017–1022
15. Watts JC, Condello C, Stohr J, Oehler A, Lee J, DeArmond SJ, Lannfelt L, Ingelsson M, Giles K, Prusiner SB (2014) Serial propagation of distinct strains of Abeta prions from Alzheimer's disease patients. Proc Natl Acad Sci U S A 111(28):10323–10328
16. Stohr J, Condello C, Watts JC, Bloch L, Oehler A, Nick M, DeArmond SJ, Giles K, DeGrado WF, Prusiner SB (2014) Distinct synthetic Abeta prion strains producing different amyloid deposits in bigenic mice. Proc Natl Acad Sci U S A 111(28):10329–10334
17. Cohen ML, Kim C, Haldiman T, ElHag M, Mehndiratta P, Pichet T, Lissemore F, Shea M, Cohen Y, Chen W, Blevins J, Appleby BS, Surewicz K, Surewicz WK, Sajatovic M, Tatsuoka C, Zhang S, Mayo P, Butkiewicz M,

Haines JL, Lerner AJ, Safar JG (2015) Rapidly progressive Alzheimer's disease features distinct structures of amyloid-beta. Brain 138 (Pt 4):1009–1022

18. Guo JL, Covell DJ, Daniels JP, Iba M, Stieber A, Zhang B, Riddle DM, Kwong LK, Xu Y, Trojanowski JQ, Lee VM (2013) Distinct α-synuclein strains differentially promote tau inclusions in neurons. Cell 154:103–117
19. Bousset L, Pieri L, Ruiz-Arlandis G, Gath J, Jensen PH, Habenstein B, Madiona K, Olieric V, Bockmann A, Meier BH, Melki R (2013) Structural and functional characterization of two alpha-synuclein strains. Nat Commun 4:2575
20. Watts JC, Giles K, Oehler A, Middleton L, Dexter DT, Gentleman SM, DeArmond SJ, Prusiner SB (2013) Transmission of multiple system atrophy prions to transgenic mice. Proc Natl Acad Sci U S A 110(48):19555–19560
21. Peelaerts W, Bousset L, Van der Perren A, Moskalyuk A, Pulizzi R, Giugliano M, Van den Haute C, Melki R, Baekelandt V (2015) alpha-Synuclein strains cause distinct synucleinopathies after local and systemic administration. Nature 522(7556):340–344
22. Prusiner SB, Woerman AL, Mordes DA, Watts JC, Rampersaud R, Berry DB, Patel S, Oehler A, Lowe JK, Kravitz SN, Geschwind DH, Glidden DV, Halliday GM, Middleton LT, Gentleman SM, Grinberg LT, Giles K (2015) Evidence for alpha-synuclein prions causing multiple system atrophy in humans with parkinsonism. Proc Natl Acad Sci U S A 112(38):E5308–E5317
23. Peretz D, Scott M, Groth D, Williamson A, Burton D, Cohen FE, Prusiner SB (2001) Strain-specified relative conformational stability of the scrapie prion protein. Protein Sci 10:854–863
24. Peretz D, Williamson RA, Legname G, Matsunaga Y, Vergara J, Burton D, DeArmond SJ, Prusiner SB, Scott MR (2002) A change in the conformation of prions accompanies the emergence of a new prion strain. Neuron 34:921–932
25. Legname G, Nguyen H-OB, Peretz D, Cohen FE, DeArmond SJ, Prusiner SB (2006) Continuum of prion protein structures enciphers a multitude of prion isolate-specified phenotypes. Proc Natl Acad Sci U S A 103:19105–19110
26. Watts JC, Giles K, Patel S, Oehler A, Dearmond SJ, Prusiner SB (2014) Evidence That bank vole PrP is a universal acceptor for prions. PLoS Pathog 10(4):e1003990
27. Angers RC, Kang HE, Napier D, Browning S, Seward T, Mathiason C, Balachandran A, McKenzie D, Castilla J, Soto C, Jewell J, Graham C, Hoover EA, Telling GC (2010) Prion strain mutation determined by prion protein conformational compatibility and primary structure. Science 328:1154–1158
28. Safar JG, Scott M, Monaghan J, Deering C, Didorenko S, Vergara J, Ball H, Legname G, Leclerc E, Solforosi L, Serban H, Groth D, Burton DR, Prusiner SB, Williamson RA (2002) Measuring prions causing bovine spongiform encephalopathy or chronic wasting disease by immunoassays and transgenic mice. Nat Biotechnol 20:1147–1150
29. Schneider CA, Rasband WS, Eliceiri KW (2012) NIH Image to ImageJ: 25 years of image analysis. Nat Methods 9(7):671–675
30. Fujiwara H, Hasegawa M, Dohmae N, Kawashima A, Masliah E, Goldberg MS, Shen J, Takio K, Iwatsubo T (2002) alpha-Synuclein is phosphorylated in synucleinopathy lesions. Nat Cell Biol 4:160–164
31. Lee BR, Kamitani T (2011) Improved immunodetection of endogenous α-synuclein. PLoS One 6:e23939

Chapter 23

Model Phospholipid Liposomes to Study the β-Amyloid-Peptide-Induced Membrane Disruption

Wei Qiang and Katelynne E. Doherty

Abstract

Model phospholipid liposomes have been utilized widely to study the molecular interactions between peptides and membrane bilayers. In the mechanistic study of Alzheimer's disease (AD), disruption of neuronal cell membranes has been considered as a major contribution for the β-amyloid (Aβ) peptides' neurotoxicity. However, clear interpretation of the Aβ-induced cellular membrane at high-resolution level is challenging because of the co-existence of multiple pathways. Here we present the generation of simplified model liposome systems that will facilitate the in-depth mechanistic studies. Protocols for the preparation of model liposomes and the characterization of individual membrane disruption effects will be described.

Key words β-Amyloid peptides, Phospholipid liposomes, Membrane disruption, Fibrillation, lipid mixing, Electrophysiology, Ion channels

1 Introduction

A major challenge that hinders the in-depth, high-resolution studies on the β-amyloid (Aβ)-peptide-induced membrane disruption process is that the Aβ-liposome systems are typically heterogeneous [1]. Such heterogeneity is mainly induced by the high tendency of aggregation for Aβ peptides in aqueous solution at the concentration of most biophysical studies (i.e., ∼ μM range) [2–4]. Even in the simplest phospholipid bilayers that contain only zwitterionic phosphatidylcholine (PC) and negatively charged phosphatidylglycerol (PG), interactions between Aβ and membranes may be initiated with monomeric, small oligomeric, and large oligomeric peptides in in vitro model systems. Distinct individual states of Aβ at the initial stage are structurally heterogeneous, and may lead to different downstream membrane disruption effects [5, 6]. The co-existence of these pathways greatly increases the complexity of model systems. For instance, two of the previously reported membrane-associated Aβ aggregation pathways, the fibrillation and the formation of Ca^{2+}-selected ion channel, involved

Bradley L. Nilsson and Todd M. Doran (eds.), *Peptide Self-Assembly: Methods and Protocols*, Methods in Molecular Biology, vol. 1777, https://doi.org/10.1007/978-1-4939-7811-3_23, © Springer Science+Business Media, LLC, part of Springer Nature 2018

membrane interactions of distinct Aβ species. The Aβ fibrillation pathway is initiated from monomeric peptides that were absorbed by the membrane bilayer [5, 7]. In this case, the bilayer surface may serve as templates, which promote the formation of specific structures on pathway of the fibrillation [5]. Along with the fibrillation process, local membrane disruption may lead to membrane leakage and/or the uptake of lipids [5, 8]. On the other hand, the ion channel activities were detected in the presence of membrane-incorporated Aβ oligomers [9]. Molecular dynamic simulations suggest that the β-barrel-like oligomer structures are stabilized by the surrounding lipids [9–11].

This protocol describes the generation and biophysical characterization of three distinct Aβ-membrane model systems [12]. These systems are produced with well-controlled initial Aβ concentrations and peptide-to-lipid (P:L) molar ratio. Biophysical techniques are utilized to determine the dominant membrane-disruption effects in each system. These model systems, because of the reduction of heterogenity, may be utilized in future works for high-resolution studies of the molecular mechanisms that contribute to Aβ-related toxicity.

2 Materials

Buffer solutions utilized in this protocol were prepared with deionized water. All organic solvents were analytical grade with the exception of HPLC-grade water and acetonitrile used for high-performance liquid chromatography (HPLC). Procedures for the manual peptide synthesis and purification are well established and considered as routine protocols, and therefore will not be described in details here.

2.1 Liposome Preparation

1. Aβ40 peptides (*see* **Note 1**).
2. HFIP.
3. Phospholipids: 1-palmitoyl-2-oleoyl-*sn*-glycero-3-phosphocholine (POPC), 25 mg/mL in chloroform (*see* **Note 2**).
4. Extruder.
5. Support filter paper.
6. 1.0 μm or 100 nm pore-size polycarbonate membranes.
7. 10 mM phosphate buffer (10 mM Na_2HPO_4, 10 mM NaH_2PO_4, 0.01% m/v NaN_3) with *p*H adjusted to 7.4.
8. Vacuum desiccator.
9. *n*-Decane.

2.2 Circular Dichroism (CD) Spectroscopy

1. Circular Dichroism Spectrometer.
2. Quartz cuvette with 1 mm path length (max. Volume 350 μL).

2.3 Fluorescence Assays

1. A fluorimeter equipped with temperature control or a water bath.
2. Fluorescent phospholipid mixture: POPC containing 1 mol% NBD-phosphatidylethanolamine (NBD-PE, 0.2 mg/mL in chloroform) and 1 mol% Rhodamine-B-labeled phosphatidylethanolamine (Rh-PE, 0.2 mg/mL in chloroform).
3. Thioflavin-T (ThT) buffer: 4 mM ThT in 10 mM phosphate buffer, pH = 7.4 (*see* **Note 3**).
4. Calcein-contained phosphate buffer: 5 mM calcein in 10 mM phosphate buffer (pH 7.4).
5. Quartz micro-scale fluorescent cuvette with 1 mm path length (max. Effective volume 120 μL).

2.4 Fluorescence Imaging

1. Rhodamine-green-labeled Aβ (Anaspec, Part # AS-61134).
2. Rh-PE.
3. Phospholipid mixture for fluorescence imaging: POPC containing 1 mol% Rh-PE (0.2 mg/mL in chloroform).

2.5 Transmission Electron Microscopy (TEM)

1. Glow-discharged carbon film, 300-mesh carbon-coated copper grid.
2. Uranyl acetate.

2.6 Electrophysiological Measurements

1. Two-compartment electrophysiological chamber with a 200 μm aperture.
2. Glutaraldehyde: ~10% *v/v* in deionized water.
3. HEPES buffer: 10 mM HEPES with 150 mM KCl and 1 mM magnesium gluconate, pH adjusted to 7.4.
4. n-Decane.

3 Methods

3.1 Preparation and Characterization of the "External Addition" Aβ-Liposome Model System with Dominant Fibrillation Pathway

Figure 1 illustrates the Aβ evolution pathway within this model system. Briefly, Aβ peptides are absorbed to membrane surface, where the enhancement of local peptide concentration promotes the initial fibrillation. Chain elongation is then continued on the rapidly formed fibril nucleus. Membrane bilayers are locally disrupted by the initial nucleation process, which results in content leakage [7, 13, 14].

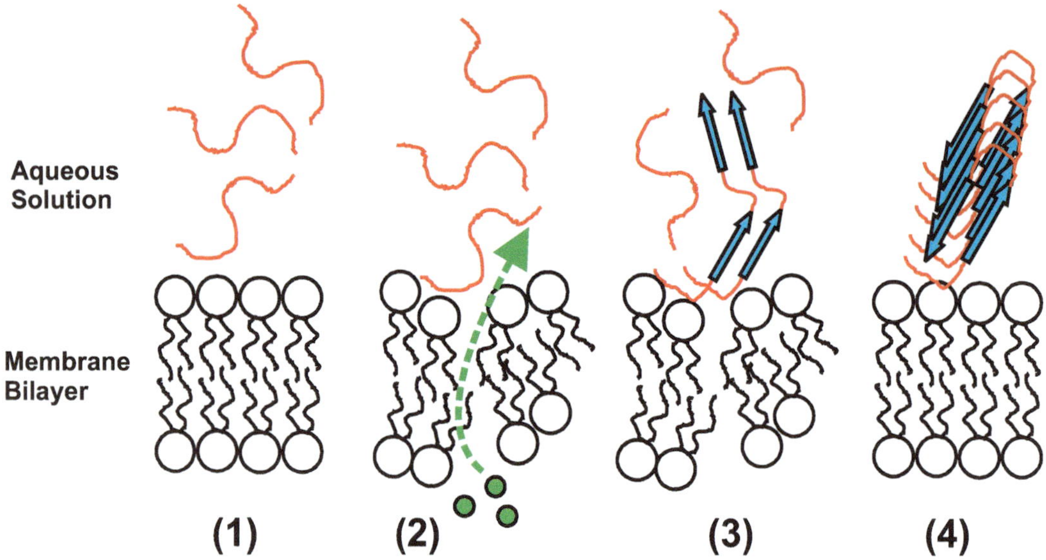

Fig. 1 Proposed Aβ fibrillation pathway in the presence of membrane. Step (1), absorption of Aβ to the membrane surface. Step (2), initial binding between Aβ and membrane bilayer, which induces local membrane disruption and cellular leakage. Step (3), initial nucleus of Aβ on membrane surface. Step (4), the fibril chain elongation

3.1.1 Liposome Preparation

1. Dry the desired amount of POPC in chloroform under N_2 flow.
2. Remove all organic solvent in a vacuum desiccator for at least 8 h.
3. Resuspend the dried lipid film in phosphate buffer (*see* **Note 4**).
4. Rotate the mixture for 1 h at room temperature.
5. Extrude the mixture using 100 nm pore-size polycarbonate membrane for at least 15 cycles at room temperature to ensure the homogeneity of vesicle size. The liposome solution can be stored at 4 °C, but extrusion should be done freshly before usage.

3.1.2 Addition of Aβ Peptides into Liposome

1. Dissolve the desired amount of Aβ peptide in 0.1 M NaOH and lyophilize to remove any pre-formed Aβ aggregates [6].
2. Dissolve the peptide powder in DMSO (2% *v/v* to the final volume of liposome solution) and transfer it into liposome solution (*see* **Note 5**).

*3.1.3 Monitoring Fibrillation Using ThT Fluorescence (see **Note 6**).*

1. Equilibrate the fluorimeter sample chamber to 37 °C and stabilized for 1 h before measurement. For example, one can use a fluorimeter that is equipped with a water-bath cuvette holder.

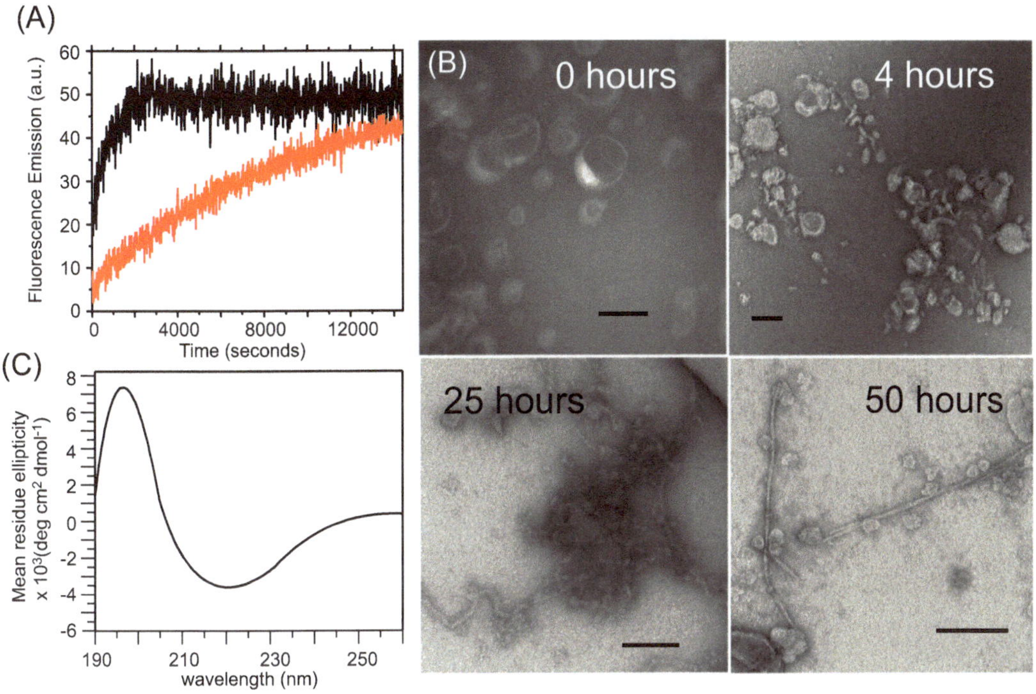

Fig. 2 Biophysical characterization of the "external addition" sample with P:L ratio 1:30. (**a**) Fluorescence assays for fibrillation (gray) and calcein content leakage (black) plotted on the same time scale. (**b**) TEM images of Aβ-membrane system taken at different time points of incubation. The scale bar represents 100 nm. (**c**) CD spectrum of Aβ-membrane system taken at 20-h incubation time

2. Mix a 100 μL aliquot of the Aβ-liposome solution with 10 μL of 4.0 mM ThT solution and transfer the mixture into the fluorescence cuvette (*see* **Note 3**) [15].
3. Cap the cuvette and seal it with parafilm.
4. Allow the sample to sit in the holder for 5 min before data collection.
5. Record the fluorescence signal with excitation and emission wavelengths at 440 nm and 490 nm, respectively, over ~14,400 s with 5 s dwell time.
6. Record control ThT fluorescence traces for liposomes without the addition of Aβ and subtract this trace from the sample trace. A typical fluorescence emission trace for this mode system was shown in Fig. 2a.

3.1.4 Monitoring of Fibrillation Using TEM

1. Prepare the sample as outlined in Subheadings 3.1.1 and 3.1.2, and incubate at 37 °C quiescently.
2. Remove a 10 μL aliquot of the Aβ-liposome solution at the desired time point and dispense onto a glow-discharged carbon

film supported by lacey carbon on a 300 mesh copper TEM grid.

3. After 2 min, blot the residual solution with tissue paper and rince the grid twice with deionized water.
4. Stain the grid with 10 μL 2% uranyl acetate for 35 s (*see* **Note** 7).
5. Blot the excess staining solution with tissue paper.
6. Dry the grid in air and tested using TEM. Representative TEM images with different incubation time was shown in Fig. 2b.

3.1.5 Monitoring of Liposome Leakage Using Calcein Fluorescence

1. Prepare the calcein-contained liposome using a protocol similar to Subheading 3.1.1, with the exception that the calcein-contained phosphate buffer is used for the resuspension of the dried lipid film in **step 3**.
2. The final concentration of calcein should be kept at 1 mM for the liposome solution (in 10 mM phosphate buffer).
3. Extrude the liposome mixture as in **step 5** Subheading 3.1.1.
4. Transfer the extruded liposome mixture to a snakeskin dialysis tube with 3.5 kDa molecular weight cutoff and dialyze against 1000× bulk phosphate buffer (10 mM, *p*H 7.4).
5. Repeat the dialysis procedure three times with buffer exchange to ensure complete removal of uncaptured calcein (*see* **Note 8**).
6. After dialysis, add Aβ peptide to the liposome solution as in Subheading 3.1.2.
7. Transfer 110 μL of Aβ-liposome (calcein contained) solution to the fluorescence cuvette and pre-warm the solution to 37 °C in the cuvette holder.
8. Record a kinetic fluorescence trace of the Aβ-liposome-calcein mixture as in Subheading 3.1.3, using excitation and emission wavelengths set to 495 nm and 515 nm, respectively.
9. Record control calcein fluorescence traces for the liposomes without Aβ and subtract the control trace from the sample. A representative trace is shown in Fig. 2a above the ThT fluorescence.

3.1.6 Monitoring of the Global Secondary Structures of Aβ Using CD Spectroscopy

1. Incubate Aβ-liposome preparations at 37 °C quiescently.
2. Dilute a 12 μL aliquot solution to 300 μL using deionized water (*see* **Note 9**), and transfer it into a 1 mm path-length CD cuvette.
3. Record the CD spectrum from 190 nm to 260 nm wavelength range with at least a 1 nm/s scanning rate and 40-scan signal averaging.

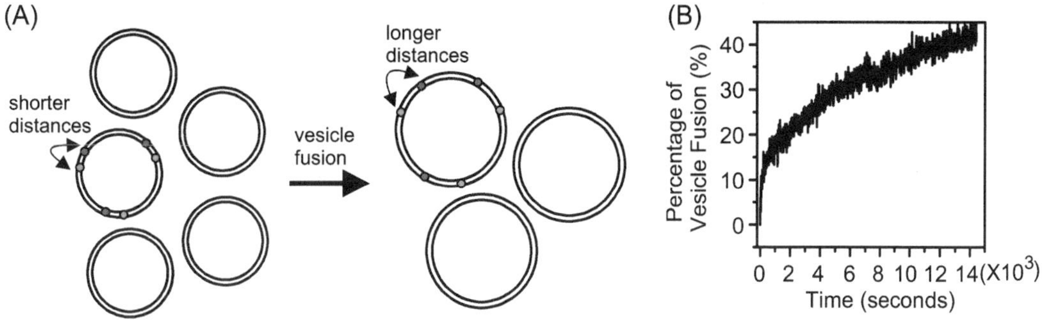

Fig. 3 Fluorescence measurements on vesicle fusion. (**a**) Fluorescence assay that has been utilized to measure the lipid mixing. The distances between fluorophore-labeled lipids (black spheres) and quenching lipids (gray spheres) become larger during vesicle fusion. (**b**) Representative fluorescence emission curve of lipid mixing for the "external addition" sample with P:L ratio 1:120

4. Control CD spectrum was recorded with only POPC liposomes in the absence of Aβ, and subtracted from the spectrum for the POPC/Aβ samples. A representative CD spectrum with 20-h incubation time for this sample was shown in Fig. 2c.

3.2 Preparation and Characterization of the "External Addition" Aβ-Liposome Model System with Dominant Lipid Mixing Pathway

The main difference between the fibrillation pathways and lipid mixing, in terms of the sample preparation protocol, is the P:L molar ratio. The identification and characterization of these two competing pathways were described in our published papers [7, 13, 14, 16]. For this protocol, we will focus on the fluorescence assay that has been utilized to characterize the vesicle fusion in the Aβ-liposome model system (sketched in Fig. 3a).

3.2.1 Liposome Preparation

1. Premix the fluorescent phospholipid mixture containing 80% *v/v* non-fluorophore-labeled liposomes (only POPC), 20% fluorophore-labeled liposomes (POPC with NBD-PE and Rh-PB) as shown in Subheading 2.3.
2. Dry the film in a similar fashion as described in **step 2** of Subheading 3.1.1.
3. Take up the mixture in 10 mM phosphate buffer and form liposomes using the same protocols in Subheading 3.1.1, **steps 4–5**.
4. Add Aβ at a concentration of 1:120 P:L ratio (*see* **Note 10**).

3.2.2 Measurements of Lipid Mixing Using Fluorescence

1. Warm up the fluorimeter to 37 °C and equilibrate for 1 h.
2. Transfer a 110 μL aliquot of the Aβ-liposome mixture to the fluorescence cuvette.
3. Incubate the mixture in the cuvette holder for 5 min and record the fluorescence trace with the excitation and emission wavelengths at 480 nm and 585 nm, respectively.
4. Record control fluorescence traces for liposomes without the addition of Aβ and subtract them from the sample.

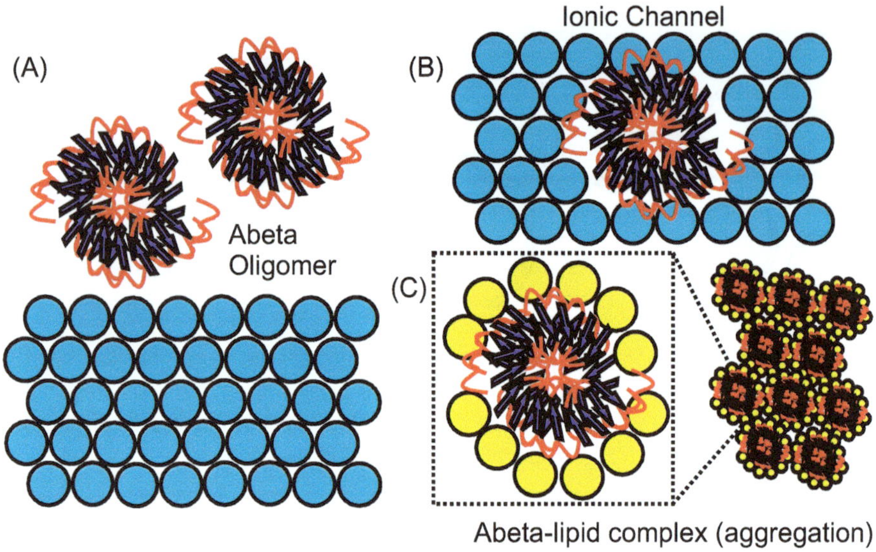

Fig. 4 Proposal of membrane disruption for the "pre-incorporation" samples. (**a**) Absorption of Aβ oligomers by membrane (blue spheres, top view). (**b**) Incorporation of Aβ oligomers in the membrane bilayer forms ion channels at low P:L ratio. (**c**) Elution of Aβ oligomers from membrane bilayer results in Aβ-lipid complex and large aggregation

5. To quantify the percentage of vesicle fusion, after the fluorescence signal reaches plateau, add a 10 μL aliquot of 20% SDS solution (SDS dissolved in deionized water) to the fluorescence cuvette.
6. Incubate the mixture for 5 min and record the fluorescence emission level. This value represents 100% lipid mixing [17, 18]. A representative fluorescence trace for this sample was shown in Fig. 3b.

3.3 Preparation and Characterization of the "Pre-incorporation" Aβ-Liposome Model System with Dominant Membrane Fragmentation Pathway

As shown in Fig. 4, using a different sample preparation protocol, named "pre-incorporation," the model system adopted a completely different membrane disruption pathway. We have identified that membranes are largely fragmentized in this model system, resulting in an aggregation composed of Aβ-lipid complex [7, 12–14]. During this process, ion channels may be temporarily formed and stabilized within the membrane bilayer at low P:L ratio such as 1:120, but not at high ratio such as 1:30. In this protocol, we will describe the "pre-incorporation" sample preparation procedure and the characterization of the formation of Aβ-lipid complex and ion channels.

3.3.1 Co-dissolving of Lipids and Aβ

1. Dissolve Aβ in HFIP (*see* **Note 11**).
2. Mix the desired amount of lipids (in chloroform) and Aβ (in HFIP).

3. Remove the organic solvents using N_2 flow.
4. Dry the film by completely removing residual solvents in a vacuum desiccator for at least 8 h.

3.3.2 Liposome Preparation

1. Resuspend the dried film in 10 mM phosphate buffer. The Aβ concentration should be kept at 2.0 mg/mL.
2. Rotate the mixture at ambient temperature for 1 h.
3. Perform 10 freeze-thaw cycles on the mixture using liquid N_2 and bath sonicator (*see* **Note 12**).
4. Extrude the mixture using a 100 nm pore-size polycarbonate membrane (*see* **Note 13**). For "pre-incorporation" samples, the Aβ-contained liposome has to be prepared freshly before measurements and cannot be stored.
5. Incubate the pre-incorporation mixture at 37 °C quiescently for further testing, such as the CD and fluorescence assays described in previous sections.

3.3.3 Visualization of the Co-localization of Lipids and Aβ Peptides

1. Prepare the "pre-incorporation" Aβ-liposomes using the protocols in Subheadings 3.3.1 and 3.3.2 using rhodamine-green-labeled Aβ and 1 mol% Rh-PE (molar percentage to the total lipids), with the exception that a 1.0 μm pore-size polycarbonate membrane was utilized for the extrusion (*see* **Note 14**).
2. Add 400 μL of Aβ-liposome suspension to glass bottom confocal dish (MatTek Corporation).
3. Capture the liposome images using a laser scanning confocal microscope (Leica TCS SP2, Germany) with HCX PL APO 63 × 1.4 oil immersion objective by 4 time digital zoom-in, with 488 nm and 543 nm excitation and emission wavelengths. A representative fluorescence image obtained on the "pre-incorporation" sample with P:L 1:30 was provided in Fig. 5a.

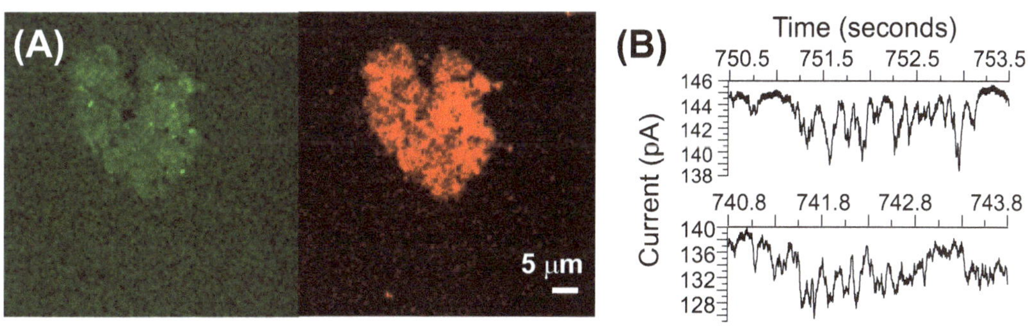

Fig. 5 Membrane fragmentation and ion channels induced by "pre-incorporation" Aβ in liposomes. (**a**) Confocal fluorescence images of Aβ-lipid aggregation (models presented in Fig. 4c) which showed the co-localization of lipid (red channel, Rh-PE) and peptides (green channel, Rh-green-labeled Aβ). (**b**) Representative electrophysiological current traces of samples prepared using protocol **steps 5–7** in Subheading 3.3

3.3.4 Measurements of Ion Channel Activities

1. Dry 20 μL, 25 mg/mL POPC in chloroform using N_2 flow.
2. Resuspend the lipid film in 75 μL, 10 mM phosphate buffer with brief sonication.
3. Add 11.7 μL, 2 mg/mL freshly dissolved Aβ in deionized water.
4. Perform 10 freeze-thaw cycles on the mixture using liquid N_2 and water bath sonicator for 10 cycles (*see* **Note 15**).
5. Sample preparation with cross-linking:
   (a) Prepare the Aβ-liposome using **steps 1–4**. After freeze-thaw, add 15.7 μL of glutaraldehyde (70% *v/v* in water) to the liposome solution to cross-link any membrane-associated Aβ oligomers [9, 19].
   (b) React at ambient temperature for 20 min.
   (c) Quench the reaction with 21.7 μL Tris buffer (1.0 M, *p*H 7).
6. Black lipid membrane (BLM) for ion channel measurements:
   (a) Dry a 50 μL aliquot of POPC in chloroform with N_2 flow.
   (b) Re-dissolve the lipid film in 75 μL *n*-decane.
   (c) Fill the electrophysiological chamber with 10 mM HEPES buffer.
   (d) Deposit a 10 μL aliquot POPC/decane solution onto the 200 μm aperture between two compartments in the electrophysiological chamber to form the BLM (*see* **Note 16**).
   (e) Place the cathode and anode of an Ag/AgCl electrode in the inner and outer compartments of the electrophysiological chamber, respectively.
   (f) Allow the BLM to stabilize for 10 min, and record the baseline.
   (g) Add the Aβ liposomes into the outer compartment of the chambers (*see* **Note 17**).
   (h) Add 3 μL, 100 mM $CaCl_2$ to both compartments to facilitate the fusion between liposomes and BLM [20].
   (i) Gently mix the solution with pipette tips and recording of current trace was continued for 30–90 min. Two representative electronic traces were shown in Fig. 5b.

4 Notes

1. For the synthesis of 40-residue Aβ peptides, it is important to use resin with relatively low substitution because the primary sequence is highly hydrophobic. We found that ~0.3 mmol/g substitution works fine for our synthesis.

2. The amount of POPC was calculated specifically for individual sample preparation based on the desired amount of Aβ and the P:L ratio.
3. The ThT solution was prepared by directly dissolving ThT powder into the deionized water, bath-sonicating for 5 min at ambient temperature and filtering through 0.22 μm syringe filter unit. This solution should be prepared fresh.
4. Phosphate buffer was added so that the final concentration of Aβ was ~ 50 μM for all our samples.
5. For practicality, it is better to dilute the concentrated DMSO/Aβ solution with phosphate buffer rather than adding the DMSO solution drop by drop into the buffer. This is to avoid instant aggregation of Aβ due to the slow dispersion of DMSO in water.
6. Technically, we have utilized ThT fluorescence to measure the initial stage of fibrillation instead of the elongation of Aβ fibrils.
7. The staining time and the concentration of uranyl acetate has been optimized to provide best contrast for the TEM imaging for the Aβ-liposome samples. These parameters must be optimized for other samples.
8. The first and second dialysis were performed at ambient temperature for 2 h, and the third round was done at 4 °C overnight. We observed that the green color (calcein) in the dialysis buffer faded over multiple rounds.
9. The dilution of Aβ-liposome solution was necessary for our CD spectrometer setup. Otherwise the driving voltage for our instrument tends to be too high. It may not be necessary for other instrumental setups, and should be optimized.
10. This particular P:L ratio gives dominant vesicle fusion effect over the fibrillation, based on our published work [13].
11. The exact volume of HFIP was not critical here. We typically use 1.0 mL to ensure all Aβ peptides are dissolved.
12. For POPC, the bath sonication was performed at ambient temperature, because this lipid has a phase transition temperature at ~ −2 °C. If other types of lipid with higher phase transition temperature are utilized. A heated water bath may be necessary.
13. For the "pre-incorporation" sample, we found it was necessary to do the extrusion with 1.0 μm pore-size membrane followed by 100 nm pore-size membrane.
14. In this case, the larger pore-size membrane was utilized because of the resolution obtained on our fluorescence imaging instrumental setup.

15. Experimentally, we observed that the mixture becomes transparent after 10 cycles of freeze-thaw.
16. Since the aperture is submerged below the surface of buffer, it is critical to insert the pipette tip under the liquid surface. Let the tip hang slightly above the aperture and slowly deposit the POPC/decane, using a microscope to visualize the process. We found that it was necessary to let the dispersed decane cover the aperture for a few seconds to obtain the best BLM. In addition, the concentration of POPC in decane seems to be critical.
17. The amount of Aβ-liposome that has been added in our experiment was 60 μL, and we found that it gave best results in the recorded electronic traces. This parameter should be optimized experimentally [21].

Acknowledgments

This work is supported by the start-up funding from the Research Foundation of the State University of New York. We acknowledge Dr. He Dong and Mr. Dawei Xu from Clarkson University, and Dr. Christof Grewer from Binghamton University for their kind helps on fluorescence imaging and electrophysiological current measurements.

Reference

1. Williams TL, Serpell LC (2011) Membrane and surface interactions of Alzheimer's Abeta peptide—insights into the mechanism of cytotoxicity. FEBS J 278:3905–3917
2. Qiang W, Kelley K, Tycko R (2013) Polymorph-specific kinetics and thermodynamics of beta-amyloid fibril growth. J Am Chem Soc 135(18):6860–6871
3. Sengupta P, Garai K, Sahoo B, Shi Y, Callaway DJ, Maiti S (2003) The amyloid beta peptide (Abeta(1-40)) is thermodynamically soluble at physiological concentrations. Biochemistry 42 (35):10506–10513
4. Hasegawa K, Ono K, Yamada M, Naiki H (2002) Kinetic modeling and determination of reaction constants of Alzheimer's beta-amyloid fibril extension and dissociation using surface plasmon resonance. Biochemistry 41 (46):13489–13498
5. Bokvist M, Lindstrom F, Watts A, Grobner G (2004) Two types of Alzheimer's beta-amyloid (1-40) peptide membrane interactions: aggregation preventing transmembrane anchoring versus accelerated surface fibril formation. J Mol Biol 335:1039–1049
6. Sciacca MF, Kotler SA, Brender JR, Chen J, Lee DK, Ramamoorthy A (2012) Two-step mechanism of membrane disruption by Abeta through membrane fragmentation and pore formation. Biophys J 103(4):702–710
7. Qiang W, Yau WM, Schulte J (2015) Fibrillation of beta amyloid peptides in the presence of phospholipid bilayers and the consequent membrane disruption. Biochim Biophys Acta Biomembr 1848(1):266–276
8. Sparr E, Engel MFM, Sakharov DV, Sprong M, Jacobs J, de Kruijff B, Hoppener JWH, Killian JA (2004) Islet amyloid polypeptide-induced membrane leakage involves uptake of lipids by forming amyloid fibers. FEBS Lett 577:117–120
9. Quist A, Doudevski I, Lin H, Azimova R, Ng D, Frangione B, Kagan B, Ghiso J, Lal R (2005) Amyloid ion channels: a common sturctural link for protein-misfolding disease. Proc Natl Acad Sci 102(30):10427–10432
10. Connelly L, Jang H, Arce FT, Ramachandran S, Kagan BL, Nussinov R, Lal R (2012) Effects of point substitutions on the structure of toxic Alzheimer's beta-amyloid

channels: atomic force microscopy and molecular dymanics simulations. Biochemistry 51 (14):3031–3038

11. Lal R, Lin H, Quist AP (2007) Amyloid beta ion channel: 3D structure and relevance to amyloid channel paradigm. Biochim Biophys Acta 1768(8):1966–1975
12. Delgado DA, Doherty KE, Cheng Q, Kim H, Xu D, Dong H, Grewer CT, Qiang W (2016) Distinct membrane disruption pathways induced by the 40-residue beta-amyloid peptides. J Biol Chem 291(23):12233–12244
13. Akinlolu RD, Nam M, Qiang W (2015) Competition between fibrillation and induction of vesicle fusion for the membrane-associated 40-residue b amyloid peptides. Biochemistry 54(22):3416–3419
14. Qiang W, Akinlolu RD, Nam M, Shu N (2014) Strcutral evolution and membrane interaction of the 40-residue beta amyloid peptides: differences in the initial proximity between peptides and the membrane bilayer studied by solid-state nulear magnetic resonance spectroscopy. Biochemistry 53 (48):7503–7514
15. Qiang W, Yau WM, Tycko R (2011) Structural evolution of Iowa mutant b-amyloid fibrils from polymorphic to homogeneous states under repeated seeded growth. J Am Chem Soc 133(11):4018–4029
16. Vestergaard MC, Morita M, Hamada T, Takagi M (2013) Membrane fusion and vesicular transformation induced by Alzheimer's amyloid beta. Biochim Biophys Acta 1828 (4):1314–1321
17. Qiang W, Sun Y, Weliky DP (2009) A strong correlation between fusogenicity and membrane insertion depth of the HIV fusion peptide. Proc Natl Acad Sci 106 (36):15314–15319
18. Yang R, Prorok M, Castellino FJ, Weliky DP (2004) A trimeric HIV-1 fusion peptide construct which does not self-associate in aqueous solution and which has 15-fold higher membrane fusion rate. J Am Chem Soc 126 (45):14722–14723
19. Lin H, Bhatia R, Lal R (2001) Amyloid beta protein forms ion channels: implications for Alzheimer's disease pathophysiology. FASEB J 15:2433–2444
20. Ostroumova OS, Schagina LV, Mosevitsky MI, Zakharov VV (2010) Ion channel activity of brain abundant protein BASP1 in planar lipid bilayers. FEBS J 278(3):461–469
21. Delgado DA, Doherty K, Cheng Q, Kim H, Xu D, Dong H, Grewer C, Qiang W (2016) Distinct membrane disruption pathways are induced by the 40-residue Î^{2}-amyloid peptides. J Biol Chem 291(23):12233–12244

Chapter 24

Using Molecular Tweezers to Remodel Abnormal Protein Self-Assembly and Inhibit the Toxicity of Amyloidogenic Proteins

Ravinder Malik, Jing Di, Gayatri Nair, Aida Attar, Karen Taylor, Edmond Teng, Frank-Gerrit Klärner, Thomas Schrader, and Gal Bitan

Abstract

Molecular tweezers (MTs) are broad-spectrum inhibitors of abnormal protein self-assembly, which act by binding selectively to lysine and arginine residues. Through this unique mechanism of action, MTs inhibit formation of toxic oligomers and aggregates. Their efficacy and safety have been demonstrated in vitro, in cell culture, and in animal models. Here, we discuss the application of MTs in diverse in vitro and in vivo systems, the experimental details, the scope of their use, and the limitations of the approach. We also consider methods for administration of MTs in animal models to measure efficacy, pharmacokinetic, and pharmacodynamic parameters in proteinopathies.

Key words Amyloidosis, Proteinopathy, Amyloid, Protein aggregation, Synaptic toxicity, Blood--brain barrier, Alzheimer's disease, Parkinson's disease

1 Introduction

Proteinopathies are characterized by abnormal folding and self-assembly of proteins into cytotoxic oligomers and aggregates. There are over 30 proteinopathies [1, 2], of which prominent examples are Alzheimer's disease (AD) and Parkinson's disease (PD). In addition, abnormal protein self-assembly contributes to the pathologic process in diverse conditions, such as human immunodeficiency virus (HIV) infection [3], traumatic brain injury [4], cancer [5], and preeclampsia [6]. The aberrant oligomerization and aggregate formation may occur intra- or extracellularly, and the resulting protein assemblies disrupt various processes in susceptible cells and organs of affected individuals [7]. The reasons for the preferential targeting of particular cells or tissues by each amyloidogenic protein are not known and the exact mechanisms by which the abnormal protein assemblies cause disease are still untangling.

Bradley L. Nilsson and Todd M. Doran (eds.), *Peptide Self-Assembly: Methods and Protocols*, Methods in Molecular Biology, vol. 1777, https://doi.org/10.1007/978-1-4939-7811-3_24,

To date, there are no therapeutic strategies available to cure proteinopathies [1]. In most cases, no therapy is available at all. For one rare disease, familial amyloidotic polyneuropathy, the drugs tafamidis and diflunisal slow the disease progression [8]. For other diseases, such as AD, PD, and amyotrophic lateral sclerosis, drugs are available that offer moderate and temporary symptomatic relief, but not disease-modifying therapy. Thus, there is a pressing need to obtain a deeper understanding of disease mechanisms and develop strategies for effective prevention and treatment of proteinopathies.

Common structural and pathogenic features of aggregating proteins in various diseases can be targeted by therapeutic strategies such as antibodies, chaperones, and small molecules [9–11]. Our chapter discusses one class of small molecules—molecular tweezers (MTs), which are unique compounds with promising activity against multiple amyloidogenic proteins [12, 13]. MTs bind to amyloidogenic proteins regardless of assembly state and remodel the assembly process into formation of nontoxic and non-amyloidogenic structures that can be degraded efficiently by the natural clearance mechanisms. The current lead MT derivative, CLR01 (Fig. 1), has been used successfully to inhibit abnormal protein aggregation and dissociate pre-formed aggregates in vitro, protect cultured cells from the toxicity of various amyloidogenic proteins, and provide therapeutic effects in multiple animal models [12, 13]. Importantly, CLR01 has been shown to have a high safety margin [14], supporting the use of MTs as attractive therapeutic drug candidates for proteinopathies.

1.1 Molecular Tweezers and Their Mechanism of Action

Lysine-specific molecular tweezers were first reported by Fokkens et al. in 2005 [15] and their ability to remodel abnormal protein aggregation and prevent the toxicity of the resulting oligomers and aggregates was discovered subsequently by the Bitan group [16]. Labile binding of MTs to lysine residues competes with a combination of key electrostatic and hydrophobic interactions that mediate the formation of toxic oligomers and fibrillation nuclei of amyloidogenic proteins. Because the binding is highly labile [17] and occurs with micromolar affinity, it does not interfere with the structure or function of normal proteins (unless substantially higher concentrations are used). This unique mechanism of action of MTs is not specific to one particular protein and thus they can be used as broad-spectrum nanochaperones that inhibit assembly and toxicity of many amyloidogenic proteins.

Recently, a new function was discovered for CLR01—disruption of viral membranes [18], which appears to be distinct from the effect of the compound on amyloidogenic proteins. Presumably, the specificity of the effect to viral membranes is related to the high concentration of cholesterol and sphingomyelin in these membranes relative to membranes of mammalian cells [18], but the mechanistic details are not known and it is not yet clear whether this activity is unique to CLR01 or common to other MTs. Additional support for the membrane activity of CLR01 also was found in a study of the effect of different assembly modulators on Aβ42 in the presence of lipid membranes [19].

1.2 In Vitro Studies

CLR01 has been the main derivative explored to date for its ability to remodel abnormal protein self-assembly and inhibit the toxicity of the resulting oligomers and aggregates. Several studies have shown that CLR01 binds with high selectivity to lysine and arginine residues in amyloidogenic proteins [16, 17, 20, 21] and does not act by a nonspecific, colloidal mechanism [18]. Mass spectrometry (MS) coupled with electron-capture dissociation, and solution-state NMR studies were used to identify the main binding sites in amyloid β-protein (Aβ) [16] and α-synuclein (α-syn) [20], whose aggregation is involved in AD and PD, respectively.

In most of the in vitro and cell culture experiments, a control derivative named CLR03, which lacks the hydrophobic side arms of the tweezer structure (Fig. 1), was used as a negative control. For example, oligomerization and aggregation of Aβ42 were disrupted effectively by CLR01, whereas CLR03 was inactive, as expected [16]. The aggregation kinetics were followed using thioflavin T (ThT) fluorescence [22] and the morphology of the aggregates characterized by electron microscopy. Formation of oligomers was studied by dot-blots using the anti-oligomer antibody, A11 [23]. More recently, ion-mobility spectroscopy-coupled mass-spectrometry experiments showed that CLR01 disrupted specifically formation of Aβ42 hexamers and dodecamers [24]. Importantly, these oligomers previously have been shown to be key toxic structures linked to Aβ42 toxicity [25–28].

CLR01, but not CLR03, inhibited the aggregation of α-syn and disaggregated pre-formed α-syn fibrils when added either in the middle of the growth phase or after mature fibrils formed

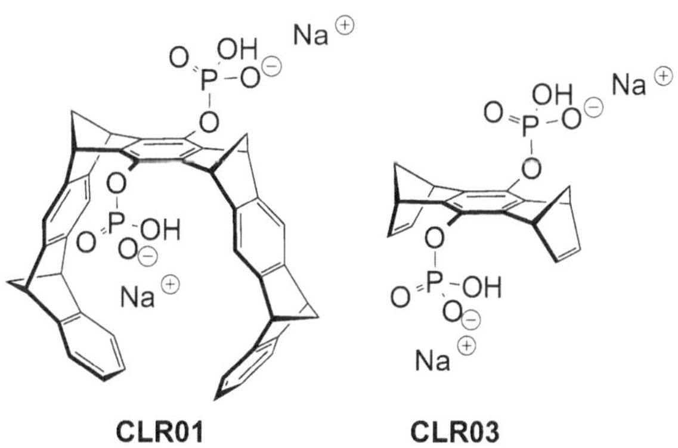

CLR01 **CLR03**

Fig. 1 Chemical structure of CLR01 and CLR03. These compounds are slightly basic and are partially protonated at physiological pH

[20]. Fibril dissociation was a slow reaction completed over ~2 months and required a tenfold excess of CLR01, consistent with the mechanism of action of MTs. Similar results were observed with Aβ40 or Aβ42 [16]. In the case of islet amyloid polypeptide, a highly amyloidogenic peptide associated with pancreatic β-cell death in type-2 diabetes, CLR01 halted fibril growth but did not dissociate preexisting fibrils [21]. In contrast, fibrils of amyloidogenic peptides derived from semen proteins, which greatly enhance HIV infection, were dissociated by CLR01dose-dependently within ~2 h [18]. These results suggest that CLR01's ability to dissociate pre-formed fibrils of amyloidogenic proteins depends strongly on the thermodynamic and kinetic stability of the fibrils.

To date, CLR01 has been found to inhibit the aggregation of 16 different amyloidogenic proteins, including tau, transthyretin (TTR), β_2-microglobulin, insulin, calcitonin [16, 29], and several others that have not yet been published. Recently, aggregation of amyloidogenic p53 mutants has been shown to be disrupted by CLR01, suggesting the potential application of MTs as anticancer agents [30].

1.3 Cell Culture Studies

Thanks to their ability to modulate the self-assembly process of amyloidogenic proteins, MTs would be expected to inhibit the toxicity of oligomers and aggregates of these proteins. Indeed, in all the cases tested to date, CLR01 showed dose-dependent inhibition of the toxicity of diverse proteins in cell viability assays using different cell lines [16, 20, 21, 29, 30].

CLR01 also was found to inhibit synaptic toxicity caused by Aβ42. When added to primary cortical or hippocampal neurons at nanomolar or low micromolar concentrations, Aβ42 caused retraction of dendritic spines and inhibition of both basal synaptic activity and long-term potentiation (LTP), a cellular correlate of learning and memory [31]. CLR01 prevented these toxic effects effectively when added at tenfold excess to the neuronal culture. Dendritic spine retraction and varicosities were alleviated (Fig. 2a) and spine density was rescued to ~80% of the baseline level in the presence of CLR01 (Fig. 2b). Similarly, basal synaptic activity was rescued and LTP was restored to ~80% of baseline [31].

A surprising activity of CLR01 was observed when it was tested with seminal amyloid proteins. CLR01 inhibited the formation of infectivity-enhancing seminal amyloids and dissociated pre-formed fibrils. Interestingly, it also inhibited viral infection directly when tested in TZM-bl cells [18]. CLR01 prevented the formation of virion-amyloid complexes and directly disrupted the membrane integrity of the viruses. Following these surprising findings, the antiviral activity of CLR01 was tested further in other viruses. Infection by the enveloped viruses herpes simplex virus type 2 and hepatitis C virus was inhibited, whereas infection by the nonenveloped human cytomegalovirus was not, supporting the direct effect

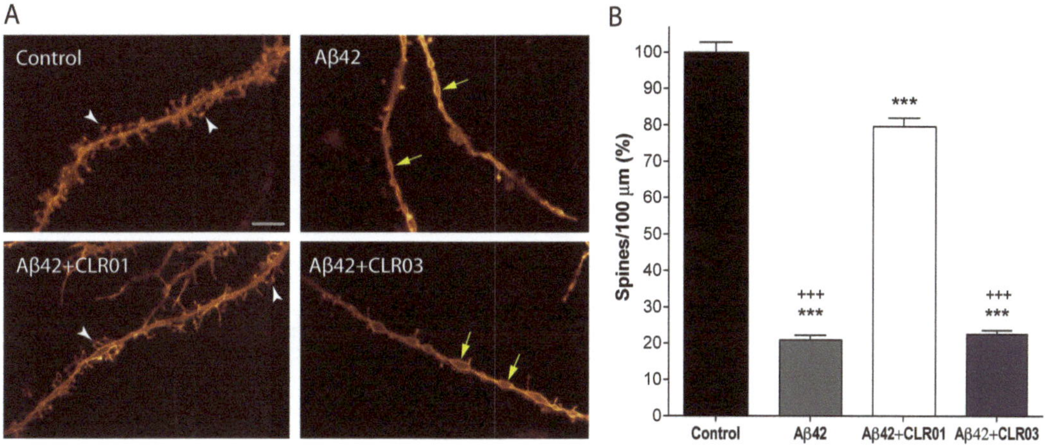

Fig. 2 CLR01 protects neurons from Aβ42-induced changes in dendritic spine number and morphology. (**a**) Rat primary hippocampal neurons incubated for 72 h with media alone or with Aβ42 in the absence or presence of CLR01 or CLR03. Arrows point to Aβ42-induced varicosities. Scale bar = 5 μm. (**b**) The number of dendritic spines per 100 μm was quantified. $^{***}p < 0.001$ compared with control; $^{+++}p < 0.001$ compared with Aβ42 + CLR01. Published originally by Oxford University Press: Brain 135(12): 3735–3748, *Protection of primary neurons and mouse brain from Alzheimer's pathology by molecular tweezers*

of CLR01 on viral membranes and suggesting that the compound could be used as a broad-spectrum antiviral microbicide [18].

1.4 In Vivo Studies

The in vivo efficacy of CLR01 was examined in animal models of several proteinopathies. The compound was evaluated in a triple-transgenic (3 × Tg) mouse model of AD, which overexpresses mutant human presenilin 1 (PS1(M146 V)), amyloid β-protein precursor (APP(KM670/671NL)), and tau(P301L). The mice develop amyloid plaques and neurofibrillary tangles in AD-relevant brain regions (hippocampus, cortex, amygdala) and deficits in synaptic plasticity and memory [32]. 15-month-old 3 × Tg mice were treated for 28 days with 0.04 mg/kg per day CLR01 in saline (vehicle), or with vehicle alone, applied subcutaneously (s.c.), continuously using osmotic minipumps [31]. Immunohistochemical (IHC) analysis of brain sections of mice showed a significant, ~33% decrease in Aβ burden in the hippocampus and cortex of CLR01-treated mice compared to vehicle-treated mice (Fig. 3a–c). Hyperphosphorylated tau (p-tau), but not total tau, also decreased substantially following CLR01 treatment (Fig. 3d–h). Presumably downstream of CLR01's effect on Aβ and p-tau burden, levels of microgliosis also decreased with CLR01 treatment, with no effect found on levels of microglia in CLR01- or vehicle-treated wild-type animals (Fig. 3i–k).

Recently, in a small study in a Tg rat model of AD, which expresses familial AD-linked mutant forms of human APP (K670N/M671L/V717I) and PS1 (M146V) [33], CLR01 was

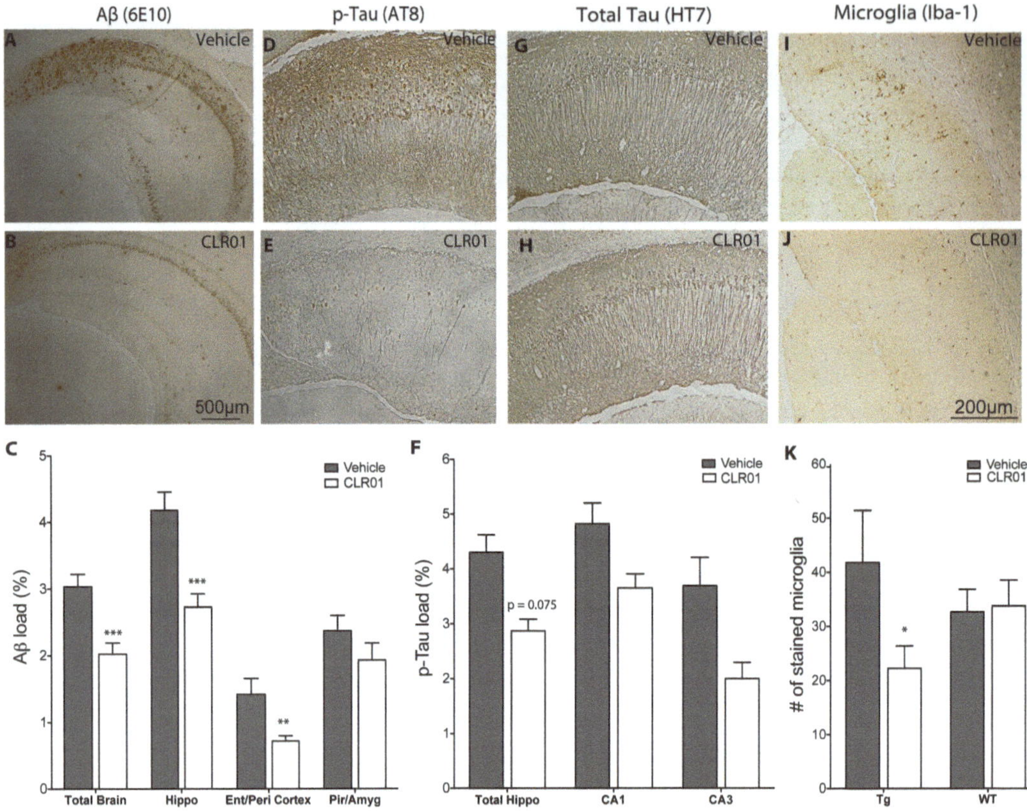

Fig. 3 CLR01 decreases abnormal Aβ and tau deposition and ameliorates microgliosis in Tg mouse brain. 3 × Tg mice were treated with 0.04 mg/kg per day CLR01 or vehicle. (**a**) Vehicle-treated Tg brain stained with monoclonal antibody (mAb) 6E10 showing amyloid plaques and intraneuronal Aβ deposition predominantly in the hippocampus. (**b**) 6E10-stained CLR01-treated Tg brain. (**c**) % Aβ burden quantified by calculating the total 6E10-stained area divided by the total area measured (Hippo—hippocampus, Ent—entorhinal, Peri—perirhinal, Pir—piriform cortex, Amyg—amygdala). (**d**) Vehicle-treated brain showing mAb AT8-positive neurofibrillary tangles in the CA1 region. (**e**) AT8-stained CLR01-treated brain. (**f**) % Aggregated p-tau load quantified by calculating the total AT8-stained area divided by the total area. (**g**) Vehicle-treated brain stained with mAb HT7 for total tau. (**h**) HT7-stained CLR01-treated brain. (**i**) Vehicle-treated brain showing Iba1-positive activated microglia in the subiculum and CA1 region. (**j**) Anti-Iba1-stained CLR01-treated brain. (**k**) Activated microglia load (# of stained microglia in a 1.14 mm^2 area) per treatment condition. The scale bar in panel B is applicable also to A and in J also to panels D–I. $^{*}p < 0.05$, $^{**}p < 0.01$, $^{***}p < 0.001$ compared to vehicle-treated mice. Published originally by Oxford University Press: Brain 135(12): 3735–3748, *Protection of primary neurons and mouse brain from Alzheimer's pathology by molecular tweezers*

administered in a similar manner to the 3 × Tg mouse experiment described above, at 0.1 or 0.3 mg/kg/day. The animals were mixed-gender and were treated at 9-months of age, an age at which they are expected to have moderate plaque pathology [34]. Due to the relatively young age of the animals, the plaque pathology found by IHC using the Aβ-specific mAb MOAB-2, which recognizes Aβ but not APP [35], was highly variable (Fig. 4d). Nonetheless, the treatment led to 45% and 52%

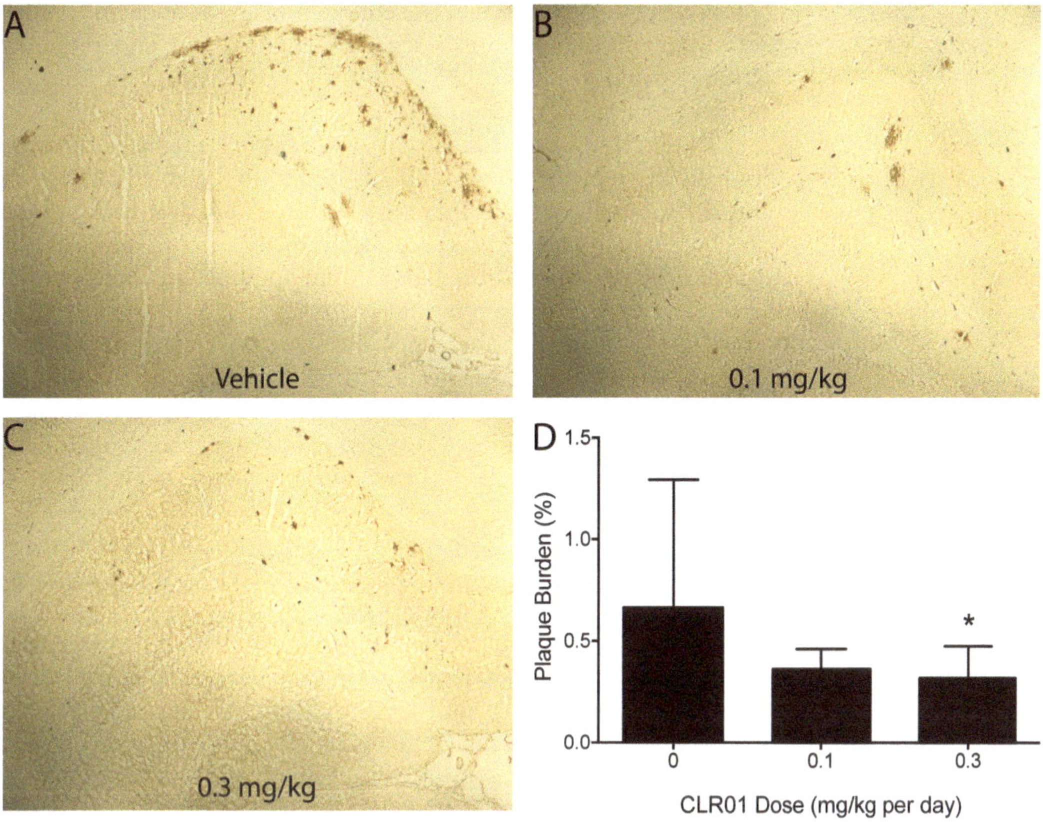

Fig. 4 CLR01 reduces amyloid plaque burden in Tg AD rats. (**a–c**) Representative brain sections. (**d**) % Plaque burden quantified by calculating the total MOAB-2-stained area divided by the total area measured. *$p < 0.05$

reduction in plaque burden in the 0.1- and 0.3-mg/kg/day treatment groups, respectively (Fig. 4, whole brain analysis).

In the context of Parkinson's disease, CLR01 was tested in a zebrafish (ZF, *Brachydanio rerio*) embryo model. In this model, neuronal expression of human, wild-type α-syn led to severe deformation and death within 48–72 hours post fertilization (hpf). Addition of 1 or 10 μM CLR01 to the water in which the embryos developed at 8 hpf caused a dramatic improvement in phenotype and survival [20]. IHC analysis showed that in untreated ZF, α-syn formed abundant cytoplasmic aggregates, whereas in CLR01-treated fish α-syn was completely soluble in the cytoplasm. Interestingly, the treatment led to ~80% reduction in total α-syn concentration levels in the ZF neurons. Additional experiments using proteasome inhibitors or a GFP-coupled degron system showed that by keeping α-syn from aggregating, CLR01 enabled its rapid clearance, predominantly by the 26S ubiquitin-proteasome system (UPS) [20]. A recent subsequent study showed that the pesticide Ziram, which increases significantly the risk of developing PD [36], caused selective aminergic neuronal death in ZF embryos, which was linked to aggregation of the endogenous ZF synuclein. CLR01 was found to significantly rescue the survival and phenotype of Ziram-treated embryos [37], similarly to its effect in the ZF model expressing human α-syn.

To determine whether CLR01 was effective against TTR amyloidosis in vivo, the compound was tested in Tg mice expressing human mutant TTR(V30M) on a mouse TTR-null background and heterozygous for deletion of the heat shock transcription factor 1 (HSF1)—a model of familial amyloidotic polyneuropathy [38]. The mice develop progressive amyloidosis in the gastrointestinal (GI) tract and peripheral nervous system. Treatment with 1.2-mg/kg/day CLR01 via s.c. osmotic minipumps for 35 days led to a significant decrease in TTR deposition in the stomach, colon, and dorsal-root ganglia, and in associated markers of disease, including apoptosis, endoplasmic reticulum stress, and protein oxidation [29].

The safety of CLR01 was evaluated in both acute and chronic administration experiments in wild-type mice [14]. Acute administration of 100 mg/kg CLR01 caused obvious signs of distress, primarily hunching and freezing, which was alleviated completely by 2 h following administration. 10 mg/kg did not appear to cause any distress. Histological and serological analysis showed expected liver injury, but not damage to other organs. No mortality was recorded in either of the groups. In follow-up chronic administration experiments, 10 mg/kg CLR01 for 30 days yielded no signs of discomfort. There were no histological findings and the only significant serum change was ~40% decrease in cholesterol [14]. These findings indicate that CLR01 has a high safety margin in mice.

2 Materials

2.1 In Vitro Studies

1. Active MTs, e.g., CLR01, in a powder form [39].
2. CLR03 in a powder form (as a negative control).
3. Appropriate buffer for dissolving MTs depending on the desired study. For details of different buffers used previously *see* **Note 1** in Subheading 4.1. For a discussion of solubility, refer to **item 1** in Subheading 3.1.
4. Protein sample under study. *See* **Note 2** in Subheading 4.1.
5. Appropriate assay for monitoring the effect of MTs (*see* **item 5** in Subheading 3.1).

2.2 Cell Culture Studies

1. Active MTs in a powder form.
2. CLR03 in a powder form (as a negative control).
3. Sterile water for making MT stocks.
4. Cell culture system for studying the effects of MTs.

5. Growth medium, growth supplements, growth factors, antibiotics, differentiation medium as appropriate for the cell line under study.
6. Protein under study, if the protein is added exogenously (*see* **Note 2** in Subheading 4.2).
7. An appropriate assay for monitoring the effect of MTs.

2.3 In Vivo Studies

The materials described below are two examples: (1) measuring blood–brain barrier (BBB) penetration of CLR01 by spiking the compound with a radiolabeled derivative following different routes of administration; and (2) administering CLR01 s.c. via osmotic minipumps for efficacy experiments. In addition to s.c. injection, several other routes of administration have been used to administer CLR01 and may be used for other MTs, including intravenous injection (i.v.), oral gavage, and intraperitoneal injection (i.p.). If osmotic pumps are used, they can be of different sizes, depending on the animal size, route of administration, delivery rate, and the duration of the experiment. The example below uses the Alzet model 1004 pump (http://www.alzet.com/downloads/1004specs.pdf), which delivers 0.11 μL/h and typically is used for up to 28 days. However, per the manufacturer's instructions, the pump use can be extended up to 35 days.

1. For efficacy studies—osmotic minipumps (model 1004; Alzet).
2. Hemostat (Kent Scientific).
3. Wound clips (7-mm Reflex clips, Alzet).
4. For BBB studies—^{3}H-CLR01 (*see* **Note 1** in Subheading 4.3).
5. A scintillation counter (*see* **Note 2** in Subheading 4.3).
6. Solvable™ (Perkin Elmer).
7. Ultima Gold™ liquid scintillation cocktail (Perkin Elmer).
8. DecapiCones (Braintree Scientific, Inc.).
9. 28 gauge × 0.5″ syringe needles (BD).
10. 30% H_2O_2.

3 Methods

3.1 In Vitro Studies

The method described here is for a broad range of experiments. Changes can be made for specific experimental designs. The high aqueous solubility of CLR01 allows dissolution in the same buffer as the protein. If other MTs are used, the concentrations may need to be adjusted based on the solubility of the specific compound. The protein:MT molar ratio should be examined and potentially optimized for specific experiments depending on the protein under study. The following are common steps for evaluating the effect of

MTs in vitro on different aspects of amyloid proteins' self-assembly, such as oligomerization, aggregation, and remodeling of oligomers or amyloid fibrils.

1. Prepare a stock solution of the MTs used in the experiment. We recommend preparing the stock solution in deionized water. This way, it is easier to recover any unused compound without any contamination of salts or buffers. Typically, 10 mM stock solutions can be made for CLR01. The limit of solubility has not been determined formally, yet based on our experience, it is between 10 and 15 mM. The solubility of other MTs may vary substantially and should be determined for each compound.
2. Prepare the protein in a buffer system in which it is to be studied. In most cases, this should be the last step and experiments should begin immediately because many amyloidogenic proteins aggregate fast (*see* **Note 3** in Subheading 4.1).
3. Add MT to the protein under study at different molar ratios to obtain a dose–response relationship (*see* **Note 4** in Subheading 4.1). The negative control MT, CLR03, may be added only at the highest concentration ratio if appropriate (*see* **Note 5** in Subheading 4.1). A positive control is the protein alone, with no MTs added.
4. Incubate the protein:MT mixtures under suitable aggregation conditions. Online monitoring of the samples may be done by measuring various parameters, such as absorbance (e.g., for turbidity assays) or fluorescence (e.g., for thioflavin T assays), using common plate readers. If desired, aliquots of the reaction mixtures may be withdrawn periodically and analyzed separately.
5. Typical methods for analysis of the aggregation reaction, conformational change, and morphology of the protein assemblies include thioflavin T fluorescence, dynamic light scattering, circular dichroism spectroscopy, and electron microscopy [40, 41] (*see* **Note 6** in Subheading 4.1).

3.2 Cell Culture Studies

The steps below are general instructions for testing the effect of MTs in cell lines. Some steps may need to be modified to suit the specific cell line being used. Specific details for PC-12 cells, the cell we have used most often, are mentioned in the Notes section.

3.2.1 Plating Cells for Measurement

1. Harvest the cells in media when cell confluence reaches 70–80% of flask surface (*see* **Note 2** in Subheading 4.2).
2. Plate in a 96-well plate at a cell density of 30,000 cells/well (90 μL/well, *see* **Notes 3** and **4** in Subheading 4.2).
3. Incubate cells in a cell culture incubator at 37 °C and 5% CO_2 for 24–48 h (*see* **Note 5** in Subheading 4.2).

3.2.2 Adding the Protein and MTs to Cells and Assay

1. Prepare stock solutions of MTs at 1–10 mM concentration by dissolving in sterile water (*see* **Note 6** in Subheading 4.2), and then dilute in media for a dose–response experiment (*see* **Notes 7** and **8** in Subheading 4.2).
2. Make stock solutions of the protein under study if the protein is added exogenously (*see* **Note 1** in Subheading 4.2).
3. Remove a small volume of the media (approximately 40 μL) and add the protein:MT mixture to make the final total volume in each well 100 μL (*see* **Note 9** in Subheading 4.2).
4. Incubate the cells in a cell culture incubator at 37 °C and 5% CO_2 for the time required for the specific assay to be performed (*see* **Notes 10–12** in Subheading 4.2).

3.3 In Vivo Studies

In vivo studies may be used to determine the safety, efficacy, pharmacokinetics, and pharmacodynamics of MTs. The species, specific disease models, and route of administration vary according to the goals of the experiment. To date, most experiments have tested CLR01 in mice or fish [13]. Below are examples of protocols suitable for measuring BBB penetration of CLR01 in mice by spiking the compound with a radiolabeled derivative, and for treating mice by s.c. administration via osmotic minipumps.

3.3.1 Measuring BBB Penetration of CLR01

1. Restrain mice using DecapiCones or other mouse restrainers.
2. Measure the body weight (BW) of mice and prepare a mixture of CLR01, 1 mg/g BW spiked with ~2 μCi/g BW of ^{3}H–CLR01. The ^{3}H–CLR01 makes up 10% of the total CLR01.
3. Inject 4 μL/g BW of the CLR01:^{3}H–CLR01 mixture into the tail vein using 28-gauge × 0.5″ syringe needles (*see* **Note 3** in Subheading 4.3).
4. After the appropriate amount of time (*see* **Note 4** in Subheading 4.3), collect the blood via cardiac puncture, perfuse the mouse thoroughly through the heart with PBS, and collect the brain.
5. Dissect the brain into the two hemispheres. Weigh out one hemisphere. Draw 100–350 μL of blood. Digest the brain hemisphere and the blood, each with 1 mL Solvable for 2 h at 60 °C.
6. To the blood sample, add 0.1 mL EDTA-Na_2 solution, then add 0.3 mL 30% H_2O_2 in small aliquots, incubate for 15–30 min at ambient temperature, cap tightly, and incubate for additional 30 min at 60 °C.
7. Add 10 mL Ultima Gold, mix well, and measure the radioactivity in the sample in a liquid scintillation counter. Take three measurements for each sample, subtract the background radioactivity, and average.
8. Brain penetration percentage is calculated as radioactivity per gram of brain divided by radioactivity per milliliter of blood.

3.3.2 Administering CLR01 to Mice Via s.c. Osmotic Minipumps

1. Anesthetize mice by intraperitoneal injection of a ketamine/xylazine cocktail (87.5 mg/kg ketamine, 12.5 mg/kg xylazine) 0.1 mL/20 g mouse BW.
2. Once the animal is anesthetized, shave and disinfect the skin over the implantation site.
3. Make an incision 1.5-times the diameter of the implant adjacent to the site chosen for pump placement and perpendicular to the long axis of the implant. If the back of the animal is the site of choice, make a mid-scapular incision across the back perpendicular to the spine.
4. Insert a hemostat into the incision and by opening and closing the jaws of the hemostat, spread the subcutaneous tissue to create a pocket for the pump. The pocket should be large enough to allow some free movement of the pump (e.g., 1 cm longer than the pump). Avoid making the pocket too large, as this will allow the pump to turn around or slip down on the flank of the animal. The pump should not rest immediately beneath the incision because this may interfere with the healing of the incision.
5. Insert a filled pump into the pocket, delivery portal first. This minimizes interaction between the compound delivered and the healing of the incision. If the pocket is not large enough to hold the implant comfortably, remove the implant and enlarge the pocket as described above.
6. Close the wound with wound clips. Two clips will normally suffice.
7. The daily dose is determined by the concentration of the CLR01 in the pump and calculated by the rate of delivery of the pump (0.11 μL/h for the Alzet 1004 model).
8. Once the treatment has been completed, anesthetize mice with pentobarbital (100 mg/kg), and collect blood via a cardiac puncture.
9. Perfuse mice with cold, fixative or non-fixative saline buffer containing protease inhibitors (*see* **Note 5** in Subheading 4.3).
10. Immediately excise and fix in 10% formalin or 4% paraformaldehyde, or flash-freeze organs of interest, by placing the sample into a 1–2 mL cryovial. Cap the vial tightly and submerge in liquid nitrogen. When the content is frozen, transfer to a − 80 °C freezer.
11. IHC, western blots, ELISA, or any other techniques may be used for analysis of the tissue.

4 Notes

4.1 In Vitro Studies

1. MTs have been used in different solvents and buffer solutions, including water [18, 19]; 10 mM sodium phosphate, pH 7.4 [16, 20]; 100 mM sodium phosphate, pH 7.7 or 9.0 [31, 42]; 100 mM sodium phosphate, pH 7.4, 10 mM tris(2-carboxyethyl)phosphine (TCEP), and 0, 10, 20, or 30% sucrose (w/w) [43]; 60 mM sodium hydroxide, 20 mM phosphate buffer; 5 mM dithiothreitol (DTT), 10 mM 4-(2-hydroxyethyl)-1-piperazineethanesulfonic acid (HEPES), 100 mM NaCl, pH 7.6; 10 mM glycine-HCl, pH 3.5; 10 mM sodium acetate pH 3.2; or 10 mM sodium acetate pH 4.4 [16, 21].
2. For best results, fresh protein solutions should be used for most experiment types. In case the protein is already present in solution in a frozen state, freeze and thaw cycles should be minimized as it may alter the state of protein and protein aggregation pattern.
3. Certain amyloidogenic proteins require removal of pre-formed aggregates, which nucleate further aggregation and reduce reproducibility substantially. For such proteins this step should be done every time to make the assay uniform and get reproducible aggregation studies. For example, we consistently treat Aβ with 1,1,1,3,3,3-hexafluoro-2-propanol (HFIP) before performing aggregation or oligomerization assays [44].
4. The concentrations of MTs to be used in an assay should be wide so as to test the minimum and maximum range of effect. If the effect on a particular protein is not known, a good rule of thumb is to start at a protein:MT concentration ratio 1:10. If no effect is observed at this ratio, one should consider whether increasing the ratio would make sense in view of future applications in biological systems, given the safety limits. If an effect is observed, the next logical step is to determine the dose–response relationship of this effect.
5. In some studies, CLR03 had unexpected effects on certain proteins [24, 30]. These effects were small relative to those of CLR01, but in those cases, CLR03 could not be considered a true negative control.
6. Molecular tweezers may interfere with the analysis by some techniques. For example, high concentrations of MTs may suppress the radical reaction required for Photo-Induced Cross-linking of Unmodified Protein (PICUP). If a technique that has not been tested before is going to be used to monitor the reaction, possible interference by MTs should be tested first.

4.2 Cell Culture Studies

1. Cell toxicity assays can be performed with peptides/proteins added exogenously [16] or plasmid transfected to express the peptide or protein endogenously, as has been described previously, e.g., for α-syn [20].
2. Mix a small volume of cells with Trypan blue (we typically mix 0.5 mL of cells in media with 0.1 mL of a Trypan blue solution) and use a hemocytometer to count viable cells. We use a Bright Line Reichert 0.1-mm deep hemocytometer. Other counting methods also may be used.
3. We typically plate PC-12 cells in 96-well plates at a cell density of 30,000 cells/well. For different cell lines, cell density may need to be optimized to avoid overcrowding of cells and to ensure proper differentiation of cells that need to be differentiated into a particular phenotype.
4. In our typical experiments, PC-12 cells are harvested with differentiation media and neuronal growth factor (NGF) is added at 150 ng/mL before plating in a 96-well plate.
5. The time of incubation may vary depending upon the requirements for preparation of cells, e.g., 80–90% confluence or differentiation into mature cells.
6. To ensure sterility, we recommend using autoclaved distilled water filtered through a 0.2-μm filter and preparing the solution inside a biological safety cabinet (laminar-flow tissue-culture hood).
7. A wide concentration range of MTs should be tested to determine the minimum concentration (lower limit) below which inhibition on peptide toxicity is not significant and up to a maximum concentration (upper limit) at which MTs show cytotoxicity [16]. For example, we tested the inhibition of toxicity induced by 5–10 μM Aβ42 using concentrations of MTs increasing by half-log steps (1, 3, 10, 30, and 100 μM). In these experiments, we used NGF-differentiated PC-12 cells and prepared serial dilutions of MTs in differentiation media containing NGF.
8. The toxicity of MTs may vary depending on the cell type used and needs to be evaluated for specific systems. For example, in NGF-differentiated PC-12 cells, CLR01 caused ~25% decrease in cell viability at 400 μM [16], whereas no toxicity was observed in TZM-bl cells at 500 μM [18].
9. The protein and MTs can be mixed together before adding to the cells [16, 20] or MTs can be added separately after adding the protein. If MTs are added after adding protein, the cells should not be left exposed to protein for more than ~20 min.
10. When adding the solution to the well, tilt the plate and let the solution trickle down along the wall of the well to avoid disrupting the cells.

11. Different cytotoxicity assays can be used such as Caspase-3, TUNEL, and other similar assays. We have used 3-(4,5-dimethylthiazol-2-yl)-2,5-diphenyltetrazolium bromide (MTT) assay to measure cell viability [16], lactate dehydrogenase (LDH) assay to assess cell death [45], and an ELISA kit which measures cytoplasmic histone-associated nucleosomes for measurement of apoptosis [21].
12. The assembly state of the proteins is important for studying cytotoxicity. For example, to induce toxicity in NGF-differentiated PC-12 cells, α-syn added exogenously had to be incubated for 24 h to allow formation of toxic oligomers, whereas freshly dissolved α-syn was not toxic [20].

4.3 In Vivo Studies

1. In our experiments, ^{3}H–CLR01 was prepared by Moravek Biochemicals (Brea, CA) using a method that provides ^{3}H incorporation into the hydrocarbon skeleton (i.e., non-labile protons) [46].
2. We use a Triathler liquid scintillation counter (model 425–034). Any other scintillation counter suitable for measuring tritium radiation may be used.
3. I.v. injection in the tail vein may be technically difficult because the veins are hard to visualize, especially for a learner. Consulting an experienced user should be considered. Injection into the exposed jugular vein may be an optional method in anesthetized mice.
4. The plasma half-life was found to be ~ 2.5 h and approximately 5% of the initial CLR01 levels were found in the plasma 8 h following i.v. administration. The radioactivity level in the brain did not change significantly up to 72 h post injection [14].
5. The buffer composition depends on the subsequent analysis and the experimental design. Protease inhibitors are always included. Phosphatase inhibitors also may be included if analysis of phosphorylated proteins is planned. Non-fixative buffer is needed if biochemical analyses are planned, whereas fixative buffer can be used for subsequent IHC analysis. Depending on the experimental design, some animals may be perfused with fixative buffer and others with non fixative buffer. Alternatively, all the animals may be perfused with non-fixative buffer and tissues may be post-fixed for IHC analysis.

Acknowledgments

We are grateful for financial support from a UCLA Alzheimer's Disease Research Center Pilot Award enabled through NIH/NIA P50AG016570, The UCLA Mary S. Easton Endowment, NIH/-NIA R01AG050721, Team Parkinson/Parkinson Alliance, RGK

Foundation grant 20143057, Michael J. Fox Foundation grant 10220, Cure Alzheimer's Fund grant 20152631, CurePSP Foundation grant 600-6-15, and MSA Coalition Grant 20170367.

References

1. Golde TE, Borchelt DR, Giasson BI, Lewis J (2013) Thinking laterally about neurodegenerative proteinopathies. J Clin Invest 123 (5):1847–1855. https://doi.org/10.1172/JCI66029
2. Buxbaum J (1996) The amyloidoses. Mt Sinai J Med 63(1):16–23
3. Munch J, Rucker E, Standker L, Adermann K, Goffinet C, Schindler M, Wildum S, Chinnadurai R, Rajan D, Specht A, Gimenez-Gallego G, Sanchez PC, Fowler DM, Koulov A, Kelly JW, Mothes W, Grivel JC, Margolis L, Keppler OT, Forssmann WG, Kirchhoff F (2007) Semen-derived amyloid fibrils drastically enhance HIV infection. Cell 131(6):1059–1071. https://doi.org/10.1016/j.cell.2007.10.014
4. Smith DH, Johnson VE, Stewart W (2013) Chronic neuropathologies of single and repetitive TBI: substrates of dementia? Nat Rev Neurol 9(4):211–221. https://doi.org/10.1038/nrneurol.2013.29
5. Xu J, Reumers J, Couceiro JR, De Smet F, Gallardo R, Rudyak S, Cornelis A, Rozenski J, Zwolinska A, Marine JC, Lambrechts D, Suh YA, Rousseau F, Schymkowitz J (2011) Gain of function of mutant p53 by coaggregation with multiple tumor suppressors. Nat Chem Biol 7 (5):285–295. https://doi.org/10.1038/nchembio.546
6. Buhimschi IA, Nayeri UA, Zhao G, Shook LL, Pensalfini A, Funai EF, Bernstein IM, Glabe CG, Buhimschi CS (2014) Protein misfolding, congophilia, oligomerization, and defective amyloid processing in preeclampsia. Sci Transl Med 6(245):245ra292. https://doi.org/10.1126/scitranslmed.3008808
7. Rahimi F, Shanmugam A, Bitan G (2008) Structure–function relationships of pre-fibrillar protein assemblies in Alzheimer's disease and related disorders. Curr Alzheimer Res 5(3):319–341
8. Adams D, Cauquil C, Theaudin M, Rousseau A, Algalarrondo V, Slama MS (2014) Current and future treatment of amyloid neuropathies. Expert Rev Neurother 14 (12):1437–1451. https://doi.org/10.1586/14737175.2014.983905
9. Cheng B, Gong H, Xiao H, Petersen RB, Zheng L, Huang K (2013) Inhibiting toxic aggregation of amyloidogenic proteins: a therapeutic strategy for protein misfolding diseases. Biochim Biophys Acta 1830(10):4860–4871. https://doi.org/10.1016/j.bbagen.2013.06.029
10. Guerrero-Muñoz MJ, Castillo-Carranza DL, Kayed R (2014) Therapeutic approaches against common structural features of toxic oligomers shared by multiple amyloidogenic proteins. Biochem Pharmacol 88(4):468–478. https://doi.org/10.1016/j.bcp.2013.12.023
11. Aguzzi A, O'Connor T (2010) Protein aggregation diseases: pathogenicity and therapeutic perspectives. Nat Rev Drug Discov 9 (3):237–248. https://doi.org/10.1038/nrd3050
12. Attar A, Bitan G (2014) Disrupting self-assembly and toxicity of amyloidogenic protein oligomers by "molecular tweezers"—from the test tube to animal models. Curr Pharm Des 20 (15):2469–2483
13. Schrader T, Bitan G, Klarner FG (2016) Molecular tweezers for lysine and arginine—powerful inhibitors of pathologic protein aggregation. Chem Commun (Camb) 52 (76):11318–11334. https://doi.org/10.1039/c6cc04640a
14. Attar A, Chan WT, Klärner FG, Schrader T, Bitan G (2014) Safety and pharmacological characterization of the molecular tweezer CLR01—a broad-spectrum inhibitor of amyloid proteins' toxicity. BMC Pharmacol Toxicol 15(1):23. https://doi.org/10.1186/2050-6511-15-23
15. Fokkens M, Schrader T, Klärner FG (2005) A molecular tweezer for lysine and arginine. J Am Chem Soc 127(41):14415–14421
16. Sinha S, Lopes DH, Du Z, Pang ES, Shanmugam A, Lomakin A, Talbiersky P, Tennstaedt A, McDaniel K, Bakshi R, Kuo PY, Ehrmann M, Benedek GB, Loo JA, Klärner FG, Schrader T, Wang C, Bitan G (2011) Lysine-specific molecular tweezers are broad-spectrum inhibitors of assembly and toxicity of amyloid proteins. J Am Chem Soc 133 (42):16958–16969. https://doi.org/10.1021/ja206279b
17. Bier D, Rose R, Bravo-Rodriguez K, Bartel M, Ramirez-Anguita JM, Dutt S, Wilch C, Klärner FG, Sanchez-Garcia E, Schrader T, Ottmann C

(2013) Molecular tweezers modulate 14-3-3 protein-protein interactions. Nat Chem 5 (3):234–239. https://doi.org/10.1038/nchem.1570

18. Lump E, Castellano LM, Meier C, Seeliger J, Erwin N, Sperlich B, Stürzel CM, Usmani S, Hammond RM, von Einem J, Gerold G, Kreppel F, Bravo-Rodriguez K, Pietschmann T, Holmes VM, Palesch D, Zirafi O, Weissman D, Sowislok A, Wettig B, Heid C, Kirchhoff F, Weil T, Klärner FG, Schrader T, Bitan G, Sanchez-Garcia E, Winter R, Shorter J, Münch J (2015) A molecular tweezer antagonizes seminal amyloids and HIV infection. eLife 4. https://doi.org/10.7554/eLife.05397
19. Malishev R, Nandi S, Kolusheva S, Levi-Kalisman Y, Klarner FG, Schrader T, Bitan G, Jelinek R (2015) Toxicity inhibitors protect lipid membranes from disruption by Aβ42. ACS Chem Neurosci 6(11):1860–1869. https://doi.org/10.1021/acschemneuro.5b00200
20. Prabhudesai S, Sinha S, Attar A, Kotagiri A, Fitzmaurice AG, Lakshmanan R, Ivanova MI, Loo JA, Klärner FG, Schrader T, Stahl M, Bitan G, Bronstein JM (2012) A novel "molecular tweezer" inhibitor of α-synuclein neurotoxicity in vitro and in vivo. Neurotherapeutics 9(2):464–476. https://doi.org/10.1007/s13311-012-0105-1
21. Lopes DH, Attar A, Nair G, Hayden EY, Du Z, McDaniel K, Dutt S, Bravo-Rodriguez K, Mittal S, Klärner FG, Wang C, Sanchez-Garcia-E, Schrader T, Bitan G (2015) Molecular tweezers inhibit islet amyloid polypeptide assembly and toxicity by a new mechanism. ACS Chem Biol 10(6):1555–1569. https://doi.org/10.1021/acschembio.5b00146
22. LeVine H 3rd (1999) Quantification of β-sheet amyloid fibril structures with thioflavin T. Methods Enzymol 309:274–284
23. Kayed R, Head E, Thompson JL, McIntire TM, Milton SC, Cotman CW, Glabe CG (2003) Common structure of soluble amyloid oligomers implies common mechanism of pathogenesis. Science 300(5618):486–489
24. Zheng X, Liu D, Klärner FG, Schrader T, Bitan G, Bowers MT (2015) Amyloid β-protein assembly: the effect of molecular tweezers CLR01 and CLR03. J Phys Chem B 119(14):4831–4841. https://doi.org/10.1021/acs.jpcb.5b00692
25. Bitan G, Kirkitadze MD, Lomakin A, Vollers SS, Benedek GB, Teplow DB (2003) Amyloid β-protein (Aβ) assembly: Aβ40 and Aβ42 oligomerize through distinct pathways. Proc Natl Acad Sci U S A 100(1):330–335. https://doi.org/10.1073/pnas.222681699
26. Gong Y, Chang L, Viola KL, Lacor PN, Lambert MP, Finch CE, Krafft GA, Klein WL (2003) Alzheimer's disease-affected brain: presence of oligomeric Aβ ligands (ADDLs) suggests a molecular basis for reversible memory loss. Proc Natl Acad Sci U S A 100 (18):10417–10422. https://doi.org/10.1073/pnas.1834302100
27. Lesné S, Koh MT, Kotilinek L, Kayed R, Glabe CG, Yang A, Gallagher M, Ashe KH (2006) A specific amyloid-β protein assembly in the brain impairs memory. Nature 440(7082):352–357
28. Bernstein SL, Dupuis NF, Lazo ND, Wyttenbach T, Condron MM, Bitan G, Teplow DB, Shea JE, Ruotolo BT, Robinson CV, Bowers MT (2009) Amyloid-β protein oligomerization and the importance of tetramers and dodecamers in the aetiology of Alzheimer's disease. Nat Chem 1(4):326–331. https://doi.org/10.1038/nchem.247
29. Ferreira N, Pereira-Henriques A, Attar A, Klärner FG, Schrader T, Bitan G, Gales L, Saraiva MJ, Almeida MR (2014) Molecular tweezers targeting transthyretin amyloidosis. Neurotherapeutics 11(2):450–461. https://doi.org/10.1007/s13311-013-0256-8
30. Herzog G, Shmueli MD, Levy L, Engel L, Gazit E, Klärner FG, Schrader T, Bitan G, Segal D (2015) The lys-specific molecular tweezer, CLR01, modulates aggregation of the mutant p53 DNA binding domain and inhibits its toxicity. Biochemistry 54 (24):3729–3738. https://doi.org/10.1021/bi501092p
31. Attar A, Ripoli C, Riccardi E, Maiti P, Li Puma DD, Liu T, Hayes J, Jones MR, Lichti-Kaiser K, Yang F, Gale GD, Tseng CH, Tan M, Xie CW, Straudinger JL, Klärner FG, Schrader T, Frautschy SA, Grassi C, Bitan G (2012) Protection of primary neurons and mouse brain from Alzheimer's pathology by molecular tweezers. Brain 135(Pt 12):3735–3748. https://doi.org/10.1093/brain/aws289
32. Oddo S, Caccamo A, Kitazawa M, Tseng BP, LaFerla FM (2003) Amyloid deposition precedes tangle formation in a triple transgenic model of Alzheimer's disease. Neurobiol Aging 24(8):1063–1070
33. Flood DG, Lin YG, Lang DM, Trusko SP, Hirsch JD, Savage MJ, Scott RW, Howland DS (2009) A transgenic rat model of Alzheimer's disease with extracellular Aβ deposition. Neurobiol Aging 30(7):1078–1090. https://doi.org/10.1016/j.neurobiolaging.2007.10.006

34. Teng E, Kepe V, Frautschy SA, Liu J, Satyamurthy N, Yang F, Chen PP, Cole GB, Jones MR, Huang SC, Flood DG, Trusko SP, Small GW, Cole GM, Barrio JR (2011) [F-18] FDDNP microPET imaging correlates with brain Aβ burden in a transgenic rat model of Alzheimer disease: effects of aging, in vivo blockade, and anti-Aβ antibody treatment. Neurobiol Dis 43(3):565–575. https://doi.org/10.1016/j.nbd.2011.05.003
35. Youmans KL, Tai LM, Kanekiyo T, Stine WB Jr, Michon SC, Nwabuisi-Heath E, Manelli AM, Fu Y, Riordan S, Eimer WA, Binder L, Bu G, Yu C, Hartley DM, LaDu MJ (2012) Intraneuronal Aβ detection in 5xFAD mice by a new A-β-specific antibody. Mol Neurodegener 7:8. https://doi.org/10.1186/1750-1326-7-8
36. Fitzmaurice AG, Rhodes SL, Cockburn M, Ritz B, Bronstein JM (2014) Aldehyde dehydrogenase variation enhances effect of pesticides associated with Parkinson disease. Neurology 82(5):419–426. https://doi.org/10.1212/WNL.0000000000000083
37. Lulla A, Barnhill L, Bitan G, Ivanova MI, Nguyen B, O'Donnell K, Stahl MC, Yamashiro C, Klarner FG, Schrader T, Sagasti A, Bronstein JM (2016) Neurotoxicity of the parkinson disease-associated pesticide ziram is synuclein-dependent in zebrafish embryos. Environ Health Perspect 124 (11):1766–1775. https://doi.org/10.1289/EHP141
38. Santos SD, Fernandes R, Saraiva MJ (2010) The heat shock response modulates transthyretin deposition in the peripheral and autonomic nervous systems. Neurobiol Aging 31 (2):280–289. https://doi.org/10.1016/j.neurobiolaging.2008.04.001
39. Bitan G, Shanmugam A, Lomakin A, Schrader T, Klärner F-G, Talbiersky P, Polkowska J, Bastkowski F, Sinha S, Frautschy SA (2009) Molecular tweezers for the treatment of amyloid-related diseases. US Patent 8,791,092
40. Li H, Murakami K, Rahimi F, Maiti P, Sinha S, Bitan G (2009) Amyloids and protein aggregation–analytical methods. Encyclopedia of analytical chemistry, September 15, 2009 edn. Wiley. https://doi.org/10.1002/9780470027318.a9038
41. Rahimi F, Bitan G (2014) Methods for studying and structure–function relationships of non-fibrillar protein assemblies in Alzheimer's disease and related disorders. In: Lahiri DK (ed) Advances in alzheimer research, vol 2. Bentham Science Publishers, Emirate of Sharjah, UAE, pp 291–374. https://doi.org/10.2174/9781608058525114020014
42. Talbiersky P, Bastkowski F, Klärner FG, Schrader T (2008) Molecular clip and tweezer introduce new mechanisms of enzyme inhibition. J Am Chem Soc 130(30):9824–9828. https://doi.org/10.1021/Ja801441j
43. Acharya S, Safaie BM, Wongkongkathep P, Ivanova MI, Attar A, Klärner FG, Schrader T, Loo JA, Bitan G, Lapidus LJ (2014) Molecular basis for preventing α-synuclein aggregation by a molecular tweezer. J Biol Chem 289 (15):10727–10737. https://doi.org/10.1074/jbc.M113.524520
44. Rahimi F, Maiti P, Bitan G (2009) Photo-induced cross-linking of unmodified proteins (PICUP) applied to amyloidogenic peptides. J Vis Exp (23). doi:https://doi.org/10.3791/1071
45. Sinha S, Du Z, Maiti P, Klärner FG, Schrader T, Wang C, Bitan G (2012) Comparison of three amyloid assembly inhibitors: the sugar *scyllo*-inositol, the polyphenol epigallocatechin gallate, and the molecular tweezer CLR01. ACS Chem Neurosci 3(6):451–458. https://doi.org/10.1021/cn200133x
46. Maegawa T, Hirota K, Tatematsu K, Mori Y, Sajiki H (2005) Facile and efficient postsynthetic tritium labeling method catalyzed by Pd/C in HTO. J Org Chem 70 (25):10581–10583. https://doi.org/10.1021/jo0517545

Chapter 25

Incorporation of an Azobenzene *β*-Turn Peptidomimetic into Amyloid-*β* to Probe Potential Structural Motifs Leading to *β*-Sheet Self-Assembly

Todd M. Doran and Bradley L. Nilsson

Abstract

Alzheimer's disease (AD) is characterized by chronic neurodegeneration and the insidious accumulation of senile plaques comprised of the amyloid-β (Aβ) peptide. An important goal in AD research is to characterize the structural basis for how Aβ aggregates exert their noxious effects on neurons. We describe herein synthetic steps to incorporate a light-controlled β-turn mimetic, 3-(3-aminomethylphenylazo)-phenylacetic acid (AMPP), into the backbone of a putative turn region within Aβ. AMPP adopts a rigid β-hairpin turn when azobenzene is in the *cis* conformation, and can adopt an extended "β-arc" turn in the *trans*-azobenzene conformation. The long lifetimes of these conformationally stable isomers permit detailed biochemical analyses that help to clarify the controversial role played by these two types of turns during the toxic misfolding pathway of Aβ. Methods to photo-nucleate the *cis*- or *trans*-AMPP isomeric turns in aqueous buffer are also described. Finally, we detail selected techniques to characterize the Aβ aggregates derived from these photoisomerized variants.

Key words Azobenzene, Photoswitch, Photoisomerization, Amyloid-β, Alzheimer's disease, β-Hairpin, β-Arc

1 Introduction

The amyloid-β (Aβ) peptide has been implicated as a causative agent in Alzheimer's disease (AD) pathology. This causal relationship has its origins in early observations that AD patients characteristically possess senile plaques comprised principally of Aβ folded into β-sheet amyloid [1]. Nevertheless, strategies aimed at reducing Aβ plaque levels in AD patients have failed to significantly delay disease progression [2]. This failure has recently been explained by observations that Aβ fibrils, though diagnostic of AD, may be pathologically benign [3]. Compelling evidence suggests that much smaller, soluble aggregates of Aβ are the functional mediators of Aβ-induced toxicity [4–7]. Indeed, in vitro preparations of these so-called pre-fibrillar species, including oligomers [8] and

Bradley L. Nilsson and Todd M. Doran (eds.), *Peptide Self-Assembly: Methods and Protocols*, Methods in Molecular Biology, vol. 1777, https://doi.org/10.1007/978-1-4939-7811-3_25, © Springer Science+Business Media, LLC, part of Springer Nature 2018

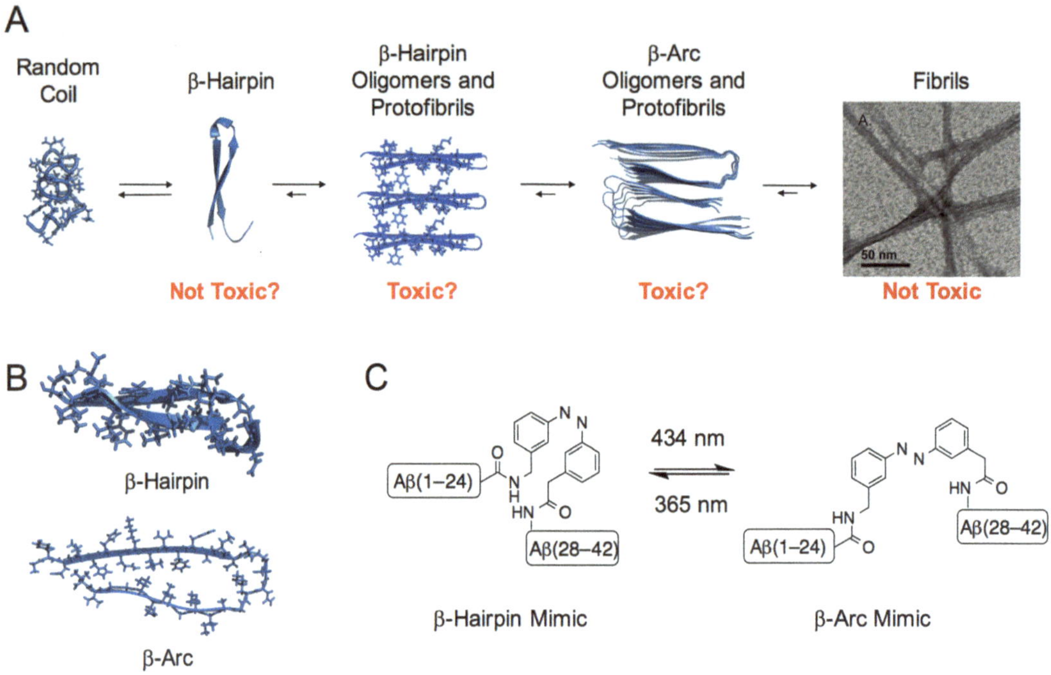

Fig. 1 The many possible β-turn structures along the Aβ42 β-sheet misfolding pathway. (**a**) Schematic of the major Aβ misfolding pathway that leads the formation of β-sheet fibrils. As the pathway progresses from a monomeric β-hairpin to a β-arc multimer, toxicity increases before decreasing again in insoluble fibrils. The difficulty in separating oligomers and protofibrils has made it difficult to unequivocally determine the structure of assemblies at the inflection point, where toxicity is highest. (**b**) The β-hairpin turn observed in monomeric Aβ (PDB:2OTK) [17] and the β-arc observed in Aβ fibrils (coordinates graciously provided by R. Tycko) (**c**) The photoswitch, *cis*-AMPP, with flanking N- and C-termini of Aβ adopts a type I′ β-hairpin turn. Upon photoisomerization at 434 nm, *cis-trans* isomerization leads to the formation of a more extended conformation, allowing it to model a β-arc [31]

protofibrils [9], are highly proteotoxic. Despite their clinical value, the precise mechanisms by which these species exert their toxicity remain elusive. Deciphering the key structure–function relationships is complicated by the dynamic and atypical folding behavior of low-order Aβ aggregates. For example, misfolding of Aβ is a multi-step, nucleation-polymerization process, and aggregation proceeds by a concentration-dependent association of soluble components into insoluble fibrils once fibrils are initially seeded (Fig. 1a) [10, 11]. Consequently, soluble preparations of Aβ oligomers and protofibrils are transient and metastable, creating an inseparable mixture of various polymorphs. Bulk analyses of Aβ secondary structures at varying degrees of self-association reveal a conformationally heterogeneous mixture of secondary structures, with each playing an uncertain role in cytotoxicity. Collectively, this behavior of Aβ soluble assemblies has obviated the use of conventional biochemical assays to unequivocally ascribe functional properties to oligomers or protofibrils. Thus, innovative methods to associate specific structural features of Aβ oligomers or protofibrils with neurotoxicity have been sought to unravel disease mechanisms and define targets for drug discovery.

Many efforts to understand the toxic behavior of discrete Aβ oligomers or protofibrils have focused on addressing the precise nature of a reverse turn motif located between residues 23 and 30 that is featured in NMR structures of Aβ fibrils [12–16]. A critical role for reverse turn nucleation on aggregate toxicity is evidenced by the acceleration of Aβ misfolding kinetics upon β-turn formation, and the significant structural reorganization of this region during key transitions associated with proteotoxicity (Fig. 1b) [17–19]. During Aβ misfolding, natively unfolded Aβ monomers adopt a β-hairpin turn, accompanied by the formation of intramolecular β-sheet secondary structure [17]. For fibrillization to occur, the β-hairpin morphs, and Aβ fibrils universally adopt a multi-residue "U-turn/β-arc" reverse turn with flanking β-strands aligned in a parallel, cross-β sheet arrangement [12–16, 20]. Conversely, structural models of this turn region in pre-fibrillar Aβ species are less definitive, having been variably described as possessing β-hairpin [21, 22] or U-turn morphology [23]. Divorcing the functional effects of these two turn elements more precisely thus remains a central problem in defining the structural basis of Aβ toxicity.

Precise structure–function analyses of the turn region of Aβ have been achieved using chemical engineering to restrict conformational space around the turn region in Aβ. For example, chemically constraining the conformation around the turn region of Aβ with disulfide bond engineering [24], lactam ring condensation [25], and a D-ProGly motif [26] have shed considerable light on the functional significance of β-turn formation. Our laboratory has extended this list to include an azobenzene turn mimetic that locks the turn region into a β-hairpin or a β-arc conformation depending on the *cis-trans* isomerization state of azobenzene [27]. Specifically, 3-(3-aminomethylphenylazo)-phenylacetic acid (AMPP) models a type I′ β-hairpin in the *cis*-azobenzene conformation or an extended turn in the *trans*-azobenzene state (Fig. 1c) [28–31]. The two states can be attained via photoisomerization using UV-visible light. By incorporating this turn mimetic into the 25–27 region of Aβ42, we demonstrated that imposing a *cis*-azobenzene β-hairpin causes Aβ to resist fibrillization via the Aβ-sheet misfolding pathway. Conversely, the *trans*-azobenzene variant readily formed toxic oligomers and nucleated fibrillization. These data reveal a less important role for assemblies containing a β-hairpin turn in toxic aggregates, and suggest that a more extended, perhaps a β-arc, turn confers toxicity to oligomers. A caveat to these findings, however, is that cis-AMPP induces a type I′ β-turn, and native Aβ appears to adopt a type II′ β-hairpin during β-sheet misfolding. Nevertheless, photo-stabilization of metastable conformers provides tantalizing new evidence about the folding behavior of Aβ, and may prove useful in screens that stabilize β-hairpins that cannot be accessed easily by other means.

In this protocol, we outline steps to investigate the morphological properties of Aβ42 containing the AMPP photoswitch in the *cis* and *trans* conformation, which we will refer to as *cis*-AMPP-Aβ42 and *trans*-AMPP-Aβ42, respectively. First, we describe the incorporation of AMPP into the Aβ42 peptide using solid-phase peptide synthesis (SPPS). We also outline a procedure for the purification and disaggregation of Aβ42 variants for use in biochemical assays. Steps to photomodulate between *cis/trans* aggregation states are detailed. Finally, this protocol describes several methods to characterize the morphology of the Aβ42 variant aggregates formed by these conformational variants. Azobenzene peptidomimetics offer exceptional spatiotemporal control over the formation of β-hairpin and β-arc turn in the Aβ42 peptide, and other peptides containing this multipurpose photoswitch may be of use as probes to investigate the folding behavior of other highly dynamic and structurally ambiguous proteins.

2 Materials

2.1 Synthesis and Purification of AMPP-Containing Peptides

1. 0.1 mmol equivalents of low-loading Rink amide resin coated in polyethylene glycol (PEG), such as NovaPEG Rink Amide resin.
2. Automated peptide synthesizer with a microwave reaction vessel for 0.1 mmol scale peptide synthesis, such as a CEM Liberty microwave-equipped peptide synthesizer running on PepDriver software (*see* **Note 1**).
3. *N,N*-Dimethylformamide (DMF).
4. Dichloromethane (DCM).
5. Amino acid solutions: 0.2 M Fmoc-protected amino acids dissolved in DMF.
6. 9-Fluorenylmethoxycarbonyl-N-3-(3-aminomethylphenylazo)-phenylacetic acid (Fmoc-AMPP-OH, *see* **Note 2**) [29].
7. Activating solution: 0.5 M O-(Benzotriazol-1-yl) *N,N,N′,N*-′-tetramethyluronium hexafluorophosphate (HBTU) in DMF.
8. *N,N*-Diisopropylethylamine (DIPEA).
9. Activator base: 2 M DIPEA in DMF.
10. 1-Hydroxybenzotriazole (HOBT).
11. Deprotection solution: 20% piperidine in DMF containing 0.1 M HOBt.
12. Cleavage solution: 95% TFA, water, 2.5% triisopropylsilane, 2.5% water (*v/v/v*).
13. Fritted syringe.

14. Diethyl ether.
15. Liquid nitrogen.
16. 50 mL conical tube.
17. Centrifuge capable of 2500 × *g*.
18. 0.2 μm syringe tip filter.
19. Lyophilizer.
20. Water purified to a resistance of 18.2 MΩ or HPLC-grade water.
21. High-performance liquid chromatography pump with a semi-preparative C4 column, such as one with an approximate diameter of 21.2 mm and length of 250 mm (*see* **Note 3**).
22. Organic HPLC solution: HPLC-grade acetonitrile containing 0.5% TFA.
23. Aqueous HPLC solution: HPLC-grade water, or ultra-purified water, containing 0.5% TFA (*v/v*).
24. Dimethylsulfoxide (DMSO).
25. MALDI-TOF mass spectrometer.
26. MALDI-TOF matrix solution: 10 mg/mL α-cyano-4-hydroxycinnamic acid in 50:50 acetonitrile:water (*v/v*).

2.2 Peptide Disaggregation

1. Freshly distilled 1,1,1,3,3,3-hexafluoroisopropanol (HFIP).
2. Anhydrous DMSO, such as from a newly opened ampule or DMSO dried over molecular sieves.
3. Liquid nitrogen.
4. 1 dram glass vials.
5. Water bath sonicator.
6. Dry nitrogen source.

2.3 Photo-isomerization and Self-Assembly of Peptides

1. Ultra-purified water, 18.2 MΩ resistance, such as obtained with a Barnstead water purifier.
2. 0.1 cm quartz cuvette.
3. 500 W mercury lamp (*see* **Note 4**).
4. Phosphate-buffered saline (PBS): 10 mM Na_2HPO_4, 1.8 mM KH_2PO_4, 137 mM NaCl, 2.7 mM KCl.
5. 100–1000 μL micropipettes.
6. Gel-loading micropipette tips.
7. 1.5 mL low-bind microcentrifuge tubes.
8. 400 nm longpass filter (*see* **Note 5**).
9. 365 nm bandpass filter (*see* **Note 5**).

2.4 Spectral and Morphological Characterization of Peptide Fibrils

1. UV-vis spectrophotometer or photo-diode array.
2. 1 cm quartz cuvette.
3. Low-bind, clear-bottom black 96-well microtiter plates.
4. Fluorometer with 485 nm excitation and 515 nm emission filters.
5. Thioflavin T stock solution: 2.5 mM thioflavin T in PBS.
6. Analytical HPLC with C18 column.
7. FT-IR spectrometer with ATR stage.
8. Deuterated water (D_2O).
9. 1 cm quartz cuvette.
10. Circular dichroism (CD) spectrophotometer.
11. TEM staining solution: 2% uranyl acetate in water.
12. Filter paper.
13. Formvar stabilized carbon on 200-mesh copper grids.
14. Filter paper.
15. Transmission electron microscope (TEM).
16. Centrifuge capable of speeds of at least 14,000 × *g*.

3 Methods

All methods employ ultrapure water with a resistance of 18.2 MΩ, obtained by filtering distilled water through a reverse osmosis ultrapurification system containing activated charcoal and UV irradiation. The protocols describe the synthesis and characterization of AMPP Aβ variants only, and any controlled experiments should include comparisons to the native peptide, Aβ42.

3.1 Synthesis of Aβ42 Containing AMPP (AMPP-Aβ42)

1. Load the resin into the automated synthesizer reaction vessel and load the vessel into an automated peptide synthesizer with microwave capabilities.
2. Using the standard protocols as defined by the manufacturer, remove the Fmoc group protecting the Rink linker with the deprotection solution (*see* **Note 1**).
3. Wash the resin with 5 mL DMF, and evacuate the reaction vessel.
4. Program the instrument to perform the following iterative steps for amino acid chain extension using command prompts provided by the manufacturer (*see* **Note 6**). In general, we use the following protocol and agitating the resin in each step using nitrogen gas bubbled through the reaction vessel.

(a) Add 5 equivalents of the appropriate amino acid solution to the reaction vessel, followed by the activator solution and the base solution.

(b) Agitate the coupling mixture at high temperatures (75 °C) for approximately 1–2 min. Following the coupling reaction, evacuate the reaction vessel and wash the resin three times with 5 mL of DMF.

(c) Deprotect the terminal Fmoc protecting group by adding 5 mL deprotection solution for 0.5 min at a microwave temperature of 75 °C. Evacuate the reaction vessel and wash the resin with 5 mL of DMF. Repeat the wash two more times.

(d) Repeat these steps iteratively until amino acids 28–42 have been incorporated into the nascent peptide.

5. Remove the resin from the reaction chamber. Coupling of Fmoc-AMPP-OH should be performed manually (*see* **Note 7**).
6. Activate Fmoc-AMPP-OH as the HOBt ester by preparing a mixture containing 0.5 mmol of each of the following in DMF: Fmoc-AMPP-OH, HOBt, and HBTU, followed by 1 mmol DIPEA.
7. Vigorously mix the activated Fmoc-AMPP-OH coupling mixture and let stand for 5 min. It will be a deep orange color.
8. Add the coupling mixture containing activated Fmoc-AMPP-OH to the resin in a separate reaction vessel.
9. Heat the activated Fmoc-AMPP-OH/resin mixture at 37–50 °C for 3 h with agitation in order to couple Fmoc-AMPP to amino acid 28 (beads will become orange, even after washing). Filter the resin and wash three times with 5–10 mL of DMF at 3 min per wash.
10. Repeat **steps 6–9** at room temperature overnight (*see* **Note 8**).
11. Wash the resin three times for 3 min with 5–10 mL of DMF.
12. Add 7 mL deprotection solution and agitate for 5 min.
13. Filter the resin and add 7 mL deprotection solution. Agitate an additional 10 min.
14. Wash the resin three times for 3 min with 5 mL of DMF.
15. Transfer the resin back into the automated peptide synthesizer reaction vessel.
16. Continue iterative amino acid chain extension using the SPPS procedures in **step 4** until residues 1–24 are incorporated (*see* **Notes 1** and **6**).
17. After coupling of the final residue is complete, deprotect the *N*-terminal Fmoc group as per **step 4c**, wash the resin with 5–10 mL of DCM three times and then with 5–10 mL of

methanol one time. Transfer the washed beads to a fritted syringe for cleavage.

18. Cleave the resulting peptide from the resin by adding cleavage solution for 1–2 h at room temperature.
19. Filter the cleavage solution and collect in a 50 mL conical tube; evaporate 90–95% of the TFA using a delicate stream of dry nitrogen gas.
20. Precipitate the product peptide by adding 10–20 mL diethyl ether that is pre-cooled in liquid nitrogen.
21. Centrifuge the crude product mixture to sediment the precipitated peptide.
22. Carefully decant the supernatant.
23. Resuspend the peptide pellet in 10–20 mL of room temperature diethyl ether and repeat **steps 21** and **22**.
24. Dissolve the pellet in 5–10 mL of DMSO.
25. Filter the solution through a 0.2 μM syringe tip filter.
26. Purify peptide by reverse-phase HPLC using a C4 column using a binary gradient solvent system that mixes water and acetonitrile (*see* **Note 9**). Inject the crude mixture onto the column and purify Aβ by eluting with a steadily increasing gradient of 25–45% organic HPLC solution in aqueous HPLC solution (*v/v*) over 15 min. Aβ will elute during this ramp, but, due to dead volume of the column, may elute after 15 min. After the initial ramp up, the gradient should be increased to 95% organic HPLC solution over 2 min and run isocratically with this concentration for at least 5 min to wash residual Aβ from the column. The column should be re-equilibrated in the aqueous HPLC solution by ramping down the organic HPLC solution to 5% over 2 min, and washing the column with this concentration for at least 5 min.
27. To ensure that a highly pure Aβ sample is obtained, lyophilize the purified Aβ42, dissolve the dried product of each purification run in 2 mL of dry DMSO. Rerun the purified Aβ42 through the column a second time using the same method as described in **step 26** of this section.
28. Confirm the identity of the peptide by MALDI-TOF MS using α-cyano-4-hydroxycinnamic acid as the matrix.
29. Pool and freeze the HPLC fractions containing the desired peptide using liquid nitrogen and lyophilize each HPLC run separately as an aliquot of pure product in glass vials (*see* **Note 9**).
30. Dissolve one aliquot of the lyophilized powder in 100 μL of DMSO.
31. Construct an HPLC concentration curve to assist in accurately quantifying the peptide in aggregation experiments. This is

done by injecting serial dilutions of the DMSO solution from **step 29** onto an analytical HPLC and plotting the integrated area of the peak absorbances at 215 nm against the amount of peptide injected. A small aliquot of known volume of the DMSO solution is saved to correlate peptide concentration by amino acid analysis. Wetzel and coworkers described in great detail the construction of HPLC concentration curves for Aβ [32, 33].

32. Once a concentration curve has been constructed, the concentration of Aβ can be quantified for any solution of Aβ stock by eluting a known volume of the stock onto the same column and HPLC system. To determine the concentration, integrate the eluted Aβ42 absorbance peak and correlate it to the concentration of Aβ using your constructed standard curve from **step 31** (*see* **Note 10**).
33. Store unused aliquots as lyophilized powders at −80 °C (*see* **Note 11**).

3.2 Disaggregation of Aβ42 and AMPP-Containing Variants

1. Dissolve 0.5–1 mg of purified, lyophilized peptide in 1 mL of TFA.
2. Sonicate the solution at room temperature for 1 min. Let the mixture stand for 10 min.
3. Evaporate the TFA using a gentle stream of dry nitrogen, leaving a thin film of peptide on the walls of the vial.
4. Dissolve the film in 500 μL HFIP and incubate at 37 °C for 1 h (*see* **Notes 12 and 13**).
5. Evaporate HFIP using a gentle stream of dry nitrogen.
6. Dissolve the peptide film in 1 mL of HFIP once again.
7. Remove a 5–10 μL of the HFIP solution and dilute it into 50 μL of DMSO to determine the concentration of peptide by HPLC analysis or UV-vis (*see* **Notes 12** and **13**). To determine concentration, inject the 40 μL of the HFIP/DMSO mixture onto an analytical HPLC column and elute.
8. Based on the estimated peptide concentration from **step** 7, aliquot the amount of HFIP stock solution corresponding to the desired amount of peptide into low-bind centrifuge tubes (*see* **Note 14**).
9. Freeze the Aβ42 variant HFIP aliquots by immersion of the tubes in liquid nitrogen.
10. Lyophilize these HFIP aliquots for 16 h to obtain a powder (*see* **Note 15**).
11. Prepare a stock solution of the peptide in dry DMSO at a concentration of 2 mM or ~10 mg/mL. These solutions should be used immediately and not stored.

3.3 Photo-isomerizing AMPP-Containing Amyloid Peptides and Monitoring Fibril Assembly by Thioflavin T Fluorescence

This protocol describes the preparation of *cis*-AMPP-Aβ42 in DMSO. DMSO is the preferred solvent because it prevents fibril nucleation of the thermally relaxed *trans*-AMPP-Aβ42 isomer that predominates following the previous steps in Subheadings 3.1 and 3.2. *Cis*-AMPP mimics a restricted type I′ β-hairpin turn providing a long-lived model of the short-lived β-hairpin turn observed in early stages of the Aβ β-sheet folding pathway (Fig. 1). This model provides a more well-behaved system to monitor the kinetics of fibrillization of β-hairpin conformers. Methods to profile cross-β fibrillization kinetics are also described.

3.3.1 Trans → Cis Isomerization of AMPP-Aβ42 in DMSO and Characterization of Fibrillization Kinetics

1. Add the freshly disaggregated peptide DMSO stock solution to a 0.1 cm pathlength quartz cuvette.
2. Obtain an absorbance spectrum and identify the maxima at ~320 nm for *trans*-azobenzene and ~430 nm for *cis*-azobenzene (Fig. 2a, *see* **Note 16**).
3. Irradiate the solution near the maximum at 320 nm using a 500 W mercury lamp with a 365 nm or similar wavelength cutoff filter (*see* **Note 5**).
4. After 5 min, obtain an absorbance spectrum to monitor the extent of isomerization at the two maxima at 320 nm and 430 nm (Fig. 2a, *see* **Note 17**).
5. Replace the cuvette in the lamp apparatus to continue irradiating the sample at 365 nm for additional 5 min.
6. Record an absorbance spectrum of the sample. If it has not changed from **step 5**, proceed to **step 8**. If the profile continues

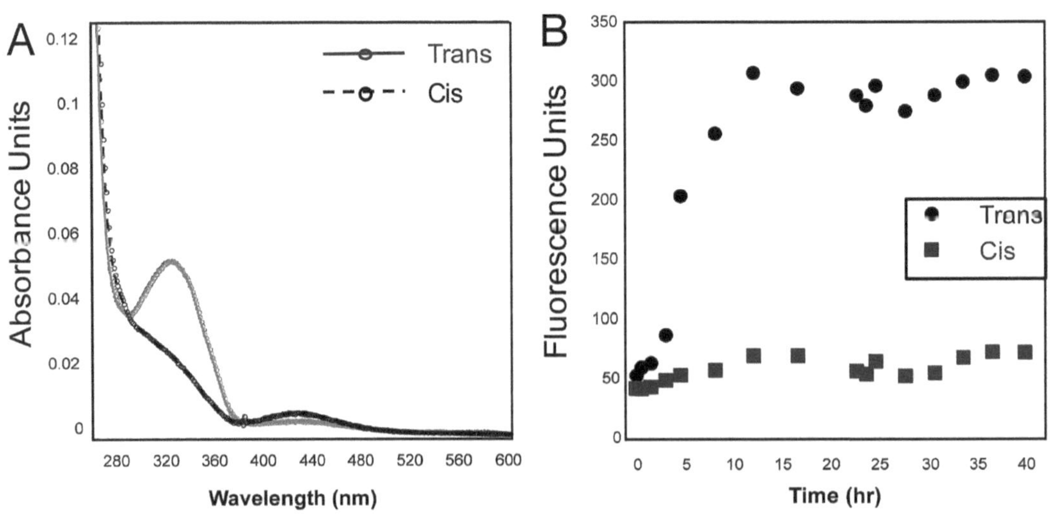

Fig. 2 Spectroscopic characterization of freshly disaggregated cis- and trans-AMPP-Aβ. (**a**) Absorbance spectra of 10 μM *cis*- and *trans*-AMPP-Aβ in PBS. (**b**) ThT fluorescence intensity at 497 nm of freshly isomerized cis- and trans-AMPP-Aβ in PBS at 10 μM monitored over 48 h

to change, repeat **steps 6** and 7 until the two absorbance maxima remain constant (*see* **Note 18**).

7. Dilute the DMSO stock into PBS to a final Aβ42 variant concentration of no more than 10 μM (*see* **Note 19**). All experiments with AMPP variants should be performed concurrently with the native peptide as a necessary comparative control.
8. Monitor ThT fluorescence by removing 148.5 μL of the Aβ solution in PBS and add 1.5 μL ThT stock solution to a final concentration of 25 μM ThT.
9. Monitor ThT fluorescence in the aliquot from **step 9** using an excitation wavelength of 465 nm and reading the emission intensity at 497 nm. Perform this measurement in triplicate for each time point (Fig. 2b, *see* **Note 20**).
10. Repeat **steps 9** and **10** until characterization has been performed, but no longer than 72 h (*see* **Note 21**). ThT fluorescence will increase proportionally to the degree of fibril formation. After an initial lag phase, you should see an exponential increase that eventually plateaus, indicating completion of fibril formation.
11. Characterize the aggregates that are formed by *cis*-AMPP-Aβ42 using the methods described in Subheading 3.4.

3.3.2 Cis → Trans Photoisomerization of AMPP-Aβ42 in Aqueous Buffer

Isomerization of *cis*- to *trans*-AMPP in Aβ42 nucleates fibrillization in the β-sheet pathway due to the opening of the β-hairpin turn to permit fibrils to form with a cross-β arrangement. The following protocol describes the *cis* → *trans* isomerization of AMPP-Aβ42 to demonstrate the importance of an extended, non-β-hairpin turn in nucleating the β-sheet fibrillization pathway. Methods to monitor the elongation kinetics of nucleated fibrillization are also described.

1. Transfer the aqueous *cis*-AMPP-Aβ42 solution from Subheading 3.3.1, **step** 7 into a 1 cm pathlength quartz cuvette.
2. Obtain an absorbance spectrum by UV-vis spectrophotometry to record the initial absorbance reading for the isomerization process (*see* **Note 22**).
3. Remove 148.5 μL of the cis-AMPP-Aβ42 solution and add 1.5 μL ThT stock to a final concentration of 25 μM ThT and record a baseline thioflavin T fluorescence measurement by exciting the mixture containing the fluorophore at 465 nm and monitoring fluorescence emission at 497 nm.
4. Transfer the aqueous solution containing the *cis* isomer into a 0.1 cm quartz cuvette. Depending on the volume required for the kinetics and characterization experiments, two lamp setups may be necessary. Alternatively, two isomerizations can be performed on one apparatus sequentially.

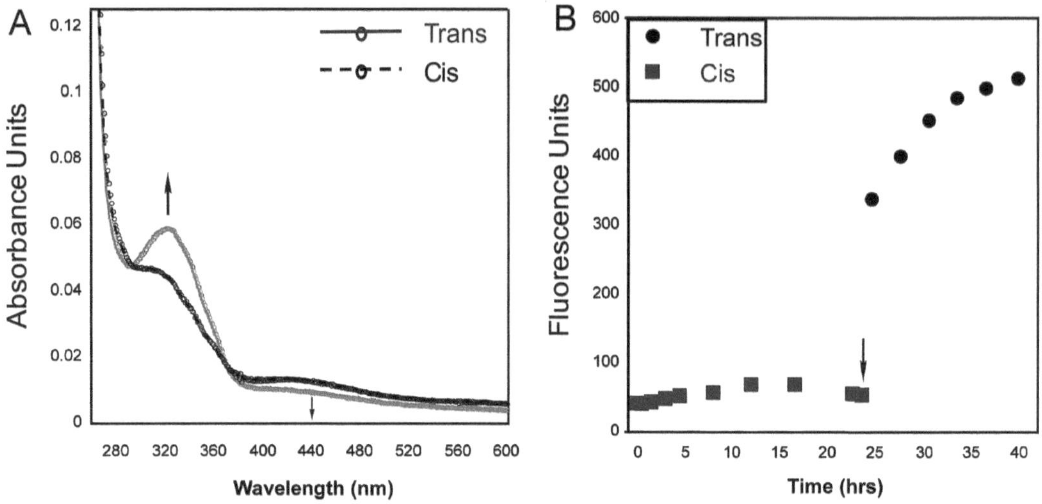

Fig. 3 Spectroscopic characterization of AMPP-modified Aβ42 before and after *cis* → *trans* isomerization in aqueous buffer. (**a**) Absorbance spectra of cis-AMPP-Aβ (black line). After 24 h, the cis-AMPP-Aβ was photoisomerized to trans-AMPP-Aβ and the absorbance spectrum recorded again (gray line). The arrows indicate the movement of the maxima after irradiation. (**b**) Kinetics of *cis*-AMPP-Aβ fibrillization were monitored using the ThT assay for 24 h. After photoisomerization (indicated by the arrow), the fibrillization kinetics assay was resumed for *trans*-AMPP-Aβ. All experiments were performed in PBS using 10 μM starting concentration of the AMPP-modified Aβ42 variants

5. Irradiate the sample in the cuvette with light of wavelength ~434 nm using a 400 nm longpass filter, or similar, for 1–2 h using a 500 W mercury lamp (*see* **Note 5**).
6. Mix the solution by inserting a gel-loading pipette tip and aspirating the solution in and out of the pipette tip at least 15–20 times. Do this every 10–15 min.
7. Move the cuvette to the UV-vis spectrophotometer and measure the absorbance spectrum in a 1 cm cuvette as in **step 2** (Fig. 3a, *see* **Note 22**).
8. Remove the sample from the UV-vis spectrophotometer and continue irradiating the sample at 434 nm for 30 min in the lamp apparatus.
9. Repeat **steps 6–8** until the absorbance of the peptide at 430 nm and 320 nm remains unchanged, indicating that conversion has reached its upper limit.
10. Monitor fibril formation by ThT as in **steps 9–10** of Subheading 3.3.1 (Fig. 3b, *see* **Note 17**)
11. Once fibrillization is complete, remove ~500 μL of sample to perform structural characterization experiments (*see* Subheading 3.4).

3.3.3 Trans → Cis Photoisomerization of AMPP-Aβ42 in Aqueous Buffer

Once fibrillization of the *trans*-AMPP-Aβ42 peptide is complete, it is possible to reverse cross-β sheet aggregation by isomerizing the *trans*-AMPP to *cis*-AMPP-Aβ42. This section describes the conversion of *trans*-AMPP-Aβ42 aggregates to *cis*-AMPP-Aβ42 aggregates and describes monitoring of cross-β sheet aggregates.

1. Dilute the *trans*-AMPP-Aβ42 variant DMSO stock from Subheading 3.2 into PBS to a final concentration of no more than 10 μM (*see* **Note 19**).
2. Transfer 148.5 μL of the *trans*-AMPP-Aβ42 solution and add 1.5 μL of stock ThT to a final ThT concentration of 25 μM.
3. Mix the sample by aspirating the sample into a pipette tip and expelling it 10 times. Then, measure ThT fluorescence over time by exciting the dye at 465 nm and recording the emission intensity at 497 nm.
4. Repeat **steps 2–3** every 15 min to monitor fibril formation as indicated by an increase in ThT fluorescence at 497 nm.
5. Once the elongation phase of fibrillization is complete the ThT fluorescence emission intensity at 497 nm will begin to plateau. At this stage, immediately transfer the AMPP-Aβ42 variant solution to a 0.1 cm quartz cuvette.
6. Irradiate the sample at 365 nm for 2 h to convert the *trans* isomers to a *cis* conformation (*see* **Note 23**).
7. Determine the extent of isomerization by monitoring the absorbance at 365 nm and 434 nm on a UV-vis spectrophotometer (Fig. 4a, *see* **Note 22**).
8. After 2 h of irradiation, monitor the level of fibrillization by ThT fluorescence. Transfer 148.5 μL to a fluorimeter sample vessel and add 1.5 μL ThT stock solution to a final concentration of 25 μM ThT.
9. Measure the ThT fluorescence at 497 nm using an excitation wavelength of 465 nm. If a reduction in fluorescence is observed, continue to **step 8**. If no reduction in fluorescence is observed, repeat **step 6**, but this time irradiate for 0.5–1 h each time (Fig. 4b).
10. Continue to monitor absorbance and ThT fluorescence over time by repeating **steps 6–9** until no additional change is observed.

3.4 Characterization of Fibrils Derived from Aβ42 and AMPP-Containing Variant

The following section describes methods to characterize the morphology of aggregated fibrils comprised of Aβ42 and the Aβ42 *cis/trans* variants.

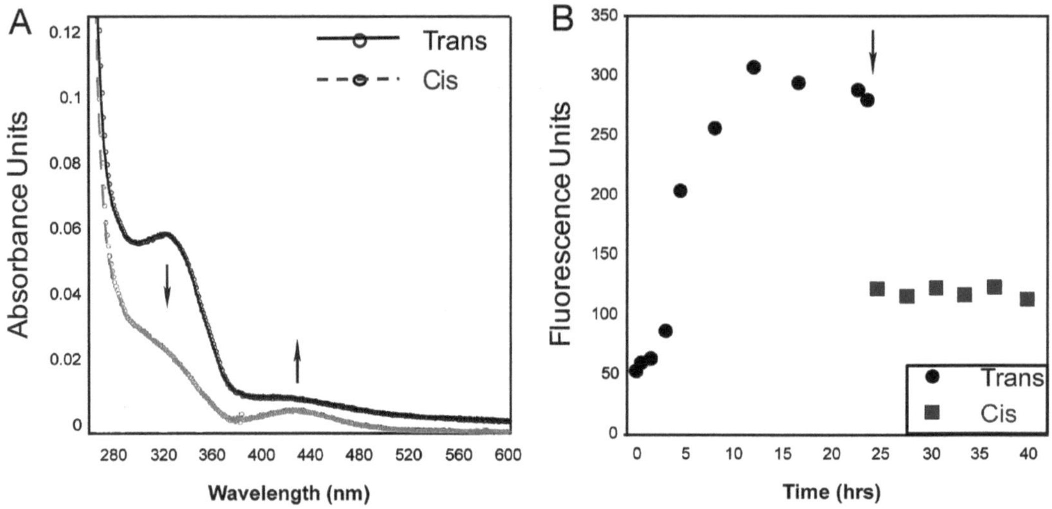

Fig. 4 Spectroscopic characterization of AMPP-modified Aβ42 before and after trans → cis isomerization in aqueous buffer. (**a**) Absorbance spectra of trans-AMPP-Aβ42 (black line). After 24 h, the cis-AMPP-Aβ was photoisomerized to cis-AMPP-Aβ and the absorbance spectrum recorded again (gray line). The arrows indicate the movement of the maxima after irradiation. (**b**) Fibrillization kinetics were monitored using the ThT assay for 24 h. After photoisomerization (indicated by the arrow), the fibrillization kinetics assay was resumed for *cis*-AMPP-Aβ42. All experiments were performed in PBS using 10 μM starting concentration of the AMPP-modified Aβ42 variants

3.4.1 Fourier Transform Infrared (FT-IR) Spectroscopy

1. Use the leftover insoluble aggregates from the kinetic experiments or prepare aggregates of the *cis*- and *trans*-Aβ42 conformers by irradiating a 2 mM DMSO stock solution at 365 nm or 434 nm to produce either the *cis*- or *trans*-Aβ42, respectively (*see* **Note 24**).
2. Dilute the DMSO stock solution into PBS to a final Aβ42 concentration of 10 μM.
3. Allow fibrils to assemble in the dark for 2 days.
4. Sediment the insoluble fibrils at 14,000 × *g*, and carefully remove the supernatant by decanting.
5. Wash the Aβ42 fibrils in three successive washes of 500 μL D_2O by suspending the fibrils, vortexing thoroughly, and sedimenting them, as in **step 4**. This D_2O exchange is necessary to avoid interference from water in the diagnostic region of the IR spectra.
6. After the final wash, resuspend the fibrils, freeze them by immersion in liquid nitrogen, and lyophilize the samples.
7. Apply the lyophilized fibrils to an ATR stage for FT-IR analysis.
8. Record the FT-IR spectrum of the sample to obtain diagnostic amide carbonyl stretching frequencies between 1600 and 1700 cm^{-1} (Fig. 5a).

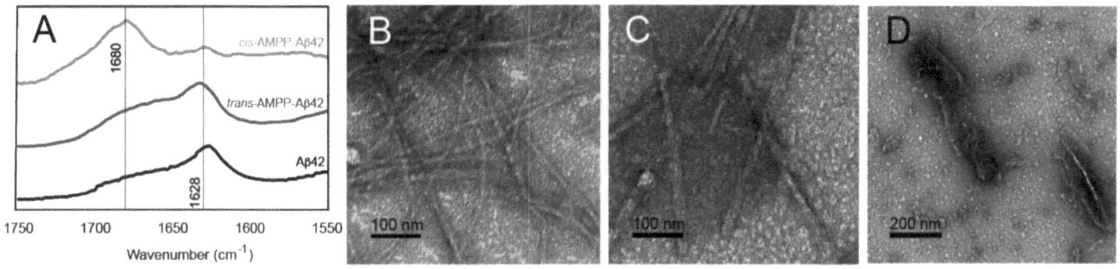

Fig. 5 Structural characterization of aggregates formed by Aβ42, *trans*-AMPP-Aβ42, and *cis*-AMPP-Aβ42. (**a**) FT-IR spectra obtained by examining aggregates comprised of Aβ42, *cis*-AMPP-Aβ42, or *trans*-AMPP-Aβ42. Wavenumbers for the strongest absorption peaks are indicated at 1628 cm^{-1} and 1680 cm^{-1} with vertical lines. (**b**) TEM images obtained from insoluble Aβ42 (**c**) TEM images obtained from insoluble *trans*-AMPP-Aβ42 aggregates. (**d**) TEM images of *cis*-AMPP-Aβ42 aggregates. Figures are reproduced with permission from Ref. 27

3.4.2 Transmission Electron Microscopy (TEM)

1. Prepare fibrils as described in **steps 1–3** of Subheading 3.4.1.
2. Apply 10 μL of a suspension of the mature fibrils onto a 200 mesh, carbon-coated copper TEM grid for 5 min.
3. Wick off excess buffer by gently touching filter paper to the edge of the grid.
4. Apply 10 μL H_2O to the grid for 10 s.
5. Remove excess H_2O by capillary action by touching a piece of filter paper gently to the edge of the grid.
6. Repeat **steps 4–5**.
7. Stain the grid with 10 μL TEM staining solution for 10 min.
8. Remove excess staining solution by gently touching filter paper to the edge of the grid.
9. Allow the grid to air-dry for at least 10 min prior to imaging.
10. Record digital TEM images using an electron microscope at 100,000–200,000× magnification (Fig. 5b, c). We typically use an accelerating voltage of 120 kV.

4 Notes

1. Alternatively, manual iterative SPPS peptide couplings may be performed, but we have observed that peptides prepared in this way suffer from poor product yield and quality.
2. Rigidity of the peptide backbone can be enhanced using azobenzene peptidomimetics containing fewer methylene units adjacent to the diazobenzene rings [34, 35]. Chain extension onto an aminophenyl nitrogen is difficult, however, and use of such moieties would require optimization of the chain extension step.
3. A C18 column can be used, though we obtained better results using a C4 column.

4. A lower power light source may be used, but longer irradiation times will be required. Optimization will be necessary.
5. Filters should be chosen judiciously, and tested to make sure that light is passable with a wavelength at or near the maxima of 320 nm or 430 nm. We typically use a longpass filter to block all incident light below 400 nm, a wavelength sufficiently high enough not to excite the *trans*-azobenzene chromophore around 330 nm. Any longpass filter can be used that has a cutoff between ~400 and 430 nm. A bandpass filter can also be used if available, but any of these changes will require optimizations (described in the Methods section). Similarly, though we employ a bandpass filter to narrow irradiating light wavelengths to 365 nm, a shortpass filter should sufficiently block light that would excite the *cis*-azobenzene chromophore.
6. Certain amino acids require special coupling conditions. You should consult with the manufacturer of your automated synthesizer about special coupling conditions that may not be listed here. More forcing conditions will be particularly important during manual synthesis. Some considerations are:
   (a) Always perform two identical coupling steps in sequence without washing or deprotecting (double couple) after β-branched amino acids Ile, Thr, and Val.
   (b) To reduce histidine and cysteine racemization during coupling, the CEM standard protocol is to reduce the temperature of coupling to no more than 50 °C after a 2 min RT reaction.
   (c) To reduce arginine lactam formation, couple arginine at room temperature for 25 min and then heat to 75 °C for 2 min. Double couple each arginine.
   (d) Double couple the last four residues onto the growing Aβ42 chain, which corresponds to the N-terminal DAEF sequence.
7. It is advised not to use automated synthesizers to deliver azobenzene peptidomimetics to the reaction vessel, as synthesizers often use spectral properties of the amino acid solution in DMF to determine the quantity delivered to the reaction vessel. Thus, care should be exercised if using automation for this step.
8. The beads will turn bright orange after a successful coupling of the AMPP chromophore onto the growing peptide chain. For a more precise qualitative test, unreacted amines can be detected using a ninhydrin test after the first and second coupling [36].
9. The product will give the acetonitrile-water mixture an orange tint. The lyophilized powder will be orange.
10. Your standard concentration curve must be a linear line. Due to detector limitations, high concentrations of peptide cannot be

measured accurately, and the curve will begin to plateau at these high concentrations. These points should not be used when constructing your best-fit line that will have the form $y = mx + b$. A good check to make sure you are working within the linear range is to ensure the square of your fitted linear line's Pearson's correlation coefficient is >95%. Also note that when measuring concentration, ensure that your experimental concentration falls within this linear range of your standard curve. If your absorbance is higher than the data points used to construct your standard curve, simply dilute your test sample and repeat HPLC-assisted quantification.

11. It is best to aliquot the desired amount of material at this pre-disaggregation stage and store under dry conditions below −20 °C. We typically dispose of unused stocks after 6 months of storage.
12. HFIP is volatile, and pipetting will be imprecise. Therefore, the concentration determined at this step should only be used as an estimation of the final peptide concentration.
13. HFIP must be dry at this stage, since any residual water will evaporate more slowly than HFIP, creating an aqueous environment that will seed fibrillization of the highly concentration Aβ42. For this reason, we always freshly distill our HFIP prior to use in disaggregation studies.
14. It is best to start with more peptide than needed at this stage, since the characterization techniques described here destroy the sample. If you run out of sample during the experiment, you must start over. It is advised that an initial pilot experiment be performed with a small amount of peptide to determine fibrillization kinetics under your experimental conditions. This will inform the number of time points that will be needed during the kinetics experiments.
15. Lyophilization is the preferred drying method here because a dry film can trap HFIP. A lyophilized powder also dissolves more quickly, whereas solvating a dry film takes more time and introduces high local concentrations of peptide that may seed aggregation.
16. It is convenient to perform photoisomerizations near a UV-vis spectrophotometer to monitor kinetics of photoswitching by observing the absorbance maxima at 320 nm and 430 nm. *Cis*-azobenzene absorbs at 430 nm and *trans*-azobenzene absorbs at 320 nm in Aβ42. Prior to photoisomerization, >80% of the peptide will contain AMPP in the trans conformation. HPLC is an inappropriate method to monitor the ratio of *cis*-AMPP:*trans*-AMPP in Aβ42 because the two isomers are difficult to resolve. In our hands, the Aβ42 chromatogram peak often has a "lagging tail." The predominant *trans*-AMPP-Aβ42 partially overlaps with the *cis*-AMPP-Aβ42, which elutes second.

17. The kinetics of switching can be affected by solvent polarity and local folding environment of the peptide [37].
18. We have found that typical irradiation times vary, and isomerization can be achieved between 15 and 30 min in the DMSO solution. However, this will vary depending on many experimental conditions such as cuvette pathlength, lamp power and filter strength, the peptide concentration, etc.
19. Amyloid self-assembly is concentration-dependent, and the preparation of Aβ42 may vary slightly due to synthetic purity, etc. Therefore, each laboratory should optimize the appropriate concentration to use for these experiments.
20. The alternative method (to add the ThT directly to the Aβ42 solution and read the fluorescence of this one solution throughout the entirety of the kinetics experiment) is inappropriate. ThT absorbs near 434 nm, and may be affected by irradiation with high-power lamps (e.g., photobleaching). We also suggest using a narrow bandwidth when exciting ThT to minimize absorption of the chromophore at wavelengths near the maximum of 430 nm.
21. The *cis*-AMPP chromophore will thermally isomerize back to the thermodynamically stable trans state over time. This will lead to fibril formation once the trans state is populated sufficiently enough to nucleate the misfolding pathway. Thermal cis-trans isomerization occurs in these peptides in water after ~48 h, though this time frame will depend on a multitude of experimental parameters.
22. The concentration of AMPP-modified Aβ42 may be too low to get an accurate absorbance using a 0.1 cm cuvette in a standard spectrophotometer. On the other hand, a 1 cm pathlength may require too large of a solution volume to obtain an accurate reading. If the volume is too low in a 1 cm pathlength cuvette for UV-vis, and a low-volume adaptor is unavailable, then multiple isomerizations must be performed in parallel as described in Subheading 3.3.2 in **step 4**. Each one should then be combined into a single 1 cm pathlength cuvette to obtain an accurate absorbance reading.
23. Presumably, almost all photoisomerization occurs when the Aβ42 variant is fully solvated. We have adopted the practice of isomerizing aggregates immediately when ThT begins to reach its maximum and mixing the solution by pipetting it in and out of the pipette tip inside the cuvette frequently. This way, all of the sedimented fibrils and protofibrils can be exposed to the light source.
24. To produce trans-AMPP-Aβ42 fibrils from a freshly disaggregated sample, you do not have to irradiate at 434 nm because the species in the sample will exist predominantly in the *trans-*

AMPP state due to thermal relaxation during synthesis and purification. However, the absorbance at 365 nm should still be recorded at 365 nm on a photodiode array spectrophotometer to estimate the ratio of *trans:cis* conformers in the thermally relaxed state and to note change in absorbance at 320 nm during isomerization.

Acknowledgments

This work was made possible by a grant from the Alzheimer's Association (NIRG-08-90797). We thank Professor Joseph P. Dinnocenzo for helpful discussions regarding photoisomerization methods and Karen Bentley of the University of Rochester Medical Center Electron Microscopy Core for assistance with transmission electron microscopy.

References

1. Braak H, Braak E (1991) Neuropathological stageing if Alzeimer-related changes. Acta Neuropathol 82:239–259
2. Qian X, Hamad B, Dias-LAlcaca G (2015) The Alzheimer disease market. Nat Rev Drug Discov 14:675–676
3. Jan Bieschke MH, Wiglenda T, Friedrich RP, Boeddrich A, Schiele F, Kleckers D, del Amo JML, Grüning BA, Wang Q, Schmidt MR, Lurz R, Anwyl R, Schnoegl S, Fändrich M, Frank RF, Reif B, Günther S, Walsh DM, Wanker EE (2012) Small-molecule conversion of toxic oligomers to nontoxic β-sheet–rich amyloid fibrils. Nat Chem Biol 8:93–101
4. Walsh DM, Klyubin I, Fadeeva JV, Cullen WK, Anwyl R, Wolfe MS, Rowan MJ, Selkoe DJ (2002) Naturally secreted oligomers of amyloid-β protein potently inhibit hippocampal long-term potentiation in vivo. Nature 416:535–539
5. Walsh DM, Selkoe DJ (2007) Aβ oligomers—a decade of discovery. J Neurochem 101:1172–1184
6. Sylvain Lesne´ MTK, Kotilinek L, Kayed R, Glabe CG, Yang A, Gallagher M, Ashe KH (2006) A specific amyloid-β protein assembly in the brain impairs memory. Nature 440:352–357
7. Liu P, Reed MN, Kotilinek LA, Grant MK, Forster CL, Qiang W, Shapiro SL, Reichl JH, Chiang AC, Jankowsky JL, Wilmot CM, Cleary JP, Zahs KR, Ashe KH (2015) Quaternary structure defines a large class of amyloid-β oligomers neutralized by sequestration. Cell Rep 11:1760–1771
8. Kayed R, Head E, Thompson JL, McIntire TM, Milton SC, Cotman CW, Glabe CG (2003) Common structure of soluble amyloid oligomers implies common mechanism of pathogenesis. Science 300:486–489
9. Brian O'Nuallain DBF, Nicoll AJ, Risse E, Ferguson N, Herron CE, Collinge J, Walsh DM (2010) Amyloid-protein dimers rapidly form stable synaptotoxic protofibrils. J Neurosci 30:14411–14419
10. Chiti F, Dobson CM (2006) Protein misfolding, functional amyloid, and human disease. Annu Rev Immunol 75:333–366
11. Liang Y, Lynn DG, Berland KM (2010) Direct observation of nucleation and growth in amyloid self-assembly. J Am Chem Soc 132:6306–6308
12. Petkova AT, Ishii Y, Balbach JJ, Antzutkin ON, Leapman RD, Delaglio F, Tycko R (2002) A structural model for Alzheimer's β-amyloid fibrils based on experimental constraints from solid state NMR. Proc Natl Acad Sci U S A 99:16742–16747
13. Paravastu AK, Leapman RD, Yau W-M, Tycko R (2008) Molecular structural basis for polymorphism in Alzheimer's β-amyloid fibrils. Proc Natl Acad Sci U S A 105:18349–18354
14. Lührs T, Ritter C, Adrian M, Riek-Loher D, Bohrmann B, Döbeli H, Schubert D, Riek R (2005) 3D structure of Alzheimer's amyloid-β (1–42) fibrils. Proc Natl Acad Sci U S A 102:17342–17347
15. Yiling Xiao BM, McElheny D, Parthasarathy S, Long F, Hoshi M, Nussinov R, Ishii Y (2015) Aβ(1–42) fibril structure illuminates self-

recognition and replication of amyloid in Alzheimer's disease. Nat Struct Mol Biol 22:499–505
16. Petkova AT, Yau W-M, Tycko R (2006) Experimental constraints on quaternary structure in Alzheimer's β-amyloid fibrils. Biochemistry 45:498–512
17. Hoyer W, Grönwall C, Jonsson A, Ståhl S, Härd T (2008) Stabilization of a β-hairpin in monomeric Alzheimer's amyloid-β peptide inhibits amyloid formation. Proc Natl Acad Sci U S A 105:5099–5104
18. Lazo ND, Grant MA, Condron MC, Rigby AC, Teplow DB (2005) On the nucleation of amyloid β-protein monomer folding. Protein Sci 14:1581–1596
19. Tomaselli S, Esposito V, Vangone P, NAJv N, Bonvin AMJJ, Guerrini R, Tancredi T, Temussi PA, Picone D (2006) The α-to-β conformational transition of Alzheimer's Aβ-(1–42) peptide in aqueous media is reversible: a step by step conformational analysis suggests the location of β conformation seeding. Chembiochem 7:257–267
20. Kajava AV, Baxa U, Steven AC (2010) β arcades: recurring motifs in naturally occurring and disease-related amyloid fibrils. FASEB J 24:1311–1319
21. Cerf E, Sarroukh R, Tamamizu-Kato S, Breydo L, Derclaye S, Dufrene YF, Narayanaswami V, Goormaghtigh E, Ruysschaert J-M, Raussens V (2009) Antiparallel β-sheet: a signature structure of the oligomeric amyloid β-peptide. Biochem J 421:415–423
22. Zimmerman DHI, Martin PK, Nix AJ, Rosenberry TL, Paravastu AK (2015) Antiparallel β-sheet structure within the C-terminal region of 42-residue Alzheimer's amyloid-β peptides when they form 150-kDa oligomers. J Mol Biol 427:2319–2328
23. Ahmed M, Davis J, Aucoin D, Sato T, Ahuja S, Aimoto S, Elliott JI, Sostrand WEV, Smith SO (2010) Structural conversion of neurotoxic amyloid-β(1-42) oligomers to fibrils. Nat Struct Mol Biol 17:561–567
24. Sandberg A, Luheshi LM, Söllvander S, TPd B, Macao B, Knowles TPJ, Biverstål H, Lendel C, Ekholm-Petterson F, Dubnovitsky A, Lannfelt L, Dobson CM, Härd T (2010) Stabilization of neurotoxic Alzheimer amyloid-β oligomers by protein engineering. Proc Natl Acad Sci U S A 107:15595–15600
25. Sciarretta KL, Gordon DJ, Petkova AT, Tycko R, Meredith SC (2005) Aβ40-Lactam (D23/K28) models a conformation highly favorable for nucleation of amyloid. Biochemistry 44:6003–6014
26. Doran TM, Anderson EA, Latchney SE, Opanashuk LA, Nilsson BL (2012) Turn nucleation perturbs amyloid β self-assembly and cytotoxicity. J Mol Biol 421:315–328
27. Doran TM, Anderson EA, Latchney SE, Opanashuk LA, Nilsson BL (2012) An azobenzene photoswitch sheds light on turn nucleation in amyloid-β self-assembly. ACS Chem Neurosci 3:211–220
28. Aemissegger A, Kräutler V, Gunsteren WFV, Hilvert D (2005) A Photoinducible β-Hairpin. J Am Chem Soc 127:2929–2936
29. Aemissegger A, Hilvert D (2007) Synthesis and application of an azobenzene amino acid as a light-switchable turn element in polypeptides. Nat Protoc 2:161–167
30. Kräutler V, Aemissegger A, Hünenberger PH, Hilvert D, Hansson T, Gunsteren WFV (2005) Use of molecular dynamics in the design and structure DETERMINATION of a photoinducible β-hairpin. J Am Chem Soc 127:4935–4942
31. Dong S-L, Loweneck M, Schrader TE, Schreier WJ, Moroder L, Renner C (2006) A photocontrolled β-hairpin peptide. Chem Eur J 12:1114–1120
32. O'Nuallain B, Thakur AK, Williams AD, Bhattacharyya AM, Chen S, Thiagarajan G, Wetzel R (2006) Kinetics and thermodynamics of amyloid assembly using a high-performance liquid chromatography-based sedimentation assay. Methods Enzymol 413:34–74
33. O'Nuallain B, Shivaprasad S, Kheterpal I, Wetzel R (2005) Thermodynamics of Aβ(1–40) amyloid fibril formation. Biochemistry 44:12709–12718
34. Ulysse L, Cubillos J, Chmielewski J (1995) Photoregulation of cyclic peptide conformation. J Am Chem Soc 117:8466–8467
35. Behrendt R, Renner C, Schenk M, Wang F, Wachtveitl J, Oesterhelt D, Moroder L (1999) Photomodulation of the conformation of cyclic peptides with azobenzene moieties in the peptide backbone. Angew Chem Int Ed 38:2771–2774
36. Donald A, Wellings EA (1997) Standard Fmoc protocols. Methods Enzymol 289:44–67
37. Chi L, Sadovski O, Woolley GA (2006) A blue-green absorbing cross-linker for rapid photoswitching of peptide helix content. Bioconjug Chem 17:670–676

Chapter 26

Solid-State NMR Studies of Amyloid Materials: A Protocol to Define an Atomic Model of Aβ(1–42) in Amyloid Fibrils

Yiling Xiao, Dan McElheny, Minako Hoshi, and Yoshitaka Ishii

Abstract

Intense efforts have been made to understand the molecular structures of misfolded amyloid β (Aβ) in order to gain insight into the pathological mechanism of Alzheimer's disease. Solid-state NMR spectroscopy (SSNMR) is considered a primary tool for elucidating the structures of insoluble and noncrystalline amyloid fibrils and other amyloid assemblies. In this chapter, we describe a detailed protocol to obtain the first atomic model of the 42-residue human Aβ peptide Aβ(1–42) in structurally homogeneous amyloid fibrils from our recent SSNMR study (Nat Struct Mol Biol 22:499–505, 2015). Despite great biological and clinical interest in Aβ(1–42) fibrils, their structural details have been long-elusive until this study. The protocol is divided into four sections. First, the solid-phase peptide synthesis (SPPS) and purification of monomeric Aβ(1–42) is described. We illustrate a controlled incubation method to prompt misfolding of Aβ(1–42) into homogeneous amyloid fibrils in an aqueous solution with fragmented Aβ(1–42) fibrils as seeds. Next, we detail analysis of Aβ(1–42) fibrils by SSNMR to obtain structural restraints. Finally, we describe methods to construct atomic models of Aβ(1–42) fibrils based on SSNMR results through two-stage molecular dynamics calculations.

Key words Aβ(1–42), Amyloid fibrils, Seeding method, SSNMR, Structure, Atomic model

1 Introduction

Misfolded fibrillar aggregates of amyloid β (Aβ) peptide are the primary components of amyloid plaques in Alzheimer's disease (AD) [1]. While monomeric Aβ is not toxic, misfolded Aβ fibrils exhibit toxicity [2, 3]. Aside from Aβ fibrils, several diffusible Aβ oligomers are reported to be highly neurotoxic [4–10]. A handful of mutations of Aβ are associated with early onset of AD, and many of the mutations promote formation of Aβ amyloid plaques [11]. Therefore, misfolded Aβ is believed to play a vital role in the development of AD.

Aβ is produced via proteolysis of amyloid precursor protein (APP) in the brain [12]. Two of the major isoforms of Aβ are 40-residue Aβ (Aβ(1–40)) and 42-residue Aβ (Aβ(1–42)). Despite

Bradley L. Nilsson and Todd M. Doran (eds.), *Peptide Self-Assembly: Methods and Protocols*, Methods in Molecular Biology, vol. 1777, https://doi.org/10.1007/978-1-4939-7811-3_26,

minor sequence differences at the C-terminus, Aβ(1–40) and Aβ (1–42) vary significantly in their propensity to aggregate, and they are considered to play different roles in AD [13–17]. The more soluble Aβ(1–40) is the abundant form of Aβ in plasma and cerebrospinal fluid in AD [18]. It is also the predominant species in vascular amyloid found in cerebral amyloid angiopathy (CAA) [16, 19]. In contrast, Aβ(1–42) is generally considered to be more pathogenically relevant to AD. Although Aβ(1–42) constitutes only approximately 20% of the total Aβ in plasma [20], fibrillar Aβ(1–42) is often the predominant species in AD plaques [14, 15, 21]. Previous studies indicated a lowered diffusible Aβ(1–42)/Aβ (1–40) ratio as a risk factor for AD [22, 23], presumably through selective aggregation of Aβ(1–42). In addition, Aβ(1–42) fibrils show higher neurotoxicity than Aβ(1–40) fibrils [11, 24]. Thus, Aβ(1–42), rather than Aβ(1–40), has been linked to the onset of AD.

Intensive efforts have focused on structural studies of amyloid aggregates of Aβ [8, 25–33] and other amyloid proteins [34–37], as molecular structures of the amyloids may give insight into the pathological mechanisms and possible means for preventing amyloid aggregation in AD [38, 39]. Due to their insoluble and non-crystalline nature, solid-state NMR (SSNMR) has become the primary tool for studying structures of amyloid and other proteins at atomic resolution [25, 40–43]. Multiple fibril structures of Aβ(1–40) prepared in vitro have been characterized by SSNMR and electron microscope (EM) techniques [26, 28–30]. However, despite numerous attempts, only a few crude structural models of Aβ(1–42) fibrils exist [9, 27, 31, 32, 44], with limited structural restraints and little assessment of the structural homogeneity of the analyzed fibrils. Because misfolding of Aβ(1–42) often leads to the formation of heterogeneous aggregates or polymorphs of Aβ fibrils [31, 32], a critical challenge remains in the preparation of homogeneous Aβ(1–42) fibrils that are suitable for structural investigation.

In the following protocol, we describe detailed procedures for the sample preparation and structural analysis used in our recently published work on the first elucidation of an atomic model for structurally homogenous Aβ(1–42) fibril based on SSNMR data [45]. In the structural study, we first prepared Aβ(1–42) fibrils with a "seeding" method [25]. The structural homogeneity of the seeded fibril sample was validated by transmission electron microscopy (TEM), Thioflavin T (ThT) fluorescence, and SSNMR results. Well-resolved two-dimensional (2D) ^{13}C, ^{15}N-SSNMR spectra allowed for a detailed structure analysis. Our SSNMR-based structure modeling yielded a unique "S"-shaped triple β-sheet motif with an in-register parallel β-sheet arrangement. The triple-β motif of the fibril is stabilized with a Lys28-Ala42 salt bridge, which is linked with unique biological properties of Aβ(1–42).

The protocol can be utilized for structural analyses of other misfolded amyloid peptides by SSNMR.

2 Materials

2.1 Aβ(1–42) Peptide Preparations

The following reagents and materials are prepared for solid-phase peptide synthesis (SPPS) and purification of Aβ(1–42). If any specific reagents and materials listed below, they were used in our past research and are given as an example. Although equivalent products may be usable, it should be noted that SPPS and purification of Aβ (1–42) are both generally very difficult unless reagents and conditions are highly optimized. All aqueous solutions are prepared with double deionized and distilled water (DDI water) purified with a High Q water purification system (High-Q Inc. Wilmette IL). Buffers described below are used without 0.22 μM filtration unless otherwise mentioned. All of the reagents used for SPPS are peptide synthesis grade.

1. 0.1 mmol equivalence (~500 mg) 9-fluorenyl-methyl-oxycarbonyl-(Fmoc)-Ala-Novasyn TGT resin (Novabiochem, 8.56125.0005; 0.2 mmol/g substitution).
2. Automated peptide synthesizer capable of a 0.1 mmol scale synthesis protocol, such as Applied Biosystems, Model ABI 433A (*see* **Note 1**).
3. An 8 mL reaction vessel (Applied Biosystems, 401571).
4. Amino acid cartridges (AnaSpec, AS-60881) packed with 1 mmol of unlabeled Fmoc-protected amino acids. (Peptides International).
5. 0.4 mmol of ^{13}C, ^{15}N-uniformly labeled amino acid (Sigma-Aldrich or Cambridge Isotope) with Fmoc-protection.
6. 400 mL of Fmoc deprotection reagent: 99.5% piperidine (Sigma, 571261-450 mL).
7. 400 mL of capping reagent: 19 mL of acetic anhydride, 25 mL of 2.0 M N-diisopropylethylamine (DIEA) in 356 mL N-methylpyrrolidone (NMP, Applied Biosystems, 401517).
8. 6 mL of 1.0 M 1-hydroxybenzotriazole (HOBt) in NMP (ReagentWorld, ABCA-0).
9. 350 mL NMP (Applied Biosystems, 400580).
10. 400 mL of unlabeled amino acid activation reagent: 0.5 M 1H-benzotriazolium-1-[bis (dimethylamino) methylene]-5-chloro-hexafluorophosphate-(1-), 3-Oxide (HCTU, Peptides International, KHC-1018-PI) in N, N-dimethylformamide (DMF, Applied Biosystems, Gen002007).

11. 100 mL of labeled amino acid activation reagent: 0.2 M HCTU in DMF.
12. Coupling reagent for coupling unlabeled amino acids: 2.0 M DIEA in NMP.
13. Coupling reagent for coupling labeled amino acids: 0.8 M DIEA in NMP.
14. Washing reagent: NMP.
15. Final washing reagent: dichloromethane (DCM, Applied Biosystems, 400142).
16. Cleavage mixture B: containing 0.375 g of crystalline phenol (Fisher Scientific, A92-100), 5.000 mL of trifluoroacetic acid (TFA) (AmericanBIO, Ab02010-00500), 0.25 mL of water, 0.25 mL of thioanisole (Fluka, 88470), and 0.125 mL of 1,2-ethanedithiol (Fluka, 02390).
17. PTFE syringe filter, 13 mm in diameter, with 0.5 μm pore size (GE Healthcare, 6784-1350).
18. Peptide extraction solvent: ethyl ether (Fisher Scientific, E134-4).
19. Analytical balance, max: 200 g; $d = 0.1$ mg.
20. 10% acetonitrile (ACN) solution: 10% ACN in DDI water with 0.1% TFA.
21. 30% ACN solution: 30% ACN in water with 0.1% TFA.
22. HPLC system (Shimadzu, CBM-20A), equipped with a UV-vis detector (Shimadzu, SPD-20A), a degassing unit (Shimadzu, DGU-20A), and a pump (Shimadzu, LC-20AT).
23. HPLC column (Agilent Technologies, 5mic Zorbax 300 - Extend-C18, 9.4 × 250, 899999-555).
24. HPLC mobile water phase: 70 mM NH_4OH in 1 L water.
25. HPLC mobile organic phase: 70 mM NH_4OH in 200 mL of water and 800 mL of acetonitrile.
26. Lyophilizer (Labconco, Model FreeZone 4.5 Plus, 7386021) for freeze-drying purified Aβ(1–42) peptide and fibril samples harvested for SSNMR.
27. Purified Aβ(1–42) peptide with purity >95% (*see* **Note 2**), stored under −20 °C (*see* **Note 3**).
28. Hexafluoroisopropanol (HFIP, Sigma-Aldrich, 105228).
29. Lyophilizer (EYELA, Model FDU-1200) for removing HFIP from an Aβ(1–42) solution in HFIP (*see* **Note 4**).

2.2 SSNMR Sample Preparation of Seeded Aβ(1–42) Fibrils

1. Peptide dissolution solution: 10 mM NaOH in DDI water, stored at 4 °C until use.
2. Peptide dilution buffer: 10 mM sodium, potassium phosphate buffer containing 7.74 mM Na_2HPO_4, 2.26 mM KH_2PO_4,

and 0.02% (w/v) NaN_3 at pH 7.5, filtered through a 0.22 μm filter. Use 0.1 M HCl in DDI water to adjust pH. The buffer and HCl solution were stored at 4 °C until use.

3. pH meter equipped with a micro and semimicro size glass combination electrodes.
4. 0.22 μm sterile syringe filters with PES membrane (EMD Millipore, SLG033RS).
5. Amicon ultra-4 and ultra-15 centrifugal filters (EMD Millipore, UFC805096 and UFC905096). They have regenerated cellulose membrane of 50 kDa nominal molecular weight limit (NMWL).
6. Refrigerated centrifuge (Eppendorf, Model 5810R) equipped with a fixed angle rotor (Eppendorf, FA-45-6-30) with capacity of 6 × 15/50 mL conical tubes, and a maximum spinning frequency of 12,100 rpm (20,130 × *g*).
7. UV-vis spectrometer (Jasco, Model V-660) assembled with a cell holder for a 50 μL micro cell (Jasco, EMC-709).
8. Fluorescence spectrometer (Hitachi, Model F-7000).
9. Thioflavin T (ThT) solution: 300 μM ThT in water, stored in a light-proof container at 4 °C until use.
10. Glycine buffer: 0.05 M glycine and 4 μM NaOH in water at pH 9.0, stored at RT.
11. Ultrasonic cleaner (Fisher Scientific, Model FS20H), used as a sonication bath with ice water.
12. Rotating shaker (Barnstead, C415110), for agitating the Aβ solution during incubation for fibril growth.
13. Staining solution for TEM and scanning TEM (STEM): 2% (w/v) uranyl acetate in water, stored at room temperature, in a light-proof container. The solution was filtered through a 0.22 μm filter before every use (*see* **Note 5**). Fixing solution for STEM: 2% glutaraldehyde in DDI water, stored at 4 °C, and filtered through a 0.22 μm filter right before use (*see* **Note 6**).
14. TEM grids: formvar/carbon 300 copper mesh (Electron Microscopy Sciences, FCF300-CU-50).
15. STEM grids: carbon 400 copper mesh (Electron Microscopy Sciences, CF400-CU-50).
16. TEM instrument (JEOL, Model JEOL-1200) and STEM instrument (JEOL, Model JEM-ARM200CF).

2.3 SSNMR Spectroscopy

1. 2.5 mm MAS rotor (11 uL sample volume) with Torlon caps and zirconia sleeve (Revolution NMR, AMP4085-001).
2. Fluorosilicone (fs) O-ring, with an inside diameter of 0.036″ and a cross section of 0.016″ (Apple Rubber Products, R0003601660FSE).

3. Custom-made Kel-F cylindrical spacer of 0.062″ in diameter and 0.191″ in height. The surface of the spacer has a machine cut circular groove across the cylindrical axis (depth 0.0135″, width 0.025″) to accommodate the O-ring for better seal. The upper edge of the groove is positioned at 0.025″ from one end of the cylinder.
4. Teflon tape.
5. Packing tools: A Teflon 2.5 mm funnel, a brass packing tool, a removal tool, and a stainless steel packing stick.
6. SSNMR spectrometer. In our previous studies, all the SSNMR experiments were performed on a 9.4 T magnetic field (^{1}H frequency of 400.2 MHz) with a Bruker Avance III or Varian Infinity-Plus SSNMR spectrometer with a home built ^{1}H, ^{13}C, ^{15}N triple-resonance 2.5 mm MAS probe that was capable of stable spinning at 10–30 kHz. A triple-resonance SSNMR system at a ^{1}H NMR frequency of 400 MHz or higher is recommended.

2.4 Software for SSNMR Data Analysis and Structure Calculations

1. NMRPipe [46].
2. TALOS-N [55].
3. Cyana 2.1 [47].
4. SpinEvolution [48].
5. AMBER 12 [49].
6. SHIFTX2 [50].
7. STRIDE [51].
8. VMD 1.9.1.
9. PROCKECK-NMR [52].

3 Methods

3.1 Peptide Synthesis and Purification of Aβ (1–42) in Various ^{13}C- and ^{15}N-Labeling Schemes

Automated peptide syntheses of Aβ(1–42) are performed with an ABI 433 (or equivalent) synthesizer following the manufacture's protocol using "fast Fmoc" chemistry (detailed operation protocols are available at online user manuals) with minor modifications.

1. Pack the 8 mL reaction vessel with 0.1 mmol Fmoc-Ala-Novasyn TGT resin.
2. Set the reaction vessel at the synthesizer and swell the resin with 3 mL of DCM for 30 min using the synthesizer's standard function. Repeat the swelling process twice.
3. Extend the Aβ(1–42) peptide chain (DAEFR-HDSGY-EVHHQ-KLVFF-AEDVG-SNKGA-IIGLM-VGGVV-IA)

from the C-terminus one amino acid per cycle using the following coupling cycle:

(a) Deprotect Fmoc using Fmoc deprotection reagent.

(b) Activate Fmoc amino acid using activation reagent.

(c) Peptide bond formation with coupling reagent.

(d) Cap unreacted N-terminal residue with capping reagent.

4. Remove the final Fmoc group from the resin after the last coupling cycle.
5. Wash the deprotected resin with NMP and then DCM.
6. Transfer the wet resin from the reaction vessel to a conical tube.
7. Vacuum dry the resin for 1 day to completely remove DCM.
8. The dried resin of Aβ(1–42) (*see* Table 1) was stored at −20 °C.
9. Remove the Aβ(1–42) peptide from the resin beads with a cleavage mixture B (15 μL of reagent per 1 mg of resin). Gently

Table 1
Labeled Aβ(1–42) samples described in this study

Sample Number	Seeding generation	Labeling Scheme[a]
1	Unseeded 1st generation	F20, A21, V24, G25, L34
2[b]	Seeded 4th generation	F20, A21, V24, G25, L34
3	Seeded 4th generation	A2, G9, F20, V39, I41
4	Seeded 13th generation	F4, V12, L17, A21, G29
5	Seeded 4th generation	H13, V18, I32, G37
6	Seeded 4th generation	V18, D23, S26, K28
7[b]	Seeded 4th generation	F19, A30, I31, G33, V36
8	Seeded 4th generation	F19, A30, I32, G38, V40
9	Seeded 4th generation	F19, I31, L34, G38, V40
10	Seeded 4th generation	A21(^{15}N), A30(^{13}CO)
11	Seeded 4th generation	L17(^{15}N), L34(^{13}CO)
12	Seeded 4th generation	I31, V39
13[b]	Seeded 4th generation	G29, I41
14[b]	Seeded 4th generation	K28(^{15}N), A42
15	Seeded 4th generation	H14, L17, V36, G38
16	Seeded 4th generation	V39(^{13}CO)

[a]Unless otherwise mentioned, ^{13}C-, ^{15}N-uniformly labeled amino acids were introduced at the residue noted above

[b]Isotope diluted samples (50%) were also prepared

stir the resin in the cleavage mixture for 2.5 h at room temperature.

10. Remove the resin beads with a 0.5 μm PTFE syringe filter and keep the flow-through containing Aβ(1–42) at 4 °C.
11. Wash the resin beads in the syringe with TFA three times (total TFA amount was 30 μL per mg of resin), and combine all flow-through at 4 °C.
12. Extract Aβ(1–42) from the cleavage mixture from **step 10** by adding ice-cold ethyl ether (5–10 mL) to the cleavage mixture.
13. Isolate the Aβ(1–42) pellets by discarding the supernatant after centrifugation at 6000 rcf at −9 °C for 5 min.
14. Wash the Aβ(1–42) pellets two more times with ice-cold ether (5–10 mL for each wash), and discard the ether after pelleting down Aβ(1–42) under the same centrifugation conditions (*see* **Note 6**).
15. Vacuum dry the ether-extracted Aβ(1–42) for 1 h.
16. Add 5 mL of 30% ACN solution to dissolve the peptide at 4 °C.
17. After 30 min of quiescent dissolution at 4 °C, gently tap the conical tube containing the sample against the bench top for 10 min.
18. After centrifugation of the Aβ solution at 3000 rpm for 1.3 min at 4 °C, transfer the supernatant containing dissolved Aβ to a weighed clean conical tube.
19. Add a 5 mL aliquot of fresh 30% ACN solution to the undissolved Aβ(1–42) pellet and collect the soluble Aβ by combining all supernatants after tapping-dissolution.
20. Repeat **steps 17** and **18** until most of the Aβ(1–42) pellets are dissolved. The combined supernatant was lyophilized for 2 days before purification.
21. Estimate the amount of crude Aβ(1–42) by measuring weights of the conical tube with and without the dry Aβ(1–42) protein.
22. Dissolve the Aβ(1–42) with 10% ACN solution to a concentration of 3 mg/mL on ice for 40 min and filter the solution with a 0.22 μm syringe filter before HPLC separation.
23. Purify Aβ(1–42) by reverse-phase HPLC using a gradient aqueous and organic phase. The mobile phase flow rate is 4 mL/min. A water-ACN gradient (containing 70 mM NH_4OH) is applied. The elution of Aβ(1–42) takes place when the ACN concentration reaches about 25%.
24. Lyophilize eluates containing pure Aβ(1–42) overnight. The purity of the Aβ samples should be approximately 85% and 95% before and after the HPLC purification (*see* **Note 2**).

25. Weigh the lyophilized peptide after HPLC purification and redissolve it in 30% ACN solution at a 2 mg/mL peptide concentration.
26. Lyophilize the solution again overnight. The lyophilized peptides were stored with drying reagents at −20 °C.

3.2 Preparation of Aβ (1–42) Fibrils for SSNMR Analysis

1. Warm the purified Aβ(1–42) peptide from −20 °C to room temperature (RT) inside a desiccator for 0.5–1 h.
2. Dissolve the peptide in HFIP at a concentration of approximately 2 mg/mL at RT (*see* **Note 3**).
3. After an hour of dissolution in HFIP, flash freeze the Aβ(1–42) solution in liquid nitrogen and lyophilize it for 3–4 h to remove HFIP.
4. Redissolve the lyophilized Aβ(1–42) peptide in HFIP at a concentration of approximately 2 mg/mL for 1 h and then freeze-dry again. The second lyophilization should be performed overnight to ensure complete removal of HFIP. The two successive HFIP treatments disaggregate preformed aggregates in the Aβ(1–42) sample and the pretreated Aβ (1–42) peptide is ready for further experiments [6].
5. Dissolve the Aβ(1–42) peptide from **step 4** in an ice-cold 10 mM NaOH solution to a concentration of 0.6 mM, and then dilute the Aß solution further to 60 μM with 10 mM phosphate buffer at 4 °C. Adjust the pH of the Aβ solution to pH 7.4 with 0.1 M HCl (about 1/10 in volume of the total Aβ solution) on ice.
6. Remove any undissolved peptide or preformed aggregates from the ice-cold fresh Aβ(1–42) solution by centrifuge filtration with a 50 kDa molecular mass cutoff filter at $4.8 \times 10^3 \times g$ for 3 min at 4 °C.
7. Use the ThT assay and TEM analysis (described below at **steps 9** and **12**) to confirm that no aggregated Aβ remains in the solution at the beginning of the incubation.
8. Determine the concentration of the recovered Aβ monomer spectrophotometrically by measuring the absorbance at 280 nm and 400 nm in triplicate. Calculate the concentration of Aβ monomer (typically about 50 μM) with an equation based on the Beer-Lambert law as:

$$c_{A\beta} = \frac{\left[\left(\mathrm{Abs}_{A\beta(280\ \mathrm{nm})} - \mathrm{Abs}_{A\beta(400\ \mathrm{nm})}\right) - \left(\mathrm{Abs}_{\mathrm{blank}(280\ \mathrm{nm})} - \mathrm{Abs}_{\mathrm{blank}(400\ \mathrm{nm})}\right)\right]}{b_{\mathrm{cell}} \cdot \varepsilon_{A\beta(280\ \mathrm{nm})}},$$

where the cell length of the cuvette bcell was 1.0 cm in our previous study, and the blank was the phosphate buffer used for dilution of Aβ monomer. The extinction coefficient of Aβ (1–42) peptide is εAβ(280 nm) = 1280 (M−1 cm−1). The minimal baseline absorption is measured at 400 nm, confirming that there is no scattering due to aggregated Aβ.

9. Monitor the time course for Aβ(1–42) fibril formation using a ThT fluorescence assay: Dilute a 10 μL aliquot of the Aβ(1–42) solution with a premixed ThT solution containing 990 μL of glycine buffer and 10 μL of ThT solution. The unseeded fresh monomeric Aβ solution should not emit fluorescence above background level in this assay.
10. For the non-seeding incubation, place the Aβ(1–42) solution on a rotator for agitation at RT for 3–7 days (typically 3–4 days). We term the fibril formed under such a condition *generation one* (G_1). For the seeding incubation, sonicate the previously formed fibrils from the *n*-th generation in an ice water bath for 2 min and then introduce the sonicated sample as seed fibrils into a freshly prepared Aβ(1–42) monomer solution from the **steps 1–9** at a concentration of 5% (w/w). Then agitate the Aβ solution on a rotator for seeded fibril formation at room temperature for 1–7 days (typically 3–4 days). The seeded fibril is named as generation $(n + 1)$ (G_{n+1}). See a schematic illustration of the protocol in Fig. 1 with typical electron microscope images of Aβ(1–42) fibrils for different generations.
11. Monitor fibril formation with the ThT fluorescence assay as described in **step 9**. The fluorescence curve represents the β-sheet formation. Fit the sigmoidal curves using an equation of:

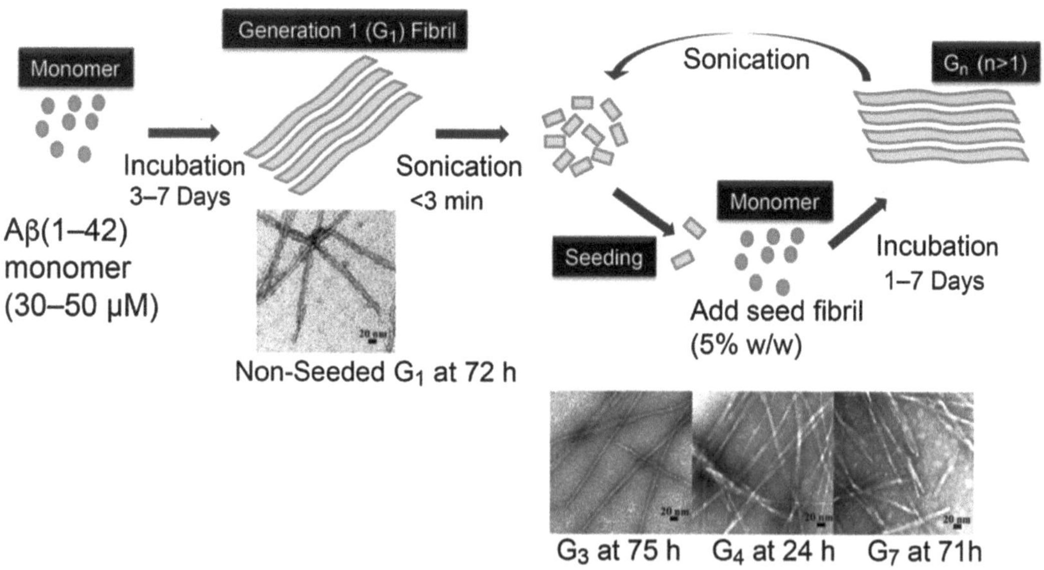

Fig. 1 A schematic illustration of preparation of Aβ(1–42) fibril samples used for SSNMR analysis. The electron microscope images were obtained for the fibrils sampled at the incubation times indicated in the figure

$$y(t) = a/[1 + \exp(-k(t - t_0))]$$

where $y(t)$ denotes the ThT fluorescence at the incubation time t, a, k, and t_0 are fitting parameters, and $t0$ and k define a lag time of tL as tL = t_0–2/k (see Ref. 45). Curve fitting is performed using χ^2 analysis, and the error range for the lag time was estimated at the 90% confidence level. For the Aβ (1–40)- and Aβ(1–42)-seeded data, which show no lag time, curve fitting was performed with an equation of:

$$y(t) = a[1 - \exp(-kt)]$$

as the data typically follow first order kinetics (*see* **Note 8**).

12. Capture the morphologies of Aβ(1–42) fibrils and the initial monomers at a nanometer scale using EM. For the TEM grid preparation, load 10 μL of Aβ solution, sampled at 24–72 h into the incubation time, onto a 300 mesh copper formvar/carbon grid for 1 min. Blot the excess solution with filter paper to remove excess. Stain the sample attached to the grid with a 10 μL solution of 2% (w/v) uranyl acetate for 1.5 min. Blot off the excess staining solution and air-dry the grid. Store the grid in a desiccating chamber before use. Perform TEM on a standard electron microscope using magnifications greater than 120,000×.
13. Obtain high-resolution images of fibrils at 400,000× using STEM. For the grid preparation, collect a 10 μL aliquot of fibrils at 24–72 h during the incubation time and load it onto a 400 mesh copper carbon grid. Blot excess solution with filter paper after 1 min. Wash the sample on the grid by placing 5 μL of DDI water onto the grid for 30 s and blotting it away and repeating the DDI water wash. Fix the sample with 10–20 μL of 2% (w/v) glutaraldehyde solution for 30 min under a fume hood, and then blotting away the excess glutaraldehyde; wash the fixed sample with DDI water again twice. Negatively stain the fixed sample with 10 μL of 2% (w/v) uranyl acetate solution for 2 min and blot the excess solution. Dry the grid in air and store it in a desiccating chamber.
14. Harvest the ^{13}C-, ^{15}N-labeled Aβ fibril samples after incubation for 1 day to 1 week at G_4 or at a later generation. This seeding protocol improves structural homogeneity as described below. Pellet the fibril samples by centrifugation at 9000 × g for 45 min at RT, remove the supernatant and then lyophilize the fibril in the pellet. Pack the lyophilized fibril samples (5–10 mg) into 2.5 mm SSNMR MAS rotors (10 μL volume) and subsequently rehydrate with about 0.5 μL of water per mg of peptide. The samples used for the SSNMR analysis are listed in Table 1.

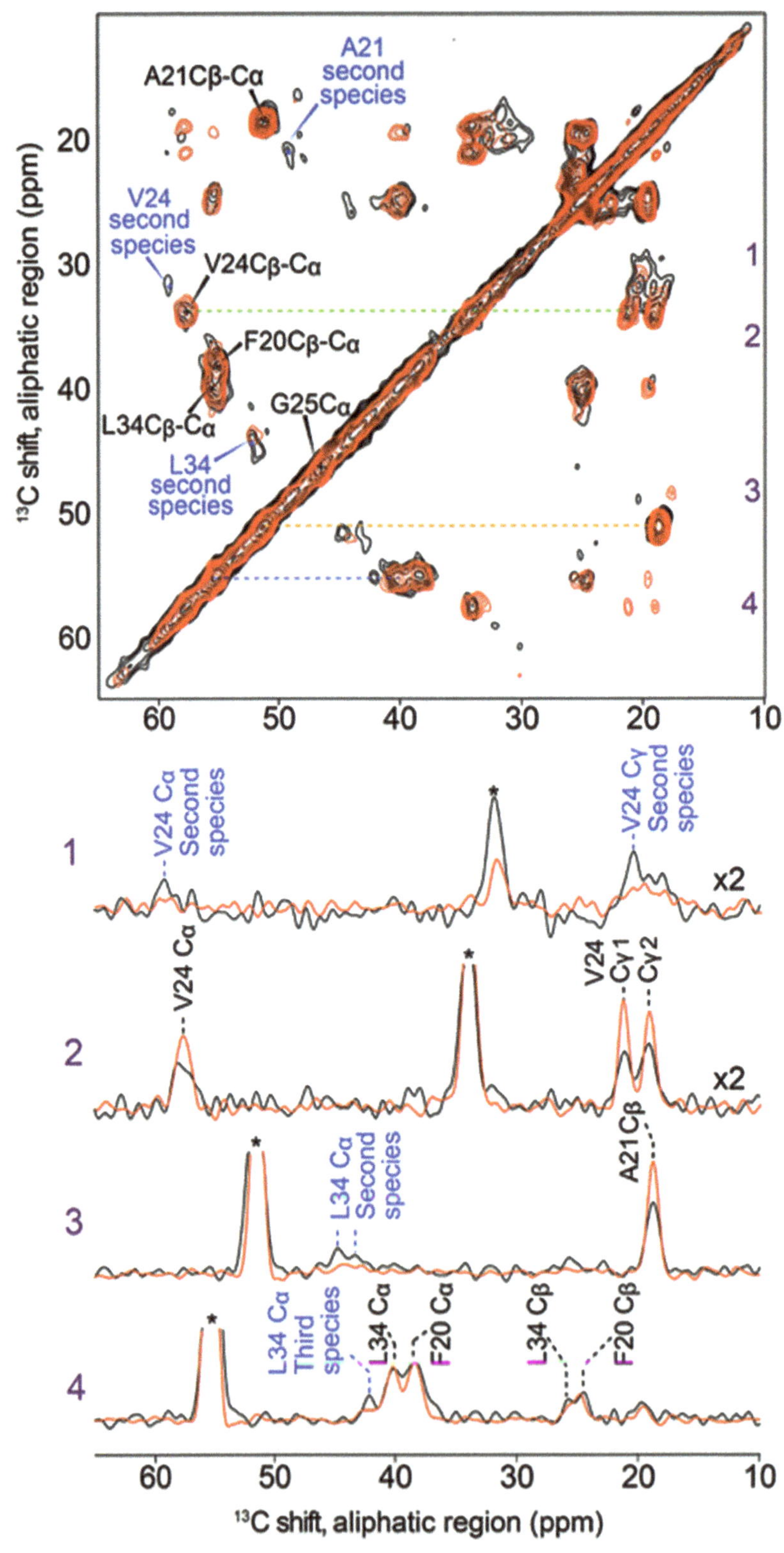
A21Cβ-Cα
A21 second species
V24 second species
V24Cβ-Cα
F20Cβ-Cα
G25Cα
L34Cβ-Cα
L34 second species
¹³C shift, aliphatic region (ppm)
1
2
3
4
V24 Cα Second species
V24 Cγ Second species
x2
V24 Cα
V24 Cγ1 Cγ2
x2
L34 Cα Second species
A21Cβ
L34 Cα Third species
L34 Cα
F20 Cα
L34 Cβ
F20 Cβ
¹³C shift, aliphatic region (ppm)

3.3 Structure Analysis Based on SSNMR Spectroscopy

Perform all SSNMR experiments in a 9.4 T magnetic field (^{1}H frequency of 400.2 MHz) using MAS at spinning speeds of 10–20 kHz with a ^{1}H, ^{13}C, ^{15}N triple-resonance 2.5 mm MAS probe.

All the detailed NMR experimental conditions can also be found in previously published work [45].

1. Validate structural homogeneity of the Aβ(1–42) fibril samples by comparing 2D ^{13}C-^{13}C chemical shift correlation (2D ^{13}C) SSNMR spectra of unseeded and seeded fibril samples (G_4), particularly for Samples 1 and 2 listed in Table 1. For both seeded and unseeded samples, 2D ^{13}C spectra were collected with an fp-RFDR sequence of a 1.6 ms mixing time, as in our previous study [53]. If the unseeded fibril for Sample 1, which is ^{13}C, ^{15}N-uniformly labeled at five residues, shows more than one set of corresponding cross peaks in a 2D ^{13}C spectrum (the black spectrum in Fig. 2), it indicates polymorphism. However, the seeded fibril for Sample 2, which is labeled at the same sites as Sample 1, should yield a corresponding 2D ^{13}C spectrum with only one set of cross peaks (the red spectrum in Fig. 2), indicating a single structurally homogeneous conformer in the fibril. It is worth mentioning that distinguishing between the seeded and unseeded fibril samples used for Fig. 2 is difficult from standard TEM data. SSNMR offers an excellent means to assess structural homogeneity of an amyloid fibril sample in a quantitative manner. The seeded Aβ(1–42) fibril samples up to G_{13} were confirmed to show identical spectra with that of the G_4 fibrils.
2. Complete ^{13}C, ^{15}N backbone and side chain chemical shift assignments by SSNMR. Perform SSNMR experiments on Samples 2–9 and 14–15 in Table 1 with 2D ^{13}C experiments with a 50 ms DARR mixing [54] and 2D ^{13}C-^{15}N chemical shift correlation (2D ^{13}C-^{15}N) experiments using a double CP scheme. The observed chemical shifts should be reproducible for different batches of the samples. All the backbone and $^{13}C_{\beta}$ assignments are used for secondary structure analysis by TALOS-N prediction [55] and secondary chemical shift calculation.

Fig. 2 (**a**) A comparison of the aliphatic spectral regions of 2D ^{13}C–^{13}C correlation spectra for non-seeded Aβ (1–42) fibrils (G_1 black) and Aβ(1–42) fibrils obtained by three successive seeded incubations (G_4 red) with slices along the indicated positions 1–4, which display cross peaks. The non-seeded fibril sample shows another set of cross peaks corresponding to the second conformer (black dotted circles) for Ala21, Val24, and Leu34 (assigned with blue arrows and labels). The seeded fibrils sample shows a single set of cross peaks for Phe20, Ala21, Val24, and Gly25 (assigned with black arrows and labels). Although Leu34 still displays a weak peak for the second set of cross peaks, the secondary species was nearly completely suppressed by seeding. The base contour levels were set to (black) 5% and (red) 4% of the diagonal signal for $^{13}C_{\alpha}$ of Ala21, which both correspond to 4 times the root-mean-squared noise level

3. Collect inter-residue restraints on Samples 2–9, 12–13, and 15 (Table 1) with 2D ^{13}C DARR experiments and a 200 ms mixing time. The 2D ^{13}C DARR spectra with a 200 ms mixing time display both cross peaks representing intra- and inter-residue contacts. Corresponding DARR 2D spectra with a 50 ms mixing time is a reference to identify intra-residue contacts, which enables the identification of long-range restraints in the data with a 200 ms mixing time.
4. Classify long-range intra-residue interactions into inter- or intramolecular contacts using an isotope dilution approach [28]. Mix the ^{13}C-, ^{15}N-labeled sample with unlabeled Aβ (1–42) at a ratio of 1:1 (Samples 2, 7, 13, and 14). The isotope diluted sample is studied using the procedure described in **step 3**. This dilution does not affect relative cross peak intensities for intramolecular contacts. On the other hand, *inter*-molecular cross peaks are attenuated as the chance of having a labeled molecule next to another labeled molecule is reduced by 50%. Thus, when peak intensities are compared between the isotope diluted and undiluted Aβ(1–42) samples, unaffected intra-residue signals can be identified as signals due to intramolecular contacts.
5. Perform inter-strand ^{13}CO-^{13}CO distance measurements on Sample 10, 11, and 16 using fp-RFDR-CT experiment [56].
6. For MD-assisted structural modeling efforts, systematically change peptide dihedral angles to minimize the deviation of experimental chemical shifts and those calculated from SHIFTX2 [50] (*see* **Note 9**).
7. Obtain inter-residue restraints between ^{13}C and ^{15}N using frequency-selective REDOR on Samples 6 and 14. Perform an isotope dilution experiment (described in **step 4**) on Sample 14 to confirm long-range interactions within the Aβ molecule.

3.4 Computational Simulation Assisted Structure Finalization

Perform a two-step structural optimization so that the final atomic model satisfies all the distance, dihedral angle, and chemical shift constraints from the SSNMR experiments.

3.4.1 Structural Modeling by CYANA 2.1

1. At the first stage, generate an ensemble of 1000 structures with CYANA 2.1 program [48] by adopting a similar approach used for the Het-s prion fibril [41]. The initial model of a 12-mer for Aβ(1–42) is built. If needed, remove N-terminus residues that are likely to be flexible and disordered. In our previous study for Aβ(1–42), the residues 1–10 were removed for this reason [45].
2. From the intermolecular ^{13}CO-^{13}CO distance measurements, estimate supramolecular arrangements and generate intermolecular restraints. Many amyloid fibrils are known to misfold

into in-register parallel β-sheet arrangements. In our previous study on Aβ(1–42) fibril, neighboring strands formed in-register parallel β strands throughout the well-structured residues at positions 17–42 [45]. Thus, internuclear ^{13}CO-^{13}CO distance restraints were included at residues 20, 24, 30, and 34 as lower and upper restraints of 4.6 and 5.0 Å, respectively. Amyloid fibrils often have stacks of β strands all with identical structure [41]. This is the case with Aβ(1–42) as evidenced by single sharp peaks observed in Fig. 2 and from internuclear ^{13}CO-^{13}CO distance measurements. In such a case, a quasi 1D-symmetry restraint can be implemented for all heavy atoms between neighboring strands [41] during both AMBER and CYANA refinements below. The upper and lower limits are set to 5.0 and 4.6 Å, respectively. It should be noted that such restraints and their force constants should not be overly weighted so as to artificially drive down the rmsd or should not be too loose to let the observed symmetry broken. We found that the force constants given below provide for a decent compromise for deriving the models. Neighboring strands of the Aβ molecules were connected by virtual atom linkers, each of which was 210 residues in length.

3. Create a list of upper-limit restraints set to 6.5 Å from the long-range DARR cross peaks. Additionally, generate a list of lower limit restraints for a pair of well-structured residues (residues 17–42) for which no cross peaks are identified in DARR experiments with a 200 ms mixing period. This should be done only when the S/N of the relevant experiments are sufficiently high and all the signals are well separated. The lower limit restraints are implemented at the C_β atoms of nonglycine residues and set to 6.5 Å.
4. Set the upper and lower limits for intramolecular contacts from quantitative REDOR distance measurements. For example, our REDOR experiment for the Aβ(1–42) fibril identified a unique contact between the $^{13}CO_2{}^-$ terminus of Ala42 and the $^{15}NH_3{}^+$ side chain of Lys28; for such a contact, the distance constraints with the lower and upper limits are set to 3.9 and 4.1 Å, respectively.
5. Set restraints on dihedral angles to define conformations. To elucidate likely dihedral angles (φ, ψ) from the obtained ^{13}C and ^{15}N chemical shifts, TALOS-N software [55] is used, and these dihedral angles are used as restraints only when the program determined the prediction as a consistent match to the database (i.e., "strong" or "generous"). Other details are discussed in our previous published paper [45].

6. Perform thermal annealing routines for the given sequence and structural restraints defined above. A total of 1000 structures is calculated with CYANA with the standard anneal.cya method included in the program, with a slight modification. The annealing procedure by CYANA is modified to include three rounds of high-temperature annealing instead of one for each molecule because this provided structures that better satisfied overall experimental restraints [45]. Of the 1000 structures, the 100 with the lowest target energies are retained for optimization at the next stage by AMBER software.

3.4.2 AMBER/ShiftX2 Refinement and Structural Validation

1. Perform a refinement by thermal annealing with AMBER 12 [49], as previously reported for globular proteins [57]. All of the distance and dihedral restraints used in CYANA are transferred to AMBER 12.
2. Minimize the energy of the CYANA structures for 10,000 steps without experimental restraints. The first 5000 steps invoke the steepest decent algorithm and the last 5000 steps are conjugate gradient minimization.
3. Carry out thermal annealing with the structural restraints determined in **step 2**. For the distance, torsional, chirality, angular (bond), and symmetry restraints, set force constants to 10 kcal/(mole $Å^2$), 540 kcal/(mole rad^2), 100 kcal/(mole rad^2), 40 kcal/(mole rad^2), and 1 kcal/(mole $Å^2$), respectively.
4. Perform three rounds of simulated annealing from 0 to 1000 K and then back to 0 K using a Berendsen thermostat to regulate temperature; each round of the annealing process is implemented for a 20 ps period. The temperature ramping and restraint weighting are the same as those previously reported [51]. A standard pairwise generalized Born solvation [58] is used with a cutoff of 12 Å. The time step of the MD simulation is set to 1 fs with a total of 60,000 steps or 60 ps for the three rounds of annealing. The last step of the simulated annealing involves a final energy minimization, including NMR restraints and implicit solvation, for 10,000 steps similar to the initial energy minimization with the steepest descent and conjugate gradient algorithms.
5. Keep the 20 structures from AMBER with the lowest nonrestraint energies for structural analysis. SHIFTX2 [50] is used to further assess structural quality by back-calculating the $^{13}C_{\alpha}$, $^{13}C_{\beta}$, ^{13}CO, and amide ^{15}N chemical shifts from the determined structures and comparing these results to the experimentally measured shifts. The shift prediction is performed for each Aβ molecule, and the ensemble average of the ^{13}C or ^{15}N was obtained for each site. The top ten models that show the lowest rms deviations between the predicted and experimental

shifts are selected as representative structural models. Figure 3a, b shows an example of the best-fit model from our previous study of Aβ(1–42) fibril. For the models, the average rms deviation shift of $^{13}C_{\alpha}$, $^{13}C_{\beta}$, ^{13}CO, and amide ^{15}N was 1.29 ppm, indicating reasonable fitting. The average rms deviation value obtained for our model is comparable to the rms deviation value of 1.38 ppm that was obtained from the

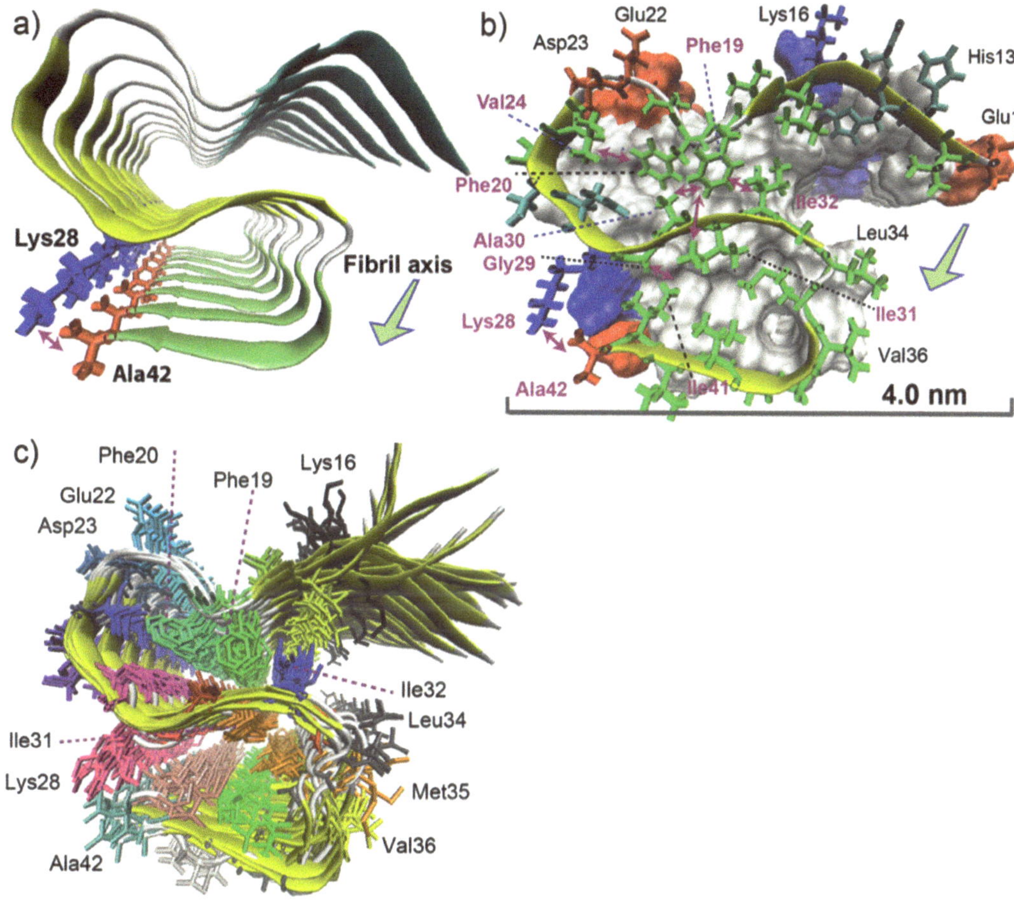

Fig. 3 A structural model of the amyloid fibril of Aβ(1–42) based on the SSNMR data. Disordered residues 1–10 were omitted for clarity. (**a**) View from the fibril axis shows three β strand regions (arrows) connected by short coil (white) or turn (silver) regions (tube); the three β strands are represented by arrows in cyan (residues 12–20), yellow (24–33), and green (36–41). The unique salt bridge between Lys28 (blue) and Ala42 (red) is shown (purple arrow). (**b**) Side chain contacts for a single Aβ chain in a skeletal and a ribbon diagram with van der Waals surface and polarity diagram for the rest of the Aβ chains. Hydrophobic, polar, acidic, and basic residues are represented by green, cyan, red, and blue, respectively. Observed long-range side chain intramolecular contacts (purple arrows) are shown. All β-sheet regions are presented in yellow. The surface plot indicates positively charged (Lys; blue) and negatively charged (Glu, Asp; red) side chains, and Ala42 that has a negatively charged carboxyl group (red). (**c**) Overlay of the ten best fit atomic structural models of the Aβ (1–42) fibril. The molecules at the edge and the N-terminal side chains at residues 11–15 were omitted for clarity. The structural images were reproduced from Ref. [45]

amyloid fibril structure for the Het-s prion protein (PDB 2RNM) and its experimental shifts [41]. Further discussion about the protocol and comparisons is given in ref. [45].

6. Perform structural validation for the obtained structures. Perform an analysis by PROCKECK-NMR [52] to examine whether nearly all the residues for the fibril model reside in allowed φ/ψ space. Perform distance and dihedral-restraint violations within PSVS and AMBER for the final structures (*see* Table 1 in ref. [45] for an example). The minimal number of violations confirms that the structures are consistent with all the SSNMR structural constraints. Confirm that overlaid structures show similar structural folds and calculate the rms deviation from the mean structure. For example, the overlaid ensemble structures for Aβ(1–42) fibril in Fig. 3c indicates that all ten models show very similar tertiary folds except for a few side chains near the loop regions and the dynamic N-terminus residues 11–16. The average rms deviation from the mean structure is estimated for the main chain and all the heavy atoms, respectively [45]. The obtained structures in the example were displayed by VMD 1.9.1 with the secondary structures elucidated by STRIDE [51].

4 Notes

1. Warm up Fmoc-amino acids and Fmoc-Ala-TGT resin to room temperature (RT) from −20 °C, and warm up 1.0 M HOBt in NMP from 4 °C to RT before use.
2. MALDI-TOF mass spectroscopy. The HPLC purified peptide was dissolved in a 50% acetonitrile solution with 0.01% TFA (0.1 mg/10 μL), and mixed (1:1 (v/v)) with a MALDI matrix solution (Sigma-Aldrich, 5 mg in 200 μL of 70% acetonitrile solution with 2% TFA). The mixture of 0.5–1 μL was loaded onto a MALDI chip (model ABI 01-192–6-AB, Life Technologies) and air dried before MALDI-TOF analysis. The peptides used in this study showed high purity (>95%).
3. Warm up the peptide in a desiccator from −20 °C to RT. Always handle (in transferring or dissolving) the dry peptide after it is warmed up. Always operate treatments of the dry peptide inside of a clean, dry box (humidity <15%).
4. A separate lyophilizer is used to remove HFIP from an Aβ (1–42) solution in HFIP for reproducible Aβ incubation. Note that an additional cold trap with dry ice (−78.5 °C) is connected between the first cold trap and a pump as HFIP is harmful to a pump.

5. Uranyl acetate has minimal radiological hazard, however, uranyl acetate-contaminated waste still must be separated from regular experimental waste for disposal.
6. Apply glutaraldehyde to the EM grids under safety hood.
7. Ice-cold ether should be cooled by dry ice; vent off ether in a fume hood for disposal.
8. Preparation of Aβ(1–40) fibrils utilizes the same method mentioned in **steps 1–7**.
9. It is recommended to start a preliminary structural modeling at an early stage of the procedure since this will provide possible contacts to be measured for additional SSNMR analysis.

Acknowledgments

This work was supported by the National Institutes of Health R01 and U01 programs (GM 098033) and Alzheimer's Association IIRG grant (08-91256) for YI.

References

1. Selkoe DJ (1994) Alzheimer's disease: a central role for amyloid. J Neuropathol Exp Neurol 53 (5):438–447
2. Lorenzo A, Yankner B (1996) Amyloid fibril toxicity in Alzheimer's disease and diabetesa. Ann N Y Acad Sci 777(1):89–95
3. Lorenzo A, Yankner BA (1994) Beta-amyloid neurotoxicity requires fibril formation and is inhibited by congo red. Proc Natl Acad Sci 91 (25):12243–12247
4. Chimon S, Ishii Y (2005) Capturing intermediate structures of Alzheimer's β-amyloid, Aβ (1−40), by solid-state NMR spectroscopy. J Am Chem Soc 127(39):13472–13473
5. Hoshi M, Sato M, Matsumoto S, Noguchi A, Yasutake K, Yoshida N, Sato K (2003) Spherical aggregates of β-amyloid (amylospheroid) show high neurotoxicity and activate tau protein kinase I/glycogen synthase kinase-3β. Proc Natl Acad Sci 100(11):6370–6375
6. Noguchi A, Matsumura S, Dezawa M, Tada M, Yanazawa M, Ito A, Akioka M, Kikuchi S, Sato M, Ideno S, Noda M, Fukunari A, Muramatsu S-i, Itokazu Y, Sato K, Takahashi H, Teplow DB, Nabeshima Y-i, Kakita A, Imahori K, Hoshi M (2009) Isolation and characterization of patient-derived, toxic, high mass amyloid β-protein (Aβ) assembly from Alzheimer disease brains. J Biol Chem 284 (47):32895–32905
7. Parthasarathy S, Inoue M, Xiao Y, Matsumura Y, Nabeshima Y-i, Hoshi M, Ishii Y (2015) Structural insight into an Alzheimer's brain-derived spherical assembly of amyloid β by solid-state NMR. J Am Chem Soc 137 (20):6480–6483
8. Olofsson A, Sauer-Eriksson AE, Öhman A (2006) The solvent protection of Alzheimer amyloid-β-(1–42) fibrils as determined by solution NMR spectroscopy. J Biol Chem 281 (1):477–483
9. Ahmed M, Davis J, Aucoin D, Sato T, Ahuja S, Aimoto S, Elliott JI, Van Nostrand WE, Smith SO (2010) Structural conversion of neurotoxic amyloid-beta(1-42) oligomers to fibrils. Nat Struct Mol Biol 17(5):561–567
10. Barghorn S, Nimmrich V, Striebinger A, Krantz C, Keller P, Janson B, Bahr M, Schmidt M, Bitner RS, Harlan J (2005) Globular amyloid β-peptide1− 42 oligomer− a homogenous and stable neuropathological protein in Alzheimer's disease. J Neurochem 95(3):834–847
11. Murakami K, Irie K, Morimoto A, Ohigashi H, Shindo M, Nagao M, Shimizu T, Shirasawa T (2003) Neurotoxicity and physicochemical properties of Aβ mutant peptides from cerebral amyloid angiopathy. J Biol Chem 278 (46):46179–46187

12. Selkoe DJ (2001) Alzheimer's disease: genes, proteins, and therapy. Physiol Rev 81 (2):741–766
13. Fryer JD, Simmons K, Parsadanian M, Bales KR, Paul SM, Sullivan PM, Holtzman DM (2005) Human apolipoprotein E4 alters the amyloid-β 40: 42 ratio and promotes the formation of cerebral amyloid angiopathy in an amyloid precursor protein transgenic model. J Neurosci 25(11):2803–2810
14. Gravina SA, Ho L, Eckman CB, Long KE, Otvos L Jr, Younkin LH, Suzuki N, Younkin SG (1995) Amyloid β protein (Aβ) in Alzheimer's disease brain. Biochemical and immunocytochemical analysis with antibodies specific for forms ending at Aβ40 or Aβ42(43). J Biol Chem 270(13):7013–7016
15. Iwatsubo T, Odaka A, Suzuki N, Mizusawa H, Nukina N, Ihara Y (1994) Visualization of A-beta-42(43) and A-beta-40 in senile plaques with end-specific A-beta monoclonals—evidence that an initially deposited species is A-beta-42(43). Neuron 13(1):45–53
16. Van Dorpe J, Smeijers L, Dewachter I, Nuyens D, Spittaels K, Van den Haute C, Mercken M, Moechars D, Laenen I, Kuiperi C (2000) Prominent cerebral amyloid angiopathy in transgenic mice overexpressing the London mutant of human APP in neurons. Am J Pathol 157(4):1283–1298
17. Burdick D, Kosmoski J, Knauer MF, Glabe CG (1997) Preferential adsorption, internalization and resistance to degradation of the major isoform of the Alzheimer's amyloid peptide, Aβ (1–42), in differentiated PC12 cells. Brain Res 746(1–2):275–284
18. Mehta PD, Pirttilä T, Mehta SP, Sersen EA, Aisen PS, Wisniewski HM (2000) Plasma and cerebrospinal fluid levels of amyloid β proteins 1–40 and 1–42 in Alzheimer disease. Arch Neurol 57(1):100–105
19. Suzuki N, Iwatsubo T, Odaka A, Ishibashi Y, Kitada C, Ihara Y (1994) High tissue content of soluble beta 1-40 is linked to cerebral amyloid angiopathy. Am J Pathol 145(2):452
20. Seppälä TT, Herukka S-K, Hänninen T, Tervo S, Hallikainen M, Soininen H, Pirttilä T (2010) Plasma Aβ42 and Aβ40 as markers of cognitive change in follow-up: a prospective, longitudinal, population-based cohort study. J Neurol Neurosurg Psychiatry 81 (10):1123–1127
21. Roher AE, Lowenson JD, Clarke S, Woods AS, Cotter RJ, Gowing E, Ball MJ (1993) beta-Amyloid-(1-42) is a major component of cerebrovascular amyloid deposits: implications for the pathology of Alzheimer disease. Proc Natl Acad Sci 90(22):10836–10840
22. van Oijen M, Hofman A, Soares HD, Koudstaal PJ, Breteler MMB (2006) Plasma A beta (1-40) and A beta(1-42) and the risk of dementia: a prospective case-cohort study. Lancet Neurol 5(8):655–660
23. Graff-Radford NR, Crook JE, Lucas J, Boeve BF, Knopman DS, Ivnik RJ, Smith GE, Younkin LH, Petersen RC, Younkin SG (2007) Association of low plasma A beta 42/A beta 40 ratios with increased imminent risk for mild cognitive impairment and Alzheimer disease. Arch Neurol 64(3):354–362
24. Davis J, VanNostrand WE (1996) Enhanced pathologic properties of Dutch-type mutant amyloid beta-protein. Proc Natl Acad Sci U S A 93(7):2996–3000
25. Petkova AT, Leapman RD, Guo Z, Yau W-M, Mattson MP, Tycko R (2005) Self-propagating, molecular-level polymorphism in Alzheimer's ß-amyloid fibrils. Science 307 (5707):262–265
26. Lu J-X, Qiang W, Yau W-M, Schwieters Charles D, Meredith Stephen C, Tycko R (2013) Molecular structure of β-amyloid fibrils in Alzheimer's disease brain tissue. Cell 154 (6):1257–1268
27. Lührs T, Ritter C, Adrian M, Riek-Loher D, Bohrmann B, Döbeli H, Schubert D, Riek R (2005) 3D structure of Alzheimer's amyloid-β (1–42) fibrils. Proc Natl Acad Sci U S A 102 (48):17342–17347
28. Petkova AT, Yau WM, Tycko R (2006) Experimental constraints on quaternary structure in Alzheimer's beta-amyloid fibrils. Biochemistry 45(2):498–512
29. Paravastu AK, Leapman RD, Yau W-M, Tycko R (2008) Molecular structural basis for polymorphism in Alzheimer's β-amyloid fibrils. Proc Natl Acad Sci 105(47):18349–18354
30. Bertini I, Gonnelli L, Luchinat C, Mao J, Nesi A (2011) A new structural model of Aβ40 fibrils. J Am Chem Soc 133(40):16013–16022
31. Masuda Y, Uemura S, Ohashi R, Nakanishi A, Takegoshi K, Shimizu T, Shirasawa T, Irie K (2009) Identification of Physiological and Toxic Conformations in Aβ42 Aggregates. Chembiochem 10(2):287–295
32. Schmidt M, Sachse C, Richter W, Xu C, Fändrich M, Grigorieff N (2009) Comparison of Alzheimer Aβ(1–40) and Aβ(1–42) amyloid fibrils reveals similar protofilament structures. Proc Natl Acad Sci 106(47):19813–19818
33. Lopez del Amo JM, Schmidt M, Fink U, Dasari M, Faendrich M, Reif B (2012) An asymmetric dimer as the basic subunit in Alzheimer's disease amyloid beta fibrils. Angewandte Chemie-International Edition 51 (25):6136–6139

34. Dobson CM (2003) Protein folding and misfolding. Nature 426(6968):884–890
35. Hyung S-J, DeToma AS, Brender JR, Lee S, Vivekanandan S, Kochi A, Choi J-S, Ramamoorthy A, Ruotolo BT, Lim MH (2013) Insights into antiamyloidogenic properties of the green tea extract (−)-epigallocatechin-3-gallate toward metal-associated amyloid-beta species. Proc Natl Acad Sci U S A 110(10):3743–3748
36. Fawzi NL, Ying J, Ghirlando R, Torchia DA, Clore GM (2011) Atomic-resolution dynamics on the surface of amyloid-beta protofibrils probed by solution NMR. Nature 480 (7376):268–U161
37. Sarkar B, Mithu VS, Chandra B, Mandal A, Chandrakesan M, Bhowmik D, Madhu PK, Maiti S (2014) Significant structural differences between transient amyloid-β oligomers and less-toxic fibrils in regions known to harbor familial Alzheimer's mutations. Angew Chem Int Ed 53(27):6888–6892
38. del Amo JML, Fink U, Dasari M, Grelle G, Wanker EE, Bieschke J, Reif B (2012) Structural Properties of EGCG-Induced, Nontoxic Alzheimer's Disease A beta Oligomers. J Mol Biol 421(4–5):517–524
39. Tycko R (2006) Molecular structure of amyloid fibrils: insights from solid-state NMR. Q Rev Biophys 39(01):1–55
40. Lange A, Giller K, Hornig S, Martin-Eauclaire MF, Pongs O, Becker S, Baldus M (2006) Toxin-induced conformational changes in a potassium channel revealed by solid-state NMR. Nature 440(7086):959–962
41. Wasmer C, Lange A, Van Melckebeke H, Siemer AB, Riek R, Meier BH (2008) Amyloid Fibrils of the HET-s(218–289) Prion Form a β Solenoid with a Triangular Hydrophobic Core. Science 319(5869):1523–1526
42. Cady SD, Schmidt-Rohr K, Wang J, Soto CS, DeGrado WF, Hong M (2010) Structure of the amantadine binding site of influenza M2 proton channels in lipid bilayers. Nature 463 (7281):689–U127
43. Loquet A, Sgourakis NG, Gupta R, Giller K, Riedel D, Goosmann C, Griesinger C, Kolbe M, Baker D, Becker S, Lange A (2012) Atomic model of the type III secretion system needle. Nature 486(7402):276–279
44. Ma B, Nussinov R (2011) Polymorphic triple beta-sheet structures contribute to amide hydrogen/deuterium (H/D) exchange protection in the Alzheimer Amyloid beta 42 peptide. J Biol Chem 286(39):34244–34253
45. Xiao Y, Ma B, McElheny D, Parthasarathy S, Long F, Hoshi M, Nussinov R, Ishii Y (2015) A[beta](1-42) fibril structure illuminates self-recognition and replication of amyloid in Alzheimer's disease. Nat Struct Mol Biol 22:499–505
46. Delaglio F, Grzesiek S, Vuister GW, Zhu G, Pfeifer J, Bax A (1995) Nmrpipe—a Multidimensional Spectral Processing System Based on Unix Pipes. J Biomol NMR 6(3):277–293
47. Herrmann T, Güntert P, Wüthrich K (2002) Protein NMR Structure Determination with Automated NOE Assignment Using the New Software CANDID and the Torsion Angle Dynamics Algorithm DYANA. J Mol Biol 319 (1):209–227
48. Veshtort M, Griffin RG (2006) SPINEVOLUTION: a powerful tool for the simulation of solid and liquid state NMR experiments. J Magn Reson 178(2):248–282
49. Case D, Darden T, Cheatham TE III, Simmerling C, Wang J, Duke R, Luo R, Walker R, Zhang W, Merz K (2012) AMBER 12. University of California, San Francisco
50. Han B, Liu Y, Ginzinger SW, Wishart DS (2011) SHIFTX2: significantly improved protein chemical shift prediction. J Biomol NMR 50(1):43–57
51. Heinig M, Frishman D (2004) STRIDE: a web server for secondary structure assignment from known atomic coordinates of proteins. Nucleic Acids Res 32:W500–W502
52. Laskowski RA, Rullmann JAC, MacArthur MW, Kaptein R, Thornton JM (1996) AQUA and PROCHECK-NMR: Programs for checking the quality of protein structures solved by NMR. J Biomol NMR 8(4):477–486
53. Ishii Y (2001) 13C–13C dipolar recoupling under very fast magic angle spinning in solid-state nuclear magnetic resonance: applications to distance measurements, spectral assignments, and high-throughput secondary-structure determination. J Chem Phys 114 (19):8473–8483
54. Takegoshi K, Nakamura S, Terao T (2003) C-13-H-1 dipolar-driven C-13-C-13 recoupling without C-13 rf irradiation in nuclear magnetic resonance of rotating solids. J Chem Phys 118(5):2325–2341
55. Shen Y, Bax A (2013) Protein backbone and sidechain torsion angles predicted from NMR chemical shifts using artificial neural networks. J Biomol NMR 56(3):227–241
56. Ishii Y, Balbach JJ, Tycko R (2001) Measurement of dipole-coupled lineshapes in a many-spin system by constant-time two-dimensional solid state NMR with high-speed magic-angle spinning. Chem Phys 266(2–3):231–236

57. Dames SA, Martinez-Yamout M, De Guzman RN, Dyson HJ, Wright PE (2002) Structural basis for Hif-1 alpha/CBP recognition in the cellular hypoxic response. Proc Natl Acad Sci U S A 99(8):5271–5276

58. Tsui V, Case DA (2000) Molecular dynamics simulations of nucleic acids with a generalized born solvation model. J Am Chem Soc 122 (11):2489–2498

Chapter 27

Experimental and Computational Protocols for Studies of Cross-Seeding Amyloid Assemblies

Baiping Ren, Rundong Hu, Mingzhen Zhang, Yonglan Liu, Lijian Xu, Binbo Jiang, Jie Ma, Buyong Ma, Ruth Nussinov, and Jie Zheng

Abstract

Alzheimer's disease (AD) and type 2 diabetes (T2D) are two common protein aggregation diseases. Compelling evidence has shown a link between AD and T2D, which may derive from interspecies cross-sequence interactions between amyloid-β peptide (Aβ), associated with AD, and human islet amyloid polypeptide (hIAPP), associated with T2D. Herein, we present experimental and computational protocols and tools to study the aggregate structures and kinetics, conformational conversion, and molecular interactions of Aβ-hIAPP mixtures. These protocols could be generally applied to other cross-seeding behaviors of amyloid peptides.

Key words Amyloid peptides, Aβ, hIAPP, Cross-seeding, Alzheimer disease, Diabetes

1 Introduction

Amyloid peptides can misfold and aggregate into conformationally similar amyloid fibrils consisting of characteristic cross-β-sheet structures. Fibrils formed by different amyloid peptides are associated with more than 20 neurodegenerative and other protein aggregation diseases [1]. Among them, Alzheimer disease (AD) and type 2 diabetes (T2D) are the two common age-related chronic disorders [2, 3], both affecting millions of people globally. Accumulating evidence has shown that ~81% of AD patients often have a higher risk of developing type II diabetes (T2D), and vice versa [4, 5]. Although the precise mechanisms linking these two disorders are still unclear, the evidence suggests a potential pathological link between AD and T2D.

Considering that amyloid peptides share a common aggregation process and their aggregates share similar structural hallmarks,

Baiping Ren and Rundong Hu contributed equally to this work.

Bradley L. Nilsson and Todd M. Doran (eds.), *Peptide Self-Assembly: Methods and Protocols*, Methods in Molecular Biology, vol. 1777, https://doi.org/10.1007/978-1-4939-7811-3_27, © Springer Science+Business Media, LLC, part of Springer Nature 2018

the AD-T2D link could arise from cross-species interactions between the amyloid-β peptide (Aβ) [6, 7] and the human islet amyloid polypeptide (hIAPP or amylin) [8–11], which are associated with AD and T2D, respectively. Several lines of evidences appear to support the hypothesis of the cross-species interaction between Aβ and hIAPP: (1) Aβ and hIAPP coexist in the serum and cerebrospinal fluids with similar concentrations [6]; (2) Aβ co-localizes with hIAPP in pancreatic islet amyloid deposits of T2D patients [7]; and (3) hIAPP and Aβ show high degrees of sequence identity (25%) and similarity (50%), making it possible for them to interact with each other to form hybrid amyloid fibrils [6, 12]. Similar cross-seeding behavior between different amyloid species (Aβ and α-synuclein [13], Aβ and tau [14], and hIAPP and insulin [15]) and even between bacterial curli and amyloid peptides of SEVI, Aβ, and hIAPP have been reported [16]. This fact at least suggests that amyloid cross-seeding sets the scene to reveal some general principles governing heterogeneous protein misfolding and aggregation.

Only a few studies have shown that Aβ and hIAPP may cross-seed fibrillization, but with different cross-seeding efficiencies, strongly depending on experimental conditions (seeding concentrations, sequence specifity, and even agitation) [17–20]. Different seeding abilities of Aβ vs. hIAPP aggregates suggest the existence of different cross-species barriers for different amyloid species to interact with the other and form hybrid amyloids [21]. Given the current state of knowledge of the cross-seeding of amyloid peptides, it still remains unclear how Aβ and hIAPP interact with each other to induce cross-seeding behaviors. Herein, we detail computational and experimental protocols, obtained from our recent works, to probe the aggregation kinetics of Aβ/hIAPP mixtures by fluorescent dye thioflavin-T (ThT), the fibril morphologies by atomic force microscopy (AFM), the secondary structure transformation by circular dichroism (CD), and atomic-scale structures and association by molecular dynamics (MD) simulations. This chapter will hopefully provide robust and general strategies to study the cross-species amyloid aggregation by different amyloid peptides, which may help to reveal a possible link between different neurodegenerative diseases.

Understanding the cross-species aggregation mechanisms between Aβ and hIAPP peptides is critical for revealing a potential pathological link between AD and T2D. In this chapter, we report the details of computational and experimental tools, protocols, and conditions to study how Aβ and hIAPP peptides interact with each other and how such cross-sequence interactions affect the structures of their aggregates and pathways. Collective data demonstrate the existence of cross-seeding interaction and aggregation between Aβ and hIAPP, yielding hybrid Aβ – hIAPP amyloid fibrils morphologically similar to pure Aβ or hIAPP fibrils. Further molecular simulations reveal different atomic structures of hybrid Aβ – hIAPP fibrils, indicating that the complex assembly of Aβ and hIAPP produces polymorphic amyloid cross-seeding species through different aggregation pathways. The cross-seeding effects in this work may provide a molecular explanation for a potential link between AD and T2D and hopefully lead to therapeutic advances that could treat different neurodegenerative diseases. Technically, the tools, protocols, and conditions we used for Aβ-hIAPP cross-seeding could be readily applied to other amyloid peptides.

2 Materials

Prepare ultrapure water by purifying deionized water to obtain a sensitivity of 18 MΩ cm at 25 °C; use this ultrapure water for all aqueous solutions. All chemical reagents were purchased at the analytical grade and stored at room temperature (unless indicated otherwise).

2.1 Experimental Materials

1. Beta-amyloid ($A\beta_{1-42}$, ≥95.0%) and human islet amyloid polypeptide ($hIAPP_{1-37}$, ≥95.0%) (American Peptide Inc., Sunnyvale, CA).
2. 1,1,1,3,3,3-Hexafluoro-2-propanol (HFIP, ≥99.9%).
3. NaOH solution: 10 mM NaOH in water (*see* **Note 1**).
4. PBS buffer: 10 mM, pH 7.4. Phosphate buffered saline (pH 7.4) dry powder was purchased in foil pouches. Dissolve each pouch in 1 L of water to yield 10 mM phosphate buffered saline, NaCl 0.138 M, KCl 0.0027 M. Store at 25 °C (*see* **Note 2**).
5. Thioflavin T (ThT) Tris-buffer: 10 μM ThT, pH 7.4. Add about 100 mL water to a 1 L amber bottle. Weigh 1.22 g Tris and transfer it to the bottle. Add water to a volume of 900 mL. Mix and adjust pH to 7.4 by carefully adding concentrated HCl. Make up to 1 L with water to produce a 1 L 10 mM Tris-buffer.
6. ThT-water solution (2 mM): add 33 mg of ThT powder into 50 mL of water. Dilute 250 μL of this ThT-water solution into 50 mL of Tris-buffer resulting in a final concentration of 10 μM ThT Tris-buffer (*see* **Note 3**).

3 Methods

3.1 Purification and Preparation of Peptides

1. Obtained both Abeta and hIAPP as a lyophilized form and each peptide of 1 mg was stored separately in a small amber glass bottle in a −20 °C refrigerator.

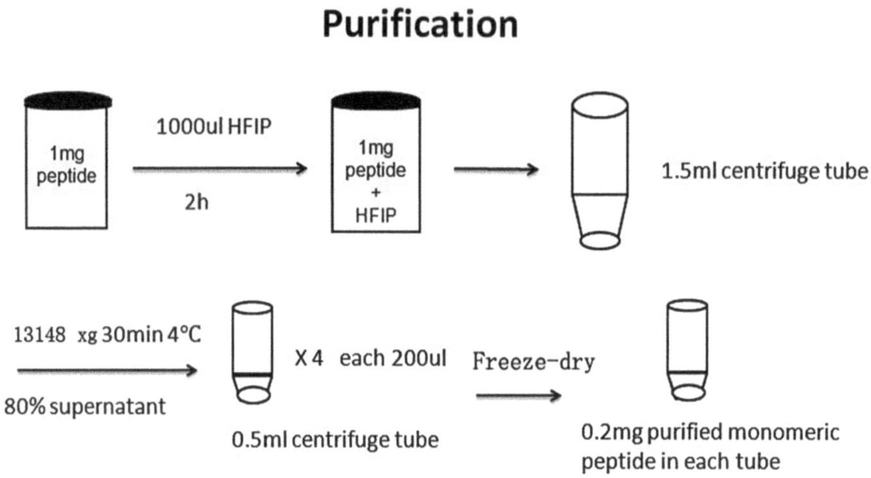

Fig. 1 Preparation procedure of amyloid peptide purification

2. Add 1 mL HFIP into each glass bottle with 1.0 mg Aβ or hIAPP peptide inside and seal the bottle cap with parafilm. Allow the peptide to fully dissolve in HFIP solution for 2 h, followed by 30 min sonication to make peptides in a monomeric state in the solution (*see* **Note 4**).
3. Transfer the 1 mL peptide-HFIP mixture to a 1.5 mL centrifuge tube, and centrifuge the mixture solution at 13,148 × *g* for 30 min at 4 °C.
4. Extract 80% (0.8 mL) of the top solution, save it in four 500-μL micro tubes, each containing 0.2 mL of the top solution.
5. Freeze the extracted solutions in a −80 °C refrigerator and freeze-dry at −110 °C and 10 mTorr.
6. Extract 0.2 mg dry peptide powder from the micro tube and lyophilize in a −80 °C refrigerator [22] (*see* **Note 5**) (Fig. 1).

3.2 Peptide Incubation

1. Add 25 μL of 10 mM NaOH solution to 0.2 mg purified Aβ powder in a microfuge tube, and sonicate the mixture for 1 min to obtain a homogeneous Aβ-NaOH solution.
2. Add 25 μL of 10 mM NaOH solution into 0.2 mg purified hIAPP powder in the micro tube and sonicate the mixture for 1 min at 13,148 × *g* to obtain a homogeneous hIAPP-NaOH solution (*see* **Note 6**).
3. Add the 25 μL Aβ-NaOH solution into 2 mL of 10 mM PBS buffer solution in 2 mL centrifuge tube at 4 °C, and mix using vortex. Right after mixing, add 25 μL hIAPP-NaOH solution to the previous Aβ-NaOH-PBS solution, and mix using vortex.
4. Centrifuge the Aβ-hIAPP mixture solution at 13,148 × *g* for 30 min at 4 °C to remove any potential oligomers, followed by

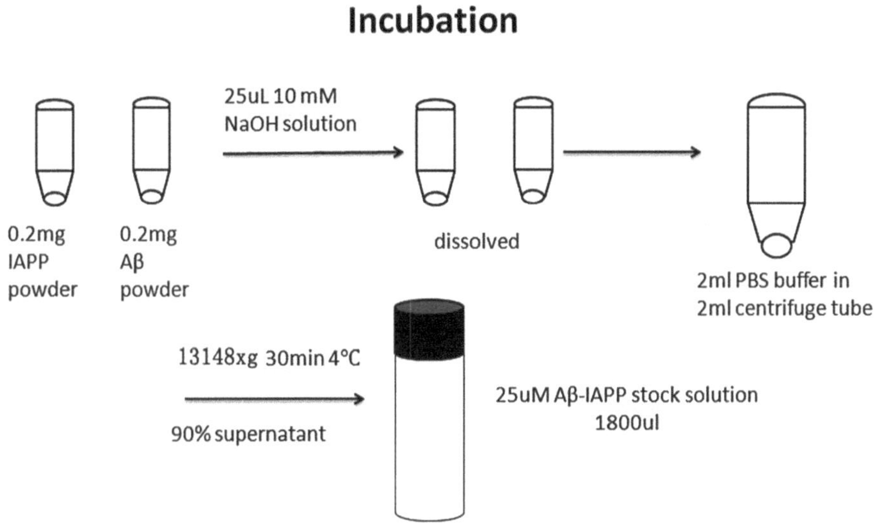

Fig. 2 Procedure of amyloid peptide incubation

extracting 90% of the top Aβ-hIAPP solution (1.8 mL) in the monomeric state of both peptides for cross-seeding tests [23] (Fig. 2).

3.3 Monitor Amyloid Formation Using Thioflavin T Fluorescence Assay (ThT)

1. Add 3 mL ThT Tris-buffer to a 1 cm × 1 cm quartz cuvette and install the cuvette in the sample chamber. Measure the emission curve between 470 and 500 nm with an excitation wavelength of 450 nm. The emission curve is used as a background control (*see* **Note 7**).
2. Add 60 μL of incubated peptide solution, obtained from "peptide incubation," to the ThT Tris-buffer in the cuvette and mix well. Measure the emission curve and subtract the background curve to obtain a true-fluorescence intensity at this time point [24]. Each peptide sample is tested in triplicate (*see* **Note 8**).
3. Repeat **steps 1–2** at different time points, collect all the data to do the analysis of variance, use graphics software (Excel, Origin et al.) to plot a time-dependent ThT curve to quantify the aggregation kinetics of the cross-seeding of Aβ and hIAPP (Fig. 3a–c).

3.4 Circular Dichroism Spectroscopy (CD)

1. Obtain CD spectra by scanning between 250 and 190 nm at a 0.5 nm resolution and 50 nm/min scan rate. Correct all spectra by subtracting the baseline and report the data as the average of three successive scans for each sample.
2. Add 160 μL incubated peptide solution into a rectangular quartz cuvette with a 1 mm path length without dilution (*see* **Note 9**).

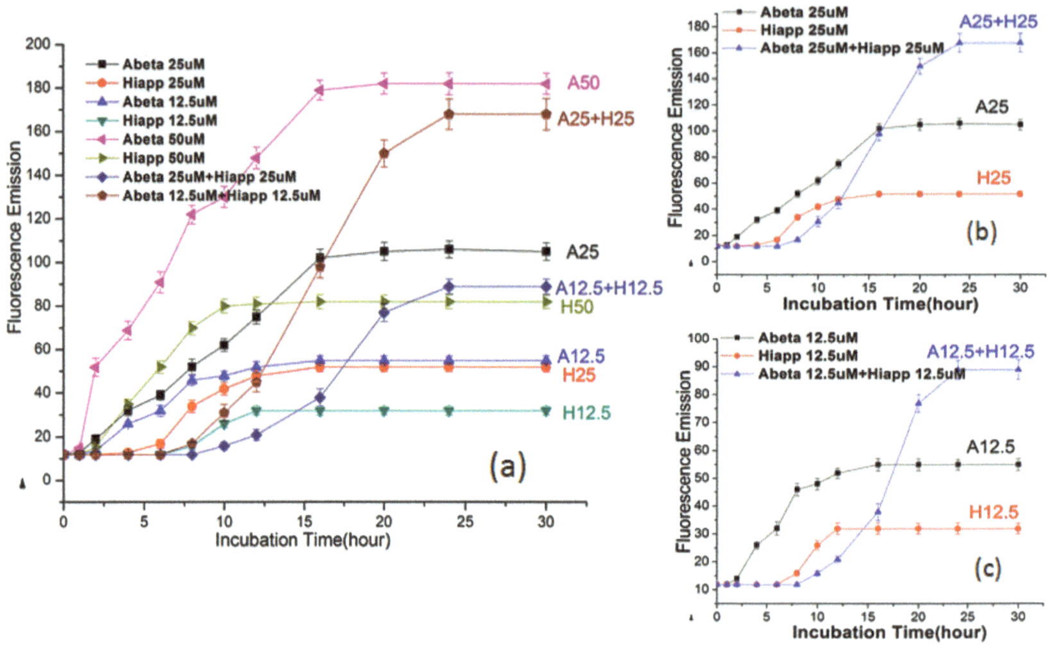

Fig. 3 Time-dependent ThT fluorescence curves for pure Aβ, pure hIAPP, and mixed Aβ/hIAPP at different concentration of 12.5 mM, 25 mM, and 50 mM. Reprinted with permission from ref. [23]. (Copyright 2015 ACS Publication)

3. Repeat **steps 1–2** at different time points to obtain a time-dependent CD curve to quantify the conformational change of the cross-seeding of Aβ and hIAPP.
4. The secondary structure contents from the CD spectra can be estimated using the self-consistent method (CDSSTR program) in the CD proanalysis software [25, 26] (Fig. 4a–f).

3.5 Tapping-Mode Atomic Force Microscopy (AFM)

1. All images are acquired in tapping mode as 512 × 512 pixel images at a typical scan rate of 1.0–2.0 Hz with a vertical tip oscillation frequency of 250–350 kHz [27].
2. Take 20 μL incubated peptide solution and deposit the solution onto a freshly cleaved mica substrate for 1 min.
3. Gently rinse the surface with 10 mL of ultrapure water to remove salts and loosely bound peptides, dry the surface with compressed air for 1 min, and leave the sample to air-dry before AFM imaging.
4. Representative AFM images are collected by scanning at least six different locations of different samples (Fig. 5).

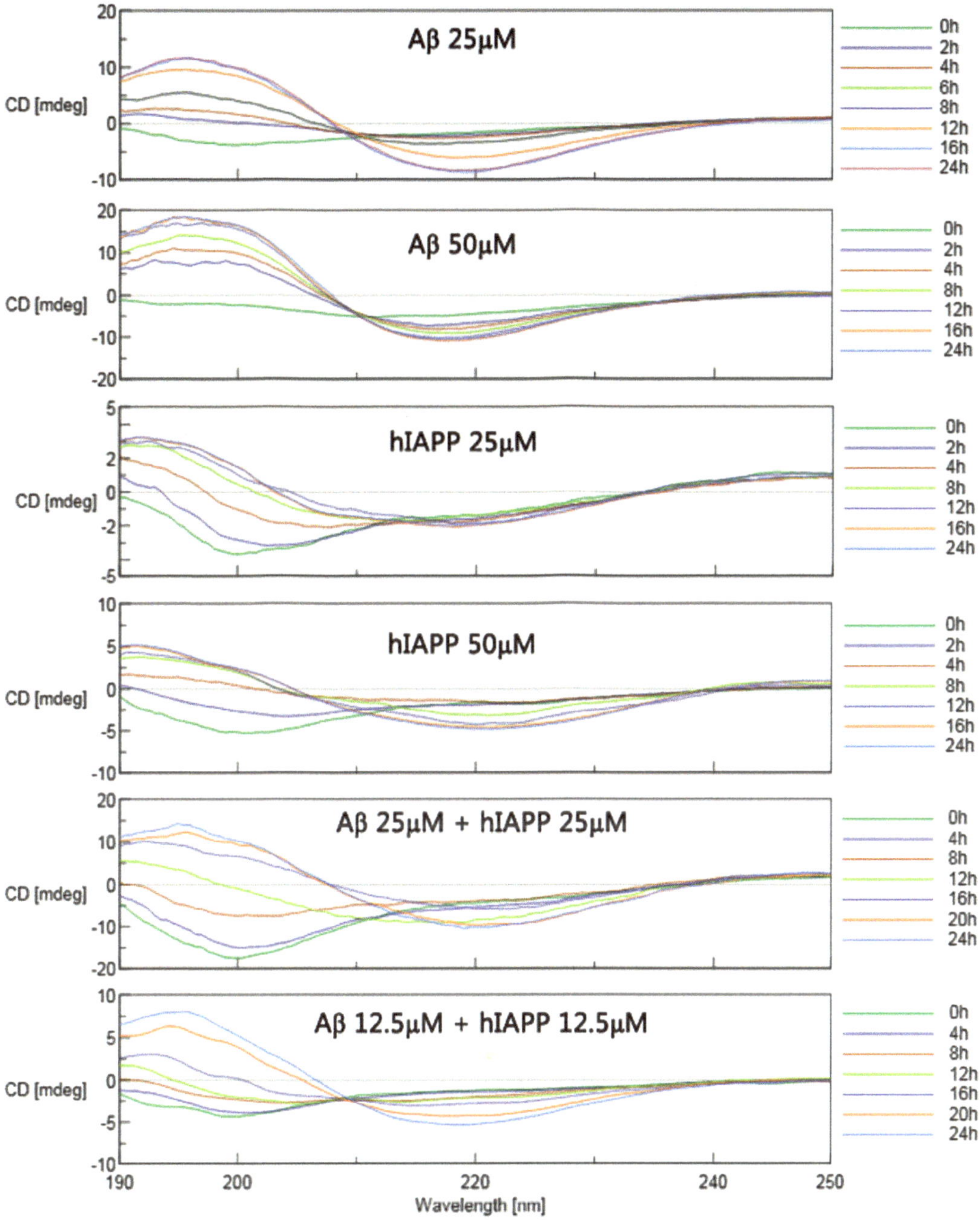

Fig. 4 Time-dependent far-UV CD spectra for (**a**) pure Aβ (25 μm), (**b**) pure Aβ (50 μm), (**c**) pure hIAPP (25 μm), (**d**) pure hIAPP (50 μm), (**e**) mixed Aβ/hIAPP (25/25 μM), and (**f**) mixed Aβ/hIAPP (12.5/12.5 μm). Reprinted with permission from ref. [23]. (Copyright 2015 ACS Publication)

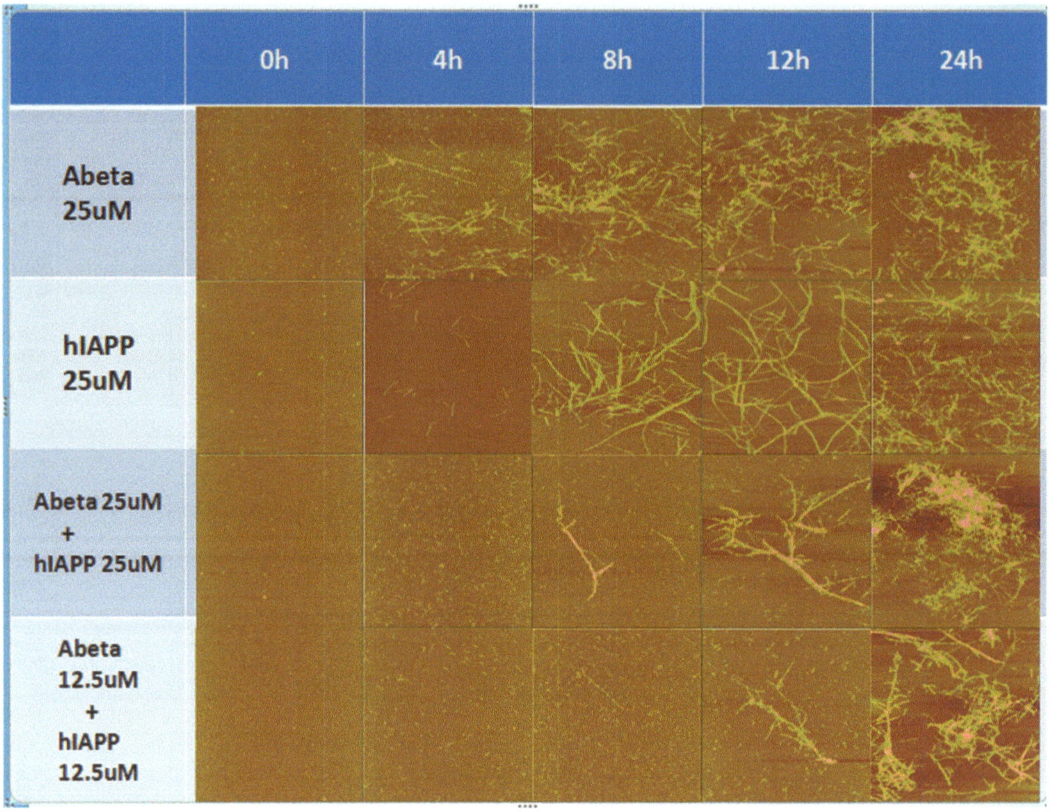

Fig. 5 Representative AFM images for pure Aβ, pure hIAPP, and mixed Aβ/hIAPP aggregates after incubation for 0, 4, 8, 12, and 24 h. Reprinted with permission from ref. [23]. (Copyright 2015 ACS Publication)

3.6 Simulation Protocols

3.6.1 Aβ Pentamer

1. Obtain the atomic structure of Aβ pentamer from the Protein Data Bank (PDB) database (ID:2EBG) [28].
2. Add the N-terminal missing residues in an in-register manner in the β-sheet conformation using the GaussView program (*see* **Notes 10** and **11**, Fig. 6).
3. Cap the N- and C-termini with NH_3^+ and COO^- groups. Assign the tautomeric state of His residues based on local environment.
4. Use a 500-step steepest descent (SD) minimization with backbone atoms fixed to relax the initial structure of the Aβ pentamer (*see* **Note 12**).

3.6.2 hIAPP Pentamer

1. The initial coordinates of hIAPP monomer were kindly provided by the Tycko lab [29].
2. Five hIAPP monomers are parallel-packed on top of each other in an in-register manner with inter-peptide distance of 4.7 Å (*see* **Note 11**).

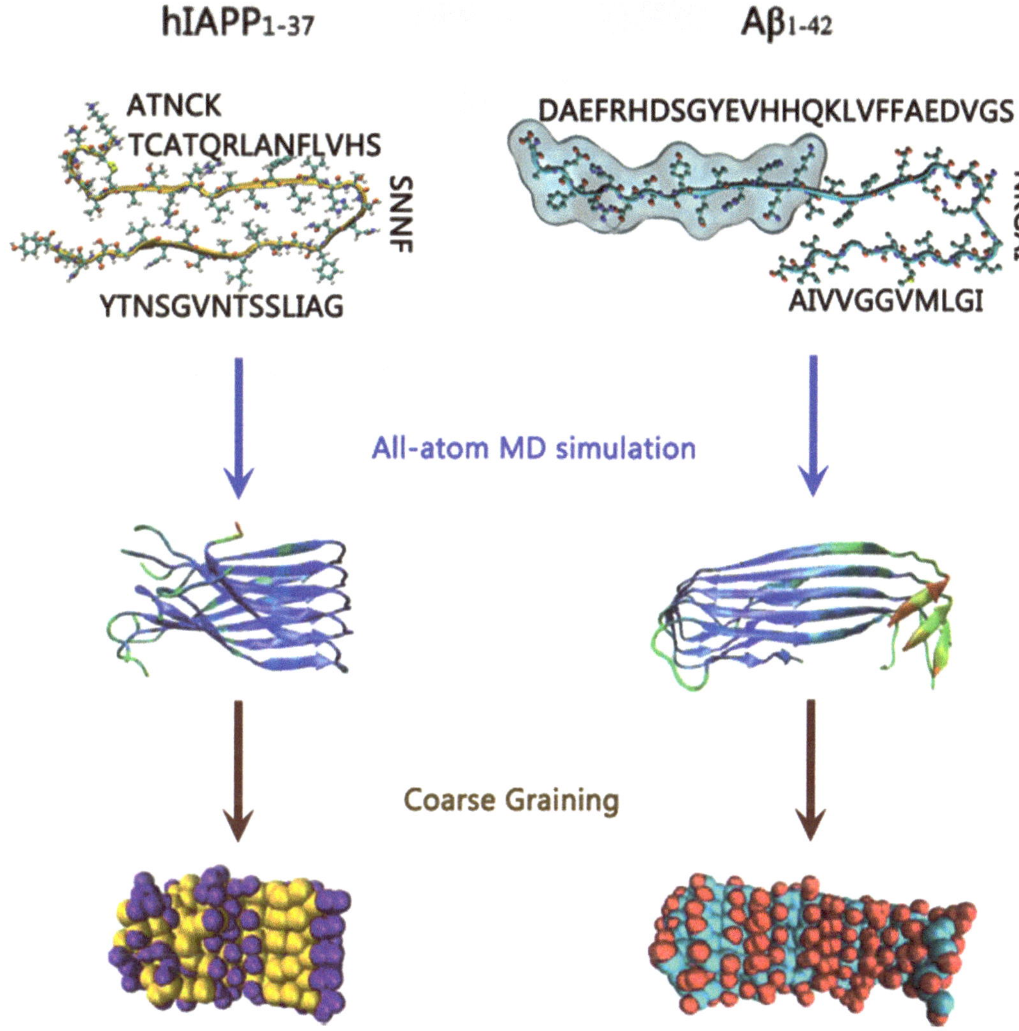

Fig. 6 All-atom and CG-based structures and sequences of hIAPP and Aβ peptides

3. Construct the intramolecular disulfide bond between Cys2 and Cys7 at the N-termini of hIAPP peptides. The tautomeric state of His residues is assigned based on local environment in the hIAPP monomer.
4. Perform a 500-step SD minimization with backbone atoms fixed to refine the overall structure (*see* **Note 12**, Fig. 6).

3.6.3 Coarse-Grained Model of Aβ Pentamer

1. First, perform a 100-ns all-atom explicit-solvent simulation of Aβ to obtain the equilibrium structure of Aβ pentamer in bulk water.
2. Extract 500 all-atomic structures of the Aβ pentamer from the last 20-ns MD trajectory excluding water molecules. Each Aβ

structure is energy minimized with 500-step SD minimization to remove any structural clashes.

3. Obtain an average structure of Aβ pentamer by averaging the 500 structures above, followed by 500-step SD minimization to refine the average structure.
4. Use the elastic network method with the Martini force fields (version 2.4) to convert the all-atom average structures of Aβ pentamer into the CG structure of the Aβ pentamer (*see* **Note 13**, Fig. 6).

3.6.4 Coarse-Grained Model of hIAPP Pentamer

1. First, perform 100-ns all-atom explicit-solvent simulation of hIAPP to obtain the equilibrium structure of the hIAPP pentamer in bulk water.
2. Extract 500 all-atomic structures of the hIAPP pentamer from the last 20-ns MD trajectory excluding water molecules. Energy minimize each hIAPP structure with 500-step SD minimization to remove any structural derivations.
3. Obtain an average structure of the hIAPP pentamer by averaging the 500 structures above, followed by a 500-step SD minimization to refine the average structure.
4. Use the elastic network method with the Martini force fields (version 2.4) to convert the all-atom average structures of the hIAPP pentamer into the CG structure of the hIAPP pentamer (*see* **Note 14**, Fig. 6).

3.7 Simulation Methods

3.7.1 Coarse-Grained Replica Exchange Molecular Dynamics (CG-REMD) Simulations of Hybrid Aβ-hIAPP Assemblies

1. Separate the CG Aβ and hIAPP pentamers from each other by random intermolecular distances and orientations.
2. Place the Aβ-hIAPP system in a simulation box of $100 \times 100 \times 100$ Å^3 with the three-dimensional periodical boundary condition (3D-PBC).
3. Fully solvate the simulation system with coarse-grained water molecules with minimal margins of 15 Å from any protein atom to any edge of water box.
4. Add Cl^- and Na^+ counterions into the systems to mimic an ionic strength of ~150 mM.
5. Structurally optimize and relax the resulting system using 5000 steps of the steepest decent minimization.
6. Heat the energy minimized system gradually from 0 to 310 K by 1-ns MD simulations, constraining the peptide backbones using the NPT ensemble ($T = 310$ K and $P = 1$ atm) and under periodic boundary conditions.
7. For the production run, treat the Aβ-hIAPP system under NPT ensemble and 3D-PBC conditions using CG-REMD simulations with the Gromacs-4.6.5 software package. A total of

40 independent replicas with temperatures ranging from 305 to 410 K should be simulated simultaneously for each Aβ-hIAPP system (*see* **Note 15**).

8. The exchange rate between two replicas is attempted every 1000 integration steps and the acceptance ratio for exchanging replicas varied between 0.20 and 0.25.
9. Control the temperatures and pressure (1 atm) by the V-rescale and Parrinello-Rahman methods with coupling constants of 1.0 and 12.0 ps, respectively. Conduct the temperature coupling separately for protein and nonprotein atoms (i.e., ions and water molecules).
10. Leapfrog integrator was utilized to allow an integration time step of 4 fs (Fig. 7).

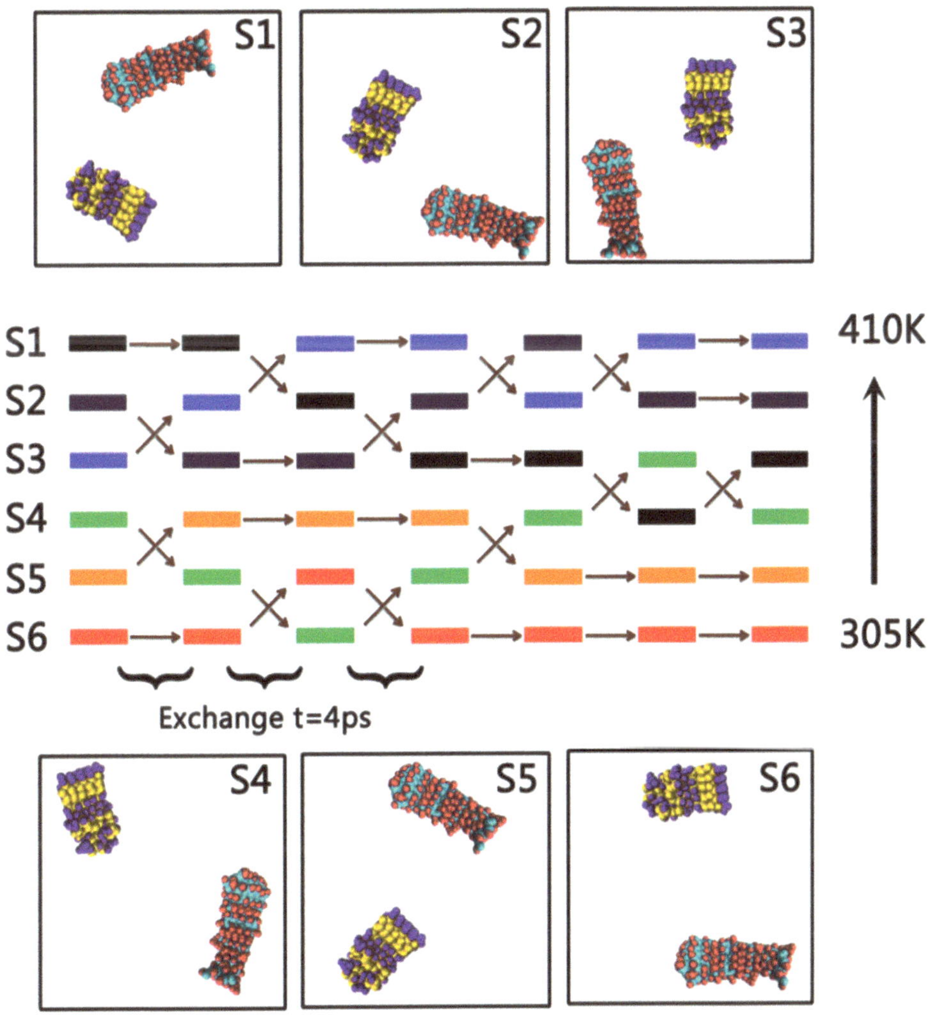

Fig. 7 Schematic illustration of the CG-REMD simulations

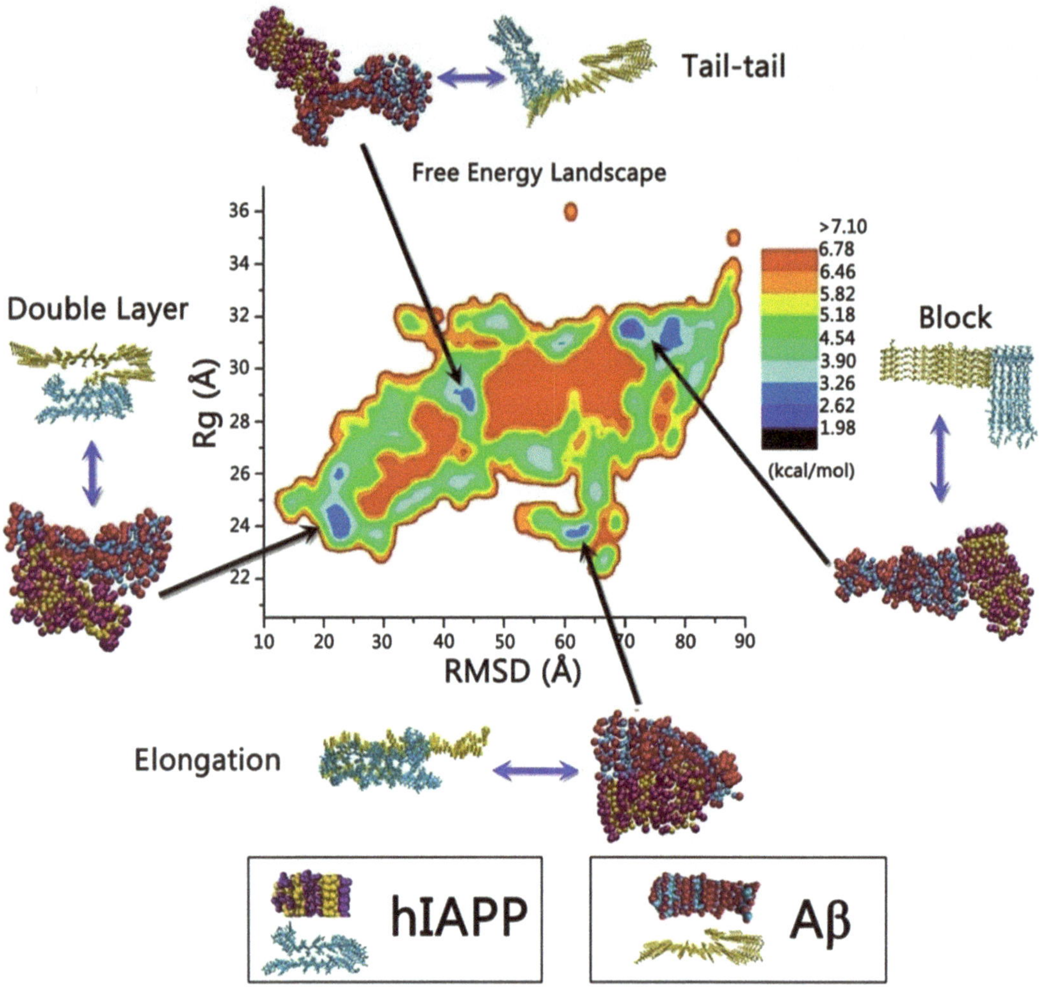

Fig. 8 CG-based free energy landscape along Aβ and hIAPP aggregation process as a function of two reaction coordinates of RMSD and Rg. Reprinted with permission from ref. [30]. (Copyright 2015 ACS Publication)

3.7.2 CG Hybrid Aβ hIAPP Assemblies

1. From the CG-REMD simulations, construct the free energy landscape that describes the interaction between hIAPP and Aβ using −RT log(H(x,y)), where H(x,y) is the histogram of two selected reaction coordinates: x is root-mean-square derivation (RMSD) and y is the radius of gyration (Rg) (*see* **Notes 16** and **17**).
2. Based on the free energy landscape, identify the four most populated CG Aβ-hIAPP hetero-assemblies (double-layer, elongation, tail-tail, and block modes) at different energy potential wells (Fig. 8).

3.7.3 Conversion of CG Aβ-hIAPP Models to All-Atom Aβ-hIAPP Structures

1. Based on the most populated CG Aβ-hIAPP hybrids obtained from the CG free energy landscape, we developed a two-step strategy to convert the CG Aβ-hIAPP structures to all-atom Aβ-hIAPP ones.
2. The first step of backmapping is to generate an initial, all-atom structure. In the CG model, the center of mass of the groups of atoms are represented by a single CG bead as the center of mass location of backbone (Bcom), side chains (Scom), and/or other peripheral groups (Pcom) for each residue. Then, the centers of mass of all-atom backbones, sidechains, and/or peripheral groups of each residue of the same hIAPP or Aβ peptides are superimposed into Bcom, Scom, and/or Pcom, and all atoms are assigned coordinates according to standard geometries using GROMACS around these mapped atoms. This backmapping step produces an initial all-atom of A-β-hIAPP structure from the CG one. The resulting structure is then optimized by energy minimization to remove any bad contacts and to correct internal coordinates for all bonded atoms (**steps 1** and **2** in Fig. 9, *see* **Note 18**).
3. The second step is to optimize interfacial contacts between Aβ and hIAPP. The resulting all-atom structure obtained from the first stage is further recast into the initial all-atom Aβ or hIAPP models to maximize residue-residue contacts by minor structural adjustments through structural superimposition and rotation, followed by energy minimization. In this way, the structural disorder of the resulting all-atom structures, caused by the high temperatures in CG-based REMD simulation, is minimized (**steps 2** and **3** in Fig. 9).

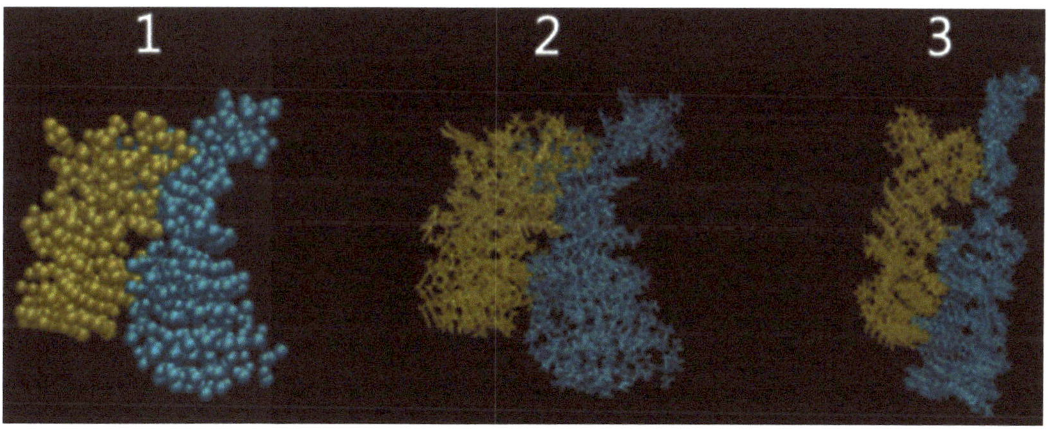

Fig. 9 Schematic illustration of the conversion of Aβ-hIAPP cross-seeding assemblies from coarse-grained to all-atom models

3.7.4 All-Atom MD Simulations of Aβ-hIAPP Models

1. The obtained all-atom Aβ-hIAPP models from the CG models are then subjected to all-atom explicit-solvent simulations to validate the models; perform these simulations using the Gromacs-4.6.5 program with the CHARMM27 force field with CMAP correction and SPCE solvent models [31, 32].
2. Solvate all-atom Aβ-hIAPP systems in a cubic water box with minimal margin of 15 Å from any protein atom to any edge of the water box. Add Cl^- and Na^+ ions into the systems to mimic ~150 mM ionic strength.
3. The resulting systems are structurally optimized and relaxed by 50,000 steps of the steepest decent minimization.
4. Gradually heat the energy minimized systems from 0 to 310 K by 1-ns MD simulations with constrained peptide backbones.
5. Conduct MD simulations for 60 ns using the NPT ensemble ($T = 310$ K and $P = 1$ atm) and under periodic boundary conditions. The Parrinello-Rahman and the V-rescale methods are used to maintain a constant pressure of 1 atm and a temperature of 310 K.
6. Constrain all covalent bonds involving hydrogen atoms by the LINCS method. Short-range Van der Waals (VDW) interactions should be described by the smoothly truncated method via potential shift at 14 Å. Long-range electrostatic interactions should be calculated by the particle mesh Ewald (PME) method with a grid spacing of 0.16 Å and a real-space cutoff of 12 Å.
7. Integrate the equations of motion using the Leapfrog integrator with a time step of 2 fs. Structures should be saved every 2 ps for analysis.

3.7.5 Structural and Population Analysis of All-Atom Aβ-hIAPP Models

1. Conduct in-house Monte Carlo simulations to calculate the relative populations for the Aβ-hIAPP models. First, 500 structures per each all-atom Aβ-hIAPP model are extracted from the last 10-ns explicit-water MD trajectories, excluding water molecules [30].
2. All extracted structures (i.e., all conformers) from all Aβ-hIAPP models are used to construct an energy landscape of Aβ-hIAPP assemblies for evaluating the probability of conformers using an in-house Monte Carlo program as reported in our previous studies [33–35].
3. Calculate the conformational energies of all conformers using the generalized Born method with molecular volume (GBMV) with the CHARMM force field, where the dielectric constant of 80 is used for water and the hydrophobic solvent accessible surfaces area (SAAS) term factor is set to 0.00592 kcal/mol Å [2]. This GBMV free energy calculation generates a Markov chain of all conformers [E_1, E_2... E_i, ...E_n].

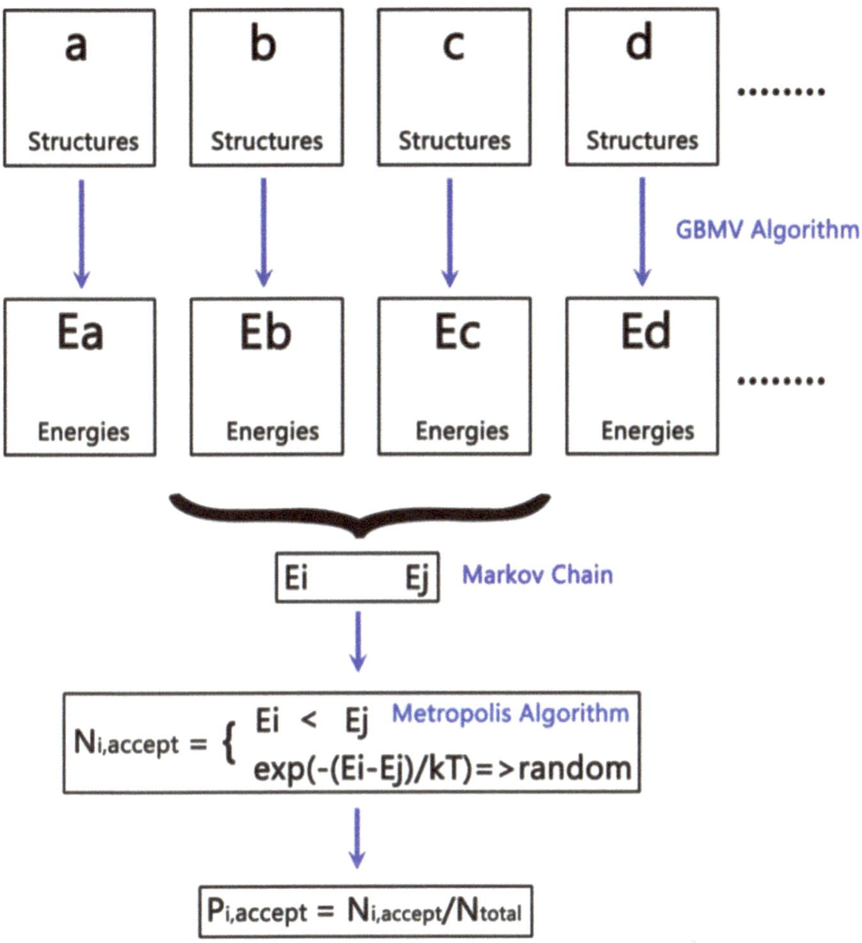

Fig. 10 Schematic illustration of the calculations of the populations for different Aβ-hIAPP hybrid assemblies

4. For conformer population analysis, randomly select any two structures from conformers *i* and *j* and calculate their conformational energies using the GBMV model. Then, the metropolis algorithm is applied to determine the transition probability (i.e., acceptance probability) from conformer *i* to conformer *j* using $P_{\text{accept, move}} = \min\left[1, \exp\left(\frac{E_i - E_j}{k_B T}\right)\right]$. After one million step samplings, a total number of "accepted" structures (N_i) for any conformer *i* is counted, and the relative probability of conformer *i* occurring in the energy landscape is evaluated as $p_i = N_i / N_{\text{total}}$ (Fig. 10).

4 Notes

1. Prepare fresh NaOH solutions on the same day as Aβ and hIAPP incubation.

2. PBS solution needs to be filtered by a hydrophilic filter of 0.22 μm pore size prior to use for peptide incubation.
3. Both dissolution and dilution of peptides use sonication to ensure uniformity of the ThT buffer solution.
4. HFIP is a powerful solvent to dissolve peptide aggregates into monomers; thus we use HFIP to remove any preexisting peptide aggregates from the originally purchased peptides.
5. This procedure allows the peptides to be maintained in monomeric, random coil structures at the initial stage of incubation.
6. Use vortex mixing to ensure the complete dissolution of peptide powders in the micro tube.
7. The ThT fluorescence assay is considered as a standard method to detect the formation of amyloid fibrils, because ThT can specifically bind to the β-sheet structure of amyloid fibrils, typically yielding a strong fluorescence emission around a wavelength of 485 nm.
8. For ThT measurement at a given time point, the fluorescence intensity difference between a peptide sample and a blank sample indicates the amount of β-sheet structure in the peptide sample solution.
9. Selection of cuvette path length is important for CD spectra, because appropriate light absorption is required to obtain accurate CD signals. Strong light adsorption increases the photomultiplier voltage resulting in the increase of signal noises, while weak light adsorption often leads to the loss or uncertainty of signal peaks. Both effects will cause inaccurate and unreliable signals.
10. The N-terminal 1–16 residues ($_{1}$DAEFRHDSGYEVHHQK$_{16}$) are added to the NMR structure of Aβ. The backbone of the added residues is aligned with the initial N-terminal strand of the Aβ structure so that the inter-peptide backbone hydrogen bonds are constructed.
11. The inter-peptide distance within both Aβ and hIAPPβ-sheets is set to 4.7 Å, which well reproduces the peptide-peptide hydrogen bonds and the β-sheet structure.
12. The initially constructed structures may contain some unreasonable steric overlaps between sidechains and backbones. The SD minimization with CHARMM27 force field is required to eliminate any bad atom contacts. To minimize the structural disruption of the β-sheet, the backbone atoms in the structure are harmoniously constrained during the SD minimization.
13. The obtained all-atom structures are converted into the coarse-grained models using the Martini force field (version 2.4) using the martinize.py script (http://md.chem.rug.nl/images/

tools/martinize/martinize-2.4/martinize.py). In the coarse-grained models, the secondary structure of the Aβ or hIAPP pentamer is determined by the DSSP algorithm.

14. Replica exchange simulation requires each replica to contain exactly the same atoms. Thus, caution should be exercised when solvating the system so that the number of water molecules in the simulation box will be strictly controlled by the "genbox" plugin in the Gromacs4.6.5 program with the "−nmol" flag.
15. The replica temperature distribution is calculated using the web-server of "A temperature predictor for parallel tempering simulations" (http://folding.bmc.uu.se/remd/). The tolerance is 1e-4, the exchange probability is 0.20, the lowest temperature is 305 K, the highest temperature is 410 K.
16. The trajectories used to calculate the free energy landscape are extracted from the previous CG-REMD simulations. The equilibrium of the systems is confirmed by studying the Rg distribution profiles of the Aβ-hIAPP cross-seeding assemblies at different simulation intervals.
17. For cluster analysis, RMSD and Rg data for each structure in the equilibrium trajectories are calculated. The structures with the RMSD and Rg difference less than 1.5 Å are assigned as the same cluster.
18. The AutoPsf plugin in the VMD program is used to reconstruct all covalent bonds when converting the CG Aβ-hIAPP models to the corresponding all-atom models.

Acknowledgments

J.Z. thanks the financial support from NSF (CBET-1510099 and DMR-1607475), Alzheimer Association (2015-NIRG-341372), and National Natural Science Foundation of China (NSFC-21528601). The high-performance computational facilities of the Biowulf PC/Linux cluster at the NIH were mainly used for the simulations. This project has been funded in whole or in part with Federal funds from the National Cancer Institute, National Institutes of Health, under contract number HHSN261200800001E. The content of this publication does not necessarily reflect the views or policies of the Department of Health and Human Services, nor does mention of trade names, commercial products, or organizations imply endorsement by the U.S. Government. This research was supported (in part) by the Intramural Research Program of the NIH, National Cancer Institute, Center for Cancer Research.

References

1. Chiti F, Dobson CM (2006) Protein misfolding, functional amyloid, and human disease. Annu Rev Biochem 75:333–366
2. Shankar GM, Li S, Mehta TH, Garcia-Munoz-A, Shepardson NE, Smith I, Brett FM, Farrell MA, Rowan MJ, Lemere CA, Regan CM, Walsh DM, Sabatini BL, Selkoe DJ (2008) Amyloid-[beta] protein dimers isolated directly from Alzheimer's brains impair synaptic plasticity and memory. Nat Med 14(8):837–842
3. DeToma AS, Salamekh S, Ramamoorthy A, Lim MH (2012) Misfolded proteins in Alzheimer's disease and type II diabetes. Chem Soc Rev 41(2):608–621
4. Janson J, Laedtke T, Parisi JE, O'Brien P, Petersen RC, Butler PC (2004) Increased risk of type 2 diabetes in Alzheimer disease. Diabetes 53(2):474–481
5. Nicolls MR (2004) The clinical and biological relationship between type II diabetes Mellitus and Alzheimers disease. Curr Alzheimer Res 1 (1):47–54
6. Andreetto E, Yan LM, Tatarek-Nossol M, Velkova A, Frank R, Kapurniotu A (2010) Identification of hot regions of the A beta-IAPP interaction interface as high-affinity binding sites in both cross- and self-association. Angewandte Chemie Int Ed 49 (17):3081–3085
7. Miklossy J, Qing H, Radenovic A, Kis A, Vileno B, Laszlo F, Miller L, Martins RN, Waeber G, Mooser V, Bosman F, Khalili K, Darbinian N, McGeer PL (2010) Beta amyloid and hyperphosphorylated tau deposits in the pancreas in type 2 diabetes. Neurobiol Aging 31(9):1503–1515
8. Luca S, Yau W-M, Leapman R, Tycko R (2007) Peptide conformation and supramolecular organization in amylin fibrils: constraints from solid-state NMR. Biochemistry 46 (47):13505–13522
9. Brender JR, Salamekh S, Ramamoorthy A (2012) Membrane disruption and early events in the aggregation of the diabetes related peptide IAPP from a molecular perspective. Acc Chem Res 45(3):454–462
10. Zhang MZ, Hu RD, Chen H, Chang Y, Ma J, Liang GZ, Mi JY, Wang YR, Zheng J (2015) Polymorphic cross-seeding amyloid assemblies of amyloid-beta and human islet amyloid polypeptide. Phys Chem Chem Phys 17 (35):23245–23256
11. Zhang M, Hu R, Chen H, Chang Y, Gong X, Liu F, Zheng J (2015) Interfacial interaction and lateral association of cross-seeding assemblies between hIAPP and rIAPP oligomers. Phys Chem Chem Phys 17:10373–10382
12. Nicolls MR (2004) The clinical and biological relationship between Type II diabetes Mellitus and Alzheimer's disease. Curr Alzheimer Res 1 (1):47–54
13. Mandal PK, Pettegrew JW, Masliah E, Hamilton RL, Mandal R (2006) Interaction between Aβ peptide and α synuclein: molecular mechanisms in overlapping pathology of Alzheimer's and Parkinson's in dementia with Lewy body disease. Neurochem Res 31(9):1153–1162
14. Stancu I-C, Vasconcelos B, Terwel D, Dewachter I (2014) Models of β-amyloid induced Tau-pathology: the long and "folded" road to understand the mechanism. Mol Neurodegener 9(1):1–14
15. Liu P, Zhang S, Chen M-s, Liu Q, Wang C, Wang C, Li Y-M, Besenbacher F, Dong M (2012) Co-assembly of human islet amyloid polypeptide (hIAPP)/insulin. Chem Commun 48(2):191–193
16. Hartman K, Brender JR, Monde K, Ono A, Evans ML, Popovych N, Chapman MR (2013) Ramamoorthy, A., Bacterial curli protein promotes the conversion of PAP248-286 into the amyloid SEVI: cross-seeding of dissimilar amyloid sequences. PeerJ 1:e5
17. Andreetto E, Yan L-M, Tatarek-Nossol M, Velkova A, Frank R, Kapurniotu A (2010) Identification of hot regions of the Aβ–IAPP interaction interface as high-affinity binding sites in both cross- and self-association. Ange Chem Inter Ed 49(17):3081–3085
18. O'Nuallain B, Williams AD, Westermark P, Wetzel R (2004) Seeding specificity in amyloid growth induced by heterologous fibrils. J Biol Chem 279(17):17490–17499
19. Yan LM, Velkova A, Tatarek-Nossol M, Andreetto E, Kapurniotu A (2007) IAPP mimic blocks Aβ cytotoxic self-assembly: cross-suppression of amyloid toxicity of Aβ and IAPP suggests a molecular link between Alzheimer's disease and type II diabetes. Angew Chem Int Ed 46(8):1246–1252
20. Seeliger J, Evers F, Jeworrek C, Kapoor S, Weise K, Andreetto E, Tolan M, Kapurniotu A, Winter R (2012) Cross-amyloid interaction of Aβ and IAPP at lipid membranes. Angew Chem Int Ed 51(3):679–683
21. Ma B, Nussinov R (2012) Selective molecular recognition in amyloid growth and transmission and cross-species barriers. J Mol Biol 421 (2–3):172–184

22. Hu R, Zhang M, Patel K, Wang Q, Chang Y, Gong X, Zhang G, Zheng J (2014) Cross-sequence interactions between human and rat islet amyloid polypeptides. Langmuir 30 (18):5193–5201
23. Hu R, Zhang M, Chen H, Jiang B, Zheng J (2015) Cross-seeding interaction between β-amyloid and human islet amyloid polypeptide. ACS Chem Neurosci 6(10):1759–1768
24. Biancalana M, Makabe K, Koide A, Koide S (2009) Molecular mechanism of thioflavin-t-binding to the surface of [beta]-rich peptide self-assemblies. J Mol Biol 385(4):1052–1063
25. Sreerama N, Woody RW (2000) Estimation of protein secondary structure from circular dichroism spectra: comparison of CONTIN, SELCON, and CDSSTR methods with an expanded reference set. Anal Biochem 287 (2):252–260
26. Juszczyk P, Kolodziejczyk A, Grzonka Z (2005) Circular dichroism and aggregation studies of amyloid beta (11-28) fragment and its variants. Acta Biochim Pol 52(2):425
27. Wang Q, Yu X, Patal K, Hu R, Chuang S, Zhang G, Zheng J (2013) Tanshinones inhibit amyloid aggregation by amyloid-β peptide, disaggregate amyloid fibrils, and protect cultured cells. ACS Chem Neurosci 4(6):1004–1015
28. Luhrs T, Ritter C, Adrian M, Riek-Loher D, Bohrmann B, Doeli H, Schubert D, Riek R (2005) 3D structure of Alzheimer's amyloid-beta(1-42) fibrils. Proc Natl Acad Sci U S A 102(48):17342–17347
29. Luca S, Yau WM, Leapman R, Tycko R (2007) Peptide conformation and supramolecular organization in amylin fibrils: constraints from solid-state NMR. Biochemistry 46 (47):13505–13522
30. Zhang MZ, Hu RD, Chen H, Gong X, Zhou FM, Zhang L, Zheng J (2015) Polymorphic associations and structures of the cross-seeding of A beta(1-42) and hIAPP(1-37) polypeptides. J Chem Inf Model 55(8):1628–1639
31. Van der Spoel D, Lindahl E, Hess B, Groenhof G, Mark AE, Berendsen HJC (2005) GROMACS: fast, flexible, and free. J Comput Chem 26(16):1701–1718
32. Buck M, Bouguet-Bonnet S, Pastor RW, MacKerell AD (2006) Importance of the CMAP correction to the CHARMM22 protein force field: dynamics of hen lysozyme. Biophys J 90(4):L36–L38
33. Yu X, Zheng J (2011) Polymorphic structures of Alzheimer's β-amyloid globulomers. PLoS One 6(6):e20575
34. Zhao J, Hu R, Sciacca MFM, Brender JR, Chen H, Ramamoorthy A, Zheng J (2014) Non-selective ion channel activity of polymorphic human islet amyloid polypeptide (amylin) double channels. Phys Chem Chem Phys 16 (6):2368–2377
35. Zhang MZ, Hu RD, Liang GZ, Chang Y, Sun Y, Peng ZM, Zheng J (2014) Structural and energetic insight into the cross-seeding amyloid assemblies of Human IAPP and Rat IAPP. J Phys Chem B 118(25):7026–7036

INDEX

Bradley L. Nilsson and Todd M. Doran (eds.), *Peptide Self-Assembly: Methods and Protocols*, Methods in Molecular Biology, vol. 1777, https://doi.org/10.1007/978-1-4939-7811-3, © Springer Science+Business Media, LLC, part of Springer Nature 2018

N

O

P

R

S

FSC
www.fsc.org
MIX
Papier | Fördert
gute Waldnutzung
FSC® C083411

Kolibri 360 GmbH
Ferdinand-Jühlke-Straße 7
99095 Erfurt, Deutschland
produktsicherheit@kolibri360.de